VOLUME EIGHTY THREE

ADVANCES IN FOOD AND NUTRITION RESEARCH

New Research and Developments of Water-Soluble Vitamins

VOLUME EIGHTY THREE

ADVANCES IN FOOD AND NUTRITION RESEARCH

New Research and Developments of Water-Soluble Vitamins

Edited by

N.A. MICHAEL ESKIN

Department of Food and Human Nutritional Sciences, University of Manitoba, Winnipeg, MB, Canada

ELSEVIER

ACADEMIC PRESS

An imprint of Elsevier

Academic Press is an imprint of Elsevier
50 Hampshire Street, 5th Floor, Cambridge, MA 02139, United States
525 B Street, Suite 1800, San Diego, CA 92101-4495, United States
The Boulevard, Langford Lane, Kidlington, Oxford OX5 1GB, United Kingdom
125 London Wall, London, EC2Y 5AS, United Kingdom

First edition 2018

Notices
Knowledge and best practice in this field are constantly changing. As new research and experience broaden our understanding, changes in research methods, professional practices, or medical treatment may become necessary.

Practitioners and researchers must always rely on their own experience and knowledge in evaluating and using any information, methods, compounds, or experiments described herein. In using such information or methods they should be mindful of their own safety and the safety of others, including parties for whom they have a professional responsibility.

To the fullest extent of the law, neither the Publisher nor the authors, contributors, or editors, assume any liability for any injury and/or damage to persons or property as a matter of products liability, negligence or otherwise, or from any use or operation of any methods, products, instructions, or ideas contained in the material herein.

ISBN: 978-0-12-811803-0
ISSN: 1043-4526

Publisher: Zoe Kruze
Acquisition Editor: Sam Mahfoudh
Editorial Project Manager: Joanna Collett
Production Project Manager: Denny Mansingh
Cover Designer: Christian J. Bilbow

Typeset by SPi Global, India

CONTENTS

CONTRIBUTORS

Marziyeh Ashoori
School of Nutrition Sciences and Food Technology, Shahid Beheshti University of Medical Sciences, Tehran, Iran

Ranjana P. Bird
School of Health Sciences, University of Northern British Columbia, Prince George, BC, Canada

Peter Eck
Department of Food and Human Nutritional Sciences, University of Manitoba, Winnipeg, MB, Canada

Matthew Granger
Department of Food and Human Nutritional Sciences, University of Manitoba, Winnipeg, MB, Canada

James D. House
Department of Food and Human Nutritional Sciences, University of Manitoba, Winnipeg, MB, Canada

James B. Kirkland
Department of Human Health and Nutritional Sciences, University of Guelph, Guelph, ON, Canada

Derrick Lonsdale
Cleveland Clinic, Cleveland, OH, United States

Mirella L. Meyer-Ficca
Utah State University, Logan, UT, United States

Nassim Naderi
Institute of Nutrition and Functional Foods (INAF), Université Laval, Quebec City, QC; Department of Food and Human Nutritional Sciences, University of Manitoba, Winnipeg, MB, Canada

Helga Refsum
Department of Nutrition, University of Oslo, Oslo, Norway

Ahmad Saedisomeolia
School of Nutrition Sciences and Dietetics, Tehran University of Medical Sciences, Tehran, Iran; School of Medicine, Western Sydney University, Sydney, NSW, Australia

A. David Smith
Department of Pharmacology, University of Oxford, Oxford, United Kingdom

Martin J. Warren
School of Biosciences, University of Kent, Canterbury, Kent, United Kingdom

Jean-Marc Zingg
Miller School of Medicine, University of Miami, Miami, FL, United States

PREFACE

Water-soluble vitamins cannot be stored in the body like fat-soluble vitamins, and therefore, they require daily replenishment. This volume covers the latest research and developments of seven important water-soluble vitamins plus the recently discovered water-soluble form of vitamin E, tocopheryl phosphate. Chapter 1 describes the chemistry and laboratory methods for depicting thiamin (B_1) deficiency. In addition to diet, this is caused by a number of genetically determined conditions as well as following illness and surgery. The potential therapeutic role of thiamin is also discussed. Chapter 2 provides the latest evidence linking riboflavin (B_2) with specific health conditions. The neuroprotective effects of riboflavin are discussed in relation to Parkinson's disease, migraine, and MS. Chapter 3 highlights the importance of niacin (B_3) as nutritional precursors of NAD and NADP and their close relation to epigenetic regulatory enzymes. Niacin is a central regulator of physiological processes maintaining genetic stability and in epigenetic control mechanisms modulating metabolism and aging. Chapter 4 discusses the growing evidence of a role for B_6 in inflammation and inflammation-related chronic diseases including cancer. Folate (B_9) is the subject of Chapter 5 in which its prevention of neural tube defects is discussed as well as public health concerns surrounding the use of folic acid in fortifying foods. The importance of vitamin B_{12} in the formation of methionine in humans is highlighted in Chapter 6 by providing methyl groups for over 100 methylation processes. Of particular concern is the suboptimal B_{12} status of around 30%–60% of the population, particularly pregnant women in less-developed countries. A continuous inverse relationship exists between serum B_{12} and neural tube defects, stroke, and dementia. Chapter 7 provides the latest information on the role of vitamin C in human health including molecular mechanism involved in its maintenance and modifying gene–nutrient interactions. The final chapter (Chapter 8) discusses the chemical and biological characteristics of water-soluble analogs of vitamin E, in particular α-tocopheryl phosphate, which was recently discovered to occur naturally.

I am particularly grateful for the important contributions to this volume by outstanding scientists working on these vitamins. This volume will provide useful information to scientists, both researchers and clinicians, to better understand the unique contributions made by these vitamins to human health. Finally, I would like to acknowledge the outstanding editorial assistance provided by the staff of Elsevier/Academic Press in completing this volume.

N.A. Michael Eskin

CHAPTER ONE

Thiamin

Derrick Lonsdale[1]
Cleveland Clinic, Cleveland, OH, United States
[1]Corresponding author: e-mail address: derricklonsdale@hotmail.com

Contents

Advances in Food and Nutrition Research, Volume 83
ISSN 1043-4526
https://doi.org/10.1016/bs.afnr.2017.11.001

Abstract

Starting with a brief history of beriberi and the discovery that thiamin deficiency is its cause, the symptoms and signs are reviewed. None are pathognomonic. The disease has a low mortality and a long morbidity. The appearance of the patient can be deceptive, often being mistaken for psychosomatic disease in the early stages. The chemistry of thiamin and the laboratory methodology for depicting its deficiency are outlined. The diseases associated with thiamin deficiency, apart from malnutrition, include a number of genetically determined conditions where mutations, either in the cofactor relationship or a transporter, provide the etiology. It is emphasized that such mutations are often epigenetically responsive to megadoses of thiamin or one of its derivatives. The use of thiamin in clinical practice requires a high index of suspicion on the part of the clinician since it has a part to play in eating disorders, diabetes, neurodegenerative disease, and cancer. A high rate of critical illness and postsurgery thiamin deficiency have been reported, particularly those associated with gastrointestinal bypass. Emphasis is placed on thiamin deficiency as a major cause of asymmetric dysautonomia, because of the high degree of sensitivity to thiamin deficiency in the brainstem, cerebellum, and hypothalamus. The relationship of thiamin with regional pain syndrome, eosinophilic esophagitis, its analgesic capacity, and its preventive use in obstetrics is raised as a potential issue. The role of thiamin in SIDS and autism is outlined. It is emphasized that megadose thiamin is being used as a drug, either in stimulating the damaged cofactor/enzyme combination, or mitochondria.

1. INTRODUCTION

Although we know today that thiamin deficiency is the major cause of beriberi, unraveling its etiology was a bafflingly long and difficult period of medical history. Because the disease was prevalent in Eastern cultures where rice was a staple part of the diet, major studies were carried out in Japan. After the discovery of thiamin deficiency as its major cause, the "Vitamin B research committee of Japan" was formed from a group of university-based researchers. This culminated in a book entitled "Beriberi and Thiamine," written in English and published in 1965, ostensibly to try to educate Western physicians concerning the destructive nature of thiamin deficiency. The symptoms, signs, and pathophysiology are relevant today because of their extreme variability and effect in many organ systems. Some of the salient features are extracted and reported from this book (Shimazono & Katsura, 1965).

2. DIET AND ENVIRONMENTAL STRESS

The word "Kakke" was the term used in Japan as early as 808 (Inouye & Katsura, 1965a). The majority of the population in Japan took unpolished rice as the staple food until the 17th century when polished rice became associated with increased affluence. The peasants would take their rice to the rice mill because white rice looked better when served. As the consumption of white rice became nationwide, the incidence of beriberi increased in the form of epidemics, reaching their peak in August and September every year. It was not surprising therefore that the investigators at that time considered that beriberi was due to an infection. Factory workers would take their lunch between factory buildings in the summer months. As the sun began to shine into the corridor, a group of the workers would develop their first symptoms of the disease. This and the incidence of epidemics in the summer months is extremely important to us in the modern world, introducing the role of environmental stress in precipitating the disease from an asymptomatic marginal deficiency. The first national statistics on mortality were published in 1899, showing a death rate of only 20 per hundred thousand. After the discovery of thiamin, the mortality decreased to 0.5 per hundred thousand. The morbidity from beriberi is prolonged.

3. SYMPTOMS AND SIGNS IN BERIBERI

Although a number of symptoms have been described, none is pathognomonic. The three cardinal signs include edema, neurologic, and cardiocirculatory symptoms. Clinically, it can be classified into (a) "wet," (b) paralytic, (c) dry (atrophic), and (d) "shoshin" (an acute fulminating form in adults). The three cardinal signs appear early in the disease, followed by motor disturbances including paralysis of the peripheral nerves and muscle atrophy. The early investigators recognized aberrant forms: (1) breaking out in groups; (2) appearing during the course of a febrile illness or puerperium; (3) complicating cachexia due to neoplasm; (4) a variant characterized by amblyopia, hearing difficulty, and ataxia; (5) infantile beriberi; (6) childhood beriberi. It is clear that the representation of symptoms depends on the metabolic rate, reflecting the age of the patient.

3.1 Subjective Symptoms

These include general malaise, a heavy sensation in the legs with loss of strength in the knee joints, peripheral paresthesiae, heart palpitations, a sense of tightness in the chest, anorexia, nausea, and vomiting, common symptoms in America today.

3.2 Objective Symptoms

The appearance of the patient did not suggest nutritional deficiency. The disease was more common in robust appearing manual laborers. The mental state is not affected until late in the disease with symptoms that are extremely variable. Edema is one of the important signs but might be demonstrable only slightly over the pretibial area. Tenderness can be elicited by gripping the calf muscle. Fluid can be found in the pleural, abdominal, and pericardial cavities with a protein content of less than 0.5%. Slight fever was noted in 50% of 159 beriberi patients, hypothesized to be due to suspected alterations in the autonomic nervous system (ANS) or diencephalon.

3.2.1 Cardiovascular System

Heart palpitations on exertion or mental excitement (stress) are important symptoms in the early stages and occur, even at rest, later in the disease. Heart sounds are intensified and may be associated with a murmur. The second pulmonic sound may be accentuated and more intense in the supine position. X-ray examination reveals bilateral cardiac enlargement, more marked on the right. A diastolic blood pressure below 60 mm/Hg is characteristic and may reach zero, particularly in children, when the femoral pulse can be heard by auscultation over the inguinal ligament. Oxygen concentration is low in arterial blood and high in venous blood, indicating the importance of thiamin in oxygen transport and delivery. The labile excitability of the ANS reveals increased vagal tone in the early stages and increased sympathetic tone in the later stages.

3.2.2 Electrocardiogram

Mild to moderately advanced cases show no changes but as the disease advances, high P and QRS waves, abnormal T waves, and prolongation of the PQ interval reveal myocardial damage. There can be tachycardia or bradycardia according to the state of the ANS.

3.2.3 The Nervous System

Polyneuritis and paralysis of the peripheral nerves dominate the neurological picture. Superficial sensation is affected first, involving touch, then pain and temperature perception. Hypesthesia ascends from the toes and the fingertips and the index finger and thumb are affected earlier and more severely than the other fingers, like carpal tunnel syndrome. The loss of sensation affects the abdomen more than the back and the boundary between affected and unaffected areas is transitional and vague. Paresthesiae are common in the early stages. Paralysis of motor nerves occurs later and follows the same pattern of development as in sensory loss. Deep tendon reflexes in the lower extremity are lost. Vertigo and instability are common and Romberg sign may be positive. There may be nystagmus. Diminished visual acuity, amblyopia, central scotomas, and narrowing of the visual field are common. Hyperemia and cloudiness in the papilla are followed by pallor occurring in the latter half of the papilla, mindful of a salient observation in the diagnosis of multiple sclerosis. There may be optic neuritis.

3.2.4 The Autonomic Nervous System

The ANS becomes dysfunctional early in the disease and is reflected most closely by changes in vasomotor function. The labile excitability is responsible for increased sympathetic or parasympathetic tone occurring at different stages of the disease. Injection of 0.1% adrenaline produces an exaggerated response, such as tachycardia, elevated blood pressure, distressing palpitation, substernal oppression, nausea, and vomiting. On the other hand in some cases injection of adrenaline produces bradycardia and lowering of blood pressure. Paradoxically, bradycardia and hypotension can occur after the administration of 0.1% atropine.

3.2.5 Respiratory System

Vital capacity is reduced and there is a decrease in both carbon dioxide and oxygen in the alveolar air as well as in the blood.

3.2.6 Digestive System

Heartburn and constipation are common. Marked anorexia, thirst, nausea, and vomiting occur in "shoshin." Diastase activity in saliva is reduced. Decreased gastric secretion and low acid in the stomach can be followed by hyperacidity during treatment before a return to normoacidity occurs. In some cases the opposite can occur. Hyperacidity is followed by hypoacidity before returning to normal. The Japanese investigators considered that this reflected the labile instability of the ANS. The activity of trypsin and lipase is reduced.

3.2.7 Blood and Hematopoietic System

Although usually normal, normoblasts could be seen in advanced cases and lymphocytosis, neutropenia, and eosinophilia are not uncommon. The bone marrow can show an increase of nucleated red blood cells and increases of eosinophils and monocytes, as well as abnormal platelets. The erythrocyte sedimentation rate is accelerated in many patients.

3.2.8 Infantile and Childhood Beriberi

Although infants with beriberi are usually pale, edematous, and ill tempered, the symptoms most frequently seen include loss of appetite, vomiting, sometimes of coffee ground appearance, diarrhea, and green-colored feces. A hoarse cry is due to paralysis of the vocal cords. On examination, failure of head fixation is common. A convulsion, followed by a comatose state may occur, encouraging a diagnosis of encephalitis. An infant may appear healthy and succumb to sudden death, usually between 3 and 4 months of age, due to thiamine deficiency in breast milk (Fehily, 1944).

Juvenile beriberi was uncommon and apt to be misdiagnosed because blepharoptosis, strabismus, nystagmus, nasal voice, hoarseness, and ataxic gait are common symptoms unlike those in adults. Sometimes, rigidity of the neck and Kernig's sign may be mistaken for meningitis.

Beriberi affects many systems, is polysymptomatic, and can produce a confusing array of laboratory studies. Often, particularly in the early stages, standard modern laboratory procedures may be negative and the symptoms are often classified as psychosomatic. Thiamin deficiency is the great imitator of many puzzling conditions because of its effect in energy metabolism. It may affect different cells according to the distribution of the deficiency. Contrary to present concepts, thiamin deficiency beriberi is surprisingly common in the United States. In spite of vitamin enrichment of many foods, the thiamin content is inadequate because of the heavy consumption of empty calories.

4. THE DISCOVERY OF THIAMIN

Takaki, a Japanese naval surgeon, studied in England from 1875 to 1880. The diet in the British Royal Navy was quite different from that consumed by Japanese naval personnel and Takaki noted that beriberi was much less common than in Japanese ships. In 1882, 61% of the crew of a Japanese naval vessel had succumbed to beriberi on a 272 day voyage. Two years later, Takaki provisioned a ship with an ample supply of dried milk and meat, giving a carbon to nitrogen ratio of 16:1, and sent it on the same voyage. Only 14 members of the crew had succumbed to beriberi. Takaki concluded that a lack of nitrogenous food was the cause of the disease (Inouye & Katsura, 1965a). This was a notable achievement long before the discovery of thiamin and the nature of the caloric intake is important.

In the closing years of the 19th century, Eijkman had found that pigeons fed on polished rice would develop neurological lesions similar to that seen in humans with beriberi. Pigeons became the animal model for studying the disease after an "antiberiberi factor" was discovered in rice polishings. In 1926 this substance was crystallized and called Vitamine (Jansen & Donath, 1926). In 1936, Williams synthesized the elusive molecule and called it thiamin (Williams, 1936). The work of Peters (1936) exposed its role with glucose metabolism. He compared respiration of cells in brain brei from thiamin-deficient pigeons with those from thiamin-sufficient pigeons. There was no difference until glucose was added to the preparation. Cells with thiamin sufficiency immediately began to respire, whereas those that were deficient did not respond at all. Peters called this the catatorulin effect. He also found that this was more marked in cells from the lower brain. Both of these observations are extremely important in understanding the mechanisms of thiamin deficiency in human brain. This work led to the association of thiamin with what was later to become the science of oxidative metabolism.

4.1 Thiamin and Its Esters

Thiamin consists of a pyrimidine ring (2,5-dimethyl-6-aminopyrimidine) and a thiazolium ring (4-methyl-5-hydroxy ethyl thiazole) joined by a methylene bridge. The naturally occurring vitamin is found in lean pork and other meats, wheat germ, liver and other organ meats, poultry, eggs, fish, beans and peas, nuts, and whole grains. Dairy products, fruit, and vegetables are not good sources. The recommended dietary allowance (RDA) is 0.5 mg per 1000 kcal. Although this may be adequate for a healthy individual consuming a healthy diet, it may only be a marginal consumption. Considerable losses occur during cooking or other heat-processing of food. Polyphenolic compounds in coffee and tea can inactivate thiamin so that heavy use of these beverages could compromise thiamin nutrition.

4.2 Thiamin Monophosphate

Thiamin monophosphate (TMP) is an intermediate in the hydrolysis of thiamin pyrophosphate (TPP) to thiamin, not in its synthesis. TPP is directly formed by thiamin pyrophosphorylation from ATP. Excess thiamin may be stored to a limited extent in liver and erythrocytes in the form of TPP. When circulating thiamin levels decrease, the stored TPP can be hydrolyzed to TMP which is then hydrolyzed to thiamin and released into the circulation. No specific thiamin pyrophosphatases or thiamin monophosphatases have been described and no biological role of TMP is known. By the action of alkaline phosphatase, TMP can also be formed in the intestine from TPP present in the ingested food, (Bettendorff L. Personal communication).

4.3 Thiamin Pyrophosphate

The role of TPP in energy metabolism is legion and often considered to be the only important thiamin ester. It is the rate-limiting cofactor in the decarboxylase of pyruvate dehydrogenase complex, a group of enzymes and cofactors forming acetyl Co-A that condenses with oxaloacetate to form citrate, the first component of the citric acid cycle. Since pyruvate is derived from glucose by the Emden–Meyerhof pathway, the energy drive from oxidation of glucose is highly dependent upon TPP. It has the same role in the decarboxylating component of alpha-ketoglutarate dehydrogenase, a link in the sequential metabolism of pyruvate through the citric acid cycle. It is also required in the decarboxylating enzyme that makes up the branched-chain amino acid dehydrogenase for leucine isoleucine and valine. Genetic deletions that relate to all three of these dehydrogenase complexes have been described and will be considered later. A fourth enzyme that requires TPP is transketolase. It occurs twice in the hexose monophosphate shunt (HMPS). This pathway provides pentose phosphate for nucleotide synthesis and supplies reduced NADP for various synthetic pathways. Because of the importance of reducing equivalents, TPP might be described as an antioxidant as well as an oxidant. Since the HMPS occurs in erythrocytes, measurement of the activity of transketolase is an important method for depicting TD. All of the enzymes dependent on thiamin also require magnesium as additional cofactor. Recently, it has been found that TPP is cofactor for 2-hydroxyacyl-CoA lyase (HACL1) in the peroxisome. This enzyme is involved in the alpha oxidation of phytanic acid and 2-hydroxy straight-chain fatty acids. The effect of TD on HACL1 in cells cultured under low thiamin conditions indicated a clear thiamin dependence of total alpha oxidation. The HACl1 gene has been mapped to chromosome 3P25 but its deletion has not been reported, so the phenotype is unknown (Casteels, Snieckers, Fraccasia, Mannaerts, & Van Veldhoven, 2007). This discovery may make a big difference to an assessment of clinical TD. An upstream effect would be an increase in phytanic acid in blood and urine. Milk and meat are the major sources of phytol, the precursor of phytanic acid, accumulation of which has been associated with risk for some malignancies (Ollberding et al., 2013). The downstream effect would be in the processing of 2-hydroxy straight-chain fatty acids for beta oxidation and production of plasmalogens (Braverman & Moser, 2012).

4.4 Thiamin Triphosphate

In 1938, Minz first suggested a relationship between thiamin and nervous excitation. He had observed that thiamin was released into the bathing medium when the pneumogastric nerve from an ox was stimulated (Minz, 1938). Cooper and Pincus found that nerve stimulation in experimental animal systems resulted in a decline in the concentration of TPP and thiamin triphosphate (TTP) from the stimulated nerve. The released metabolites were in the form of TMP and free thiamin, making it difficult to interpret the function of the vitamin in nerve conduction. They found that thiamin appeared to be uniformly distributed in nervous tissue and was highly localized in membrane structures (Cooper & Pincus, 1979).

4.4.1 Electrophoresis

The electric organ of the electric eel (*Electrophorus electricus*) is an adaptation of a cranial nerve similar to that in humans. It is particularly rich in TTP, suggesting a role in electrophoresis (Bettendorff, Grandfils's, Wins, & Schoffeniels, 1989). It activates an ion channel of large unit conductance in neuroblastoma cells (Bettendorff, Kolb, & Schoffeniels, 1993). In addition to TPP at least three other thiamin phosphates occur naturally in most cells, TMP, TTP, and the recently discovered adenosine thiamin triphosphate (Bettendorff & Wins, 2009). TTP is present in rat brain and its synthesis in mitochondria is strongly inhibited by respiratory chain inhibitors (Gangolf, Wins, Thiry, EWl Moualij, & Bettendorff, 2010). Thiamin metabolism in vivo was studied by intracerebroventricular injection of labeled thiamin. Thiamin, TMP, TPP, and TTP were 8%–12%, 12%–14%, 72%–74%, and 2%–3%, respectively, in cerebral cortex (Iwata, Yabushita, Doi, & Matsuda, 1985). Although TTP was discovered over 60 years ago, its presence and action in most cell types from bacteria to mammals remain elusive, although available information suggests a role in cellular energetics (Bettendorff, Lakaye, Kohn, & Wins, 2014). The role of TTP is still incompletely understood.

5. THIAMINASE

Thiaminase (EC 2.5.1.2) and thiaminase II (EC 3.5.99.2) are naturally occurring enzymes that destroy thiamin. Thiaminase I is produced by *Clostridium thiaminolyticum*, an anaerobic organism found in human small intestine and by *Bacillus thiaminolyticus*, an aerobic organism that occurs in human colon. It splits the pyrimidine ring of thiamin from the thiazolium ring at the methylene bridge and adds a base compound to the pyrimidine, creating an analogue inhibitor of thiamin. *Bacillus aneurinolyticus* is aerobic, occurs in the human colon, and produces thiaminase II. It also attacks the methylene bridge but does not add a base compound. Murata reported a potent thiaminase activity in the feces of a Japanese patient with beriberi and although it was called "thiaminase disease," dietary deficiency was also apparently present (Murata, 1965). Their role in human disease is conjectural. Cell-free extracts of *Saccharomyces cerevisiae* contains thiaminase II (Kimura & Iwashima, 1987), a yeast that is used in the construction of some vaccines (Bill, 2015). It had been noted that the symptoms of seasonal ataxia in Nigeria were similar to those of acute thiamin deficiency in 37 patients, all of whom had consumed a stew containing the roasted larvae of *Anaphe venata*. (Adamolekun & Ndububa, 1994). Subsequently, a fairly high activity of a relatively heat-resistant thiaminase was detected and characterized in the pupae of an African silkworm *Anaphe* spp. The thiaminase in the buffer extract was thiaminase I strongly suggesting the etiology (Nishimune, Watanabe, Okazaki, & Akai, 2000).

6. THIAMIN TRANSPORTERS

The SLC solute carrier group of membrane transport proteins includes over 300 members organized into 52 families. Thiamin transporter 1 (ThTr1 SLC 19A2) and thiamin transporter-2 (ThTr2 SLC 19 A3) deliver their substrates to systemic tissues. SLC 19 A3 mediates intestinal thiamin absorption (Zhao & Goldman, 2013).

7. GUANYLATE CYCLASE

Unlike other B vitamins, thiamin activates the guanylate cyclase (GC)/cyclic guanosine monophosphate (cGMP) system but not the adenolyte cyclase system (Baumgartner, 1991). GC is a lyase enzyme that synthesizes cGMP from GTMP, allowing entry of calcium into the cell like cAMP (Sakurai, Chen, & Kefalov, 2011). In smooth muscle, cGMP is the signal for relaxation, including regulation of vascular and airway tone, insulin secretion, and peristalsis. This might point to some of the clinical effects of thiamin deficiency that are not explicable by defective conventionally considered enzymes. It is also important to remember that magnesium is, as far as we know to date, always a necessary cofactor with thiamin. Thus, the symptoms of thiamin deficiency are the same as magnesium deficiency. Even though magnesium is by far the least abundant serum electrolyte, it is extremely important for the metabolism of Ca, K, P, Zn, Cu, Fe, Na, Pb, Cd, HCL, acetylcholine, and nitric oxide. It is therefore important for a wide gamut of crucial body functions as is thiamin (Johnson, 2001).

8. LABORATORY STUDIES FOR THIAMIN DEFICIENCY

Perhaps the commonest method of depicting thiamin deficiency is by an analysis of its concentration in the blood. Since, like magnesium, thiamin is an intracellular enzyme cofactor, it is entirely possible that this may be normal in the presence of thiamin deficiency. Blood cell thiamin diphosphate, monophosphate, and thiamin levels were measured using high-performance

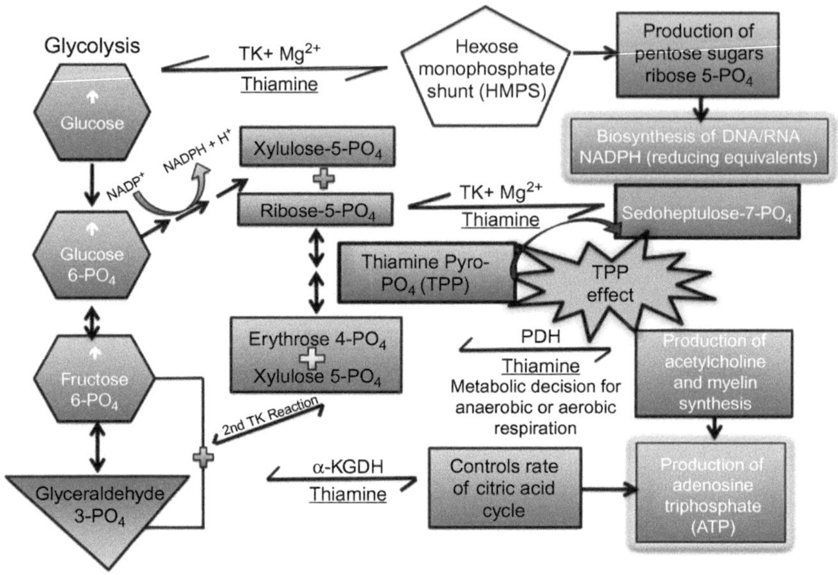

Fig. 1 The hexose monophosphate shunt, showing the positions of transketolase.

liquid chromatography (HPLC) to diagnose Alzheimer's disease (AD). The authors concluded that the measurement of blood thiamin metabolites by HPLC is an ideal diagnostic tests for AD with inexpensive, easy to perform, noninvasive merits (Pan et al., 2015) (Fig. 1).

8.1 Erythrocyte Transketolase

Erythrocyte transketolase (ETKA) was accessed by incubating hemolyzed whole blood with ribose-5-phosphate, both in the presence and absence of TPP, by measuring the amount of pentoses produced from the HMPS in a given time period (Boni, Kieckens, & Hendricz, 1980). For many years our group used erythrocyte transketolase activity (TKA, normal range 42–86 mU/L/m in.) with thiamin pyrophosphate effect (TPPE) as described by Massod, McGuire, and Werner (1971). The first part of the test is the baseline transketolase activity (TKA). The reaction is repeated after the addition of TPP. If the baseline reaction shows increased activity, it is reported as the percentage increase over baseline and referred to as the TPPE ("acceptable" range 0%–18%). Three case reports illustrated the clinical use of the test and its response to therapeutic thiamin (Lonsdale, 2007). Our experience was that TKA activity is rarely below the normal range when the TPPE is grossly abnormal. If the TKA is subnormal, it may indicate severe prolonged deficiency or a genetic effect in the enzyme (Blass & Gibson, 1977). After administration of therapeutic thiamin, the TKA usually increased within the normal range while the TPPE decreased, usually to zero, indicating full saturation with the cofactor. It may well be that although the TKA has a normal range of baseline activity, the TPPE represents a gradual change from thiamin sufficiency to deficiency. If the enzyme is fully saturated with its cofactor there should be no acceleration in its activity with the addition of TPP. The question arises as to how much acceleration in any individual indicates clinical deficiency. Since the normal range of TPPE in the laboratory that we used was calibrated from ostensibly healthy individuals, it might be that some of them had unrecognized functional changes from mild thiamin deficiency, or perhaps an asymptomatic degree of acceleration is allowable. The major factor that we noted was that any acceleration of the TPPE almost invariably decreased to zero with thiamin supplementation.

9. DISEASES RELATED TO THIAMIN METABOLISM

9.1 Dietary Thiamin Deficiency

The highly processed, calorie-dense, nutrient-depleted diet favored in the current American culture frequently leads to exaggerated supraphysiological postprandial dysmetabolism. A transient increase in free radicals triggers atherogenic changes including inflammation, endothelial dysfunction, and sympathetic hyperactivity (O'Keefe, Gheewala, & O'Keefe, 2008). Recent evidence points to a prominent role of the hypothalamus in triglyceride metabolism through innervating white adipose tissue in the liver and brown adipose tissue, mainly via sympathetic branches of the ANS (Geerling et al., 2014). Limited data are available on the relation between thiamin requirements and the intake of carbohydrates. A study showed the influence of stepwise increases of carbohydrate intake on the status of thiamin in healthy volunteers under isocaloric conditions. Starting at 55% of total energy it was gradually increased to 75%. Thiamin decreased significantly in plasma and urine. The erythrocyte transketolase and thiamin content of feces remained unchanged (Elmadfa, Majchrcak, Rust, & Genser, 2001). There is abundant evidence linking diets that are high in carbohydrate content with hypertriglyceridemia. Dietary sugar and refined carbohydrates are associated with the rising prevalence of the metabolic syndrome and type II diabetes (Morgantini, Xiao, Dash, & Lewis, 2014). Beriberi, long thought to have been conquered in developed cultures, is the prototype for dysautonomia in its early stages. It is the best known nutritional deficiency disease caused by an imbalance between ingested calories and the vitamins required for their oxidation, particularly thiamin (Lonsdale, 2007). Erythrocyte transketolase activation was studied in elderly patients with moderate to severe chronic dementia, acute alcoholic admissions, and chronic alcoholics with evidence of brain damage. The modified transketolase activation may warn of marginal thiamin deficiency, or susceptibility to brain damage in patients at risk (Jeyasingham et al., 1987). The thiamin status of 42 physically healthy nonalcoholic psychiatric inpatients was evaluated by measuring TKA and TPPE. Sixteen patients (38%) showed evidence of thiamin deficiency, five of whom had received thiamin-containing vitamins for a period of from 2 to 47 days prior to testing, suggesting that poor nutrition may not fully account for the thiamin deficiency observed (Schwartz, Gross, Lonsdale, & Shamnberger, 1979). It is hypothesized that the massive consumption of carbohydrate and fat results in a high calorie/thiamin ratio as a major cause of disease. Because mild to moderate thiamin deficiency results in pseudohypoxia in the limbic system and brainstem, emotional and stress reflexes of the ANS are stimulated and exaggerated, producing symptoms often diagnosed as psychosomatic disease. If the biochemical lesion is recognized at this stage, the symptoms are easily reversible. If not, and the malnutrition continues, neurodegeneration may follow, resulting in a variety of chronic brain diseases. If such a hypothesis were to be accepted, it could easily be tested by performing erythrocyte transketolase and TPP uptake studies when "psychosomatic" symptoms are present. If dietary correction is ignored, supplementary thiamin and magnesium could be provided, together with a well-rounded multivitamin (Lonsdale, 2015a). The physical components of disease are as much related to the brain as they are to the body. For example, a study assessed heart rate variability in healthy subjects, individuals considered to be at risk of developing arthritis and patients with diagnosed rheumatoid arthritis. The data supported the notion that autonomic dysfunction precedes the development of rheumatoid arthritis (Koopman et al., 2016). Dysautonomia, in association with a number of different diseases, has been reported (Low, 1994). It has been hypothesized that the dysautonomia is really the etiological forerunner of the associated disease (Lonsdale, 1981).

10. THIAMIN DEFICIENCY INDUCES GENETIC CHANGES

Thiamin deficiency can occur from multiple causes, many of which are genetic in origin. It is well established that thiamin deficiency results in an excess of metabolic intermediates that include lactate and pyruvate. An in vitro study, using thiamin-deficient neuroblastoma cells revealed that both pyruvate and lactate can increase the stabilization of the hypoxia-inducible 1-alpha (HIF-1α) transcription factor, resulting in the trans-activation of HIF-1α-regulated genes, independent of low oxygen. This is termed pseudohypoxia. Similar to cells exposed to hypoxia, there was a corresponding increase in HIF-1α from thiamin deficiency. Therefore, the resulting dysfunction in cellular metabolism from this deficiency may facilitate a pseudohypoxic state (Sweet & Zastre, 2013). Thiamin deficiency symptoms can imitate those from hypoxia, particularly in the brain.

11. NITRIC OXIDE

Nitric oxide (NO) plays a wide variety of roles in biological organisms. Nitric oxide synthases (NOSs) synthesize the metastable free radical nitric oxide (NO). There are three isoforms of the NOS enzyme: endothelial (eNOS), neuronal (nNOS), and inducible (iNOS). A study exploring the physiological relevance of NO signaling under conditions of thiamin deficiency was reported. It was found that thiamin deficiency modulates eNOS-dependent NO production, leading to a decreased phase of relaxation and an increased contractile response in the rat aorta (Gioda, Capettini, Cruz, & Lemos, 2014).

12. THIAMIN IN CLINICAL PRACTICE

In addition to its coenzyme roles, TPP plays a role in nerve structure and function as well as in brain metabolism. Signs and symptoms of its deficiency include peripheral neuropathy, ataxia, various forms of dysautonomia and nystagmus, in which the clue may be lactic acidosis. More advanced symptoms include confabulation, memory loss, psychosis, Wernicke encephalopathy, and/or Wernicke–Korsakoff syndrome (Frank, 2015). The clinical presentation of 20 patients with functional symptoms, previously diagnosed as psychosomatic, had abnormal red cell transketolase activity revealing thiamin deficiency. Although in most patients the etiology was due to heavy consumption of "junk" foods, carbonated beverages, and candy, this was not so in all of them and the etiology in these cases remains obscure. The symptoms in all 20 patients were due to early beriberi and all were clinically improved by the administration of thiamin. In only two cases were other vitamins administered. The improvement was slow and there was no dramatic recovery. In 10 patients, improved transketolase activity was correlated with clinical well-being. The other 10 patients did not keep their return appointments and telephone contact revealed that they were asymptomatic (Lonsdale & Shamberger, 1980).

12.1 Anorexia Nervosa

A developmentally appropriate adolescent boy presented with upper and lower extremity glove-and-stocking parasthesias, weakness, vertigo, inattention, ataxia, and binocular diplopia after a voluntary 59-kg weight loss over 5 months. Serum thiamin levels were extremely low and nerve conduction studies demonstrated polyneuropathy. Eating disorders can be an uncommon cause of rapidly progressive thiamin deficiency (Renthal, Marin-Valencia, & Evans, 2014) and other vitamin deficiencies (Rock & Vasantharajan, 1995).

Thiamin deficiency requires clinical suspicion. Apart from the genetic aspects, risk factors include patients with malnutrition due to one or more nutrition-related etiologies. Unexplained heart failure, renal failure with dialysis, alcoholism, starvation, hyperemesis gravidarum, and following bariatric surgery are at increased risk. The differential diagnosis seldom includes the possibility of thiamin deficiency. Patients who are critically ill may also be at risk, especially those given intravenous dextrose without adequate thiamin repletion. Furthermore, the clinician needs to understand the role of thiamin as a potential therapeutic agent in diabetes, some inborn errors of metabolism, and neurodegenerative diseases (Frank, 2015). Glucose-derived pyruvate is the source of acetyl-CoA in all brain cells through pyruvate dehydrogenase complex action. Cholinergic neurons, like other transmitter systems, utilize acetyl-CoA for energy production in mitochondria as well as in diverse synthetic pathways. However, cholinergic neurons require additional amounts of acetyl-CoA, thus making them more susceptible to thiamin deficiency (Szutowicz, Bielarczyk, Jankowska-Kulawy, Pawelczyk, & Ronowska, 2013). Thiamin sufficiency may be an essential component in prevention of neurodegenerative disease.

Erythrocyte TKA and TPPE were performed on 1011 patients between 1983 and 1986. TKA or TPPE or both were abnormal in 283 (28%). Of these, 36 patients were selected as representative of the clinical presentation and this was correlated with subsequent laboratory testing in order to draw attention to the practical value of the test in the clinical setting (Lonsdale, 1988). The data strongly suggested that there is widespread marginal nutritional deficiency in the United States and that it represents a clinical problem that deserves far more attention than it is presently receiving.

12.2 Marginal Thiamin Deficiency in the Adult

Seventeen adults presented with symptoms of insomnia, night cough (17), fatigue, emotional lability, panic attacks, joint, muscle, or limb pain (15), sugar craving (14), gastro esophageal reflux, abdominal pain, daily headache or migraine, nasal congestion, alternating recumbent nasal congestion, dyspnea at rest, numbness, pins and needles, cold extremities (13), heart palpitations, recurrent chest pain (11), diarrhea alternating with constipation, bowel urgency, mega stools (10), intolerance to ambient temperature (9), and recurrent nausea and vomiting (8). Three patients had sleep apnea. Of eight young women, five had premenstrual syndrome. In all of these patients the blood pressure was significantly different in the two arms. The mean blood pressure in the right arm was 150/70 and in the left arm 192/78 (Lonsdale, Shamberger, & Obrenovich, 2011a, 2011b). Thiamin deficiency, by the use of erythrocyte TKA and TPPE, was not proven in all of these patients and other vitamin deficiencies were not sought. It is hypothesized that all of them represented defective mitochondrial function and that thiamin deficiency is a major culprit in the symptomology.

Laboratory changes in thiamin deficiency often include increased triglycerides, fibrinogen, and hypersensitive CRP, usually ascribed to other causes, because a connection between inflammation and thiamin deficiency is easily induced (Lonsdale, 2015a, 2015b). Sometimes it is the social activities that do most harm because they often include carbonated beverages and simple carbohydrates. Laboratory changes recorded in an 84-year-old man with insomnia and tenosynovitis in one finger are shown in Figs. 2 and 3. Because of the increase in TPPE, the only treatment given to this patient was a complete absence of sugar in the diet. His dietary ingestion of sugar had not been excessive. It took 6 months for the TPPE to decrease to zero. There was concomitant symptomatic improvement and his weight decreased by 11 pounds. Notice the slight increase of the TPPE in September after ingestion of a very small intake of simple carbohydrate the day before. This gentleman may have been unusually hypersensitive to sugar.

12.3 Thiamin Deficiency in Children and Adolescents

As a pediatrician in a multispecialty clinic, emotional problems in children were a common referral. There was a frequent history of a variety of clinical

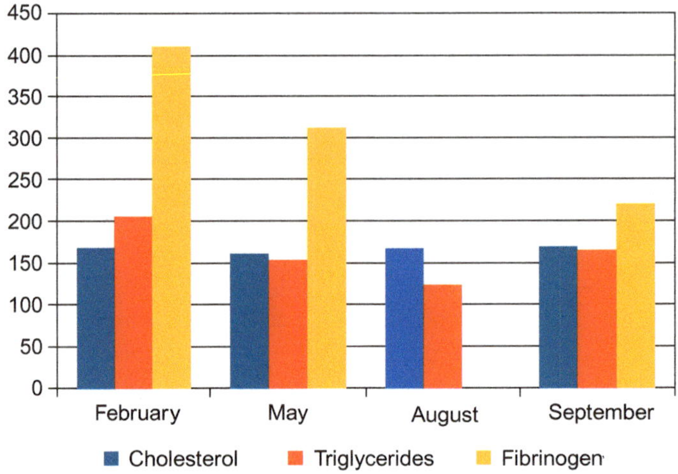

Fig. 2 Repeated laboratory data from an 84-year-old man with insomnia and tenosynovitis. Cholesterol, <200 mg/dL; Triglycerides, *N*<150 mg/dL; Fibrinogen, *N*. 180–350 mg/dL.

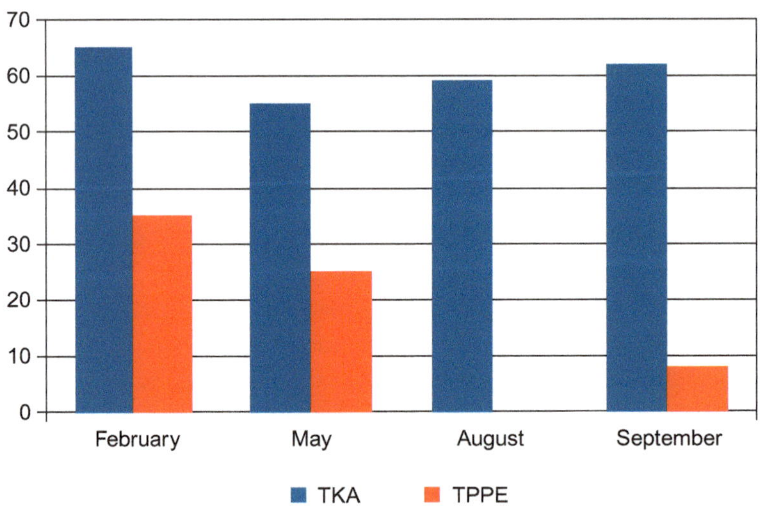

Fig. 3 Repeated measurements of transketolase activity (TKA) and thiamin pyrophosphate effect (TPPE) from an 84-year-old man whose sole treatment was complete discontinuation of sugar. TKA, 42–86 mU; TPPE, acceptable 0%–18%. Note that the TPPE was zero in August and rose to 8% in September after ingestion of a small amount of simple carbohydrate 1-day previously.

problems in pregnancy, together with variable evidence of hypoxia at birth. Neonatal jaundice and recurrent ear infections, both now known to be associated with oxidative stress (Asher & Guiford, 2016; Hansen, 2000) were common and reflected pregnancy as well as neonatal diet. Temper tantrums, hyperactivity, attention deficit, and learning disability were the commonest complaints. Adolescents had a variety of symptoms that had been usually diagnosed elsewhere as psychosomatic in origin (Lonsdale & Shamberger, 1980). The children often had a characteristic appearance with bright red cheeks, circumoral pallor, and red spots on the tongue due to inflammation of the filiform papillae. Examination of the throat often produced gagging before the tongue depressor touched the tongue and the blood pressure was abnormal. Deep tendon reflexes were invariably abnormal and stroking the inside of the thigh and leg with the tip of a finger often produced blanching dermographia in the line of the stroke. The normal blood pressure of a child around 5–8 years is about 90/60 and I have seen blood pressures of 120/0 (wide pulse pressure) or 60/40 (narrow pulse pressure). This is entirely compatible with what we know about the symptoms of classic beriberi, depending on the stage of the disease. We were always able to prove or disprove thiamin deficiency by the use of TKA and TPPE.

13. THIAMIN-RESPONSIVE MUTATIONS

13.1 Pyruvic Dehydrogenase

The pyruvic dehydrogenase complex (PDHC) consists of three enzymes. E1, pyruvate decarboxylase or dehydrogenase, is the thiamine-dependent decarboxylating element of the complex and is the rate-limiting component. E2, dihydrolipoyl transacetylase, produces acetyl-CoA, that is released into the citric acid cycle. E3, dihydrolipoyl dehydogenase, produces NADH. Mutations can occur in any of the enzymes or in the binding of their respective cofactors, producing slowing of the citric acid cycle. The clinical consequence is that of inadequate mitochondrial energy, commonly due to thiamin deficiency from diet or, rarely, from genetically determined cause. The biochemical clue is lactic acidosis. Decrease of pyruvate and ketoglutarate dehydrogenase complex activities represents the main causes of energy and acetyl-CoA deficits in thiamine deficiency-evoked cholinergic encephalopathies. Thiamine deficiency was induced in rats, yielding deficits of acetyl-CoA in mitochondrial and cytoplasmic compartments of nerve terminals in the brain (Jankowska-Kulawy, Bielarczyk, Pawelczyk, Wroblewska, & Szutowicz, 2010).

A 6-year-old boy with intermittent episodes of cerebellar ataxia was reported. The episodes of ataxia were always initiated by some form of physical stress, a mild infection being the commonest, but on one occasion related to a mild head injury and on another, an inoculation. Details derived from studies of a single untreated episode revealed abnormal, fluctuating day/night urinary metabolites, suggesting reversed transamination of pyruvate to alanine. He responded to megadoses of thiamin hydrochloride that had to be increased dramatically if he showed signs of impending infection (Lonsdale, Faulkner, Price, & Smeby, 1969). This was the first example of thiamine dependency due to a mutation in pyruvate dehydrogenase (PDHC) E1 (Blass, Avigan, & Uhlendorf, 1970). He went through high school but was lost to follow-up in his late 1920s. His brother had the same defect in the enzyme but had not experienced ataxia. A mild head injury induced unconsciousness, illustrating the risk of initiating the clinical effects of PDHC deficiency through some form of stress.

The clinical presentation of PDHC deficiency ranges from fatal neonatal lactic acidosis to chronic neurologic dysfunction without lactic acidosis. Paroxysmal neurologic problems include intermittent ataxia, episodic weakness, exercise-induced dystonia, and recurrent demyelination (Kara et al., 2017). These authors describe an 8-year-old boy who complained of acute episodic proximal muscle weakness of the upper and lower extremities with normal mental status. He had a history of Guillain–Barre-like syndrome at the age of 2 years. Genetic analysis revealed a hemizygous novel mutation of the PDH A1 gene (p.A 353T/c. 1057G > A) that encodes the E1α subunit of PDHC. Dietary carbohydrates were restricted and he was treated with thiamin (15 mg/kg/day), resulting in clinical improvement within a few weeks.

The medical records of 19 PDHC-deficient patients were reviewed. Seventeen patients received thiamin treatment. Eight of them did not respond. Four showed sustained response and the others responded temporarily or questionably. Responsiveness to thiamin was based on clinical parameters only (Van Dongen, Brown, Brown, Thorburn, & Boneh, 2015). The clinical manifestations of mutations in this enzyme complex evidently vary and few have so far been described.

13.2 Thiamin Pyrophosphokinase Deficiency

Thiamin is taken up by specific transporters into the cell and converted to the active TPP by thiamin pyrophosphokinase (TPK). Five individuals from three families presented with variable degrees of ataxia, psychomotor retardation, dystonia, and lactic acidosis. A reduced concentration of TPP was found in muscle and blood. Mutation analysis uncovered three missense mutations resulting in decreased TPK protein levels (Mayr et al., 2011). Two patients with novel homozygous TPK-1 mutations were reported. One of the patients presented with global developmental delay. In the other patient, thiamin supplementation prevented encephalopathic episodes and improved developmental progression (Banka et al., 2014).

13.3 Leigh Disease

Also known as subacute necrotizing encephalomyelopathy, it was reported that this disease responded partially to thiamine derivatives (Pincus et al., 1973). Some early research showed that tissue extracts from a patient with this disease inhibited thiamin pyrophosphate-adenosine triphosphate phosphotransferase (Cooper, Itokawa, & Pincus, 1969). To my knowledge, this was the only reported research that applied to clinical appreciation of TTP deficiency. The research was terminated for lack of funding and there still is no information concerning this important ester of thiamin in clinical medicine.

A 15-year-old girl developed a rapid onset ptosis and ophthalmoplegia, evolving into a subacute encephalopathy. Neuroimaging demonstrated symmetrical basal ganglia in midbrain lesions. Oral biotin and thiamin were administered and symptoms improved dramatically the next day (Fassone et al., 2013). Biotin-responsive basal ganglia disease is caused by mutations in the SLC19A3 gene (Distelmaier et al., 2014). These authors reported a 6-year-old child who responded to thiamine and biotin treatment. They recommend this treatment for any patient with symmetrical basal ganglia lesions and neurological symptoms. Based on current data, it is suggested that severe ATP depletion, gliosis, hyperlactic acidemia, and excitotoxicity all contribute to the pathology of Leigh syndrome (Lake, Bird, Isohanni, & Paetau, 2015). Thiamine transporter-2 deficiency is caused by mutations in the SLC 19 A3 gene, as opposed to other causes of Leigh syndrome. CSF fluid samples from patients with the syndrome due to mutations in the SLC 19 A3 gene had reduced free-thiamin. Early treatment with thiamine provided clinical improvement and restored thiamine values in fibroblasts and CSF (Ortigoza-Escobar et al., 2016). The use of Leigh syndrome is no longer a diagnostic category. It is finding the biochemical lesion that counts and clinical suspicion that leads to its discovery. In an emergency situation, clinical suspicion may be enough to justify a trial with high-dose thiamin therapy, since there is no known toxicity in therapeutic doses.

13.4 Maple Syrup Urine Disease (Branched-Chain Ketoaciduria)

A neonate was diagnosed with branched-chain ketoaciduria (Lonsdale & Barber, 1964). In 2016, I received a communication concerning a follow-up of this patient. He had not had a crisis event in 30 years and was 52 years old. He had been picked up by my informant when he was in a coma following a strep throat about 20 years previously. In May 2016 the leucine level was 544 nmol/mL (nl 68–133), valine 497 (136–309), and isoleucine 205 (36–107) without symptoms. At one time the alloisoleucine was 92 (nl <5), but he had no unusual odor and was his "usual self." He was receiving a protein-restricted diet, could carry on a reasonable conversation, and was able to mow the lawn at his local church. He had developed gout and type II diabetes. He was found to be homozygous for a mutation in the BCKDHB, not previously described (Personal communication, W.G. Wilson, M.D. University of Virginia Children's Hospital, Charlottesville, VA). Thiamin is emerging in the treatment of diabetes (Frank, 2015), and sugar-sweetened beverages have a relationship with the etiology of gout (Dalbeth et al., 2015). It would seem therefore to be reasonable to conclude that this example of Maple Syrup Urine Disease (MSUD) might have been responsive to thiamin. It has been proposed that each child with MSUD be assessed for response to thiamin by quantifying the concentration of branched-chain amino acids in plasma before and after vitamin supplementation (Fernhoff et al., 1985). There are presently five known clinical phenotypes for MSUD; classic, intermediate, intermittent, thiamin-responsive, and dihydrolipoamide dehydrogenase (E3) deficient. This is based on the severity of the disease, the gene locus affected, and the response to thiamin. To date, approximately one hundred mutations have been identified and there is a strong correlation between the presence of mutant E2 proteins and the thiamin-responsive phenotype (Chuang, Chuang, & Wynn, 2006). There is a growing body of literature implicating BCAA metabolism in more common disorders such as metabolic syndrome, cancer, and hepatic disease. Whether this implies genetic involvement or dietary deficiency of thiamin affecting branched-chain dehydrogenase may have some clinical importance (Burrage, Nagamani, Campeau, & Lee, 2014).

13.5 Thiamin Transporter Disease

Thiamin is transported into cells by two carriers, THTR 1 and THTR 2. Deficiency of these results in thiamin-responsive megaloblastic anemia and biotin-responsive basal ganglia disease, respectively (Brown, 2014).

ThTR1 (SLC 19 A2) and ThTR2 (SLC 19 A3) both transport thiamin. SLC 19 A2 delivers its substrate to systemic tissues and SLC 19 A3 mediates intestinal thiamin absorption (Zhao & Goldman, 2013). Thiamin-responsive megaloblastic anemia syndrome is the association of diabetes mellitus, anemia, and deafness due to mutations in the SLC 19 A2 gene. In 13 patients from 7 families, the anemia and diabetes responded to oral thiamine hydrochloride initially but this later became ineffective (Ricketts et al., 2006). A 30-day-old patient presented with an acute encephalopathy and increased level of lactate in the blood and CSF due to a mutation in the SLC 19 A3 gene. There was a high content of alpha-ketoglutarate in the urine and increased concentrations of the branched-chain amino acids leucine and isoleucine in the plasma. MRI detected bilateral and symmetrical cortical–subcortical lesions. After treatment with thiamin and biotin, irritability and opisthotonus disappeared and the patient recovered consciousness. The biochemical disturbances disappeared within 48 h (Perez-Duenas et al., 2013). A 3-year-old boy was reported with a novel homozygous SLC 19 A2 gene mutation. Insulin-dependent diabetes and respiratory illness had their onset at 11 months of age. He had bilateral profound hearing loss, refractory anemia, decreased visual acuity, and photophobia. The patient markedly improved several days after initiating 100 mg of thiamin a day (Mikstiene et al., 2015). A girl, born to consanguineous parents, was found with a novel mutation in the SLC 19 A2 gene. In addition to megaloblastic anemia, diabetes, and deafness, she had short stature, hepatosplenomegaly, retinal degeneration, and a brain MRI lesion. Biochemical analysis showed a severe deficiency of pyruvate dehydrogenase and complex one of the respiratory chain. In the patient's fibroblasts, supplementation with high doses of thiamin resulted in restoration of complex I activity (Scharfe et al., 2000). Thiamin-responsive megaloblastic anemia syndrome can manifest at any time between infancy and adolescence, and not all cardinal findings are present initially. The anemia typically improves significantly with pharmacological doses of thiamin. Variable improvement in diabetes is also noted but the hearing loss is apparently irreversible. However, a 2-year-old girl with a novel mutation in the SLC 19 A2 gene did not suffer from hearing loss (Yilmaz Agladioglu, Aycan, Bas, Peltek Kendirci, & Onder, 2012).

13.6 Thiamin in Diabetes

Low plasma thiamin concentration is prevalent in patients with type I and type II diabetes, associated with increased thiamin urinary clearance.

The conventional assessment of thiamin status was masked by increased thiamin transporter content of erythrocytes (Thornalley et al., 2007). The growing prevalence of both type I and type II diabetes and its related morbidity due to vascular complications creates a large burden on medical care worldwide. The physiological role of thiamin and metabolism of glucose and amino acids were reviewed and published data indicate that thiamin metabolism in diabetes is compromised. Experimental studies on the molecular mechanisms of thiamin deficiency in diabetes are critically needed (Pacal, Kurikova, & Kankova, 2014). The accumulation of triosephosphates and the increased formation of the potent glycating agent methylglyoxal in intracellular hyperglycemia are implicated in the development of diabetic complications. A strategy to counter this is to stimulate the anaerobic pentose phosphate pathway by maximizing transketolase activity by thiamin supplementation. An in vitro study showed that there is a characteristic increase in the concentration of the triosephosphate pool and formation of methylglyoxal in hyperglycemic conditions, both of which decreased by the addition of thiamin by stimulation of transketolase activity. This provides a biochemical basis for high-dose thiamine therapy for the prevention of diabetic complications (Thornalley, Jahan, & Ng, 2001). A strong association between low blood thiamine levels and type I diabetes was detected in a Saudi population (Al-Daghri et al., 2015). Both type I and type II diabetes have been associated with reduced performance on multiple domains of cognitive function. Although the pathophysiology is complex, including autonomic function, neuroinflammatory pathways, mitochondrial metabolism, and Tau signaling (Zilliox, Chadrasekaran, Kwan, & Russell, 2016), the role of thiamine deficiency might well be the important factor. The number of Americans with multiple chronic diseases is projected to increase from 57 million in 2000 to 80 million in 2020. A heuristic behavioral framework for the etiology of multimorbidity has been proposed (Suls, Green, & Davidson, 2016). A revision of the medical model has been suggested, based on the prevalence of dietetically induced dysautonomia. Widespread high calorie malnutrition, particularly with the prevalence of simple carbohydrates and fats, results in pseudohypoxia in brain cells. It is hypothesized that this is the forerunner of many organic brain diseases (Lonsdale, 2009).

The human gut is a complex ecosystem. Countless bacteria colonize it. A way has been developed to study this system in detail. The researchers can identify the proteins that are subsequently synthesized by bacteria that catalyze metabolic reactions. The gut bacteria have to adapt to changes in their environment. They do this by adjusting their metabolism, changing the amounts of vitamins they produce such as thiamin. The methodology enables the researchers to learn how functional differences in the biosynthesis of thiamin relate to type I diabetes (Heintz-Buschart et al., 2016). If bacteria are indeed capable of synthesizing thiamin, it raises the question of the need of a thiaminase enzyme. Although they can bind the thiazolium and pyrimidine rings together if present in their free form, the equilibrium of the reaction favors destruction of the thiamin molecule rather than its synthesis (Murata, 1965).

13.7 Diabetic Autonomic Neuropathy

Diabetic autonomic neuropathy is highly prevalent in the diabetes population and increases the risk of morbidity and mortality in older adults. It often goes undiagnosed and untreated (Scheinberg, Salbu, Goswami, & Cohen, 2016). Cholesterol, glucose, and triglycerides were significantly increased in 77 patients with type I diabetes. HDL-cholesterol, thiamine mono- and diphosphate were significantly decreased. These results support the role of thiamin in preventing the biochemical changes and complications in this disease (Al-Daghri et al., 2015) including autonomic neuropathy as a manifestation of thiamin deficiency.

13.8 Thiamin in Obesity

After excluding 101 individuals with prior bariatric surgery or heavy alcohol consumption, 400 patients with a body mass index of at least 35 kg/M^2 were evaluated. Sixty six patients (16.5%) were found to have clinical thiamin deficiency. Nine had gastrointestinal symptoms, 46 had cardiac manifestations, 39 had peripheral neurologic manifestations, and 3 had neuropsychiatric manifestations (Nath, Tran, Shope, & Koch, 2017).

13.9 Mitochondrial Function and Seizures

Seizures are often the key manifestation of neurological diseases caused by pathogenic mutations in 169 of the genes that have so far been identified to affect mitochondrial function. Irrespective of their origin, however, they represent an excessive acute energy demand in the brain. Accordingly, secondary mitochondrial dysfunction has been described in various epileptic disorders (Zsurka & Kunz, 2015).

When I was working at a multispecialty clinic as a pediatrician, many of the childhood seizure problems could not be controlled. A child might be admitted to hospital by a pediatric neurologist for a trial on a new drug to be replaced if unsuccessful. A 12-year-old boy was in hospital for such a drug trial. The current drug was inadvertently discontinued and, as expected, he went into *status epilepticus*. As emergency duty physician, I was called. I was experimenting with the clinical use of thiamine tetrahydrofurfuryl disulfide (TTFD) that I was receiving in both intravenous and tablet form from Takeda Chemical Industries in Japan under an independent investigator license. Each vial of TTFD for intravenous use contained 25 mg of the derivative. I injected two or three vials, amounting to 50–75 mg of thiamin. Not only did the child recover from status, he began to receive the TTFD in tablet form and was seen on the following day walking around the ward and talking to the other patients, seizure free. The neurologist normally responsible for the care of this patient never asked me a question about the use of this derivative. He discontinued the TTFD tablets and went back to his drug trial. This could not be called vitamin replacement. It was the use of thiamin as a drug (Lonsdale, 2004).

13.10 Hypoxia and Pseudohypoxia

Every 2 weeks an elderly woman participated in square dancing. After each episode she would develop a febrile illness for a few days. An abnormal transketolase indicated thiamin deficiency and in addition her blood showed hemoconcentration and polycythemia. It was concluded that thiamin deficiency had caused partial maladaptation to the equivalent of a high altitude, leading to inefficient brain oxidation and compensatory hemoconcentration. It was hypothesized that the febrile episodes were imitations of mountain sickness precipitated by the exercise stress and increased energy requirement of square dancing. With thiamin therapy the febrile episodes ceased and normal hemoconcentration and normocythemia followed. A rise in body temperature has long been associated with acute mountain sickness (Roggla, Moser, Wagner, & Roggla, 2000). Exposure to high altitude has been observed to result in a hypercoagulable state, thus predisposing to thromboembolic events. One of the predisposing factors is hemoconcentration (Gupta & Ashraf, 2012). The hypoxia in altitude sickness stimulates hypoxia-inducible factor (HIF) release (Hartman-Ksyclnska, Kluz-Zawadzka, & Lewandowski, 2016).

A study found an increase in intracellular pyruvate and extracellular lactate after thiamin deficiency exposure to a neuroblastoma cell line. Similar to cells exposed to hypoxia, there was a corresponding increase in HIF-1α stabilization and activation of target gene expression during thiamin deficiency.

Both hypoxia and thiamin deficiency exposure resulted in an increase in the expression of the thiamin transporter SLC 19 A3. Therefore, the resulting dysfunction in cellular metabolism and accumulation of pyruvate and lactate during thiamin deficiency may facilitate a pseudohypoxic state similar to that observed in hypoxic stress (Sweet & Zastre, 2013). Nine chronic mountain sickness subjects with neurological deterioration showed diffuse cerebral edema on imaging and more deranged cerebral hemodynamics. A contributing factor was exaggerated polycythemia (Bao et al., 2017). Since edema is a characteristic of thiamin deficiency, it begs the question, could thiamin deficiency be the underlying lesion?

13.11 Anxiety and Panic Attacks

Inhalation of carbon dioxide-enriched air has been used as a laboratory model for a number of anxiety disorders. Two studies were performed; in one, 35 healthy volunteers inhaled 7.5% carbon dioxide for 2 min and in study 2, 20 volunteers inhaled 20% CO_2 for 30 s. Both carbon dioxide mixtures were associated with increases in skin conductance levels and self-reported anxiety (Pappens, De Peuter, Vansteenwegen, Van den Bergh, & Van Diest, 2012). Inhalation of 35% carbon dioxide induces panic and anxiety in females with bulimia nervosa as well as panic disorder. Carbon dioxide reactivity was similar in both groups (Woznica, Vickers, Koerner, & Fracalanza, 2015). Panic disorder patients chronically hyperventilate and remain clinically unwell between attacks. Evidence from provocative studies using lactate, but particularly carbon dioxide mixtures, suggests that panic patients may have hypersensitive carbon dioxide chemoreceptors. Studies in the nonpanic state have shown EEG abnormalities as well as abnormal cerebral blood flow and cerebral glucose metabolism. These abnormalities can be interpreted as signs of cerebral hypoxia (Dratcu, 2000). Although the biological basis of uncued panic attacks is not yet understood, an important theory is the "suffocation false alarm theory." Patients showed increased brainstem activation in response to hypercapnia compared with controls and experienced divers (Goossens et al., 2014). Converging evidence suggests that patients with panic disorder have a metabolic disturbance. The consistent finding of elevated brain lactate responses to various metabolic challenges in panic disorder appears to support this model (Maddock, Buonocore, Copeland, & Richards, 2009). Hypoxia and pseudohypoxia in brain are a threat to the organism and the highly processed, calorie-dense, nutrient-depleted diet favored in the current American culture frequently leads to exaggerated supraphysiological postprandial spikes in blood glucose and lipids. This postprandial dysmetabolism induces immediate oxidant stress, including sympathetic hyperactivity (O'Keefe et al., 2008). Thus, it is hypothesized that panic attacks are fragmented fight-or-flight reflexes and may well be due to thiamin deficiency in many cases.

13.12 Neurodegenerative Disease

Emerging evidence suggests that thiamin deficiency, the cause of Wernicke's encephalopathy, produces alterations in brain function and structural damage that closely model a number of maladies in which neurodegeneration is a characteristic feature. These include Alzheimer's disease, amyotrophic lateral sclerosis, Parkinson's disease, and multiple sclerosis (Jahla & Hazell, 2011). Experimental autoimmune encephalomyelitis (EAE) has been widely used to investigate the mechanisms underlying the pathogenesis of multiple sclerosis. Chemokines, including CCL2, are involved in the development of EAE and it has been shown that thiamine deficiency induces CCL2 in neurons (Ji et al., 2014). Whether this represents a model for multiple sclerosis remains to be seen.

More than 20 syndromes, among the increasing number of neurodegenerative diseases, are known to be associated with diabetes mellitus: increased insulin resistance, obesity, disturbed insulin sensitivity, and excessive or impaired insulin secretion. Other associated diseases include Alzheimer's disease (AD), ataxia-telangiectasia, Prader–Willi syndrome, mitochondrial disorders affecting oxidative phosphorylation, and thiamin deficiency (Ristow, 2004). Reductions in brain glucose metabolism and increased oxidative stress invariably occur in Alzheimer's disease (AD) and thiamin deficiency. Both conditions cause irreversible cognitive impairment and their behavioral consequences overlap but are not identical. The activities of thiamin-dependent enzymes are characteristically diminished in AD and abnormalities in thiamin-dependent processes can be plausibly linked to its pathology (Gibson & Blass, 2007). In animal models, thiamin deficiency exacerbates plaque formation, promotes phosphorylation of tau, and impairs memory. Benfotiamin, an *S*-acyl derivative of thiamin, diminishes plaques, decreases phosphorylation of tau, and reverses memory deficits. It decreases the effects of peripheral neuropathy in diabetic humans and animals (Gibson et al., 2013). The earliest and perhaps best example of an interaction between nutrition and dementia is related to thiamin deficiency, long associated with neurological problems, including cognitive deficits and encephalopathy.

Multiple similarities exist between classical thiamin deficiency and AD (Gibson et al., 2016). Fursultiamine (thiamin tetrahydrofurfuryl disulfide, TTFD), another derivative of thiamin, has been used at an oral dose of 100 mg/day. It produced a mild beneficial effect in patients with AD. Only mildly impaired subjects showed cognitive improvement (Mimori, Katsuoka, & Nakamura, 1996). Five patients with mild to moderate AD received 300 mg a day of oral benfotiamine. It significantly improved cognitive abilities in all five patients (Pan et al., 2016).

Parkinson's disease (PD) is the second most common neurodegenerative disease. Although its cause is still unclear, numerous studies have shown that dysfunction of mitochondria plays key roles (Hu & Wang, 2016). Parenteral thiamin was used in three patients with newly diagnosed PD, leading to considerable improvement as measured by the Unified Parkinson's Disease Rating Scale (Constantini, Pala, Compagnoni, & Colangeli, 2013). Fifty patients were assessed with the Unified Parkinson's Disease Rating Scale and the Fatigue Severity Scale before being treated with 100 mg of intramuscular thiamine twice a week. They were reevaluated after 1 month and then every 3 months during treatment. This treatment led to significant improvement of motor and nonmotor symptoms and some patients with a milder phenotype had complete clinical recovery (Constantini et al., 2015). Nineteen patients with nonarthritic anterior ischemic optic neuropathy (NAION), compared with 31 age- and sex-matched control participants, were evaluated for obstructive sleep apnea syndrome (OSAS) hypertension, diabetes, hyperlipidemia, ischemic heart disease, and tobacco consumption. Of the NAION patients, 18 had OSAS, whereas in the control group only 13 had OSAS. Also the frequency of diabetes and hypertension was significantly higher in the NAION patients (Ghaleh Bandi et al., 2015). Of 57 patients with severe chronic kidney disease, only 2 did not have sleep apnea (Huang et al., 2017). These data illustrate the common incidence of OSAS and its overlap with several other common disease phenomena, suggesting that compromised oxidative metabolism is at the root. Perhaps in our present concept of disease, we are looking at the forest and not seeing the trees. To have at least two diseases with a common biochemical etiology, and possibly linked to nutrition, strongly suggests that the neurodegeneration involved in both is the endpoint of long-term nutritional deprivation, possibly linked with genetic risk. Since thiamin is the common denominator, it is hypothesized that polysymptomatic disease, usually classified as psychosomatic or trivial, represents the early deficiency of thiamin, leading to irreversible neurodegeneration years later. If this is true, it would really mean that most neurodegenerative disease is preventable. By testing for thiamin deficiency in the common polysymptomatic and emotional diseases that haunt a modern physician's office, the ideal of preventive medicine might well be established.

13.13 Thiamin in Cancer

Thiamin deficiency is increasingly recognized in medically ill patients. Its prevalence among cancer patients is unknown, however. Thiamin deficiency was found in 119 (55.3%) of 217 patients with various types of cancer. Risk factors included effects of chemotherapy, or undergoing active treatment. The majority of patients were normal in weight, overweight, or obese and 17% were receiving a multivitamin supplement. Measurement of serum thiamin concentration preceded psychiatric consultation in only 10% of cases, strongly suggesting that symptoms were merely being accepted as evidence of psychologic stress associated with the diagnosis (Isenberg-Grzeda et al., 2016). A 4-year-old boy with medulloblastoma presented with lactic acidosis in the setting of septic shock. This persisted even after his hemodynamic status improved. After administration of intravenous thiamin, the lactate level rapidly normalized and remained stable (Shah & Wald, 2015). Malignancy-associated primary thiamin deficiency has been documented in several experimental tumors, sporadic clinical case reports, and in a number of patients with fast-growing hematological malignancies. Thiamine status was assessed in 14 untreated patients with chronic lymphocytic leukemia and in 14 age-matched control patients with nonmalignant disease. An elevated TPPE, indicating thiamin deficiency, was found in five of 14 patients and none of the controls (Seligmann, Levi, Konijn, & Prok, 2001).

Thiamin supplementation in cancer cell lines stimulates growth at low doses and suppresses it at high doses. Both thiamin and dichloroacetate reduced the extent of pyruvic dehydrogenase phosphorylation, reduced glucose consumption, lactate production, and mitochondrial membrane potential (Hanberry, Berger, & Zastre, 2014). In mice with Ehrlich's ascites tumor, thiamine supplementation in doses between 12.5 and 250 times, the RDA for mice was started on day 4 of tumor inoculation. A high stimulatory effect on tumor growth of 164% was observed. At a dose of 2500 times the RDA resulted in 10% inhibition of tumor growth. The inhibitory effect at high doses of thiamin is unexplained and merits further study (Comin-Anduix et al., 2001).

14. THIAMIN ADMINISTERED WITH ANALGESICS

A study was performed to assess the influence of vitamins B1, B6, and B12 on the analgesia success achieved by diclofenac in subjects with acute lumbago. One group received 50 mg of diclofenac, +50 mg thiamin, 50 mg pyridoxine, and 1 mg of cyanocobalamin given by mouth twice daily. The other group received 50 mg diclofenac alone. The combination therapy was reported to yield superior results in pain reduction, improvement of mobility, and functionality than those receiving diclofenac alone (Mibielli et al., 2009). An article in Germany claimed that high doses of vitamins B1, B6, and B12, administered separately or in combination, can alleviate acute pain and potentiate the analgesia achieved by nonopioid analgesics when given concurrently. Therapeutic effects were observed in neuropathic and musculoskeletal pain (Jurna, 1998). A study in rats reported that TPP has biochemical and histological preventive effects on the oxidative damage induced in liver by paracetamol (acetominophen) (Mibielli et al., 2009).

14.1 Cardiac Pain

Pain related to cardiac disease has been recognized for centuries. The precise mechanisms of angina pectoris pain remain bafflingly obscure. Management of angina concentrates on methods to improve oxygen delivery. Improved imaging and other physiological assessments have demonstrated important central nervous system responses to myocardial ischemia, including activation of central nervous system areas known to be vital in chronic pain syndromes (Leach & Fisher, 2013). A review considers how ischemia at the tissue level triggers activation of afferent nociceptive pain fibers. The afferent projections of sympathetic and vagal afferent fibers are described. A difficult aspect is the wide variability of symptoms experienced by an individual patient. Neuroimaging applied at multiple levels of neuraxis-peripheral nerves, spinal cord, and brain, invites exploitation, whether pharmacological or electrical, for the benefit of the cardiac patient in pain (Rosen, 2012). In Indonesia, beriberi is still endemic but subclinical cases are not uncommon. Three patients were presented with different clinical manifestations, all of whom responded to TTFD. The authors state that a good case could be made for routine administration of thiamin to all patients in whom heart failure is present without clear evidence of the cause (Djoenaidi, Notermans, & Dunda, 1992). Since Jurna (1998) reported the alleviation of neuropathic and musculoskeletal pain using B complex, this could be a simple and completely nontoxic clinical trial.

14.2 Thiamin Supplementation in the Critically Ill

Different studies have shown that critical illness in adults and children is characterized by absolute or relative thiamin depletion, associated with as much as 50% increase in mortality. Symptoms and signs lack sensitivity and specificity. Consequently, depletion is frequently unrecognized and underdiagnosed by clinicians (Manzanares & Hardy, 2011). Three cases of hospitalized patients who had received parenteral nutrition without vitamin supplementation were reported. All three patients were found to have lactic acidosis and each presented with a different clinical condition. Intravenous administration of thiamin restored homeostasis and disappearance of neurological disturbance (Giacolone et al., 2015). The classical peracute form of beriberi, known as Shoshin, once considered an uncommon finding, is now encountered among the critically ill. Four critically ill, nonseptic, nonalcoholic patients were reported with lactic acidosis and refractory cardio-circulatory collapse caused by acute fulminant beriberi that responded to thiamin administration (Dabar, Harmouche, Habr, Riachi, & Jaber, 2015).

15. THIAMIN DEFICIENCY AFTER SURGERY

Bariatric surgery was first introduced in the 1950s and follow-up data are now available. Most of the patients will develop some nutritional deficiency after this procedure (Alvarez-Leite, 2004). One hundred twenty eight postbariatric surgery patients were subjected to the oral glucose tolerance test. There was a high incidence of adverse effects that included nausea, dizziness, weakness, diarrhea, hypoglycemia, tachycardia, tremor, and profuse sweating (Andrade, Pedrosa, Diniz, & Passos, 2016). The symptoms are those that would be expected from thiamin deficiency. A 27-year-old woman developed Wernicke encephalopathy 10 months after laparoscopic Roux-en-Y gastric bypass for morbid obesity. This complication was initially misdiagnosed, delaying treatment (Iannelli, Addeo, Novellas, & Gugenheim, 2010). The same complication occurred in a 13-year-old boy who developed acute cerebellar findings while receiving total parenteral nutrition after gastroduodenostomy for duodenal stenosis. The authors emphasize the importance of magnetic resonance imaging for diagnosis (Arslan, Ekinci, Akkus, Gocmen, & Haliloglu, 2014). The epidemic growth of morbid obesity has led to an increase in the number of bariatric interventions. It was reported that, up to December 2013, 255 patients had been diagnosed with postoperative thiamin deficiency. The majority of these patients develop dry beriberi with peripheral neuritis, ataxia, and paraplegia approximately 4–12 weeks postoperatively. These authors indicate that it can occur in the first 1–3 postoperative months (Stroh, Mayer, & Manger, 2014).

16. THIAMIN DEFICIENCY DYSAUTONOMIA

Beriberi, the classical thiamin deficiency disease is the prototype of autonomic dysfunction. It is the best known nutritional deficiency disease caused by an imbalance between ingested calories and vitamins required for their oxidation, particularly thiamin (Lonsdale, 2009). Cardiovascular autonomic imbalance is a cardinal phenotype of human heart failure and has adverse implications for cardiac, renal, immune function, for exercise capacity, for lifespan, and mode of death (Floras & Ponikowski, 2015). It suggests that thiamin treatment is virtually mandatory as clinical trial.

16.1 Complex Regional Pain Syndrome, Postural Orthostatic Tachycardia Syndrome, and Fibromyalgia Following HPV Vaccination

Complex regional pain syndrome (CRPS), postural orthostatic tachycardia syndrome (POTS), and fibromyalgia have been reported in some subjects after HPV vaccination. Sympathetic nervous system dysfunction appears to play a major role in the pathogenesis of these syndromes (Martinez-Lavin, 2015). Recent evidence points to a prominent role of the hypothalamus, operating through sympathetic activity (Geerling et al., 2014). Nineteen patients with CRPS were tested by skin capillary hemoglobin oxygenation (HBo2) and compared with two control groups, healthy subjects and surgery patients with edema. As compared to controls CRPS patients had reduced HBo2 on both sides. These results indicated that skin hypoxia occurs in CRPS. Since the second control group did not differ from the first, the authors concluded that edema could not be the reason for hypoxia (Koban, Leis, Schultze-Mosgau, & Birklein, 2003). CRPS, formally known as reflex sympathetic dystrophy, or causalgia, is a difficult therapeutic problem. Despite many divergent and often conflicting theories, the cause of severe pain, alterations in regional blood flow, and edema is unknown (Hogan & Hurwitz, 2002). To investigate oxidative metabolism in CRPS, skin lactate was measured by dermal microdialysis in 11 patients. Although venous lactate was unaltered, skin lactate was increased in patients with CRPS, suggesting enhanced anaerobic glycolysis (Birklein, Weber, & Neundorfer, 2000). It is hypothesized that B complex given to these patients might well be beneficial.

In 2013 I received an email from the mother (SF) of an 18-year-old girl who had received HPV vaccination 4 years previously and who was crippled with POTS. She had seen many physicians without any benefit and the mother had done her own library research. She had come to the conclusion that her daughter had a typical example of beriberi, a fact that made no sense at all under the circumstances. At that time, our clinical laboratory was able to do the erythrocyte transketolase test (TKA) and TPPE. The TKA was in the low normal range and the TPPE clearly showed thiamin deficiency. She has since responded partially to TTFD. Two other girls and a boy, known to SF were also crippled with POTS and all three were shown to have thiamin deficiency. Because the TKA was also abnormal in the boy, I asked his mother whether there was a history of alcoholism in the family. She replied that his father had Wernicke encephalopathy and there was a strong family history of alcoholism on the father's side, suggesting a genetic component (Lonsdale, 2015a, 2015b).

16.2 Thiamin Deficiency Eosinophilic Esophagitis

A 14-year-old boy with eosinophilic esophagitis had many symptoms, beginning in infancy, treated elsewhere as psychosomatic until the age of 8 years when upper endoscopy revealed the esophagitis. Recurrent ear infections, ADD, ADHD, and OCD marked early childhood. He complained of light and sound sensitivity and, on examination, he demonstrated CRPS by screaming on touch. An abnormal TPPE revealed abnormal thiamin homeostasis. He had eosinophilia, one of the laboratory signs of beriberi. He was treated with intravenous water-soluble vitamins that included thiamine hydrochloride. Because the erythrocyte transketolase became worse for unknown reasons, TTFD was substituted, with improvement in the TPPE and an increase in transketolase activity. At the beginning of treatment his body weight was 105 pounds, placing him in the 25th percentile. At the end of 1 year of treatment his body weight was 122 pounds, raising him to the 50th percentile. During the same time period, his stature increased from 64.5 in. to 68.5 in., raising him from the 50th to the 75th percentile. Addiction to sugar was responsible for abnormal thiamin metabolism. It is hypothesized that dysautonomia was the underlying cause of his "psychosomatic" symptoms and that his growth spurt was due to overall improvement in mitochondrial function, responsible for the dysautonomia. The widespread ingestion of sugar suggests that it might be a relatively common, but by no means unique, cause of the disease (Lonsdale, 2016).

17. THIAMIN IN OBSTETRICS

Hyperemesis gravidarum-induced Wernicke's encephalopathy is an underestimated condition. In four of five cases, ataxia and ocular signs were reported. Papilledema was observed in one patient. All five patients improved with thiamin treatment but fetal death had occurred for one patient (Ashraf, Prijesh, Praveenkumar, & Saifudheen, 2016). Thiamin (100 mg daily) proved to be 100% effective in preventing toxemia of pregnancy and other complications (Irwin, 2008) in over 1000 prenatal patients. Since there are 30,000 fetal deaths a year in the United States due to toxemia, the number of toxemia patients would be expected to be over 150 cases. The actual occurrence was zero.

17.1 Sudden Infant Death Syndrome

A medical officer of health was sent out from Britain to Hong Kong to investigate a common cause of sudden death in infants of Chinese mothers. They were often thought to be the healthiest infants in the family when they suddenly died at about 3 to 4 months, almost invariably during the night. It was found to be due to thiamine deficiency in breast milk from the rice-consuming mothers (Fehily, 1944). A historical note indicates that when the Japanese invaded Hong Kong, the Chinese mothers were severely rice rationed and although both mothers and infants were near starvation, the sudden infant death disappeared from the colony. After the Japanese retired from China, the mothers had as much rice as they wanted. The sudden death began to reappear. This suggests the catatorulin effect (Peters, 1936).

In the 1970s, our group became interested in Sudden Infant Death Syndrome (SIDS), recognizing that "near-miss" infants could appear as a clinical issue and who responded to large doses of thiamin. Fifteen such infants were tested by brainstem auditory evoked potential (BAEP) and all of them demonstrated abnormalities in two or more of the seven criteria employed in assessing results (Nodar, Lonsdale, & Orlowski, 1979). Thirty six infants identified as infant apnea syndrome (IAS) and 25 controls with a comparable age distribution were subjected to BAEP. There was a significant predilection for left-sided abnormalities in IAS patients (Orlowski, Lonsdale, Nodar, & Williams, 1982). BAEP responses were recorded from

63 near-miss SIDS infants (NMSIDS), 26 of their siblings (SSIBS), and 67 control infants. The distribution of interpeak intervals for all NMSIDS and SSIBS infants was skewed toward longer times compared to control infants, suggesting that maturation of neural processing in the brainstem of these infants may be delayed (Pettigrew & Rahilly, 1985). Auditory brainstem responses of thiamine-deficient rats were reported to be abnormal. They were corrected in 2 to 4 days after the initiation of daily intraperitoneal thiamin injections (Shigematsu, Nakai, Kuriyama, et al., 1990). A wide range of clinical findings had been found in 58 near-miss SIDS infants. Thiamin deficiency was found in 12 of 43 tested infants. Five of the deficient infants had a familial history of SIDS. Gastroesophageal reflux was found to occur in active and indeterminate sleep but not in quiet sleep (Jeffery, Rahilly, & Read, 1983). Central hypoventilation syndrome has recently been reported as a disorder of autonomic regulation and more recent evidence suggests that SIDS is also a disorder of autonomic regulation (Rand, Patwari, Carroll, & Weese-Mayer, 2013). Of 110 SIDS cases submitted to postmortem, in-depth, histological examination of the ANS, 13 cases (11.8%) had died in temporal association with administration of the hexavalent vaccine (Matturri, Del Corno, & Lavezzi, 2014). Although it has been generally accepted that moving the infant from the prone to the supine position solved the problem of SIDS, the data above suggested that this was an insufficient explanation. It has been hypothesized that some form of stressful incident (e.g., prone position, vaccination) is coupled with genetic risk and energy deficiency as three distinct components, each of which may dominate or simply contribute to the final event (Lonsdale, 2015a, 2015b). A group in Australia had measured serum thiamin concentration in 233 infants dying from SIDS and had compared them with 46 infants dying from other explicable causes. The SIDS victims had a mean concentration of 144.3 μg/L compared with a mean of 26.5 μg/L in the other group (Davis, Icke, & Hilton, 1982). TTP was found to be deficient in the phrenic nerve of a SIDS victim (Barker & Jordan, 1982).

18. AUTISM

According to the Autism Society of America, autism is now considered to be in epidemic form. Toxicity and oxidative stress may be a cause of neuronal insult and the loss of the Purkinje cells indicates their vulnerability (Kern & Jones, 2006). Postmortem studies have revealed neurobiological abnormalities in cerebellar cellular architecture (Hampson & Blatt, 2015).

It has been shown that a significant decrease in Purkinje cell density occurs in the *Cerebellar vermis* in all subjects in a group of thiamin-deficient chronic alcoholics, indicating that cerebellar Purkinje cells are selectively vulnerable to thiamin deficiency (Baker, Harding, Halliday, Kril, & Harper, 1999). Diencephalic regions and in particular, the cerebellum, demonstrate lesions in cases of prolonged thiamin deficiency, such as that observed in alcohol-dependent individuals or in patients with cancer or AIDS (Mulholland, 2006). Significant respiratory complex activity deficiencies were found in 39 patients with autistic spectrum disorder, more prevalent in severe cases (Goldenthal et al., 2015). Plasma thiamin and TMP concentrations were similar in a group of autistic children compared with healthy controls. TPP was decreased, however, by 24% in autistic children compared to the controls. Plasma protein dityrosine content was increased by as much as 88% in autistic spectrum disorder. Other oxidative markers were unchanged (Anwar et al., 2016).

18.1 Thiamin in Autism

Autism rates in the United States are increasing at a rate of 15% a year. A study was reported that uses nutritional epidemiology and an ecologic study design. It was found that the States with the highest participation in the Women, Infants, and Children (WIC) program had significantly lower autism rates. In contrast, there was a direct correlation with the increasing percentage of women exclusively breast-feeding between the years 2000 and 2004. Infants who were solely breast-fed had diets that contained less thiamin, riboflavin, and vitamin D than the minimal daily requirements. There was no correlation with MMR inoculations with the incidence of autism (Shamberger, 2011). The case histories of a mother and her two children were reported.

The mother was a recovered alcoholic. She and her two children, both of whom had symptoms that are typical of autistic spectrum disorder, had dysautonomia. All of them had intermittently abnormal erythrocyte transketolase studies indicating abnormal TPP homeostasis. Both children had unusual concentrations of urinary arsenic. All had symptomatic improvement with diet restriction and supplementary vitamin therapy but quickly relapsed after ingestion of sugar, milk, or wheat. The stress of a heavy metal burden, superimposed on existing genetic or epigenetic risk factors, may be important in the etiology of ASD when in combination. Dysautonomia has been associated with several diseases, including autism (Lonsdale et al., 2011b).

ASD is marked by a complex interaction between environmental factors and genetic predisposition, including epistasis. Evidence for abnormal thiol metabolism, marked by significant alteration in the deposition of several trace heavy metal species was presented, suggesting the biochemical alterations in transketolase, abnormal thiamin homeostasis, and the clinical presentation of dysautonomia (Obrenovich, Shola, Schroedel, Agrahari, & Lonsdale, 2015). Because of this accumulated evidence, a pilot study was carried out treating 10 autistic children with thiamin tetrahydrofurfuryl disulfide (TTFD) administered as rectal suppositories. Diagnosis was confirmed through the use of form E2, a computer assessed symptom score and symptomatic responses were determined through the use of the computer assessed Autism Treatment Evaluation Checklist (ATEC). Both forms had been developed by the Autism Research Center in San Diego and had been widely tested for authenticity. TTFD appears to have a beneficial clinical effect in autism and ASD since 8 of the 10 children improved clinically. A full-scale interinstitutional placebo-controlled study was planned, but required a new IND for the application of suppositories and this was prohibitive (Lonsdale, 2002).

I performed a survey of records from the autistic children that I had seen in my practice. This was not a protocol study. Of 157 records surveyed there were 128 boys and 29 girls. Ninety-one of these children (58%) had erythrocyte transketolase tests, TKA and TPPE. Of the 91 tests for TKA, 82 (90%) were in the normal range: only two (2%) were below the normal range, suggesting a genetic influence (Blass & Gibson, 1977) and seven (8%) were above the normal range, suggesting the possibility of vitamin B 12 deficiency (Markannen & Kalliomaki, 1966). Of the 91 studies of TPPE, 19 (21%) were over the acceptable range of 18%, indicating abnormal thiamin homeostasis. Only six (7%) were zero, indicating thiamin saturation of the enzyme and 66 (72%) were in the acceptable range of up to 18% acceleration, indicating incomplete cofactor saturation. This obviously supports the concept that autism and ASD have multiple biochemical etiologies, each of which affects energy metabolism. The evidence for abnormal thiamin homeostasis is emphasized.

19. THIAMIN DERIVATIVES

Allicin, a compound possessing the odor characteristic of garlic is produced from alliin by the action of alliinase during the grinding of fresh bulbs of garlic. Allicin conjugates with thiamin in an alkaline medium, yielding allithiamin (2′methyl-4′ amino pyrimidy-(5′) methylformamino-5-hydroxy-2-pentenyl-(3) allyl disulfide). A number of allithiamin derivatives

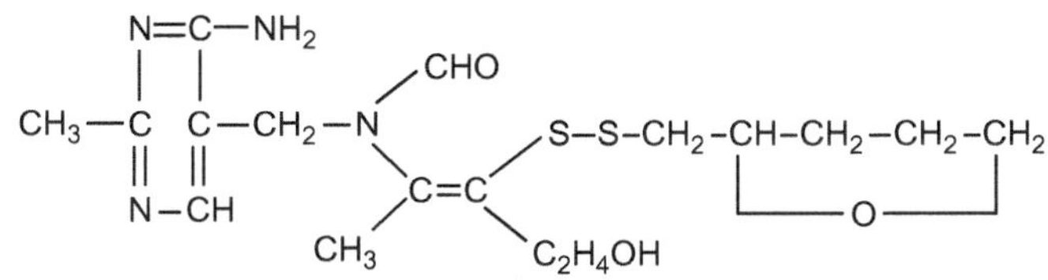

Fig. 4 Thiamin tetrahydrofurfuryl disulfide, an open ring derivative of thiamin.

were synthesized by substituting the *S*-alkyl radicals of allithiamin by different alkyl groups. Each was tested for its chemical and biological properties by the members of the Vitamin B Research Committee of Japan.

The hitherto completely unknown characteristics of allithiamin, vastly different from thiamin became clear. When allithiamin is administered by mouth, urinary excretion occurs exclusively in the form of free thiamin, because allithiamin is easily reduced in the presence of cysteine or glutathione. The open thiazolium ring form of thiamin passes through the cell membrane, and the ring closes inside the cytosol. It is this that has given rise to the concept that disulfide derivatives are fat-soluble. It refers only to the ability to pass through the lipid barrier of the cell membrane. An early derivative was thiamin propyl disulfide (TPD). This was eventually abandoned since it gave a powerful garlic odor to the patient. It was eventually replaced by TTFD and sold under the trade name of Alinamin (odorless) in Japan. It is soluble in water and can be given intravenously (Fig. 4).

19.1 Effect of Thiamin on Various Types of Intoxication

Fifty-one mice were pretreated with 1 mg of TPD given intravenously and then administered potassium cyanide in a dose of 150 μg per 10 g body weight by stomach tube. Of the 51 treated with TPD and KCN, only 6 died, whereas 36 of the mice treated with only KCN succumbed. The administration of TPD was reportedly effective in the prevention of trichloroethylene or lead intoxication. Thiamine hydrochloride alone or in combination with calcium edetate was used to treat experimentally induced lead toxicity in calves. The combination was more effective than use of thiamin alone (Dey, Swarup, & Kalicharan, & Singh, 1995). Since lead poisoning in America is not uncommon, this nontoxic treatment should be better known.

19.2 Stimulatory Effect of Allithiamin on Intestinal Mucosa

A segment of dog's jejunum was disconnected from its mesenteric innervation and TPD was applied topically to the intestinal mucosa or administered intravenously. It exerted a marked stimulatory effect on peristalsis, while ordinary thiamin was entirely devoid of such an effect.

19.3 *S*-Acyl Thiamin Derivatives

A large number of *S*-acyl thiamin homologues were synthesized, the best known being *S*-benzoyl thiamin monophosphate, also known as Benfotiamine. All of these derivatives were absorbed in the intestine far more readily at a rate comparable to TPD. However, it was reported that they are devoid of the marked property of disulfide derivatives in penetration into blood cells. Although Benfotiamine is emerging in thiamin therapy (Dakshinamurti, 2015), my particular interest has been with TTFD when I took out an independent investigator license from the FDA in 1973.

19.4 Clinical Use of TTFD

19.4.1 Animal Studies

Weanling mice of the DBA/2J strain are spontaneously audiogenically seizure prone between 19 and 24 days. Thereafter, susceptibility declines rapidly within the next 7 to 10 days. It was found that TTFD significantly delayed the natural disappearance of cholinergically activated seizure proneness in male animals compared with controls treated with identical doses of saline. Evidence was presented that suggested that TTFD is cholinergic in its action and has pharmacologic effects other than a simple vitamin replacement (Lonsdale, 1982). Due to genetic defects or illness, some individuals require higher amounts of thiamin than are typically provided by the diet. Lipid-soluble thiamin precursors can achieve high blood levels of thiamin and result in increased concentrations in the central nervous system. A study examined the effect of TTFD on behavior in juvenile male DBA/2J mice. The results supported the concept that TTFD provided during early development can affect a number of behavioral parameters by increasing whole brain thiamin concentrations (Hills, Golub, Bettendorff, & Keen, 2012).

Impaired energy metabolism is considered a possible cause of fatigue and TTFD is prescribed for its improvement. By using the forced swimming test on weight-loaded rats, the swimming duration from TTFD, but not of thiamin-treated rats, was significantly longer than that of the controls. Daily TTFD resulted in a characteristic distribution of thiamin and its phosphate esters in rat skeletal muscle, liver, kidney, heart, brain, and plasma and attenuated the decrease in ATP content in skeletal muscle caused by forced swimming with a weight load. These results indicated that TTFD exerts antifatigue affects by improving energy metabolism during physical fatigue (Nozaki et al., 2009). The effects of TTFD on the gradual increase in coronary blood flow, inherent in the canine heart–lung preparation, was studied. TTFD is reported to have an antiinflammatory effect in experimental animals. It was found that it could reverse the gradual increase in coronary blood flow through inhibition of the arachidonic acid cascade-line activation (Matsui et al., 1985). The therapeutic effect of glucosamine hydrochloride and chondroitin sulfate in combination with TTFD on the development of cartilage lesions was investigated in an animal model of osteoarthritis. Results showed that the protective effect of glucosamine hydrochloride and chondroitin sulfate is enhanced by the addition of TTFD (Kobayashi, Notoya, Nakamura, & Akimoto, 2005).

19.4.2 Human Subjects

A study was performed on students from a women's University. Each nine students went through three cross-over treatments: placebo, training treatment, and thiamin treatment. Bicycle, ergometer exercise was performed at 70% of maximal oxygen uptake and exercise intensity. The thiamin treatment group were given 10 mg of TTFD per kg for 4 weeks. Lactate and ammonia concentrations were significantly decreased with thiamin treatment after exercise (Choi, Baek, & Choi, 2013). Oral administration of TPD or TTFD rapidly increased thiamin in whole blood in erythrocytes, CSF and urine in normal and thiamin-deficient subjects. Both TPD and TTFD restored erythrocyte transketolase to normal in alcoholics with thiamin deficiency. Repletion equaled that produced by parenteral, water-soluble thiamine hydrochloride, or TPP. Orally administered disulfide derivatives of thiamin are therefore recommended for prophylaxis and treatment of thiamin deficits because, while having essentially the same biological properties as parenterally administered water-soluble thiamin, they have not produced any untoward effects after long-term administration and are far more efficiently utilized (Baker & Frank, 1976).

Three family members were reported with functional symptoms considered to be caused by intracellular deficiency of thiamin. Persistence of desaturation of erythrocyte transketolase in the face of megadose thiamine hydrochloride suggested a familial thiamin dependency state. Each of the three individuals responded clinically to the administration of TTFD. Erythrocyte transketolase became fully saturated with normal TPPE (Lonsdale, 1990). A similar problem developed in the treatment of a patient with eosinophilic esophagitis in whom the transketolase test showed abnormal thiamin homeostasis. The TKA and TPPE did not respond to thiamine hydrochloride, but did respond to intravenous TTFD. Although the mechanism is unknown, it was suspected that a transporter was at fault (Lonsdale, 2016).

20. UNUSUAL CASE REPORTS

Case 1

An African American 8-year-old girl had a 4 year history of increasingly severe nightly asthma and many episodes of bronchitis. Occasionally, there was associated fever, headache, intractable vomiting, and sweating. Auscultation revealed high-pitched expiratory wheezing, expiratory sibilant rhonchi, and crepitant rhales. Widespread piloerection on the trunk indicated sympathetic dominance in spite of sympathetic stimulants being a key approach to treatment of asthma. The maternal grandmother had chronic asthma and liver disease of unknown cause. Serum alpha-1 antitrypsin assay in the patient revealed a concentration of 394 μg/mL (0–400 μg/mL = homozygous range of deficiency). Because of sympathetic dominance a clinical trial with 150 mg/day of thiamine hydrochloride was started. During the next 5 months she experienced only two mild attacks of asthma and her weight had increased by 6.4 kg. The lungs were clear by auscultation and serum alpha-1 antitrypsin assay revealed a concentration of 1786 μg/mL, an increase of more than four times the former concentration. Nutritional factors, AAT phenotype and level were assessed on 1537 consecutive persons aged 16–90 years old. At all ages, persons with non-MM genotype had a significantly higher proportion of thiamin deficiency (50% in Pi MZ) and reactive hypoglycemia. In older persons, Pi MZ genotype had a significantly increased proportion of brain MRI T2 white matter abnormalities (Schmechel, 2007).

Case 2

A 6-year-old Caucasian boy was admitted to hospital with severe croup. Similar attacks had been occurring five or six times a year since the age of 1 year, all during the winter months. Each episode began between 2 and 4 AM, characterized by tachycardia, a barking cough, and repeated vomiting. On examination there was intercostal retraction, inspiratory wheezing, and stridor. The epiglottis was inflamed and petechial hemorrhages were seen on the face. A poor appetite had caused great maternal anxiety. He had frequent temper tantrums, restless sleep, and talked in his sleep.

Resolution followed symptomatic therapy and he was discharged from hospital. Two weeks later he was examined again. Rough ichthyosiform skin was noted below the knees, a clinical observation noted in many children whose symptoms responded to nutrient therapy. Deep tendon patellar reflexes were difficult to obtain without reinforcement. Blood pressure was 100/0 and he exhibited constant movement of a choreiform nature. Red cell TKA was 46.49 mU/L/min and TPPE 23.6%, demonstrating thiamin deficiency. After a 1 month trial of thiamine hydrochloride, 100 mg/day his general health had improved. He had a normal appetite for age, was sleeping restfully, and had no more croup. The supplement was continued and 1 year later he had remained completely well with no further episodes of croup. Thiamin supplement was discontinued and another episode of croup occurred 3 weeks later, requiring admission to hospital. After symptomatic resolution, the supplement was continued, but the case was lost to follow-up. Table 1 shows the results of measuring urinary creatine, creatinine, and uric acid over 1 year of treatment with thiamin. It should be noted that the creatine/creatinine ratio decreased in the day collections and increased in the night collections. Perhaps this is commensurate with the nighttime episodes so characteristic of this disease.

Urinary creatine excretion was measured in two patients with two different forms of metabolic aciduria. During periods of metabolic decompensation

Table 1 Creatine, Creatinine, and Uric Acid in Urine Before and After Thiamin Supplementation

				Day	**Night**	**Total**
Date	**3/77**	**2/78**	**3/77**	**2/78**	**3/77**	**2/78**
Creatine	117	16	57	87	174	103
Creatinine	130	184	162	175	292	359
C/Cr ratio	0.9	0.08	0.35	0.49	0.59	0.28
Uric acid	213	48	103	145	316	193
Body weight, kg					21.8	24.5
Creatine, mg/kg/24 h					7.98	4.5
Creatinine, mg/kg/24 h					13.4	15.7
Uric acid, mg/kg/24 h					14.5	8.4

Note the decrease in creatine and uric acid per unit body weight, whereas the concentration of creatinine increased.

the creatine/creatinine ratio increased and fell during recovery. The authors suggested that the increased creatine excretion during metabolic perturbations may result from loss from creatine-containing tissues such as muscle and may represent an underlying defect in energy metabolism (Davies, Iles, Stacey, & Chalmers, 1990).

There are several points of unusual clinical interest. Nighttime croup is a common pediatric problem, but this child had many such episodes as did his sister and two brothers. In fact the complaint of the mother was "why have all my children had repeated episodes of croup? Could it be hereditary?". Each episode of croup was accompanied by symptoms which were autonomic in nature, tachycardia and vomiting being the most obvious. The blood pressure of 100/0 was typical of thiamin deficiency, a phenomenon that I had witnessed many times in children whose symptoms were vague and variable. Poor appetite, the maternal anxiety that accompanied it, temper tantrums, and sleep characteristics supported a concept of "sympathetic dominance." It is hypothesized that the urinary biochemical changes after a year of treatment with thiamin represent fundamental recuperation of energy metabolism. On many occasions I had found that the urinary creatine/creatinine ratio was initially high and often differed considerably in the day/night collections. The normalization of this ratio was an invariable response that coincided with clinical improvement.

21. CONCLUSION

It is well known that a drug used to treat a child with ADD or ADHD will produce a worsening of symptoms. When megadoses of nutrients are used in treatment, the symptoms often become exaggerated and continue for a variable length of time that is unpredictable. Although the mechanism is unknown, perhaps we can take a leaf out of the book of the ancient Chinese. They had formulated the philosophy of Yin and Yang whose pictograms indicated extremes. In the practice of acupuncture they were able to recognize that the symptoms of the patient they treated would improve to a certain point. If treatment continued, the original symptoms would begin to return. This was thought of as a point of balance between extremes and is formulated in the modern world as "moderation in all things." Because exaggeration of symptoms experienced by my patients was obviously initially unexpected, I called it "paradox," the opposite of the patient's expectation. We found that "paradox" was much more common when water-soluble vitamins were given intravenously and the patient was always forewarned. It did not always occur but it was invariably a feature that forecast eventual improvement, although its extent could last from days to weeks. It would of course be natural that the patient would interpret "paradox" as side effects and discontinue treatment unless this was explained. Paradox may occur with TTFD treatment taken by mouth.

21.1 Paradox in Treatment of Dysautonomia

Here is the report of a patient who had suffered lifelong dysautonomia: "A few days after I started TTFD, my migraines were even worse than normal. I was terribly tired and could not sleep. My body aches and pains felt like "the flu." My paradox effect went on for more than 4 weeks, but eventually subsided. Not too long after, I felt better than I had in years. I have no headaches and more energy than I know what to do with. I also sleep better than I ever have. My 14-year-old daughter also had the paradox effect. She had been suffering from joint pains and TMJ (Tempero Mandibular Joint Syndrome). The doctors were suspecting juvenile rheumatoid arthritis. After she started the TTFD she felt worse for about 2 weeks. Now, she has been running her fastest cross country times ever, beating her personal best almost every time she runs.

The symptoms of the mother and her daughter were quite different but strongly suggestive of mitochondrial energy deficiency as the underlying cause in common. It is the polysymptomatic and variable nature of the symptoms that provide the major clue to mitochondrial dysfunction. One hundred and four patients with migraine, age 34 ± 10 years included 28 with and 76 without hypertension. The data revealed increased vasomotor reactivity in migraine patients, with or without concomitant hypertension and frequently associated with a family history of cardiovascular disease (Babayan, Mamontov, Amelin, Bogachev, & Kamshilin, 2017). Fifty cases of rheumatoid arthritis were investigated for dysautonomia, compared with 82 healthy subjects. Thirty of these patients with rheumatoid arthritis were found to have ANS dysfunction (Toussirot, Serratrice, & Valentin, 1993). Both of these references support the contention that the ANS is frequently involved in the etiology of disease and that defective energy metabolism is the underlying cause.

21.2 Thiamin in Treatment of Diseases Other Than Beriberi

In 1962 a symposium was held to celebrate the synthesis of thiamin in 1936. One of the manuscripts reported 696 published papers in which thiamin had been used to treat 230 different diseases, many of which responded surprisingly well (Zbinden, 1962). There was major research interest in thiamin between 1938 and 1940 and again between 1952 and 1955, so it is hard to understand why vitamin therapy in the 1970s and 1980s was regarded virtually as "quackery." It is clear from the references provided in this chapter that we are not dealing with simple vitamin replacement. Megadose use implies that the cofactor is being used as a drug in stimulating the enzyme cofactor relationship. Even beriberi, recognized as one of the major vitamin deficiency diseases, required megadoses of thiamin for months to restore the health of the patient. It is not sufficiently recognized that chronic nutritional deficiency, experienced over years, causes damage that is increasingly difficult to treat. It has been suggested in this chapter that the end result is an assortment of chronic degenerative brain diseases. If true, this makes nutritional prevention an imperative because "cure" may well be a "pipe dream." Thiamin is the central figure in energy metabolism, thus explaining the polysymptomatic presentation of beriberi in its early stages. The mortality from the disease is relatively low whereas the morbidity is prolonged and chronic. It may be the most important cause of mitochondrial energy insufficiency and perhaps explain many of the conditions that haunt the modern physician's office, masquerading as psychosomatic disease.

22. A QUESTION OF HISTORICAL SIGNIFICANCE

There are two historical figures whose mysterious illnesses are so similar that it strongly suggests that each of them suffered from lifelong dysautonomia and that thiamin deficiency may have been the central underlying cause. Both were geniuses, implying great brain energy demand.

Holmes, in "The Life of Mozart," wrote in 1845 "He [Mozart] sank over his Composition into frequent swoons, in which he remained for several minutes, before consciousness returned" (Holmes, 1845). In 1891, Jahn wrote "These fainting fits exhausted his strength and increased his depression" (Jahn, 1891). In the *Journal of the Royal Society of Medicine* in September 1983, it was reported that Mozart experienced many episodes of a combination of fever, joint pains, abdominal colic, and vomiting between August 1784 and August 1790. These symptoms are those that are recorded in beriberi and the fainting episodes are typical of syncope that often occurs in POTS. Money was a constant problem and may well have been reflected in the quality of his diet.

For over 40 years, Charles Darwin suffered intermittently from various combinations of symptoms. These included malaise, vertigo, dizziness, muscle spasms, tremors, vomiting, abdominal pain (colic), bloating, headaches severe fatigue, dyspnea, and a variety of skin problems. He experienced a sensation of impending death, fainting, tachycardia, insomnia, and tinnitus (Wikipedia). These symptoms may well have been mistakenly diagnosed as psychosomatic. Darwin was reportedly careful concerning diet. Perhaps his acknowledged brilliance consumed more energy in the brain than it could synthesize, thus explaining the intermittent nature of the symptomology and its relationship with stress.

REFERENCES

Adamolekun, B., & Ndububa, D. A. (1994). Epidemiology and clinical presentation of a seasonal ataxia in western Nigeria. *Journal of the Neurological Sciences, 124*(1), 95–98.

Al-Daghri, N. M., Alharbi, M., Wani, K., Abd-Alrahman, S. H., Sheshah, E., & Alokail, M. S. (2015). Biochemical changes correlated with blood thiamine and its phosphate esters levels in patients with diabetes type I (DMT1 E). *International Journal of Clinical & Experimental Pathology, 8*(10), 13483–13488.

Alvarez-Leite, J. I. (2004). Nutrient deficiencies secondary to bariatric surgery. *Current Opinion in Clinical Nutrition & Metabolic Care,* 7(5), 569–575.

Andrade, H. F., Pedrosa, W., Diniz, M. F., & Passos, V. M. (2016). Adverse effects during the oral glucose tolerance test in post-bariatric surgery patients. *Archives in Endocrinology & Metabolism, 60*(4), 307–313. Feb 23. pii: S2359-39972016005002110.

Anwar, A., Marini, M., Abruzzo, P. M., Bolotta, A., Ghezzo, A., Visconti, P., et al. (2016). Quantitation of plasma thiamine, related metabolites and plasma protein oxidative damage markers in children with autistic spectrum disorder and healthy controls. *Free Radical Research, 50*(Suppl. 1), S85–S90.

Arslan, E. A., Ekinci, S., Akkus, P. Z., Gocmen, R., & Haliloglu, G. (2014). Wernicke encephalopathy due to thiamine after surgery on a child with duodenal stenosis. *Pediatric Neurology, 51*(6), 840–842.

Asher, B. F., & Guiford, F. T. (2016). Oxidative stress and low glutathione in common ear, nose, and throat conditions: A systematic review. *Alternative Therapies in Health and Medicine, 22*(5), 44–50.

Ashraf, V. V., Prijesh, J., Praveenkumar, R., & Saifudheen, K. (2016). Wernnicke's encephalopathy due to hyperemesis gravidarum: Clinical and magnetic resonance imaging characteristics. *Journal of Postgraduate Medicine, 62*(4), 260–263.

Babayan, L., Mamontov, O. V., Amelin, A. V., Bogachev, M., & Kamshilin, A. A. (2017). Arterial hypertension in migraine: role of familial history and cardiovascular phenotype. *Autonomic Neuroscience, 203,* 103–107. https://doi.org/10.1016/J.autneu.2017.01.004 [Epub 2017 Jan 25].

Baker, H., & Frank, O. (1976). Absorption, utilization and clinical effectiveness of allithiamines compared to water-soluble thiamine. *Journal of Nutritional Science & Vitaminology (Tokyo), 22*(Suppl), 63–68.

Baker, K. G., Harding, A. J., Halliday, G. M., Kril, J. J., & Harper, C. G. (1999). Neuronal loss in functional zones of the cerebellum of chronic alcoholics with and without Wernicke's encephalopathy. *Neuroscience, 91*(2), 429–438.

Banka, S., de Goede, C., Yue, W. W., Morris, A. A., von Bremen, B., Chandler, K. E., et al. (2014). Expanding the clinical and molecular spectrum of thiamine pyrophosphokinase deficiency: A treatable neurological disorder caused by TPK1 mutations. *Molecular Genetics & Metabolism, 113*(4), 301–306.

Bao, H., Wang, D., Zhao, X., Wu, Y., Yin, G., Meng, L., et al. (2017). Cerebral edema in chronic mountain sickness: A new finding. *Science Reports*, 7, 43224. https://doi.org/10.1038/srep43224.

Barker, J. N., & Jordan, F. (1982). Phrenic thiamine and neuropathy in sudden infant deaths. In H. C. Sable & C. J. Gubler (Eds.), *Annals of the New York Academy of Sciences*: *Vol. 378. Thiamin: 20 years of progress* (pp. 449–452).

Baumgartner, TG. (1991). What the practicing nurse should know about thiamine. *Journal of Intravenous Nursing, 14*(2), 130–135.

Bettendorff, L., Grandfils's, C., Wins, P., & Schoffeniels, E. (1989). Thiamin triphosphatase in the membranes of the male electric organ of Electroforus electricus: Substrate-enzyme interactions. *Journal of Neurochemistry, 53*(3), 738–746.

Bettendorff, L., Kolb, H. A., & Schoffeniels, E. (1993). Thiamin triphosphate activates an anion channel of large unit conductance in neuroblastoma cells. *Journal of Membrane Biology, 136*(3), 281–288.

Bettendorff, L., Lakaye, B., Kohn, G., & Wins, P. (2014). Thiamin triphosphate: A ubiquitous molecule in search of a physiological role. *Metabolic Brain Disease, 29*(4), 1069–1082.

Bettendorff, L., & Wins, P. (2009). Thiamin diphosphate in biological chemistry: New aspects of thiamin metabolism, especially triphosphate derivatives acting other than as cofactors. *FEBS Journal, 276*(11), 2917–2925.

Bill, R. M. (2015). Recombinant proteins subunit vaccines synthesis in microbes: A role for yeast? *Journal of Pharmacy & Pharmacology, 67*(3), 319–328.

Birklein, P. E., Weber, M., & Neundorfer, B. (2000). Increased skin lactate in complex regional pain syndrome: Evidence for tissue hypoxia? *Neurology, 55*(8), 1213–1215.

Blass, J. P., Avigan, J., & Uhlendorf, B. W. (1970). A defect in pyruvate decarboxylase with an intermittent movement disorder. *Journal of Clinical Investigation, 49*(3), 423–432.

Blass, J. P., & Gibson, G. E. (1977). Abnormality of a thiamine-requiring enzyme in patients with Wernicke-Korsakoff syndrome. *New England Journal of Medicine, 297*(25), 1367–1370.

Boni, L., Kieckens, L., & Hendricz, A. (1980). An evaluation of a modified erythrocyte transketolase assay for assessing thiamin nutritional adequacy. *Journal of Nutritional Science & Vitaminology. (Tokyo), 26*(5), 507–514.

Braverman, N. E., & Moser, A. B. (2012). Functions of plasmalogen lipids in health and disease. *Biochimica et Biophysica Acta, 1822*(9), 1442–1452.

Brown, G. (2014). Defects of thiamin transport and metabolism. *Journal of Inherited Metabolic Disease, 37*(4), 577–585.

Burrage, L. C., Nagamani, S. C., Campeau, P. M., & Lee, B. H. (2014). Branched-chain amino acid metabolism: From rare Mendelian diseases to more common disorders. *Human Molecular Genetics, 23*(R1), R1–R8.

Casteels, M., Snieckers, M., Fraccasia, P., Mannaerts, G. P., & Van Veldhoven, P. P. (2007). The role of 2-hydroxy acyl-CoA lyase, a thiamin pyrophosphate dependent enzyme, in the peroxisomal metabolism of 3-methyl branched fatty acids and 2-hydroxy straight-chain fatty acids. *Biochemical Society Transactions, 35*(5), 876–880.

Choi, S. K., Baek, S. H., & Choi, S. W. (2013). The effects of endurance training and thiamin supplementation on anti-fatigue during exercise. *Journal of Exercise Nutrition & Biochemistry, 17*(4), 189–198.

Chuang, D. T., Chuang, J. L., & Wynn, R. M. (2006). Lessons from genetic disorders of branched-chain amino acid metabolism. *Journal of Nutrition, 136*(1 Suppl), 243S–249S.

Comin-Anduix, B., Boren, J., Martinez, S., Moro, C., Centelles, J. J., & Trebukhina, R. (2001). The effect of thiamine supplementation on tumor proliferation. A metabolic control analysis study. *European Journal of Biochemistry, 268*(15), 4177–4182.

Constantini, A., Pala, M. I., Compagnoni, L., & Colangeli, M. (2013). High-dose thiamine as initial treatment for Parkinson's disease. *BMJ Case Reports*. https://doi.org/bcr-2013-009289/bcr-2013-009289.

Constantini, A., Pala, M. I., Grossi, E., Mondonico, S., Cardelli, L. E., et al. (2015). Long-term treatment with high-dose thiamine in Parkinson disease: An open-label pilot study. *Journal of Alternative and Complementary Medicine*, *21*(12), 740–747.

Cooper, J. R., Itokawa, Y., & Pincus, J. H. (1969). Thiamine triphosphate deficiency in subacute necrotizing encephalomyelopathy. *Science*, *164*(3875), 74–75.

Cooper, J. R., & Pincus, J. H. (1979). The role of thiamine in nervous tissue. *Neurochemical Research*, *4*, 223–239.

Dabar, G., Harmouche, C., Habr, B., Riachi, M., & Jaber, B. (2015). Shoshin beriberi in critically-ill patients: Case series. *Nutrition Journal*, *14*, 51. https://doi.org/10.1186/s12937-015-0039-7.

Dakshinamurti, K. (2015). Vitamins and their derivatives in the prevention and treatment of metabolic syndrome diseases (diabetes). *Canadian Journal of Physiology & Pharmacology*, *93*(5), 355–362.

Dalbeth, N., Phipps-Green, A., House, M. E., Gamble, G. D., Horne, A., Stamp, L. K., et al. (2015). Body mass index modulates the relationship of sugar-sweetened beverage intake with serum urate concentrations and gout. *Arthritis Research & Therapy*, *17*, 263. https://doi.org/10.1186/S13075-015-0781-4.

Davies, S. E., Iles, R. A., Stacey, T. E., & Chalmers, R. A. (1990). Creatine metabolism during metabolic perturbations in patients with organic acidurias. *Clinica Chimica Acta*, *194*(2–3), 203–217.

Davis, R. E., Icke, G. C., & Hilton, J. M. (1982). High serum thiamin and the sudden infant death syndrome. *Clinica Chimica Acta*, *123*(3), 321–328.

Dey, S., Swarup, D., Kalicharan, & Singh, B. (1995). Treatment of lead toxicity in calves. *Veterinary & Human Toxicology*, *37*(3), 230–232.

Distelmaier, F., Huppke, P., Pieperhoff, P., Amunts, K., Schaper, J., Morava, E., et al. (2014). Biotin-responsive basal ganglia disease: A treatable differential diagnosis of Leigh syndrome. *JIMD Reports*, *13*, 53–57. https://doi.org/10.1007/8904_2013_271.

Djoenaidi, W., Notermans, S. L., & Dunda, G. (1992). Beriberi cardiomyopathy. *European Journal of Clinical Nutrition*, *46*(3), 227–234.

Dratcu, L. (2000). Panic, hypoventilation and perpetuation of anxiety. *Progress in Neuro-Psychopharmacology & Biological Psychiatry*, *24*(7), 1069–1089.

Elmadfa, I., Majchrcak, D., Rust, P., & Genser, D. (2001). The thiamine status of adult humans depends on carbohydrate intake. *International Journal for Vitamin & Nutrition Research*, *71*(4), 217–221.

Fassone, E., Wedatilake, Y., DeVile, C. J., Chong, W. K., Carr, L. J., & Rahman, S. (2013). Treatable Leigh-like encephalopathy presenting in adolescence. *BMJ Case Reports*, *2013*, 200838. https://doi.org/10.1136/bcr-2013-200838.

Fehily, L. (1944). Human milk intoxication due to B1 avitaminosis. *British Medical Journal*, *2*, 590–592.

Fernhoff, P. M., Lubitz, D., Danner, D. J., Dembure, P. P., Schwartz, H. P., Hillman, R., et al. (1985). Thiamine response in maple syrup urine disease. *Pediatric Research*, *19*(10), 1011–1016.

Floras, J. S., & Ponikowski, P. (2015). The sympathetic/parasympathetic imbalance and heart failure with reduced ejection fraction. *European Heart Journal*, *36*(30), 1974–1982.

Frank, L. L. (2015). Thiamin in clinical practice. *Journal of Parenteral & Enteral Nutrition*, *39*(5), 503–520.

Gangolf, M., Wins, P., Thiry, M., EWl Moualij, B., & Bettendorff, L. (2010). Thiamine triphosphate synthesis in rat brain occurs in mitochondria and is coupled to the respiratory chain. *Journal of Biological Chemistry*, *285*(1), 583–594.

Geerling, J. J., Boon, M. R., Kooijman, S., Parlevliet, E. T., Havekes, L. M., Romijn, J. A., et al. (2014). Sympathetic nervous system control of triglyceride metabolism: Novel concepts derived from recent studies. *Journal of Lipid Research, 55*(2), 180–189.
Ghaleh Bandi, M. F., Naserbakht, M., Tabasi, A., Marghaiezadeh, A., Riazee Esfahan, M., & Golzarian, Z. (2015). Obstructive sleep apnea syndrome and non-arthritic anterior ischemic optic neuropathy: A case-control study. *Medicinal Journal of the Islamic Republic of Iran, 22*(29), 300.
Giacolone, M., Martinelli, R., Abramo, A., Rubino, A., Pavoni, V., Iacconi, P., et al. (2015). Rapid reversal of severe lactic acidosis after thiamine administration in critically ill adults: A report of 3 cases. *Nutrition in Clinical Practice, 30*(1), 104–110.
Gibson, G. E., & Blass, J. P. (2007). Thiamine-dependent processes and treatment strategies in neurodegeneration. *Antioxid Ants & Redox Signalling, 9*(10), 605–1619.
Gibson, G. E., Hirsch, J. A., Cirio, R. T., Jordan, B. D., Fonzetti, P., & Elder, J. E. (2013). Abnormal thiamine-dependent processes in Alzheimer's disease. Lessons from diabetes. *Molecular & Cellular Neuroscience, 55*, 17–25. https://doi.org/10.1016/j.mcn.2012.09.001.
Gibson, G. E., Hirsch, J. A., Cirio, R. T., Jordan, B. D., Fonzetti, P., & Elder, J. E. (2016). Vitamin B1 (thiamine) and dementia. *Annals of the New York Academy of Sciences, 1367*(1), 21–30.
Gioda, C. R., Capettini, L. S., Cruz, J. S., & Lemos, V. S. (2014). Thiamine deficiency leads to reduced nitric oxide production and vascular dysfunction in rats. *Nutrition, Metabolism & Cardiovascular Disease, 24*(2), 183–188.
Goldenthal, M. G., Damle, S., Sheth, S., Shah, N., Melvin, J., Jethva, R., et al. (2015). Mitochondrial enzyme dysfunction in autism spectrum disorders; a novel biomarker revealed from buccal swab analysis. *Biomarkers in Medicine, 9*(10), 957–965.
Goossens, L., Leibold, N., Peeters, R., Esquivel, G., Knuts, I., Backes, W., et al. (2014). Brainstem response to hypercapnia: A symptom provocation study into the pathophysiology of panic disorder. *Journal of Psychopharmacology, 28*(5), 449–456.
Gupta, N., & Ashraf, M. Z. (2012). Exposure to high altitude: A risk factor for venous thromboembolism? *Seminar in Thrombosis & Hemostasis, 38*(2), 156–163.
Hampson, D. R., & Blatt, G. J. (2015). Autism spectrum disorders and neuropathology of the cerebellum. *Frontiers in Neuroscience, 9*, 420. https://doi.org/10.3389/fnins.2015.00420.
Hanberry, B. S., Berger, R., & Zastre, J. A. (2014). High-dose vitamin B1 reduces proliferation in cancer cell lines analogous to dichloroacetate. *Cancer Chemotherapy and Pharmacology, 73*(3), 585–594.
Hansen, T. W. (2000). Bilirubin oxidation in brain. *Molecular Genetics & Metabolism, 71*(1–2), 411–417.
Hartman-Ksyclnska, A., Kluz-Zawadzka, J., & Lewandowski, B. (2016). High altitude illness. *Przeglad Epidemiologiczny, 70*(3), 490–499.
Heintz-Buschart, A., May, P., Laczny, C. C., Lebrun, A., Bellora, C., Krishna, A., et al. (2016). Integrated multi-omics of the human gut microbiome in a case study of familial type I diabetes. *Nature Microbiology, 2*, 16180.
Hills, J. I., Golub, M. S., Bettendorff, L., & Keen, C. I. (2012). The effect of thiamin tetrahydrofurfuryl disulfide on behavior of juvenile DBA/2J mice. *Neurotoxicology & Teratology, 34*(2), 242–252.
Hogan, C. J., & Hurwitz, S. R. (2002). Treatment of complex regional pain syndrome of the lower extremity. *Journal of American Orthopedic Surgery, 10*(4), 281–289.
Holmes. (1845). *Life of Mozart*. London: Chapman and Hall.
Hu, Q., & Wang, G. (2016). Mitochondrial dysfunction in Parkinson's disease. *Transl Neurodegener, 5*, 14. https://doi.org/10.1186/s40035-016-0060-6.
Huang, H. C., Walters, G., Talaulikar, G., Figurski, D., Carroll, A., & Hurwitz, M. (2017). Sleep apnea prevalence and chronic kidney disease-association with total body water and symptoms. *BMC Nephrology, 18*(1), 125. https://doi.org/10.1186/s12882-017-0544-3.

Iannelli, A., Addeo, P., Novellas, S., & Gugenheim, J. (2010). Wernicke's encephalopathy after laparoscopic Roux-en-Y gastric bypass: A misdiagnosed complication. *Obesity Surgery, 20*(11), 1594–1596.

Inouye, K., & Katsura, E. (1965a). Etiology and pathology of beriberi. In N. Shimazono & E. Katsura (Eds.), *Beriberi and Thiamine* (pp. 1–29). Tokyo: Igaku Shoin Ltd.

Irwin, J. B. (2008). *The natural way to a trouble-free pregnancy: The toxemia/thiamine connection.* Aslan Publishing Fairfield. CT 06825.

Isenberg-Grzeda, E., Shen, M. J., Alici, Y., Wills, J., Nelson, C., & Breitbart, W. (2016). High rate of thiamine deficiency among inpatients with cancer referred for psychiatric consultation: results of a single site prevalence study. *Psychooncology, 26*(9), 1384–1389. https://doi.org/10.1002/pon.4155.

Iwata, H., Yabushita, Y., Doi, T., & Matsuda, T. (1985). Synthesis of thiamine triphosphate in rat brain in vivo. *Neurochemical Research, 10*(6), 779–787.

Jahla, S. S., & Hazell, A. S. (2011). Modeling neurodegenerative disease pathophysiology in thiamin deficiency: Consequences of impaired oxidative metabolism. *Neurochemistry International, 58*(3), 248–260.

Jahn, O. (1891) Life of Mozart. Translated from the German by P. D. Townsend. Novello, Ewer, & Co, London.

Jankowska-Kulawy, A., Bielarczyk, H., Pawelczyk, T., Wroblewska, M., & Szutowicz, A. (2010). Acetyl-CoA and acetylcholine metabolism in nerve terminal compartment of thiamine deficient rat brain. *Journal of Neurochemistry, 115*(2), 333–342.

Jansen, B. C. T., & Donath, W. F. (1926). On the isolation of the antiberiberi vitamin. *Proceedings of the Koninklijke Nederlandse Academie van Wetenschappen, 29*, 1390–1400.

Jeffery, H. E., Rahilly, B., & Read, D. J. (1983). Multiple causes of asphyxia in infants at high risk for sudden infant death. *Archives of Disease in Childhood, 58*(2), 92–100.

Jeyasingham, M. D., Pratt, O. E., Burns, A., Shaw, G. K., Thomson, A. D., & Marsh, A. (1987). The activation of red blood cell transketolase in groups of patients especially at risk from thiamin deficiency. *Psychological Medicine, 17*(2), 311–318.

Ji, Z., Fan, Z., Zhang, Y., Yu, R., Yang, H., & Zhou, C. (2014). Thiamine deficiency promotes T cell infiltration in experimental autoimmune encephalomyelitis: The involvement of CCL2. *Journal of Immunology, 193*(5), 2157–2167.

Johnson, S. (2001). The multifaceted and widespread pathology of magnesium deficiency. *Medical Hypotheses, 56*(2), 163–170.

Jurna, I. (1998). Analgesic and analgesia-potentiating action of B vitamins. *Schmerz, 12*(2), 136–141.

Kara, B., Genc, H. M., Uyur-Yalcin, E., Sakarya-Gunes, A., Yopcu, U., Mulayim, S., et al. (2017). Pyruvate dehydrogenase-E1α deficiency presenting as recurrent acute proximal muscle weakness of upper and lower extremities in an 8-year-old boy. *Neuromuscular Disorders, 27*(1), 94–97.

Kern, J. K., & Jones, A. M. (2006). Evidence of toxicity, oxidative stress, and neuronal insult in autism. *Journal of Toxicology & Environmental Health, 9*(6), 485–499.

Kimura, Y. H., & Iwashima, A. (1987). Occurrence of thniaminase II in *Saccharomyces cerevisiae. Experientia, 43*(8), 888–890.

Koban, M., Leis, S., Schultze-Mosgau, S., & Birklein, F. (2003). Tissue hypoxia in complex regional pain syndrome. *Pain, 104*(1–2), 149–157.

Kobayashi, T., Notoya, K., Nakamura, A., & Akimoto, K. (2005). Fursultiamine, a vitamin B1 derivative, enhances chondroprotective effects of glucosamine hydrochloride's and chondroitin sulfate in rabbit experimental osteoarthritis. *Inflammation Research, 54*(6), 249–255.

Koopman, F. A., Tang, M. W., Vermeij, J. M., de Hair, M. G., Choi, I. Y., Vervoordeldonk, M. J., et al. (2016). Autonomic dysfunction precedes development of rheumatoid arthritis: A prospective cohort study. *eBioMedicine, 6*, 231–237.

Lake, N. J., Bird, M. J., Isohanni, P., & Paetau, A. (2015). Leigh syndrome: Neuropathology and pathogenesis. *Journal of Neuropathology & Experimental Neurology*, *74*(6), 482–492.
Leach, A., & Fisher, M. (2013). Myocardial ischemia and cardiac pain—A mysterious relationship. *British Journal of Pain*, 7(1), 23–30.
Lonsdale, D. (1981). The syndrome of functional dysautonomia. *Medical Hypotheses*, 7(4), 495–502.
Lonsdale, D. (1982). The effect of thiamin tetrahydrofurfuryl disulfide on audioenics seizures in DBA/2J mice. *Developments in Pharmacology & Therapeutics*, *4*(1–2), 28–36.
Lonsdale, D. (1988). Red cell transketolase studies in a private practice specializing in nutritional correction. *Journal of the American College of Nutrition*, 7(1), 61–67.
Lonsdale, D. (1990). Hypothesis and case reports: Possible thiamin deficiency. *Journal of the American College of Nutrition*, *9*(1), 13–17.
Lonsdale, D. (2002). Treatment of autistic spectrum children with thiamin tetrahydrofurfuryl disulfide: A pilot study. *Neuroendocrinology Letters*, *23*(4), 303–308.
Lonsdale, D. (2004). Thiamine tetrahydrofurfuryl disulfide: A little-known therapeutic agent. *Medical Science Monitor*, *10*(9), RA199–RA203. www.medscimonit.com.
Lonsdale, D. (2007). Three case reports to illustrate clinical applications in the use of erythrocyte transketolase. *Evidence Based Complementary & Alternative Medicine*, *4*(2), 247–250.
Lonsdale, D. (2009). Dysautonomia, a heuristic approach to a revised model for etiology of disease. *Evidence Based Complementary & Alternative Medicine*, *6*(1), 3–10.
Lonsdale, D. (2015a). Thiamin and magnesium deficiencies: Keys to disease. *Medical Hypotheses*, *84*(2), 129–134.
Lonsdale, D. (2015b). Sudden infant death syndrome and abnormal metabolism of thiamin. *Medical Hypotheses*, *85*(6), 922–926.
Lonsdale, D. (2016). Is eosinophilic esophagitis a sugar sensitive disease? *Journal of Gastric Disorders & Therapy*, *2*(1). https://doi.org/10.16966/2381-8689.114.
Lonsdale, D., & Barber, D. H. (1964). Maple-syrup-urine disease. Report of a case, with a pedigree. *New Engl and Journal of Medicine*, *271*, 1338–1341.
Lonsdale, D., Faulkner, W. R., Price, J. W., & Smeby, R. R. (1969). Intermittent cerebellar ataxia associated with hyperpyruvic acidemia, hyperalaninemia, and hyperalaninuria. *Pediatrics*, *43*(6), 1025–1034.
Lonsdale, D., & Shamberger, R. J. (1980). Red cell transketolase as an indicator all of nutritional deficiency. *The American Journal of Clinical Nutrition*, *33*(2), 205–211.
Lonsdale, D., Shamberger, R. J., & Obrenovich, M. E. (2011a). Exaggerated autonomic asymmetry: A clue to nutrient deficiency dysautonomia. *Web medCentral Alternative Medicine*, *2*(4), WMC001854.
Lonsdale, D., Shamberger, R. J., & Obrenovich, M. E. (2011b). Dysautonomia in autism spectrum disorder: case reports of a family with review of the literature. *Autism Research & Treatment*, *2011*, 129795. https://doi.org/10.1155/2011/129795.
Low, P. A. (1994). Autonomic neuropathies. *Current Opinions in Neurology*, 7(5), 402–406.
Maddock, R. J., Buonocore, M. H., Copeland, L. E., & Richards, A. L. (2009). Elevated brain lactate responses to neural activation and panic disorder: A dynamic 1H-MRS study. *Molecular Psychiatry*, *14*(5), 537–545.
Manzanares, W., & Hardy, G. (2011). Thiamine supplementation in the critically ill. *Current Opinions in Clinical Nutrition & Metabolic Care*, *14*(6), 610–617.
Markannen, T., & Kalliomaki, J. L. (1966). Transketolase activity in blood cells in various clinical conditions. *The American Journal of Medical Sciences*, *252*(5), 564–569.
Martinez-Lavin, M. (2015). Hypothesis: Human papilloma virus vaccination syndrome—Small fiber neuropathy and dysautonomia could be its underlying pathogenesis. *Clinical Rheumatology*, *34*(7), 1165–1169.
Massod, M. F., McGuire, S. L., & Werner, K. R. (1971). Analysis of blood transketolase activity. *American Journal of Clinical Patholology*, *55*(4), 465–470.

Matsui, K., Nakahara, H., Watanabe, H., Tamatsu, H., Nakazawa, M., Nakagawa, Y., et al. (1985). Inhibition by thiamin tetrahydrofurfuryl disulfide (TT FD) of the arachidonic acid cascade-line activation as evidenced in the heart-lung preparation of the dog. *Japanese Journal of Pharmacology*, *39*(3), 375–379.

Matturri, L., Del Corno, G., & Lavezzi, A. M. (2014). Sudden infant death following hexavalent vaccination: A neuropathologic study. *Current Medicinal Chemistry*, *21*(7), 941–946.

Mayr, J. A., Freisinger, P., Schlachter, K., Rolinski, B., Zimmermann, F. A., Scheffner, T., et al. (2011). Thiamine pyrophosphokinase deficiency in encaephalpathic children with defects in the pyruvate oxidation pathway. *American Journal of Human Genetics*, *89*(6), 806–812.

Mibielli, M. A., Geller, M., Cohen, J. C., Goldberg, S. G., Cohen, M. T., & Nun, C. P. (2009). Diclofenac plus B vitamins versus diclofenac monotherapy in lumbago: The DOLOR study. *Current Medical Research and Opinion*, *25*(11), 2589–2599.

Mikstiene, V., Songailiene, J., Byckova, J. E., Rutkauskiene, G., Jasinskiene, E., Verkauskiene, R., et al. (2015). Thiamin responsive megaloblastic anemia syndrome: A novel homozygous SLC 19 A2 gene mutation identified. *American Journal of Medical Genetics. Part A*, *167*(7), 1605–1609.

Mimori, Y., Katsuoka, H., & Nakamura, S. (1996). Thiamine therapy in Alzheimer's disease. *Metabolic Brain Disease*, *11*(1), 89–94.

Minz, B. (1938). Sur la liberation de la vitamin par le tronc isole de nerf pneumogastrique soumis a la excitation electrique. *C.R. Soc Biol (Paris)*, *127*, 1251–1263.

Morgantini, C., Xiao, C., Dash, S., & Lewis, G. P. (2014). Dietary carbohydrates and intestinal lipoprotein production. *Current Opinions in Clinical Nutrition & Metabolic Care*, *17*(4), 355–359.

Mulholland, P. J. (2006). Susceptibility of the cerebellum to thiamin deficiency. *Cerebellum*, *5*(1), 55–63.

Murata, K. (1965). Discovery and nomenclature of thiaminase. In N. Shimazono & E. Katsura (Eds.), *Beribei and Thiamine* (pp. 220–248). Tokyo: Igaku Shoin Ltd.

Nath, A., Tran, T., Shope, T. R., & Koch, T. R. (2017). Prevalence of clinical thiamin deficiency in individuals with medically complicated obesity. *Nutrition Research*, *37*, 29–36. https://doi.org/10.1016/J.nutres.2016.11.012.

Nishimune, T., Watanabe, Y. H., Okazaki, H., & Akai, H. (2000). Thiamin is decomposed due to Anaphe spp. entomophagy in seasonal ataxia patients. *Journal of Nutrition*, *130*(6), 1625–1628.

Nodar, R. H., Lonsdale, D., & Orlowski, JP. (1980). Abnormal brainstem potentials in infants with threatened sudden infant death syndrome. *Otolaryngology and Head and Neck Surgery* (1979), *88*(5), 619–621.

Nozaki, S., Mizuma, H., Tanaka, M., Jin, G., Tahara, T., & Mizuno, K. (2009). Thiamin tetrahydrofurfuryl disulfide improves energy metabolism and physical performance during physical-fatigue loading in rats. *Nutrition Research*, *29*(12), 867–872.

O'Keefe, J. H., Gheewala, N. M., & O'Keefe, J. O. (2008). Dietary strategies for improving post-prandial glucose, lipids, inflammation, and cardiovascular health. *Journal of the American College of Cardiology*, *51*(3), 249–255.

Obrenovich, M. E., Shola, D., Schroedel, K., Agrahari, A., & Lonsdale, D. (2015). The role of trace elements, thiamin (e) and transketolase in autism and autistic spectrum disorder. *Frontiers in Bioscience (Elite Ed)*, *7*, 229–241.

Ollberding, N. J., Aschebrook-Kilroy, B., Caces, D. B., Weisenerger, D. D., et al. (2013). Phytanic acid and the risk of non-Hodgkin lymphoma. *Carcinogenesis*, *34*(1), 170–175.

Orlowski, J. P., Lonsdale, D., Nodar, R. H., & Williams, G. W. (1982). Brainstem dysfunction in the infant apnea syndrome. *Clinical Electroencephalography*, *13*(4), 226–232.

Ortigoza-Escobar, J. D., et al. (2016). Free-thiamine is a potential biomarker of thiamine transporter-2 deficiency: A treatable cause of Leigh syndrome. *Brain, 139*(Pt. 1), 31–38.
Pacal, L., Kurikova, K., & Kankova, K. (2014). Evidence for altered thiamine metabolism in diabetes: Is there a potential to oppose gluco-and lipotoxicity by rational supplementation? *World Journal of Diabetes, 5*(3), 288–295.
Pan, X., Chen, Z., Fei, G., Pan, S., Bao, W., Ren, S., et al. (2016). Long-term cognitive improvement after benfotiamine administration in patients with Alzheimer's disease. *Neuroscience Bulletin, 32*(6), 591–596.
Pan, X., et al. (2015). Measurement of blood thiamin metabolites for Alzheimer's disease diagnosis. *EBioMedicine, 26*(3), 155–162. https://doi.org/10.1016/J.ebiom.2015.11.039. eCollection 2016.
Pappens, M., De Peuter, S., Vansteenwegen, D., Van den Bergh, O., & Van Diest, I. (2012). Psychophysiological responses to CO2 inhalation. *International Journal of Psychophysiology, 84*(1), 45–50.
Perez-Duenas, B., Serrano, M., Rebollo, M., Muchart, G., Gargallo, E., Dupuits, C., et al. (2013). Reversible lactic acidosis in a newborn with thiamin transporter-2 deficiency. *Pediatrics, 131*(5), e1670–e1675.
Peters, R. A. (1936). The biochemical lesion in vitamin B1 deficiency. *Lancet,* 1162–1165.
Pettigrew, A. G., & Rahilly, P. M. (1985). Brainstem auditory evoked responses in infants at risk of sudden infant death. *Early Human Development, 11*(2), 99–111.
Pincus, J. H., Cooper, J. R., Murphy, J. V., Rabe, E. F., Lonsdale, D., & Dunn, H. G. (1973). Thiamine derivatives in subacute necrotizing encephalomyelopathy. A preliminary report. *Pediatrics, 51*(4), 716–721.
Rand, C. M., Patwari, P. P., Carroll, M. S., & Weese-Mayer, D. E. (2013). Congenital central hypoventilation syndrome and sudden infant death syndrome: Disorders of autonomic regulation. *Seminars in Pediatric Neurology, 20*(1), 44–55.
Renthal, W., Marin-Valencia, I., & Evans, P. A. (2014). Thiamine deficiency secondary to anorexia nervosa: An uncommon causal peripheral neuropathy and Wernicke encephalopathy in adolescence. *Pediatric Neurology, 51*(1), 100–103.
Ricketts, C. J., Minton, J. A., Samuel, J., Ariyawansa, I., Wales, J. K., Lo, I. F., et al. (2006). Thiamin-responsive megaloblastic anemia syndrome: Long-term follow-up and mutation analysis of seven families. *Acta Paediatrica, 95*(1), 99–104.
Ristow, M. (2004). Neurodegenerative disorders associated with diabetes mellitus. *Journal of Molecular Medicine (Berl), 822*(8), 510–529.
Rock, C. L., & Vasantharajan, S. (1995). Vitamin status of eating disorder patients: Relationship to clinical indices and effective treatment. *The International Journal of Eating Disorders, 18*(3), 257–262.
Roggla, G., Moser, B., Wagner, A., & Roggla, M. (2000). Correlation between raised body temperature and acute mountain sickness score at moderate altitude. *Wiener Klinische Wochenschrift, 112*(6), 290–292.
Rosen, S. D. (2012). From heart to brain: The genesis and processing of cardiac pain. *Canadian Journal of Cardiology, 28*(2 Suppl), S7–S19.
Sakurai, K., Chen, J., & Kefalov, V. (2011). The role of guanylyl cyclase modulation in mouse cone phototransduction. *Journal of Neuroscience, 31,* 7991–8000. https://doi.org/10.1523/jneurosci.6650-102011. PMC 3124626. PMID 21632921.
Scharfe, C., Hauschild, M., Klopsock, T., Jannsen, A. J., Heidemann, P. H., Meitinger, T., et al. (2000). A novel mutation in the thiamin responsive megaloblastic anemia gene SLC 19 A2 in a patient with deficiency of respiratory chain complex I. *Journal of Medical Genetics, 37*(9), 669–673.

Scheinberg, N., Salbu, R. L., Goswami, G., & Cohen, K. (2016). Treatment of diabetic autonomic neuropathy in older adults with diabetes mellitus. *Consultant Pharmacist, 31*(11), 633–645.
Schmechel, D. E. (2007). Art, alpha-1-antitrypsin polymorphism and intense creative energy: Blessing or curse? *Neurotoxicology, 28*(5), 899–914.
Schwartz, R. A., Gross, R., Lonsdale, D., & Shamnberger, R. J. (1979). Transketolase activity in psychiatric patients. *Journal of Clinical Psychiatry, 40*(10), 427–429.
Seligmann, H., Levi, R., Konijn, A. M., & Prok, M. (2001). Thiamine deficiency in patients with B-chronic lymphocytic leukaemia: A pilot study. *Postgraduate Medical Journal,* 77(911), 582–585.
Shah, S., & Wald, E. (2015). Type B lactic acidosis secondary to thiamin deficiency in a child with malignancy. *Pediatrics, 135*(1), e221–e224.
Shamberger, R. J. (2011). Autism rates associated with nutrition and the WIC program. *Journal of the American College of Nutrition, 30*(5), 348–353.
Shigematsu, Y., Nakai, A., Kuriyama, M., et al. (1990). Delayed auditory brainstem response in thiamin-deficient rats. *Journal of Nutritional Science and Vitaminology (Tokyo), 36*(3), 209–215.
Shimazono, N., & Katsura, E. (Eds.), (1965). *Review of Japanese Literature on Beriberi and Thiamine.* Tokyo: Igaku Shoin Ltd.
Stroh, C., Mayer, F., & Manger, T. (2014). Beriberi, a severe complication after metabolic surgery-review of the literature. *Obesity Facts,* 7(4), 246–252.
Suls, J., Green, P. A., & Davidson, K. M. (2016). A biobehavioral [corrected] framework to address the emerging challenge of multimorbidity. *Psychosomatic Medicine, 78*(3), 281–289.
Sweet, R. L., & Zastre, J. A. (2013). HIF-α-mediated gene expression induced by vitamin B1 deficiency. *International Journal of Vitamology & Nutritional Research, 823*(3), 188–197.
Szutowicz, A., Bielarczyk, H., Jankowska-Kulawy, A., Pawelczyk, T., & Ronowska, A. (2013). Acetyl CoA, the key factor for survival or death of cholinergic neurons in course of neurodegenerative diseases. *Neurochemical Research, 38*(8), 1523–1542.
Thornalley, P. J., Babaei-Jadidi, R., Al Ali, H., Rabbani, N., Antonysunil, A., Larkin, J., et al. (2007). High prevalence of low plasma thiamine to concentration in diabetes linked to a marker of vascular disease. *Diabetologia, 50*(10), 2164–2170.
Thornalley, P. J., Jahan, I., & Ng, R. (2001). Suppression of the accumulation of triosephosphates and increased formation of methylglyoxal in human red blood cells during hyperglycaemia by thiamine in vitro. *Journal of Biochemistry, 129*(4), 543–549.
Toussirot, E., Serratrice, G., & Valentin, P. (1993). Autonomic nervous system involvement in rheumatoid arthritis. 50 cases. *Journal of Rheumatology, 20*(9), 1508–1514.
Van Dongen, S., Brown, R. M., Brown, G. K., Thorburn, D. R., & Boneh, A. (2015). Thiamine-responsive and non-responsive patients with PDHC-E1 deficiency: A retrospective assessment. *JIMD Reports, 15,* 13–27. https://doi.org/10.1007/8904_2014_293.
Williams, R. R. (1936). Chemistry of thiamine (vitamin B1). *JAMA,* 110727–110731.
Woznica, A., Vickers, K., Koerner, N., & Fracalanza, K. (2015). Reactivity to 35% carbon dioxide in bulimia nervosa and panic disorder. *Psychiatry Research, 228*(3), 571–575.
Yilmaz Agladioglu, S., Aycan, C., Bas, V. N., Peltek Kendirci, H. N., & Onder, A. (2012). Thiamin-responsive megaloblastic anemia syndrome: A novel mutation. *Genetic Counseling, 23*(2), 149–156.
Zbinden, G. (1962). Therapeutic use of vitamin B1 in diseases other than beriberi. *Annals of the New York Academy of Science, 26*(98), 550–561.
Zhao, R., & Goldman, I. D. (2013). Folate and thiamin transporters mediated by facilitative carriers (SLC 19A1-3 and SLC 46A1) and folate receptors. *Molecular Aspects of Medicine, 34*(2–3), 373–385.

Zilliox, L. A., Chadrasekaran, K., Kwan, J. Y., & Russell, J. W. (2016). Diabetes and cognitive impairment. *Current Diabetes Reports*, *16*(9). https://doi.org/10.1007/s11892-016-0775.
Zsurka, G., & Kunz, W. S. (2015). Mitochondrial dysfunction and seizures: The neuronal energy crisis. *Lancet Neurology*, *14*(9), 956–966.

FURTHER READING

Hollis, F., Isgor, C., & Kabbaj, R. (2003). The consequences of adolescent chronic unpredictable stress exposure on brain and behavior. *Neuroscience*, *249*, 232–241.
Inouye, K., & Katsura, E. (1965b). Clinical signs and metabolism of beriberi patients. In N. Shimazono & E. Katsura (Eds.), *Beriberi and Thiamine* (pp. 29–63). Yokyo: Igaku Shoin Ltd.

CHAPTER TWO

Riboflavin in Human Health: A Review of Current Evidences

Ahmad Saedisomeolia[*,†,1], **Marziyeh Ashoori**[‡]
[*]School of Nutrition Sciences and Dietetics, Tehran University of Medical Sciences, Tehran, Iran
[†]School of Medicine, Western Sydney University, Sydney, NSW, Australia
[‡]School of Nutrition Sciences and Food Technology, Shahid Beheshti University of Medical Sciences, Tehran, Iran
[1]Corresponding author: e-mail address: a_saedi@tums.ac.ir

Contents

Abstract

Riboflavin is a water-soluble vitamin, which was initially isolated from milk. There are two coenzyme forms of riboflavin, flavin mononucleotide and flavin adenine dinucleotide, in which riboflavin plays important roles in the enzymatic reactions. Riboflavin is found in a wide variety of animal and plant foods. Meat and dairy products are the major contributors of riboflavin dietary intake. In this chapter, the latest evidence on the relationship between riboflavin status and specific health risks will be reviewed. Also, some of the mechanisms by which riboflavin exerts its roles will be discussed. The evidence accrued suggests that riboflavin is an antioxidant nutrient which may prevent lipid peroxidation and reperfusion oxidative injury. Moreover, riboflavin deficiency may increase the risk of

Advances in Food and Nutrition Research, Volume 83
ISSN 1043-4526
https://doi.org/10.1016/bs.afnr.2017.11.002

some cancers. Riboflavin may also exert a neuroprotective effects in some neurological disorders (e.g., Parkinson disease, migraine, and multiple sclerosis) through its role in some pathways that are hypothesized to be impaired in neurological disorders such as antioxidation, myelin formation, mitochondrial function, and iron metabolism.

1. RIBOFLAVIN AT A GLANCE

1.1 The History of Riboflavin

Riboflavin is a yellow fluorescent pigment which was named "lactochrome" (Northrop-Clewes & Thurnham, 2012) in 1872. However, its structure was not described until 1935, and finally in 1939, when riboflavin was shown to be an essential nutrient for human health and should be provided by the diet (Northrop-Clewes & Thurnham, 2012).

1.2 Structure and Functions of Riboflavin

The riboflavin molecule consists of an isoalloxazine ring to which a ribitol side chain is attached. There are two coenzyme forms of riboflavin, flavin mononucleotide (FMN) and flavin adenine dinucleotide (FAD), in which riboflavin plays important roles in the enzymatic reactions involved in energy metabolism (Fig. 1). Riboflavin is phosphorylated by flavokinase to form FMN. FMN could be converted to FAD through an enzymatic

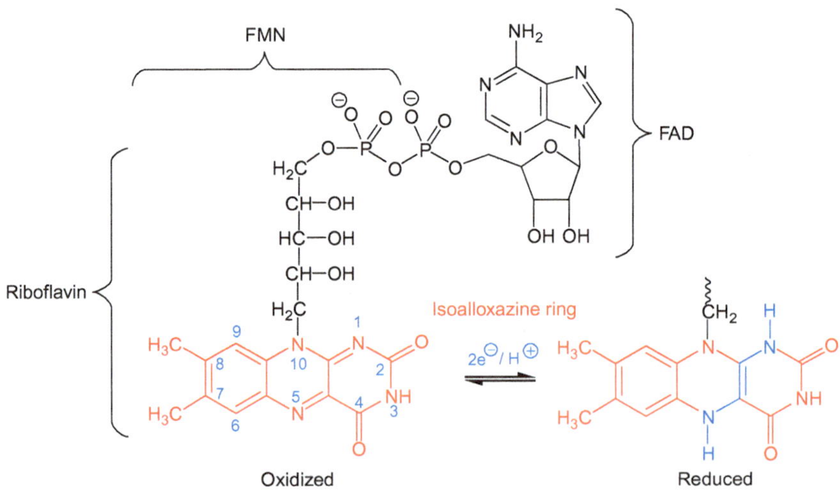

Fig. 1 Structure of riboflavin, FMN, and FAD in reduced and oxidized forms (Macheroux, Kappes, & Ealick, 2011).

reaction in which FAD synthetase adds an adenosine monophosphate (AMP) to FMN. FAD and FMN act as coenzymes for many flavoproteins in the different metabolic pathways. Some of the FAD/FMN-dependent pathways are described later.

In the electron transport chain, complex I (NADH dehydrogenase) and complex II (succinate dehydrogenase) need FMN and FAD, respectively, for their action. Riboflavin also plays a role in some other redox reactions, including fatty acid beta oxidation (acyl-CoA dehydrogenase is FAD dependent), oxidative decarboxylation of pyruvate and α-ketoglutarate (FAD acts as an electron carrier in pyruvate/α-ketoglutarate dehydrogenase), reduction of oxidized glutathione to its reduced form (FAD is required by glutathione reductase), reduction of thioredoxin catalyzed by thioredoxin reductase (FAD is an intermediate electron carrier between NADPH and thioredoxin), and choline catabolism (some dehydrogenase enzymes in this pathway require FAD). Some oxidases are also FAD dependent, including xanthine oxidase (in purine catabolism), sphinganine oxidase (in sphingosine synthesis), and aldehyde oxidase (Gropper & Smith, 2012).

Riboflavin is also involved in the metabolism of some vitamins such as pyridoxine, niacin, and folate in their synthesis and activation. *Pyridoxine phosphate oxidase*, which catalyzes the conversion of pyridoxamine phosphate and pyridoxine phosphate to pyridoxal phosphate, is an FMN-dependent enzyme. FAD is also required for kynurenine monooxygenase which catalyzes a step in niacin synthesis from its precursor, tryptophan. Riboflavin is also required for methylene tetrahydrofolate reductase (MTHFR) activity, which catalyzes the conversion of 5,10-methylene tetrahydrofolate to 5-methyl tetrahydrofolate (the active form of folate) (Gropper & Smith, 2012).

1.3 Assessment of Riboflavin

Riboflavin status is measured by several methods, but the most common method is based on the measurement of the activity of glutathione reductase (GR) which is a riboflavin-containing antioxidant enzyme (Ross, 2013). As mentioned earlier, GR is one of the enzymes which needs FAD for its activity. The activity of GR is responsive to riboflavin status in the human body. The ratio of GR activity with added FAD in vitro to GR activity without added FAD is known as erythrocyte glutathione reductase activity coefficient (EGRAC). EGRAC is one of the most sensitive indicators of riboflavin status used by many researchers these days (Gropper & Smith, 2012).

Greater increase in GR activity after an added FAD is associated with a lower GR saturation with FAD and a lower riboflavin status. Measuring cellular riboflavin concentration, activity of pyridoxamine phosphate oxidase (an FMN-dependent enzyme), and 24-h urinary excretion of riboflavin (adjusted with creatinine excretion) are the other methods used to assess riboflavin status (Gropper & Smith, 2012).

1.4 Dietary Sources and Reference Intakes of Riboflavin

Riboflavin is found in a wide variety of animal and plant foods mostly in the form of FMN or FAD attached to proteins. Meat and dairy products are the major contributors of riboflavin dietary intake in developed countries (Mahan, Escott-Stump, Raymond, & Krause, 2012; Powers, 2003). The other riboflavin-rich foods are eggs, organ meat (kidney and liver), green leafy vegetables, and legumes (Gropper & Smith, 2012; Mahan et al., 2012; Ross, 2013). Riboflavin is heat stable; however, UV light exposure or alkali can destroy riboflavin (Mahan et al., 2012). Based on the latest recommendations of Food and Nutrition Board (FNB), the daily requirement of riboflavin is 1.3 mg/day for adolescents and adults, 0.5–0.9 mg/day for children, 0.3–0.4 mg/day for infants, 1.4 mg/day during pregnancy, and 1.6 mg/day during lactation (Institute of Medicine (US) Standing Committee on the Scientific Evaluation of Dietary Reference Intakes, 1998).

1.5 Deficiency and Toxicity of Riboflavin

Riboflavin deficiency is more prevalent in low-income countries where the intake of meat and dairy product is low; however, marginal riboflavin deficiency has also been reported in some developed countries including the United Kingdom, the United States, and France (Anderson, Suchindran, & Roggenkamp, 2009; Powers et al., 2011; Preziosi et al., 1999). Riboflavin deficiency often occurs concurrently with other nutrient deficiencies. The clinical symptoms of deficiency appear after several months of inadequate dietary intake. The symptoms of severe deficiency include lesions and inflammation of lips (cheilosis), fissuring at the corner of the mouth (angular stomatitis), inflammation of the tongue and oral cavity, seborrheic dermatitis (an inflammatory skin condition), neuropathy, and anemia (Gropper & Smith, 2012; Mahan et al., 2012; Ross, 2013). Congenital heart disease, some cancers, excessive alcohol intake, diabetes mellitus, thyroid disease, trauma, and stress are the conditions which increase the risk of riboflavin deficiency (Gropper & Smith, 2012; Ross, 2013). Phototherapy in infants with hyperbilirubinemia may lead to riboflavin deficiency if riboflavin supplementation is not considered (because UV light exposure destroys riboflavin) (Gromisch, Lopez, Cole, & Cooperman, 1977; Sisson, 1987).

No toxic effects have been reported in relation to intake of high oral dose of riboflavin most probably due to its water-soluble property. The tolerable upper intake level of riboflavin has not been established by the FNB.

2. ANTIOXIDANT PROPERTIES OF RIBOFLAVIN

The antioxidant properties of some nutrients such as vitamin C (Padayatty et al., 2003), vitamin E (Traber & Atkinson, 2007), and carotenoids (Young & Lowe, 2001) have attracted lots of attention, and the effect of these antioxidants on human health has been investigated in many studies. However, the antioxidant potential of riboflavin has not been given serious consideration. In this section, two aspects of the antioxidant properties of riboflavin are discussed including its effect in preventing lipid peroxidation and its role in the attenuation of reperfusion oxidative injury will be discussed.

2.1 The Role of Riboflavin in Glutathione Redox Cycle

As mentioned previously, GR requires riboflavin (in FAD coenzyme form) for its activity (Mahan et al., 2012). GR catalyzes a reaction through which an oxidized glutathione (GSSG) is converted to its reduced form (GSH) (Fig. 2). FAD is reduced by a reduced nicotinamide adenine dinucleotide phosphate (NADPH) and the transient FADH-anion is produced. The

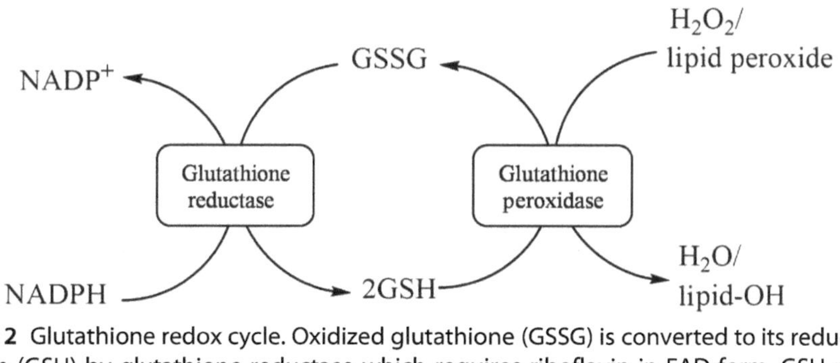

Fig. 2 Glutathione redox cycle. Oxidized glutathione (GSSG) is converted to its reduced form (GSH) by glutathione reductase which requires riboflavin in FAD form. GSH provides reduced equivalent for glutathione peroxidase activity.

FADH-anion then reduces the oxidized glutathione. Glutathione is known as a master endogenous antioxidant in the body (Pompella, Visvikis, Paolicchi, De Tata, & Casini, 2003) and mainly exists in the reduced form (about 90%) in the normal cells (Halprin & Ohkawara, 1967). An increase in oxidized glutathione to reduced glutathione ratio (GSSG/GSH) may lead to oxidative damage of the cells, a condition which is expected to occur during riboflavin deficiency.

Reduction of peroxides is one of the main antioxidant activities of glutathione; the reaction is catalyzed by glutathione peroxidase (GPx) (Hayes & McLellan, 1999). In this reaction, GPx transfers reduced equivalent from a reduced glutathione to lipid hydroperoxides producing oxidized glutathione and alcohol (Fig. 2). Considering the role of riboflavin in the reduction of oxidized glutathione, theoretically, an increase in lipid peroxidation is expected when riboflavin is deficient.

2.2 The Effect of Riboflavin Status on Glutathione Content of Tissues and Antioxidant Enzyme Activity

As the reduction of the oxidized glutathione is a riboflavin-dependent process, the effect of riboflavin status on the amount of reduced glutathione (GSH) in tissues has been examined extensively (Alam, Iqbal, & Naseem, 2015; Al-Harbi et al., 2015; Chen et al., 2015a, 2015b; Dutta, Rivlin, & Pinto, 1990; Horiuchi, Hirano, & Ono, 1984; Liang, Liu, & Xu, 1999; Taniguchi & Hara, 1983). Riboflavin deficiency leads to a decrease in GSH content in the liver of rats (Taniguchi & Hara, 1983). In another experimental study, the GSH content of gills and intestine is decreased in fishes fed with a riboflavin-deficient diet (Chen et al., 2015a, 2015b). Moreover, there was a decrease in the GSH level of the kidney, liver, and pancreas following diabetes induction in mice but was recovered following the riboflavin administration (Alam et al., 2015). Some other experimental studies have also reported similar results (Al-Harbi et al., 2015; Dutta et al., 1990; Horiuchi et al., 1984; Liang et al., 1999). However, in the only human study comparing the erythrocytes reduced glutathione status between riboflavin-deficient and healthy subjects, no significant differences were reported (Powers & Thurnham, 1981). Some limited animal studies have also shown no change in glutathione content in the tissues of riboflavin-deficient animals (Bates, 1991; Huang, Tian, Wu, Yang, & Liu, 2010). Increased de novo synthesis of glutathione from its precursor amino acids (cysteine, glutamate, glycine) or the GR ability to retain reduced glutathione concentration even in its low activity level could be the possible explanation for the results observed in these studies (Kaplowitz, Aw, & Ookhtens, 1985; Rogers & Augusteyn, 1978).

The activity of some antioxidant enzymes (such as GPx, catalase, and superoxide dismutase) is also affected by riboflavin status. In many studies, a poor riboflavin status was related to a decrease in antioxidant enzyme activity, as riboflavin administration has also been reported to enhance the activity of these enzymes (Hirano, Hamajima, Horiuchi, Niitsu, & Ono, 1983; Lee, Ye, Jones, & McCormick, 1983; Liang et al., 1999; Tumkiratiwong et al., 2003; Wang et al., 1998). However, in some other studies inconsistent results were reported. For example, an increase in GPx activity has been reported in liver (Taniguchi & Hara, 1983), lens (Rao & Bhat, 1989), and blood (Levin, Cogan, Levy, & Mokady, 1990) tissues of riboflavin-deficient animals.

2.3 Riboflavin Status and Lipid Peroxidation

The results of animal and some limited human research confirmed that riboflavin status could affect the production of lipid peroxides. In experimental studies, fish receiving a diet with a limited amount of riboflavin for 8 weeks showed significantly higher malondialdehyde (a lipid peroxidation marker) and protein carbonyl (marker of oxidative damage to proteins) content in gills and intestine tissues (Chen et al., 2015a, 2015b). An increase in oxidative stress, especially lipid peroxidation, was found in riboflavin-deficient animals, while an improvement in the oxidative stress marker was reported in some other animal studies after a riboflavin supplementation (Bates, 1991; Hirano et al., 1983; Huang et al., 2010; Levin et al., 1990; Liang et al., 1999; Rao & Bhat, 1989; Taniguchi, 1980; Taniguchi & Hara, 1983; Wang, Li, Lu, & Zhao, 2011; Yagi, Komur, Yoshino, Konishi, & Abe, 1989). Limited human studies addressing the relation between riboflavin status and oxidative stress have also confirmed these results (Das et al., 1990; George & Ojegbemi, 2009; Kodentsova et al., 2003).

Finally, according to these findings, it seems that riboflavin is a key component of the antioxidant defense system of human body and its deficiency could impair the oxidant/antioxidant balance. The mechanism by which riboflavin acts against oxidative stress and specifically lipid peroxidation could not be limited to the role of riboflavin in GR activity and glutathione redox cycle (Ashoori & Saedisomeolia, 2014). A suggested mechanism for the antioxidant action of riboflavin (independent of its role as the GR coenzyme) is the conversion of reduced riboflavin (dihydroriboflavin) to its oxidized form which provides reducing equivalents for deactivation of hydroperoxides (Toyoashi, 1992).

2.4 Riboflavin as a Protective Agent Against Reperfusion Oxidative Injury

Reperfusion injury is the tissue damage that takes place fallowing restoration of blood flow after a period of ischemia (Carden & Granger, 2000). Free radicals (Ambrosio et al., 1991; Bolli et al., 1995) and inflammatory responses (Kukielka et al., 1995) have a key role in creating reperfusion tissue damage. The results of several experimental studies showed that riboflavin could be a protective agent against oxidative tissue damage following reperfusion (Betz, Ren, Ennis, & Hultquist, 1994; Iwanaga et al., 2007; Mack, Hultquist, & Shlafer, 1995; Seekamp, Hultquist, & Till, 1999). It has been found that riboflavin administration attenuates reperfusion oxidative injury of different tissues such as heart (Iwanaga et al., 2007; Mack et al., 1995), brain (Betz et al., 1994), and lungs (Seekamp et al., 1999) in animal models. However, to our knowledge, there is no human study reported so far, suggesting that further investigations are needed.

The mechanism whereby riboflavin exerts such a protective effect may be attributed to dihydroriboflavin which is the product of flavin reductase activity of NADPH-dependent methemoglobin reductase (Hultquist et al., 1993). It has been shown that dihydroriboflavin could reduce oxidized iron in hemoproteins (Matsuki et al., 1978; Xu & Hultquist, 1991) which has been known to play a role in oxidative damage including reperfusion oxidative damage.

3. RIBOFLAVIN AND CANCER

Riboflavin has been repeatedly investigated in relation to its involvement in cancer risk and progress. The early epidemiological studies demonstrated that alcohol consumption and Plummer–Vinson syndrome are associated with upper gastrointestinal tract cancer (Wynder & Bross, 1961; Wynder, Hultberg, Jacobsson, & Bross, 1957). Both of these conditions are related to some nutritional deficiencies such as iron and riboflavin (Meulengracht & Bichel, 1941; Rosenthal, Adham, Lopez, & Cooperman, 1973; Savilahti, 1946). Considering the role of these nutrients in the maintenance of epithelial integrity, this hypothesis emerged that riboflavin deficiency may increase the risk of epithelial dysplasia particularly in upper gastrointestinal tract. Wynder and Klein (1965) described the histological changes that occurred in the epithelium of mice during the riboflavin deficiency. They showed that riboflavin deficiency could cause atrophy and hyperkeratosis of the upper gastrointestinal tract and skin hyperplasia. In further investigations, these researchers reported that riboflavin deficiency tends to increase the skin carcinogenesis induced by DMBA (7,12-dimethylbenzanthracene) (Wynder & Chan, 1970), a polycyclic aromatic hydrocarbon used as an inducer of the cancer in experimental settings. Some other experiments have also reported the effect of riboflavin status on chemically induced carcinogenesis in different tissues; most of them support the protective effect of riboflavin against carcinogens (Chan, Okamoto, & Wynder, 1972; Griffin & Baumann, 1948; Lambooy, 1970, 1976; Pangrekar, Krishnaswamy, & Jagadeesan, 1993; Webster, Gawde, & Bhattacharya, 1996; Wood et al., 1982). Some of these animal studies have been focused on hepatic carcinogenesis induced by Azo dye (Griffin & Baumann, 1948; Lambooy, 1970, 1976; Mulay & O'Gara, 1968). The results of these studies showed that the development of hepatic tumor induced by Azo dye is increased in riboflavin-deficient rat models. It has also been shown that riboflavin can modulate DNA damage caused by some carcinogens such as aflatoxin and benzopyrene (Pangrekar et al., 1993; Webster et al., 1996; Wood et al., 1982).

Along with relatively extensive animal studies addressing the relationship between riboflavin status and cancer, some observational and interventional human studies have also been conducted in this area.

Given this background that riboflavin deficiency may have a deleterious impact on esophagus epithelial, several observational and interventional studies have been focused on the possible relationship between riboflavin status and the risk of esophageal cancer. Much of this research has been reported from China (Guo et al., 1990; Lin et al., 1988; Taylor et al., 1994; Thurnham et al., 1985; Wahrendorf et al., 1988; Yang et al., 1984), which is one of the countries with a high incidence of esophageal cancer. Yang et al. (1984) evaluated the nutritional status of Linxian, a Chinese region with a high esophageal cancer incidence rate. They found that about 90% of the subjects suffer from riboflavin deficiency (EGRAC >1.2) in that area. A comparison of the zinc, vitamin A, and riboflavin status between the two populations in China at low and high risk for esophageal cancer showed that riboflavin status of high-risk population was much lower than the population with the low risk (Thurnham et al., 1985). Another ecological study from China, conducted in 65 Chinese regions, reported that the esophageal cancer mortality rate was inversely related to the plasma riboflavin levels (Guo et al., 1990). These results were consistent with results reported in other populations outside of China. Results of a case–control study conducted in Northeastern Iran (another high-incidence region of esophageal cancer) showed that esophageal cancer risk was more than double in riboflavin-deficient households compared to nondeficient households (Siassi & Ghadirian, 2005). Similar results have been reported in other case–control studies (Chen et al., 2002; Gao et al., 1994; Pottern, Morris, Blot, Ziegler, & Fraumeni, 1981). However, the results of intervention studies on the Chinese population were inconsistent. Lin et al. (1988) conducted a randomized clinical trial among 2411 subjects with mild esophageal dysplasia. The results of this study indicated that riboflavin supplementation over 3 years reduced the incidence rate of esophageal cancer by 19% compared to the control group. In another double-blind intervention trial conducted on the healthy subject in China, it was shown that supplementation with the combination of riboflavin, zinc, and vitamin A for 13.5 months did not alter the prevalence of precancerous lesions on the esophagus compared to the placebo group (Wahrendorf et al., 1988). In this study, it should be noted that zinc and riboflavin status of the control group was also improved, and further analysis indicated that those subjects with a higher increase in blood retinol, zinc, and riboflavin, independent of intervention groups, had a greater chance of having a histologically normal esophagus. Another intervention in China showed no change in esophageal cancer incidence rate following supplementation with riboflavin and niacin for 2.25 years (Taylor et al., 1994). However, as the authors mentioned in their publication, the small number of subjects and even smaller number of end points were described as the limitations for this study.

According to the evidence accrued so far, it seems that riboflavin deficiency may serve as an esophageal cancer risk factor. Further systematic reviews or meta-analysis studies in this area could provide stronger evidence by making it possible to have more accurate interpretation of the involvement of riboflavin in esophageal cancer risk, especially on intervention settings.

Several possible mechanisms can explain the protective effect of riboflavin against esophageal cancer. As mentioned earlier, the protective effect of riboflavin may be attributed to its role in maintaining esophagus epithelial integrity. Consequently, its deficiency may lead to damaged esophageal tissue followed by precancerous lesions, leading to cancer. Moreover, the results of experimental studies indicated that the riboflavin may inhibit the mutagenicity of some mutagens (Rivlin, 1973; Wood et al., 1982). It has been shown that riboflavin creates a complex with mutagen and deactivates it in vitro (Tung & Lin, 1964). Another hypothesis has been proposed that some presumed enzymes, which metabolize mutagens, may be flavin-dependent (Rivlin, 1973). As a result, decreased activity of these enzymes during riboflavin deficiency may explain the increased cancer risk among deficient subjects. We recently reviewed the latest evidence regarding the probable antioxidative properties of riboflavin (Ashoori & Saedisomeolia, 2014). This evidence indicated that riboflavin can play an important role in the maintenance of the oxidant/antioxidant balance of cells and act against oxidative stress, especially lipid peroxidation and reperfusion oxidative injury. Riboflavin exerts its antioxidant effects through its role in the glutathione redox cycle and also other probable mechanism such as oxidation of reduced form of riboflavin. Oxidative stress could cause DNA damage and enhance carcinogenesis (Sosa et al., 2013). Consequently, the antioxidant properties of riboflavin could provide another explanation for the anticancer effect by this nutrient.

The relationship between riboflavin and the risk of the other cancers including breast, lung, ovarian, cervix, gastric, and colorectal cancer has been also examined. These investigations have reported variable results. In the case of colorectal cancer, for example, some studies reported that the riboflavin intake is associated with a decreased colorectal cancer risk (La Vecchia et al., 1997; van den Donk et al., 2005). No significant association, however, has been observed in several investigations (Boyapati et al., 2004; de Vogel et al., 2008; Kabat, Miller, Jain, & Rohan, 2008; Le Marchand et al., 2002). Such an inconsistency has also seen in relation to other cancers. While several case–control and cohort studies provided evidence supporting the protective effect of riboflavin against some cancers, including breast cancer (Agnoli et al., 2016), gastric cancer (Eussen et al., 2010; Ji et al., 1998; Kweon et al., 2014), lung cancer (Bassett et al., 2012; Takata et al., 2012), ovarian cancer (Kabat et al., 2008), and cervix cancer (Liu et al., 1993), the results of some other investigations did not show any significant relationship between riboflavin and the risk of cancer (Johansson, Relton, Ueland, et al., 2010; Kabat et al., 2008).

As explained by Willett (2013), considering some inevitable errors in assessing dietary intake and the relatively low variation in diet, inconsistency in results is not surprising in the case–control studies in which the exposure is dietary intakes. Therefore, using more accurate indicators of riboflavin status (e.g., EGRAC) instead of dietary intakes and employing stronger study design (e.g., prospective cohorts) may be helpful to produce more consistent and reliable findings in this area.

4. RIBOFLAVIN AS A NEUROPROTECTIVE AGENT

Riboflavin is one of the nutrients considered as a promising neuroprotective agent in some neurological disorders (e.g., Parkinson disease (PD), migraine, and multiple sclerosis (MS)). The exact mechanism by which riboflavin acts as a neuroprotective factor is not clear yet; however, riboflavin plays an important role in some pathways that are hypothesized to be impaired in neurological disorders such as antioxidation (Hwang, 2013), mitochondrial function (Colombo, Saraceno, & Comi, 2014), iron metabolism (Bartzokis et al., 2000; Jellinger, 1999), and myelin formation (Ogunleye & Odutuga, 1989). Moreover, the results of some human and experimental studies have confirmed this action of riboflavin (Boehnke et al., 2004; Coimbra & Junqueira, 2003; Ghadirian, Jain, Ducic, Shatenstein, & Morisset, 1998; Nambiar, Aiyappa, & Srinivasa, 2011). Considering these facts, the neuroprotective effects of riboflavin do not seem to be implausible. The most probable effects of riboflavin on some selected neurological diseases are discussed as follows.

4.1 Riboflavin and Parkinson's Disease

PD is one of the most prevalent neurodegenerative disorders that mainly impairs motor function in a chronic and progressive manner. The main cause of PD remains unknown yet; however, there are a large body of evidences supporting the role of oxidative stress, inflammation, and mitochondrial dysfunction in pathogenesis of PD (Jenner, 2003; Michel, Hirsch, & Agid, 2002). The results of epidemiological and biochemical studies suggest that dietary intake may be one of the important factors affecting the risk of PD (Seidl, Santiago, Bilyk, & Potashkin, 2014).

Some evidence suggests that riboflavin may be a neuroprotective agent in PD. Lower riboflavin status has been reported in PD patients compared to the control group (Coimbra & Junqueira, 2003). In this investigation improved motor function was observed after a high dose of riboflavin supplementation (90 mg/day) along with red meat elimination from the diet for 6 months. It is suggested that a poor riboflavin status may be a specific feature of PD. Despite an adequate dietary intake of riboflavin in this study, all of the PD patients were riboflavin deficient. It has been suggested that a low absorption of riboflavin (as a result of expression of flavokinase isoforms with a low affinity for riboflavin) may explain the riboflavin deficiency in 10%–15% of the population provided with adequate riboflavin intake

(Bates, 1997). This hypothesis has emerged suggesting that PD patients may belong to the above-mentioned percentage of population. Modified flavokinase may be expressed not only in the intestinal cells but also in the brain–blood barrier of PD patients, leading to an impaired homeostatic mechanism that maintains riboflavin concentration in CNS even through there is a low dietary intake of riboflavin (DalPai et al., 2007). This hypothesis may explain the results of some case–control studies reporting no significant association between dietary intakes of riboflavin and risk of PD (Haglin, Johansson, Forsgren, & Backman, 2017; Hellenbrand et al., 1996; Johnson, Gorell, Rybicki, Sanders, & Peterson, 1999; Murakami et al., 2010).

As explained previously, riboflavin could act as an antioxidant in the human body (Ashoori & Saedisomeolia, 2014). It can protect the human body against oxidative stress damage which is known as a contributing factor in PD development (Hwang, 2013). Glutathione depletion is also considered one of the earliest neurochemical impairments in PD (Dexter et al., 1994). We know that glutathione redox cycle, in which oxidized glutathione is converted to a reduced form, is a riboflavin-dependent process (Schulz, Schirmer, & Pai, 1982). Therefore, the possible neuroprotective effect of riboflavin may, at least in part, be explained by its antioxidant activities.

The other factor considered to be involved in the pathogenesis of PD is abnormal iron metabolism in the brain. Iron accumulation in brain has been repeatedly reported in PD subjects (Weinreb, Mandel, Youdim, & Amit, 2013). Excessive iron storage may lead to an increased oxidative stress through "Fenton reaction" in which iron reacts with H_2O_2 and produces hydroxyl radicals, a highly toxic ROS. Therefore, an impaired iron metabolism could be a key event in the pathogenesis of PD. The results of several human and animal studies suggest that riboflavin may play a role in iron mobilization from its storage through a reductive process via FMN oxidoreductase (Powers et al., 2011; Sirivech, Driskell, & Frieden, 1977; Sirivech, Frieden, & Osaki, 1974; Ulvik, 1983). Regarding the suggested role of riboflavin in iron metabolism and the riboflavin depletion in PD patient (Coimbra & Junqueira, 2003), this hypothesis has emerged that brain iron accumulation in PD is probably a riboflavin-dependent mechanism.

Mitochondrial dysfunction is also known as a classic characteristic of PD. Mitochondrial DNA mutation is increased in dopaminergic neurons in PD (Di Monte, 1991; Winklhofer & Haass, 2010). Riboflavin (in FAD coenzyme form) is required in DNA replication and the repair (Nelson & Cox, 2012). Therefore, riboflavin deficiency or an impaired homeostatic mechanism involved in the maintenance of riboflavin concentration in CNS may increase the risk of PD by exacerbating the mitochondrial DNA mutation. Moreover, riboflavin is an essential component of complex I and complex II of the respiratory chain, in the form of FAD and FMN, respectively. An impaired activity of complex I and complex II has been reported to be involved in pathogenesis of PD (Bindoff, Birch-Machin, Cartlidge, Parker, & Turnbull, 1989; Blin et al., 1994). Therefore, the role of riboflavin in mitochondrial function may provide further explanation for the neuroprotective effects of riboflavin in PD.

Although all of these mechanisms support the neuroprotective effects of riboflavin in PD, there are not enough interventional or observational human studies in this regard. Some case–control studies examined the association between dietary intakes of riboflavin (but not serum riboflavin or EGRAC) and the risk of PD, reporting no significant association (Haglin et al., 2017; Hellenbrand et al., 1996; Johnson et al., 1999; Murakami et al., 2010). As mentioned earlier, poor riboflavin status in PD is probably independent of dietary intakes, but may be due to a decreased absorption/uptake of riboflavin (both in the gut and in the brain). Therefore, the results of these case–control studies are not convincing enough. Further research comparing riboflavin status (determined by serum values not dietary intakes) between PD subjects and healthy controls as well as examining the effects of high-dose riboflavin therapy through blinded controlled trials could provide stronger evidence for supporting the neuroprotective effects of riboflavin in PD.

4.2 Riboflavin and Multiple Sclerosis

MS is an autoimmune disease in which the myelin sheath of nerve cells in the brain and spinal cord is damaged. There are limited studies investigating the role of riboflavin in the pathogenesis of MS or its therapeutic effects for MS patient (Akhoondan et al., 2014; Behrooz, Asghari, Hosseini, Mirmiran, & Rashidkhani, 2014; Ghadirian et al., 1998; Naghashpour et al., 2016, 2013). However, these limited results provide evidence as well as highlight the need for further investigations.

In a case–control study investigating the role of dietary factors in the etiology of MS, it was found that a higher intake of riboflavin was associated with a decreased risk of developing MS (Ghadirian et al., 1998). In this regard, some case–control studies have reported a consistent result (Akhoondan et al., 2014; Behrooz et al., 2014). Riboflavin supplementation has been shown to suppress the neurological disability in animal models of MS (Naghashpour et al., 2016). However, in a human interventional study, riboflavin supplementation caused no improvement in the disability status of MS patient (Naghashpour et al., 2013), most likely due to the limited sample size.

Homocysteine (a nonprotein amino acid synthesized through the demethylation of methionine) has been suggested to play some roles in MS pathogenesis (Ramsaransing et al., 2006; Triantafyllou et al., 2008). Homocysteine may exert deleterious effects in the nervous system through its neurotoxic action (Ho, Ortiz, Rogers, & Shea, 2002). It is also suggested that homocysteine exacerbates the neuroinflammatory processes (Triantafyllou et al., 2008), which is known to be involved in MS development (Jadidi-Niaragh & Mirshafiey, 2011). The results of several studies suggested that riboflavin is one of the factors affecting the plasma homocysteine level (Hustad et al., 2000; Jacques et al., 2001). Riboflavin is required for MTHFR activity, which catalyzes the conversion of 5,10-methylene tetrahydrofolate to 5-methyl tetrahydrofolate. Then, 5-methyl tetrahydrofolate provides the methyl group for homocysteine remethylation to methionine. Thus, riboflavin status can affect the production of homocysteine in human body. The role of riboflavin in homocysteine metabolism provides an explanation for the protective effect of this nutrient against MS risk that has been reported in epidemiological studies (Akhoondan et al., 2014; Behrooz et al., 2014; Ghadirian et al., 1998).

In animal models, riboflavin deficiency has resulted in impaired brain lipid metabolism, similar to what is happening in the deficiency of essential fatty acids (Cai et al., 2006; Ogunleye & Odutuga, 1989). The reduction in the proportion of myelin lipids (such as cerebrosides, sphingomyelin, and phosphatidylethanolamine) has been reported in riboflavin-deficient rats (Ogunleye & Odutuga, 1989). Therefore, the other possible mechanism through which riboflavin exerts its beneficial effects in MS is the suggested role of riboflavin in essential fatty acid metabolism and myelin formation.

4.3 Riboflavin in Migraine Prophylaxis

Although the exact *pathophysiological mechanism* of migraine is not clear yet, an impaired mitochondrial energy metabolism in brain has been suggested to play a role in migraine pathogenesis (Colombo et al., 2014). In the brain of migraineurs, a decreased phosphocreatine/inorganic phosphate ratio and an increased ADP level have been observed, indicating a low intracellular energy level and mitochondrial dysfunction (Montagna, Cortelli, & Barbiroli, 1994; Welch, Levine, D'Andrea, Schultz, & Helpern, 1989). These conditions may lead to the inability of neurons to cope with the increased energy need, making the brain of migraine patients more sensitive to some stimuli which are known as triggers of migraine attacks (Colombo et al., 2014). Considering the role of riboflavin in mitochondrial respiratory chain and energy production in the cells, theoretically it may have a beneficial effect in migraine patients.

The results of several clinical trials support the prophylactic effect of high-dose riboflavin supplementation on migraine attacks. In an open-label study, a high-dose riboflavin supplementation (400 mg/day) for 6 months led to a reduction in migraine attack frequency (Boehnke et al., 2004). Nambiar et al. indicated that riboflavin supplementation (100 mg/day) is equally effective and better tolerated than 80 mg/day propranolol (a gold standard prophylactic drug) in reduction of headache frequency, duration, and severity (Nambiar et al., 2011). Some other studies have reported a consistent result (Maizels, Blumenfeld, & Burchette, 2004; Sandor, Afra, Ambrosini, & Schoenen, 2000; Schoenen, Jacquy, & Lenaerts, 1998; Schoenen, Lenaerts, & Bastings, 1994). Based on these evidences, American Academy of Neurology Guideline has considered riboflavin to be probably effective in migraine prophylaxis in adults (Holland et al., 2012). The use of riboflavin as a prophylactic drug is also strongly recommended by the Canadian Headache Society (Pringsheim et al., 2012).

The studies addressing the effectiveness of riboflavin supplementation in the prevention of pediatric migraine have reported inconsistent results. An uncontrolled open-label study reported that a high-dose supplementation of riboflavin is effective in pediatric migraine prophylaxis (Condo, Posar, Arbizzani, & Parmeggiani, 2009), whereas two randomized controlled trials failed to detect any significant improvement in riboflavin-treated patients than the placebo group (Bruijn et al., 2010; MacLennan et al., 2008). It have been indicated that the placebo response rate is higher in pediatric migraine patients than in adult patients with migraine (Ho et al., 2009; Maas, Danhof, & Della Pasqua, 2009). Therefore, inadequate sample size (MacLennan et al., 2008) or relatively low dose of supplemented riboflavin (Bruijn et al., 2010) may lead these studies not to be adequately powered to detect a small improvement of migraine between riboflavin and placebo groups.

REFERENCES

Agnoli, C., Grioni, S., Krogh, V., Pala, V., Allione, A., Matullo, G., et al. (2016). Plasma riboflavin and vitamin B-6, but not homocysteine, folate, or vitamin B-12, are inversely associated with breast cancer risk in the European Prospective Investigation into Cancer and Nutrition-Varese Cohort. *The Journal of Nutrition, 146*(6), 1227–1234.

Akhoondan, M., Behrooz, M., Naja, F., Jessri, M., Hoseini, Z., & Rashidkhani, B. (2014). Nutrient patterns and risk of multiple sclerosis: A case-control study. *Journal of Neuroscience Research, 4*(2–3), 72–80.

Alam, M. M., Iqbal, S., & Naseem, I. (2015). Ameliorative effect of riboflavin on hyperglycemia, oxidative stress and DNA damage in type-2 diabetic mice: Mechanistic and therapeutic strategies. *Archives of Biochemistry and Biophysics, 584*, 10–19.

Al-Harbi, N. O., Imam, F., Nadeem, A., Al-Harbi, M. M., Korashy, H. M., Sayed-Ahmed, M. M., et al. (2015). Riboflavin attenuates lipopolysaccharide-induced lung injury in rats. *Toxicology Mechanisms and Methods, 25*(5), 417–423.

Ambrosio, G., Flaherty, J. T., Duilio, C., Tritto, I., Santoro, G., Elia, P. P., et al. (1991). Oxygen radicals generated at reflow induce peroxidation of membrane lipids in reperfused hearts. *The Journal of Clinical Investigation, 87*(6), 2056–2066.

Anderson, J. J., Suchindran, C. M., & Roggenkamp, K. (2009). Micronutrient intakes in two US populations of older adults: Lipid research clinics program prevalence study findings. *The Journal of Nutrition, Health & Aging, 13*(7), 595.

Ashoori, M., & Saedisomeolia, A. (2014). Riboflavin (vitamin B(2)) and oxidative stress: A review. *The British Journal of Nutrition, 111*(11), 1985–1991.

Bartzokis, G., Sultzer, D., Cummings, J., Holt, L. E., Hance, D. B., Henderson, V. W., et al. (2000). In vivo evaluation of brain iron in Alzheimer disease using magnetic resonance imaging. *Archives of General Psychiatry, 57*(1), 47–53.

Bassett, J. K., Hodge, A. M., English, D. R., Baglietto, L., Hopper, J. L., Giles, G. G., et al. (2012). Dietary intake of B vitamins and methionine and risk of lung cancer. *European Journal of Clinical Nutrition, 66*(2), 182–187.

Bates, C. J. (1991). Glutathione and related indices in rat lenses, liver and red cells during riboflavin deficiency and its correction. *Experimental Eye Research, 53*(1), 123–130.

Bates, C. J. (1997). Bioavailability of riboflavin. *European Journal of Clinical Nutrition, 51*(Suppl. 1), S38–42.

Behrooz, M., Asghari, G., Hosseini, Z., Mirmiran, P., & Rashidkhani, B. (2014). Association of macro- and micro-nutrients intake with the risk of multiple sclerosis: A case control study. *Nutrition and Food Science Research, 1*(2), 27–33.

Betz, A. L., Ren, X. D., Ennis, S. R., & Hultquist, D. E. (1994). Riboflavin reduces edema in focal cerebral ischemia. *Acta Neurochirurgica. Supplementum (Wien), 60*, 314–317.

Bindoff, L. A., Birch-Machin, M., Cartlidge, N. E., Parker, W. D., Jr., & Turnbull, D. M. (1989). Mitochondrial function in Parkinson's disease. *Lancet, 2*(8653), 49.

Blin, O., Desnuelle, C., Rascol, O., Borg, M., Peyro Saint Paul, H., Azulay, J. P., et al. (1994). Mitochondrial respiratory failure in skeletal muscle from patients with Parkinson's disease and multiple system atrophy. *Journal of the Neurological Sciences, 125*(1), 95–101.

Boehnke, C., Reuter, U., Flach, U., Schuh-Hofer, S., Einhaupl, K. M., & Arnold, G. (2004). High-dose riboflavin treatment is efficacious in migraine prophylaxis: An open study in a tertiary care centre. *European Journal of Neurology, 11*(7), 475–477.

Bolli, R., Zughaib, M., Li, X. Y., Tang, X. L., Sun, J. Z., Triana, J. F., et al. (1995). Recurrent ischemia in the canine heart causes recurrent bursts of free radical production that have a cumulative effect on contractile function. A pathophysiological basis for chronic myocardial "stunning" *The Journal of Clinical Investigation, 96*(2), 1066–1084.

Boyapati, S. M., Bostick, R. M., McGlynn, K. A., Fina, M. F., Roufail, W. M., Geisinger, K. R., et al. (2004). Folate intake, MTHFR C677T polymorphism, alcohol consumption, and risk for sporadic colorectal adenoma (United States). *Cancer Causes & Control, 15*(5), 493–501.

Bruijn, J., Duivenvoorden, H., Passchier, J., Locher, H., Dijkstra, N., & Arts, W. F. (2010). Medium-dose riboflavin as a prophylactic agent in children with migraine: A preliminary placebo-controlled, randomised, double-blind, cross-over trial. *Cephalalgia, 30*(12), 1426–1434.

Cai, Z., Finnie, J. W., Blumbergs, P. C., Manavis, J., Ghabriel, M. N., & Thompson, P. D. (2006). Early paranodal myelin swellings (tomacula) in an avian riboflavin deficiency model of demyelinating neuropathy. *Experimental Neurology, 198*(1), 65–71.

Carden, D. L., & Granger, D. N. (2000). Pathophysiology of ischaemia-reperfusion injury. *The Journal of Pathology, 190*(3), 255–266.

Chan, P. C., Okamoto, T., & Wynder, E. L. (1972). Possible role of riboflavin deficiency in epithelial neoplasia. 3. Induction of microsomal aryl hydrocarbon hydroxylase. *Journal of the National Cancer Institute, 48*(5), 1341–1345.

Chen, L., Feng, L., Jiang, W. D., Jiang, J., Wu, P., Zhao, J., et al. (2015a). Dietary riboflavin deficiency decreases immunity and antioxidant capacity, and changes tight junction proteins and related signaling molecules mRNA expression in the gills of young grass carp (Ctenopharyngodon idella). *Fish & Shellfish Immunology, 45*(2), 307–320.

Chen, L., Feng, L., Jiang, W. D., Jiang, J., Wu, P., Zhao, J., et al. (2015b). Intestinal immune function, antioxidant status and tight junction proteins mRNA expression in young grass carp (Ctenopharyngodon idella) fed riboflavin deficient diet. *Fish & Shellfish Immunology, 47*(1), 470–484.

Chen, H., Tucker, K. L., Graubard, B. I., Heineman, E. F., Markin, R. S., Potischman, N. A., et al. (2002). Nutrient intakes and adenocarcinoma of the esophagus and distal stomach. *Nutrition and Cancer, 42*(1), 33–40.

Coimbra, C. G., & Junqueira, V. B. (2003). High doses of riboflavin and the elimination of dietary red meat promote the recovery of some motor functions in Parkinson's disease patients. *Brazilian Journal of Medical and Biological Research, 36*(10), 1409–1417.

Colombo, B., Saraceno, L., & Comi, G. (2014). Riboflavin and migraine: The bridge over troubled mitochondria. *Neurological Sciences, 35*(Suppl. 1), 141–144.

Condo, M., Posar, A., Arbizzani, A., & Parmeggiani, A. (2009). Riboflavin prophylaxis in pediatric and adolescent migraine. *The Journal of Headache and Pain, 10*(5), 361–365.

DalPai, J., Borges, A. A., Grassl, C., Favero-Filho, L. A., Xavier, G. F., Junqueira, V. B., et al. (2007). Dietary riboflavin restriction and chronic hemin administration does not Alter brain function in rats: The importance of vitamin homeostasis in the brain. *Current Topics in Nutraceutical Research, 5*(4), 149.

Das, B. S., Thurnham, D. I., Patnaik, J. K., Das, D. B., Satpathy, R., & Bose, T. K. (1990). Increased plasma lipid peroxidation in riboflavin-deficient, malaria-infected children. *The American Journal of Clinical Nutrition, 51*(5), 859–863.

de Vogel, S., Dindore, V., van Engeland, M., Goldbohm, R. A., van den Brandt, P. A., & Weijenberg, M. P. (2008). Dietary folate, methionine, riboflavin, and vitamin B-6 and risk of sporadic colorectal cancer. *The Journal of Nutrition, 138*(12), 2372–2378.

Dexter, D. T., Sian, J., Rose, S., Hindmarsh, J. G., Mann, V. M., Cooper, J. M., et al. (1994). Indices of oxidative stress and mitochondrial function in individuals with incidental Lewy body disease. *Annals of Neurology, 35*(1), 38–44.

Di Monte, D. A. (1991). Mitochondrial DNA and Parkinson's disease. *Neurology, 41*(5 Suppl. 2), 38–42.

Dutta, P., Rivlin, R. S., & Pinto, J. (1990). Enhanced depletion of lens reduced glutathione adriamycin in riboflavin-deficient rats. *Biochemical Pharmacology, 40*(5), 1111–1115.

Eussen, S. J., Vollset, S. E., Hustad, S., Midttun, O., Meyer, K., Fredriksen, A., et al. (2010). Vitamins B2 and B6 and genetic polymorphisms related to one-carbon metabolism as risk factors for gastric adenocarcinoma in the European prospective investigation into cancer and nutrition. *Cancer Epidemiology, Biomarkers & Prevention, 19*(1), 28–38.

Gao, Y. T., McLaughlin, J. K., Gridley, G., Blot, W. J., Ji, B. T., Dai, Q., et al. (1994). Risk factors for esophageal cancer in Shanghai, China. II. Role of diet and nutrients. *International Journal of Cancer, 58*(2), 197–202.

George, B., & Ojegbemi, O. (2009). Oxidative stress and the effect of riboflavin supplementation in individuals with uncomplicated malaria infection. *African Journal of Biotechnology, 8*(5), 849–853.

Ghadirian, P., Jain, M., Ducic, S., Shatenstein, B., & Morisset, R. (1998). Nutritional factors in the aetiology of multiple sclerosis: A case-control study in Montreal, Canada. *International Journal of Epidemiology, 27*(5), 845–852.

Griffin, A. C., & Baumann, C. A. (1948). Hepatic riboflavin and tumor formation in rat fed azo dyes in various diets. *Cancer Research, 8*(6), 279–284.

Gromisch, D. S., Lopez, R., Cole, H. S., & Cooperman, J. M. (1977). Light (phototherapy) induced riboflavin deficiency in the neonate. *The Journal of Pediatrics, 90*(1), 118–122.

Gropper, S. S., & Smith, J. L. (2012). *Advanced nutrition and human metabolism* (6th ed.). Belmont, CA: Wadsworth/Cengage Learning [chapter 9].

Guo, W. D., Li, J. Y., Blot, W. J., Hsing, A. W., Chen, J. S., & Fraumeni, J. F., Jr. (1990). Correlations of dietary intake and blood nutrient levels with esophageal cancer mortality in China. *Nutrition and Cancer, 13*(3), 121–127.

Haglin, L., Johansson, I., Forsgren, L., & Backman, L. (2017). Intake of vitamin B before onset of Parkinson's disease and atypical parkinsonism and olfactory function at the time of diagnosis. *European Journal of Clinical Nutrition, 71*(1), 97–102.

Halprin, K. M., & Ohkawara, A. (1967). The measurement of glutathione in human epidermis using glutathione reductase. *The Journal of Investigative Dermatology, 48*(2), 149–152.

Hayes, J. D., & McLellan, L. I. (1999). Glutathione and glutathione-dependent enzymes represent a co-ordinately regulated defence against oxidative stress. *Free Radical Research, 31*(4), 273–300.

Hellenbrand, W., Boeing, H., Robra, B. P., Seidler, A., Vieregge, P., Nischan, P., et al. (1996). Diet and Parkinson's disease. II: A possible role for the past intake of specific nutrients. Results from a self-administered food-frequency questionnaire in a case-control study. *Neurology, 47*(3), 644–650.

Hirano, H., Hamajima, S., Horiuchi, S., Niitsu, Y., & Ono, S. (1983). Effects of B2-deficiency on lipoperoxide and its scavenging system in the rat lens. *International Journal for Vitamin and Nutrition Research, 53*(4), 377–382.

Ho, T. W., Fan, X., Rodgers, A., Lines, C. R., Winner, P., & Shapiro, R. E. (2009). Age effects on placebo response rates in clinical trials of acute agents for migraine: Pooled analysis of rizatriptan trials in adults. *Cephalalgia, 29*(7), 711–718.

Ho, P. I., Ortiz, D., Rogers, E., & Shea, T. B. (2002). Multiple aspects of homocysteine neurotoxicity: Glutamate excitotoxicity, kinase hyperactivation and DNA damage. *Journal of Neuroscience Research, 70*(5), 694–702.

Holland, S., Silberstein, S., Freitag, F., Dodick, D. W., Argoff, C., & Ashman, E. (2012). Evidence-based guideline update: NSAIDs and other complementary treatments for episodic migraine prevention in adults: Report of the Quality Standards Subcommittee of the American Academy of Neurology and the American Headache Society. *Neurology, 78*(17), 1346.

Horiuchi, S., Hirano, H., & Ono, S. (1984). Reduced and oxidized glutathione concentrations in the lenses of riboflavin-deficient rats. *Journal of Nutritional Science and Vitaminology, 30*(4), 401–403.

Huang, J., Tian, L., Wu, X., Yang, H., & Liu, Y. (2010). Effects of dietary riboflavin levels on antioxidant defense of the juvenile grouper Epinephelus coioides. *Fish Physiology and Biochemistry, 36*(1), 55–62.

Hultquist, D. E., Xu, F., Quandt, K. S., Shlafer, M., Mack, C. P., Till, G. O., et al. (1993). Evidence that NADPH-dependent methemoglobin reductase and administered riboflavin protect tissues from oxidative injury. *American Journal of Hematology, 42*(1), 13–18.

Hustad, S., Ueland, P. M., Vollset, S. E., Zhang, Y., Bjorke-Monsen, A. L., & Schneede, J. (2000). Riboflavin as a determinant of plasma total homocysteine: Effect modification by the methylenetetrahydrofolate reductase C677T polymorphism. *Clinical Chemistry, 46*(8 Pt. 1), 1065–1071.

Hwang, O. (2013). Role of oxidative stress in Parkinson's disease. *Experimental Neurobiology, 22*(1), 11–17.

Institute of Medicine (US) Standing Committee on the Scientific Evaluation of Dietary Reference Intakes and its Panel on Folate, Other B Vitamins, and Choline. (1998). *Dietary reference intakes for thiamin, riboflavin, niacin, vitamin B6, folate, vitamin B12, pantothenic acid, biotin, and choline.* Washington, DC: National Academies Press (US). Available from: https://www.ncbi.nlm.nih.gov/books/NBK114296/.

Iwanaga, K., Hasegawa, T., Hultquist, D. E., Harada, H., Yoshikawa, Y., Yanamadala, S., et al. (2007). Riboflavin-mediated reduction of oxidant injury, rejection, and vasculopathy after cardiac allotransplantation. *Transplantation, 83*(6), 747–753.

Jacques, P. F., Bostom, A. G., Wilson, P. W., Rich, S., Rosenberg, I. H., & Selhub, J. (2001). Determinants of plasma total homocysteine concentration in the Framingham offspring cohort. *The American Journal of Clinical Nutrition, 73*(3), 613–621.

Jadidi-Niaragh, F., & Mirshafiey, A. (2011). Th17 cell, the new player of neuroinflammatory process in multiple sclerosis. *Scandinavian Journal of Immunology, 74*(1), 1–13.

Jellinger, K. A. (1999). The role of iron in neurodegeneration: Prospects for pharmacotherapy of Parkinson's disease. *Drugs & Aging, 14*(2), 115–140.

Jenner, P. (2003). Oxidative stress in Parkinson's disease. *Annals of Neurology, 53*(Suppl. 3), S26–S36.

Ji, B. T., Chow, W. H., Yang, G., McLaughlin, J. K., Zheng, W., Shu, X. O., et al. (1998). Dietary habits and stomach cancer in Shanghai, China. *International Journal of Cancer, 76*(5), 659–664.

Johansson, M., Relton, C., Ueland, P., et al. (2010). Serum b vitamin levels and risk of lung cancer. *JAMA, 303*(23), 2377–2385.

Johnson, C. C., Gorell, J. M., Rybicki, B. A., Sanders, K., & Peterson, E. L. (1999). Adult nutrient intake as a risk factor for Parkinson's disease. *International Journal of Epidemiology, 28*(6), 1102–1109.

Kabat, G. C., Miller, A. B., Jain, M., & Rohan, T. E. (2008). Dietary intake of selected B vitamins in relation to risk of major cancers in women. *British Journal of Cancer, 99*(5), 816–821.

Kaplowitz, N., Aw, T. Y., & Ookhtens, M. (1985). The regulation of hepatic glutathione. *Annual Review of Pharmacology and Toxicology, 25*, 715–744.

Kodentsova, V. M., Vrzhesinskaia, O. A., Beketova, N. A., Pereverzeva, O. G., Kharitonchik, L. A., Lavrent'eva, I. B., et al. (2003). The connection between vitamin and antioxidant status of the children with decreased hemoglobin level. *Voprosy Pitaniia, 72*(3), 3–7.

Kukielka, G. L., Smith, C. W., Manning, A. M., Youker, K. A., Michael, L. H., & Entman, M. L. (1995). Induction of interleukin-6 synthesis in the myocardium. Potential role in postreperfusion inflammatory injury. *Circulation, 92*(7), 1866–1875.

Kweon, S. S., Shu, X. O., Xiang, Y., Yang, G., Ji, B. T., Li, H., et al. (2014). One-carbon metabolism dietary factors and distal gastric cancer risk in Chinese women. *Cancer Epidemiology, Biomarkers & Prevention, 23*(7), 1374–1382.

Lambooy, J. P. (1970). Riboflavin protection against azo dye hepatoma induction in the rat. *Proceedings of the Society for Experimental Biology and Medicine, 134*(1), 192–194.
Lambooy, J. P. (1976). Influence of riboflavin antagonists on azo dye hepatoma induction in the rat. *Proceedings of the Society for Experimental Biology and Medicine, 153*(3), 532–535.
La Vecchia, C., Braga, C., Negri, E., Franceschi, S., Russo, A., Conti, E., et al. (1997). Intake of selected micronutrients and risk of colorectal cancer. *International Journal of Cancer, 73*(4), 525–530.
Lee, S. S., Ye, J. H., Jones, D. P., & McCormick, D. B. (1983). Correlation of H2O2 production and liver catalase during riboflavin deficiency and repletion in mammals. *Biochemical and Biophysical Research Communications, 117*(3), 788–793.
Le Marchand, L., Donlon, T., Hankin, J. H., Kolonel, L. N., Wilkens, L. R., & Seifried, A. (2002). B-vitamin intake, metabolic genes, and colorectal cancer risk (United States). *Cancer Causes & Control, 13*(3), 239–248.
Levin, G., Cogan, U., Levy, Y., & Mokady, S. (1990). Riboflavin deficiency and the function and fluidity of rat erythrocyte membranes. *The Journal of Nutrition, 120*(8), 857–861.
Liang, H., Liu, Q., & Xu, J. (1999). The effect of riboflavin on lipid peroxidation in rats. *Wei Sheng Yan Jiu, 28*(6), 370–371.
Lin, P. Z., Zhang, J. S., Cao, S. G., Rong, Z. P., Gao, R. Q., Han, R., et al. (1988). Secondary prevention of esophageal cancer—Intervention on precancerous lesions of the esophagus. *Zhonghua Zhong Liu Za Zhi, 10*(3), 161–166.
Liu, T., Soong, S. J., Wilson, N. P., Craig, C. B., Cole, P., Macaluso, M., et al. (1993). A case control study of nutritional factors and cervical dysplasia. *Cancer Epidemiology, Biomarkers & Prevention, 2*(6), 525–530.
Maas, H. J., Danhof, M., & Della Pasqua, O. E. (2009). Analysis of the relationship between age and treatment response in migraine. *Cephalalgia, 29*(7), 772–780.
Macheroux, P., Kappes, B., & Ealick, S. E. (2011). Flavogenomics—A genomic and structural view of flavin-dependent proteins. *FEBS Journal, 278*(15), 2625–2634.
Mack, C. P., Hultquist, D. E., & Shlafer, M. (1995). Myocardial flavin reductase and riboflavin: A potential role in decreasing reoxygenation injury. *Biochemical and Biophysical Research Communications, 212*(1), 35–40.
MacLennan, S. C., Wade, F. M., Forrest, K. M., Ratanayake, P. D., Fagan, E., & Antony, J. (2008). High-dose riboflavin for migraine prophylaxis in children: A double-blind, randomized, placebo-controlled trial. *Journal of Child Neurology, 23*(11), 1300–1304.
Mahan, L. K., Escott-Stump, S., Raymond, J. L., & Krause, M. V. (2012). *Krause's food & the nutrition care process* (13th ed.). St. Louis: Saunders [chapter 3].
Maizels, M., Blumenfeld, A., & Burchette, R. (2004). A combination of riboflavin, magnesium, and feverfew for migraine prophylaxis: A randomized trial. *Headache, 44*(9), 885–890.
Matsuki, T., Yubisui, T., Tomoda, A., Yoneyama, Y., Takeshita, M., Hirano, M., et al. (1978). Acceleration of methaemoglobin reduction by riboflavin in human erythrocytes. *British Journal of Haematology, 39*(4), 523–528.
Meulengracht, E., & Bichel, J. (1941). Riboflavin-avitaminose und das Plummer-Vinson-syndrom. *Journal of Molecular Medicine, 20*(33), 831–834.
Michel, P. P., Hirsch, E. C., & Agid, Y. (2002). Parkinson disease: Mechanisms of cell death. *Revue Neurologique (Paris), 158*, S24–32. Spec no 1.
Montagna, P., Cortelli, P., & Barbiroli, B. (1994). Magnetic resonance spectroscopy studies in migraine. *Cephalalgia, 14*(3), 184–193.
Mulay, A. S., & O'Gara, R. W. (1968). Enhancing effect of a riboflavin analog on azo-dye carcinogenesis in rats. *Journal of the National Cancer Institute, 40*(4), 731–735.
Murakami, K., Miyake, Y., Sasaki, S., Tanaka, K., Fukushima, W., Kiyohara, C., et al. (2010). Dietary intake of folate, vitamin B6, vitamin B12 and riboflavin and risk of Parkinson's disease: A case-control study in Japan. *The British Journal of Nutrition, 104*(5), 757–764.

Naghashpour, M., Amani, R., Sarkaki, A., Ghadiri, A., Samarbafzadeh, A., Jafarirad, S., et al. (2016). Brain-derived neurotrophic and immunologic factors: Beneficial effects of riboflavin on motor disability in murine model of multiple sclerosis. *Iranian Journal of Basic Medical Sciences, 19*(4), 439–448.

Naghashpour, M., Majdinasab, N., Shakerinejad, G., Kouchak, M., Haghighizadeh, M. H., Jarvandi, F., et al. (2013). Riboflavin supplementation to patients with multiple sclerosis does not improve disability status nor is riboflavin supplementation correlated to homocysteine. *International Journal for Vitamin and Nutrition Research, 83*(5), 281–290.

Nambiar, N., Aiyappa, C., & Srinivasa, R. (2011). Oral riboflavin versus oral propranolol in migraine prophylaxis: An open label randomized controlled trial. *Neurology Asia, 16*(3), 223–229.

Nelson, D. L., & Cox, M. M. (2012). *Lehninger principles of biochemistry* (6th ed.). . New York: W. H. Freeman.

Northrop-Clewes, C. A., & Thurnham, D. I. (2012). The discovery and characterization of riboflavin. *Annals of Nutrition & Metabolism, 61*(3), 224–230.

Ogunleye, A. J., & Odutuga, A. A. (1989). The effect of riboflavin deficiency on cerebrum and cerebellum of developing rat brain. *Journal of Nutritional Science and Vitaminology (Tokyo), 35*(3), 193–197.

Padayatty, S. J., Katz, A., Wang, Y., Eck, P., Kwon, O., Lee, J.-H., et al. (2003). Vitamin C as an antioxidant: Evaluation of its role in disease prevention. *Journal of the American College of Nutrition, 22*(1), 18–35.

Pangrekar, J., Krishnaswamy, K., & Jagadeesan, V. (1993). Effects of riboflavin deficiency and riboflavin administration on carcinogen-DNA binding. *Food and Chemical Toxicology, 31*(10), 745–750.

Pompella, A., Visvikis, A., Paolicchi, A., De Tata, V., & Casini, A. F. (2003). The changing faces of glutathione, a cellular protagonist. *Biochemical Pharmacology, 66*(8), 1499–1503.

Pottern, L. M., Morris, L. E., Blot, W. J., Ziegler, R. G., & Fraumeni, J. F., Jr. (1981). Esophageal cancer among black men in Washington, D.C. I. Alcohol, tobacco, and other risk factors. *Journal of the National Cancer Institute, 67*(4), 777–783.

Powers, H. J. (2003). Riboflavin (vitamin B-2) and health. *The American Journal of Clinical Nutrition, 77*(6), 1352–1360.

Powers, H. J., Hill, M. H., Mushtaq, S., Dainty, J. R., Majsak-Newman, G., & Williams, E. A. (2011). Correcting a marginal riboflavin deficiency improves hematologic status in young women in the United Kingdom (RIBOFEM). *The American Journal of Clinical Nutrition, 93*(6), 1274–1284.

Powers, H. J., & Thurnham, D. I. (1981). Riboflavin deficiency in man: Effects on haemoglobin and reduced glutathione in erythrocytes of different ages. *The British Journal of Nutrition, 46*(2), 257–266.

Preziosi, P., Galan, P., Deheeger, M., Yacoub, N., Drewnowski, A., & Hercberg, S. (1999). Breakfast type, daily nutrient intakes and vitamin and mineral status of French children, adolescents, and adults. *Journal of the American College of Nutrition, 18*(2), 171–178.

Pringsheim, T., Davenport, W., Mackie, G., Worthington, I., Aube, M., Christie, S. N., et al. (2012). Canadian headache society guideline for migraine prophylaxis. *The Canadian Journal of Neurological Sciences, 39*(2 Suppl. 2), S1–S59.

Ramsaransing, G. S., Fokkema, M. R., Teelken, A., Arutjunyan, A. V., Koch, M., & De Keyser, J. (2006). Plasma homocysteine levels in multiple sclerosis. *Journal of Neurology, Neurosurgery, and Psychiatry, 77*(2), 189–192.

Rao, P. V., & Bhat, K. S. (1989). Influence dietary riboflavin deficiency on lenticular glutathione redox cycle, lipid peroxidation, and free radical scavengers in the rat. *Journal of Clinical Biochemistry and Nutrition, 6*(3), 195–204.

Rivlin, R. S. (1973). Riboflavin and cancer: A review. *Cancer Research, 33*(9), 1977–1986.

Rogers, K. M., & Augusteyn, R. C. (1978). Glutathione reductase in normal and cataractous human lenses. *Experimental Eye Research, 27*(6), 719–721.
Rosenthal, W. S., Adham, N. F., Lopez, R., & Cooperman, J. M. (1973). Riboflavin deficiency in complicated chronic alcoholism. *The American Journal of Clinical Nutrition, 26*(8), 858–860.
Ross, A. C. (2013). *Modern nutrition in health and disease* (11th ed.). Philadelphia, PA: Lippincott Williams & Wilkins [chapter 22].
Sandor, P. S., Afra, J., Ambrosini, A., & Schoenen, J. (2000). Prophylactic treatment of migraine with beta-blockers and riboflavin: Differential effects on the intensity dependence of auditory evoked cortical potentials. *Headache, 40*(1), 30–35.
Savilahti, M. (1946). On the pathologic anatomy of the Plummer-Vinson syndrome. *Acta Medica Scandinavica, 125*(1), 40–54.
Schoenen, J., Jacquy, J., & Lenaerts, M. (1998). Effectiveness of high-dose riboflavin in migraine prophylaxis. A randomized controlled trial. *Neurology, 50*(2), 466–470.
Schoenen, J., Lenaerts, M., & Bastings, E. (1994). High-dose riboflavin as a prophylactic treatment of migraine: Results of an open pilot study. *Cephalalgia, 14*(5), 328–329.
Schulz, G. E., Schirmer, R. H., & Pai, E. F. (1982). FAD-binding site of glutathione reductase. *Journal of Molecular Biology, 160*(2), 287–308.
Seekamp, A., Hultquist, D. E., & Till, G. O. (1999). Protection by vitamin B2 against oxidant-mediated acute lung injury. *Inflammation, 23*(5), 449–460.
Seidl, S. E., Santiago, J. A., Bilyk, H., & Potashkin, J. A. (2014). The emerging role of nutrition in Parkinson's disease. *Frontiers in Aging Neuroscience, 6*, 36.
Siassi, F., & Ghadirian, P. (2005). Riboflavin deficiency and esophageal cancer: A case control-household study in the Caspian Littoral of Iran. *Cancer Detection and Prevention, 29*(5), 464–469.
Sirivech, S., Driskell, J., & Frieden, E. (1977). NADH-FMN oxidoreductase activity and iron content of organs from riboflavin and iron-deficient rats. *The Journal of Nutrition, 107*(5), 739–745.
Sirivech, S., Frieden, E., & Osaki, S. (1974). The release of iron from horse spleen ferritin by reduced flavins. *The Biochemical Journal, 143*(2), 311–315.
Sisson, T. R. (1987). Photodegradation of riboflavin in neonates. *Federation Proceedings, 46*(5), 1883–1885.
Sosa, V., Moline, T., Somoza, R., Paciucci, R., Kondoh, H., & ME, L. L. (2013). Oxidative stress and cancer: An overview. *Ageing Research Reviews, 12*(1), 376–390.
Takata, Y., Cai, Q., Beeghly-Fadiel, A., Li, H., Shrubsole, M. J., Ji, B.-T., et al. (2012). Dietary B vitamin and methionine intakes and lung cancer risk among female never smokers in China. *Cancer Causes & Control, 23*(12), 1965–1975.
Taniguchi, M. (1980). Effects of riboflavin deficiency on lipid peroxidation of rat liver microsomes. *Journal of Nutritional Science and Vitaminology (Tokyo), 26*(4), 401–413.
Taniguchi, M., & Hara, T. (1983). Effects of riboflavin and selenium deficiencies on glutathione and its relating enzyme activities with respect to lipid peroxide content of rat livers. *Journal of Nutritional Science and Vitaminology (Tokyo), 29*(3), 283–292.
Taylor, P. R., Li, B., Dawsey, S. M., Li, J. Y., Yang, C. S., Guo, W., et al. (1994). Prevention of esophageal cancer: The nutrition intervention trials in Linxian, China. Linxian nutrition intervention trials study group. *Cancer Research, 54*(7 Suppl), 2029s–2031s.
Thurnham, D. I., Zheng, S. F., Munoz, N., Crespi, M., Grassi, A., Hambidge, K. M., et al. (1985). Comparison of riboflavin, vitamin A, and zinc status of Chinese populations at high and low risk for esophageal cancer. *Nutrition and Cancer, 7*(3), 131–143.
Toyoashi, T. (1992). Antioxidant effect of riboflavin in enzymic lipid peroxidation. *Journal of Agricultural and Food Chemistry, 40*(10), 1727–1730.
Traber, M. G., & Atkinson, J. (2007). Vitamin E, antioxidant and nothing more. *Free Radical Biology & Medicine, 43*(1), 4–15.

Triantafyllou, N., Evangelopoulos, M. E., Kimiskidis, V. K., Kararizou, E., Boufidou, F., Fountoulakis, K. N., et al. (2008). Increased plasma homocysteine levels in patients with multiple sclerosis and depression. *Annals of General Psychiatry*, *7*, 17.

Tumkiratiwong, P., Tungtrongchitr, R., Migasena, P., Pongpaew, P., Rojekittikhun, W., Vudhivai, N., et al. (2003). Antioxidant enzyme levels in the erythrocytes of riboflavin-deficient and Trichinella spiralis-infected rats. *The Southeast Asian Journal of Tropical Medicine and Public Health*, *34*(3), 480–485.

Tung, T. C., & Lin, J. K. (1964). The interaction of riboflavin with 4-dimethylaminoazobenzene and 4-aminoazobenzene. *Tsa Chih Gaoxiong Yi Xue Yuan Tong Xue Hui*, *63*, 225–232.

Ulvik, R. J. (1983). Reduction of exogenous flavins and mobilization of iron from ferritin by isolated mitochondria. *Journal of Bioenergetics and Biomembranes*, *15*(3), 151–160.

van den Donk, M., Buijsse, B., van den Berg, S. W., Ocke, M. C., Harryvan, J. L., Nagengast, F. M., et al. (2005). Dietary intake of folate and riboflavin, MTHFR C677T genotype, and colorectal adenoma risk: A Dutch case-control study. *Cancer Epidemiology, Biomarkers & Prevention*, *14*(6), 1562–1566.

Wahrendorf, J., Munoz, N., Lu, J. B., Thurnham, D. I., Crespi, M., & Bosch, F. X. (1988). Blood, retinol and zinc riboflavin status in relation to precancerous lesions of the esophagus: Findings from a vitamin intervention trial in the People's Republic of China. *Cancer Research*, *48*(8), 2280–2283.

Wang, G., Li, W., Lu, X., & Zhao, X. (2011). Riboflavin alleviates cardiac failure in type I diabetic cardiomyopathy. *Heart International*, *6*(2), e21.

Wang, S., Mei, J., Chen, Q., Xu, Y., Chen, Y., & Zhu, H. (1998). Effect of beta-carotene and riboflavin on lipid peroxidation in rats. *[Ying yang xue bao] Acta Nutrimenta Sinica*, *21*(1), 22–27.

Webster, R. P., Gawde, M. D., & Bhattacharya, R. K. (1996). Modulation of carcinogen-induced DNA damage and repair enzyme activity by dietary riboflavin. *Cancer Letters*, *98*(2), 129–135.

Weinreb, O., Mandel, S., Youdim, M. B., & Amit, T. (2013). Targeting dysregulation of brain iron homeostasis in Parkinson's disease by iron chelators. *Free Radical Biology & Medicine*, *62*, 52–64.

Welch, K. M., Levine, S. R., D'Andrea, G., Schultz, L. R., & Helpern, J. A. (1989). Preliminary observations on brain energy metabolism in migraine studied by in vivo phosphorus 31 NMR spectroscopy. *Neurology*, *39*(4), 538–541.

Willett, W. (2013). *Nutritional epidemiology* (3rd ed.). New York, USA: Oxford University Press [chapter 1].

Winklhofer, K. F., & Haass, C. (2010). Mitochondrial dysfunction in Parkinson's disease. *Biochimica et Biophysica Acta*, *1802*(1), 29–44.

Wood, A. W., Sayer, J. M., Newmark, H. L., Yagi, H., Michaud, D. P., Jerina, D. M., et al. (1982). Mechanism of the inhibition of mutagenicity of a benzo[a]pyrene 7,8-diol 9,10-epoxide by riboflavin 5′-phosphate. *Proceedings of the National Academy of Sciences of the United States of America*, *79*(17), 5122–5126.

Wynder, E. L., & Bross, I. J. (1961). A study of etiological factors in cancer of the esophagus. *Cancer*, *14*, 389–413.

Wynder, E. L., & Chan, P. C. (1970). The possible role of riboflavin deficiency in epithelial neoplasia. II. Effect of skin tumor development. *Cancer*, *26*(6), 1221–1224.

Wynder, E. L., Hultberg, S., Jacobsson, F., & Bross, I. J. (1957). Environmental factors in cancer of the upper alimentary tract: A Swedish study with special reference to Plummer-Vinson (Paterson-Kelly) syndrome. *Cancer*, *10*(3), 470–487.

Wynder, E. L., & Klein, U. E. (1965). The possible role of riboflavin deficiency in epithelial neoplasia. I. Epithelial changes of mice in simple deficiency. *Cancer*, *18*, 167–180.

Xu, F., & Hultquist, D. E. (1991). Coupling of dihydroriboflavin oxidation to the formation of the higher valence states of hemeproteins. *Biochemical and Biophysical Research Communications, 181*(1), 197–203.

Yagi, K., Komur, S., Yoshino, K., Konishi, H., & Abe, H. (1989). Serum lipid peroxides and cataractogenesis in riboflavin deficiency. *Journal of Clinical Biochemistry and Nutrition, 6,* 39–48.

Yang, C. S., Sun, Y., Yang, Q. U., Miller, K. W., Li, G. Y., Zheng, S. F., et al. (1984). Vitamin A and other deficiencies in Linxian, a high esophageal cancer incidence area in northern China. *Journal of the National Cancer Institute, 73*(6), 1449–1453.

Young, A. J., & Lowe, G. M. (2001). Antioxidant and prooxidant properties of carotenoids. *Archives of Biochemistry and Biophysics, 385*(1), 20–27.

CHAPTER THREE

Niacin

James B. Kirkland*, Mirella L. Meyer-Ficca†,1
*Department of Human Health and Nutritional Sciences, University of Guelph, Guelph, ON, Canada
†Utah State University, Logan, UT, United States
[1]Corresponding author: e-mail address: mirella.meyer@usu.edu

Contents

Abstract

Nicotinic acid and nicotinamide, collectively referred to as niacin, are nutritional precursors of the bioactive molecules nicotinamide adenine dinucleotide (NAD) and nicotinamide adenine dinucleotide phosphate (NADP). NAD and NADP are important cofactors for most cellular redox reactions, and as such are essential to maintain cellular metabolism and respiration. NAD also serves as a cosubstrate for a large number of ADP-ribosylation enzymes with varied functions. Among the NAD-consuming enzymes

Advances in Food and Nutrition Research, Volume 83
ISSN 1043-4526
https://doi.org/10.1016/bs.afnr.2017.11.003

identified to date are important genetic and epigenetic regulators, e.g., poly(ADP-ribose)polymerases and sirtuins. There is rapidly growing knowledge of the close connection between dietary niacin intake, NAD(P) availability, and the activity of NAD(P)-dependent epigenetic regulator enzymes. It points to an exciting role of dietary niacin intake as a central regulator of physiological processes, e.g., maintenance of genetic stability, and of epigenetic control mechanisms modulating metabolism and aging. Insight into the role of niacin and various NAD-related diseases ranging from cancer, aging, and metabolic diseases to cardiovascular problems has shifted our view of niacin as a vitamin to current views that explore its potential as a therapeutic.

1. INTRODUCTION

Niacin (nicotinic acid or vitamin B3) is a functional group present in the coenzymes nicotinamide adenine dinucleotide (NAD) and nicotinamide adenine dinucleotide phosphate (NADP), which are essential for oxidative processes. Niacin was first identified in 1937 as the dietary factor that prevented black tongue disease in dogs, the animal model for pellagra (Koehn & Elvehjem, 1937). Pellagra is a dietary-deficiency disease whose study facilitated the identification of niacin as the pellagra-preventing factor, the discovery of niacin as the essential vitamin required for adequate NAD synthesis in humans, and the identification of multiple roles of coenzyme NAD(P) in metabolism.

2. ROLE OF DIETARY NIACIN IN DISEASE ORIGIN AND PROGRESSION

2.1 Historic Niacin-Deficiency Syndrome: Pellagra

The disease pellagra was first described in 1763 by the Spanish physician Casal, soon after maize (corn) became a major part of the European diet. In 1771, Francesco Frapolli, an Italian physician, coined the term "pellagra" (literally: sour or rough skin) to describe the crusty and inflamed skin characteristic in human patients suffering from dietary niacin wasting deficiency.

In 1870, frequent pellagra outbreaks among poor populations in southern and central Europe had followed the introduction of maize. At the end of the 19th and the beginning of the 20th centuries, pellagra reached epidemic proportions among the population of the southern United States and became the most severe deficiency disease in the history of the United States, with more than 120,000 deaths by the 1920s

(Etheridge, 1972; Rajakumar, 2000). Joseph Goldberger, working for the United States Public Health Service in the 1920s, discovered that milk and meat could provide a pellagra-preventing (P—P) factor, which was absent in corn.

In 1937, Koehn and Elvehjem used liver extracts to isolate the dietary P—P factor that prevented black tongue disease in a dog model of pellagra and subsequently identified it as nicotinic acid (Elvehjem, Madden, Strong, & Woolley, 1937; Koehn & Elvehjem, 1937). This discovery laid the foundation for more studies that confirmed lack of nicotinic acid as the underlying cause of pellagra in humans, and the rapid implementation of nicotinic acid and nicotinamide supplementation as the efficient treatment.

It was initially puzzling that Latin American populations had been spared from pellagra, despite using corn as their main staple food for centuries. This riddle was solved when differences in the traditional food processing methods were discovered. Traditional Latin American food processing practices involved presoaking corn in alkaline "lime-water" prior to cooking, which released the niacin from binding to glycoproteins, increasing its bioavailability.

In 1942, the US government mandated a niacin enrichment program for flour, and similar programs in other developed countries have diminished the occurrence of clinical pellagra to sporadic cases associated with malnutrition (e.g., neglect, abuse, or anorexia nervosa), alcoholism, iatrogenic situations (e.g., cancer treatments), malabsorption, and prolonged diarrhea.

In the 1960s and 1970s, pellagra still was considered a public health concern in developing African and Asian countries, where seasonally between 15% and 50% of the population presented with skin manifestations of pellagra (Aykroyd, 1971; Barrett-Connor, 1967; Gopalan, 2009). Pellagra outbreaks still occur frequently among refugee populations, where they continue to be of ongoing concern.

Pellagra became known as the disease of the "4 D's," referring to three commonly seen components (diarrhea, dermatitis, and dementia) of the syndrome culminating in death (the 4th D) (Hegyi, Schwartz, & Hegyi, 2004). The first D, diarrhea, the gastrointestinal presentation of pellagra, results from intestinal inflammation, when the normally metabolically active and rapidly dividing mucosal tissue is affected. Patients appear nauseous and have poor appetite and abdominal pain. Anorexia and malabsorptive diarrhea exacerbate a state of malnutrition.

The second D, dermatitis, refers to extensive skin lesions. After an initial erythematous phase resembling sunburn with blisters, the skin becomes thickened and hyperpigmented when dermatitis progresses. Typical for pellagra is a symmetrical and bilateral distribution of affected regions in sun-exposed skin, e.g., dorsal parts of hands, feet, cheeks, forehead, and sun-exposed regions of the neck, sometimes referred to as "Casal's necklace."

The third D refers to frequently observed neurological symptoms, e.g., muscle weakness, and the frequent deterioration of mental health observed in pellagra patients, progressing from lethargy, apathy, and depression to anxiety and irritability. While these symptoms are not surprising during severe malnutrition, many patients with pellagrous dementia experience visual and auditory hallucinations and display paranoid and aggressive behaviors, which can disappear within hours of niacin supplementation, suggesting impairment of brain signaling rather than structural degeneration.

The final D, death, indicates the high mortality rate observed in severe pellagra cases when depletion of the niacin-based coenzyme NAD prevents maintenance of vital body functions.

2.2 Current Risks and Causes of Niacin Deficiency and Pellagra

In addition to a diet deficient in niacin or its precursor tryptophan, the common causes of pellagra, other mechanisms can contribute to suboptimal niacin status. Primary causes of pellagra include dependence on a corn-based diet, or general malnutrition (associated with poverty, neglect, abuse, and famine, as well as anorexia nervosa) (Jagielska, Tomaszewicz-Libudzic, & Brzozowska, 2007). Some secondary causes include chronic alcoholism (Nogueira, Duarte, Magina, & Azevedo, 2009) and general malabsorptive states such as prolonged diarrhea. Alcohol-associated pellagra cases continue to be reported in countries with niacin fortification (Li et al., 2016; Terada, Kinoshita, Taguchi, & Tokuda, 2015). In scientific and medical reports of pellagra cases, alcoholism is the most frequently identified etiology (35%) compared to medications (56%), inadequate intake and malabsorption (16% and 13%, respectively), metabolic disorders (8%), and infrequent unknown etiologies (Goudarzi et al., 2016). When diarrhea sets in as part of the progression of pellagra, it tends to cause a vicious cycle of worsening niacin status.

In addition, other secondary factors induce pellagra-like symptoms and contribute to impaired NAD synthesis:

- Diets rich in leucine, like those based on *Sorghum*, can lead to pellagra due to altered tryptophan metabolism in the presence of excessive leucine (Beretich, 2005; Gopalan, 1969, 2009).

- Hartnup disease is an autosomal recessive mutation in the gene for the neutral amino acid transporter SLC6A19, leading to depletion of tryptophan and increased susceptibility for development of pellagra (Bröer, Cavanaugh, & Rasko, 2005; Kraut & Sachs, 2005; Nozaki et al., 2001). Administration of niacin reverts the pellagra symptoms and highlights tryptophan's importance as complementary NAD precursor.
- Carcinoid tumors can cause secondary tryptophan degradation due to excessive serotonin production. As tryptophan is a precursor for both, NAD and serotonin synthesis, an increased tryptophan demand via an increase in serotonin synthesis depletes tryptophan that is available for NAD synthesis resulting in niacin deficiency and occasional pellagra-like symptoms (Bouma et al., 2016; Shah et al., 2005).
- Infections with human immunodeficiency virus (HIV) have been associated with decreased plasma tryptophan levels and a pellagra-like state that responded to nicotinamide supplementation (Murray, Langan, & MacGregor, 2001).
- Additional diseases, e.g., liver cirrhosis, diabetes mellitus, and prolonged fevers, have been associated with niacin deficiency and pellagra.
- Increasing age correlates with dropping NAD^+ levels (Braidy et al., 2011). Because nicotinamide and NAD^+ levels are currently moving to the limelight as players determining longevity (Li, Chong, & Maiese, 2006; Verdin, 2015), current research aims to augment intracellular NAD^+ to prevent aging-associated disorders and decline (Srivastava, 2016).
- Women have about twice the risk of niacin deficiency compared to men (Miller, 1978). A study of sex-specific effects of tryptophan to niacin conversion in rats found a sex hormone-dependent decrease in the efficiency of tryptophan to niacin and NAD synthesis in females. This sex-specific difference puts women, and in particular pregnant females, at an increased risk for niacin deficiency and pellagra (Shibata & Toda, 1997). Indeed, a recent human study measured blood vitamin concentrations in pregnant women during the trimester of pregnancy. The women were healthy and eating a good diet supplemented with vitamins. Despite hypervitaminemic levels of folate, biotin, pantothene, and riboflavin, blood levels of vitamin A, niacin, thiamine, and vitamin B12 were deficient. The niacin hypovitaminemia was especially pronounced. A high percentage of pregnant women were deficient already in the first trimester, and this deficiency worsened in the later trimesters (Baker, DeAngelis, Holland, Gittens-Williams, & Barrett, 2002).

2.3 Niacin and Vitamin B3, Assessment of Niacin Status, and Recommended Daily Allowance

2.3.1 Vitamin B3 Structures

Pellagra is the manifestation of a deficiency in NAD. Molecules that serve as dietary precursors for NAD, other than tryptophan, constitute the vitamin B3 family. Fig. 1 gives an overview of currently known vitamin B3 molecules. The first identified "pellagra-preventing factor," i.e., vitamin B3, was nicotinic acid or pyridine-3-carboxylic acid, subsequently referred to as niacin (nicotinic acid vitamin), and this is the main source of B3 found in plant-based foods. Nicotinamide or pyridine-3-carboxamide, sometime called niacinamide, also serves as nutritional NAD precursor and prevents pellagra through a similar, albeit not completely identical pathway. Nicotinamide is formed by the cleavage of NAD^+ and is the main circulating form of B3 in blood. Despite this original distinction in nomenclature between niacin and niacinamide, today the term niacin often indiscriminately refers to both, nicotinic acid or nicotinamide, or a combination of both. In 2004, Bieganowsi and Brenner discovered that nicotinamide ribose, a molecule present, e.g., in milk, can similarly be converted to NAD and thus classifies as the newest discovered member of the vitamin B3 family (Bieganowski & Brenner, 2004).

2.3.2 Measurement of Niacin Status

Since the availability of dietary niacin in humans is directly reflected by NAD content in the body, biochemical determination of blood or tissue NAD is utilized as a measure of niacin status (Jacobson & Jacobson, 1997; Shah et al., 2005). A study determining blood NAD content in a metabolic ward demonstrated that the NAD content in red blood cells can be used as a marker for niacin status in humans. Blood NAD content directly mirrored low niacin diet or niacin supplementation in the form of either decreased or increased blood NAD levels (Fu, Swendseid, Jacob, & McKee, 1989; Jackson, Rawling, Roebuck, & Kirkland, 1995; Jacobson, Shieh, & Huang, 1999). Since blood NADP content remains relatively stable even with a niacin poor diet that causes NAD to drop (Tang, Sham, Hui, & Kirkland, 2008), the ratio of NAD to NADP (called niacin number) is sometimes used as a measure of niacin status. Interestingly, even in developed countries with niacin-fortified flour, blood niacin numbers vary to a surprising degree, and 15%–20% of the population were found niacin deficient (Jacobson, 1993).

Alternatively, quantification of urinary metabolites, i.e., *N*-methylnicotinamide and 2-pyridone, can be measured to determine

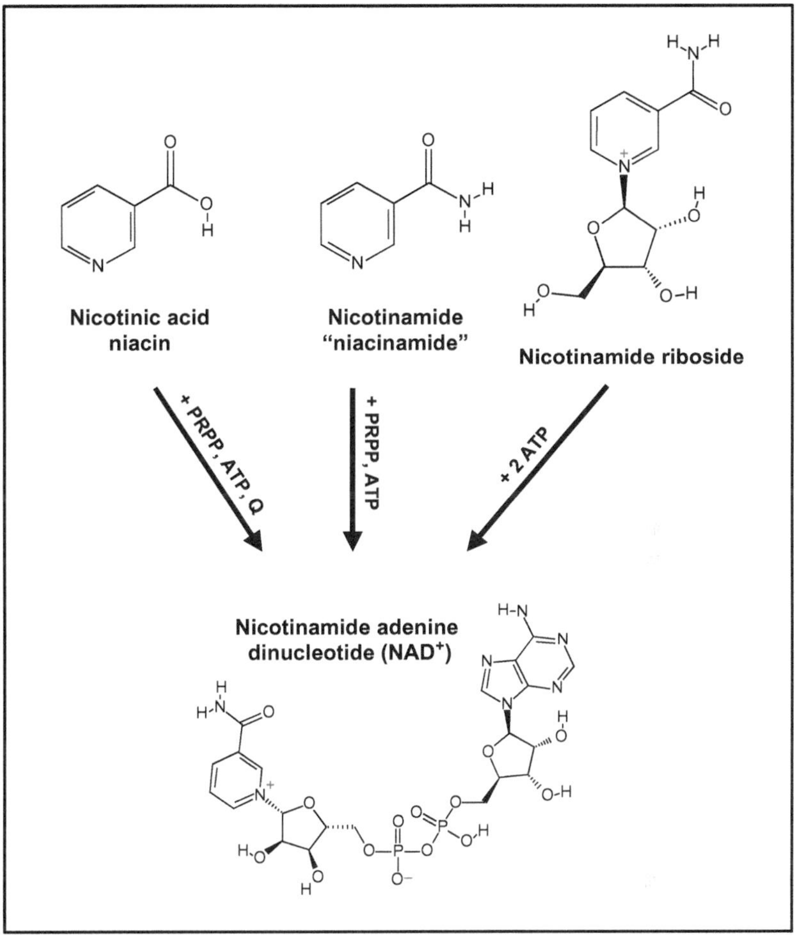

Fig. 1 Vitamin B3 molecules (nicotinic acid, nicotinamide, and nicotinamide riboside) are dietary precursors that support the formation of nicotinamide adenine dinucleotide (NAD). Cofactors for the synthesis of NAD from the respective precursor are indicated as follows: *ATP*, adenosine triphosphate; *PRPP*, 5-phosphoribosyl-ribose-1-pyrophosphate; *Q*, glutamate.

recent dietary niacin intake and niacin status at an individual and population level, with the caveat that 2-pyridone more likely indicates adequate protein intake than niacin status.

2.3.3 Dietary Need and Recommended Daily Intake

Dietary need for NAD synthesis is usually expressed as of "niacin" requirements, i.e., the direct precursors for NAD synthesis taken up as vitamin B3 molecules. A small percentage of NAD is generated from the essential amino acid tryptophan via the de novo NAD synthesis. The efficiency of the tryptophan to NAD metabolism is estimated to require 60 mg tryptophan for the generation of 1 mg niacin, so that a 60 mg tryptophan intake counts as 1 niacin equivalent (NE) (Horwitt, Harper, & Henderson, 1981).

In the United States, the Food and Drug Administration regulates vitamins as "dietary supplements" and requires vitamin manufacturers to follow current GMP regulations. To prevent and correct dietary insufficiencies in certain food products, most flour and grain products produced as "enriched products" must contain specified levels of thiamin, niacin, riboflavin, iron, and folic acids. According to 21 CFR137.165 (fda.gov; revised as of April 1, 2016), each pound of flour must contain 24 mg of niacin.

The recommended daily allowance (RDA) is the amount of nutrient or vitamin that is enough to meet the nutrient requirements of nearly all (97%–98%) healthy people. RDA for niacin in adults is 16 mg/day of NE for men and 14 mg of NE for women according to the FOA/WHO, status 1996 (see Table 1). For children, these values are adjusted to age groups and gradually increase from 6 NEs daily in toddlers to 12 in teenagers, until they reach the adult requirements. Recent studies estimate the median daily intake of preformed dietary niacin in the United States to be 28 mg for men and 18 mg for women. In Canada, median daily intake was approximately 41 mg in men and 28 mg in women (Institute of Medicine, 1998). Chronic high oral doses of nicotinic acid can induce hepatotoxicity and dermatological problems in the form of skin flushing, so that the US Food and Nutrition Board has recommended a 35 mg tolerable upper limit (UL) for daily intake (Institute of Medicine, 1998).

2.3.4 Factors That Influence Dietary Niacin Requirement

One important factor influencing the dietary niacin requirements is the bioavailability of the niacin that is present in available food. In mature cereal grains, the majority of niacin is bound and only about 30% is bioavailable. Beans and liver are among the foods that contain niacin in the free form. As mentioned, tryptophan as an NAD precursor in humans is thought to be converted to NAD at an efficiency of 60:1 to 70:1 (Horwitt et al., 1981). This conversion rate has been found to vary widely (30%) between individuals and tends to increase with increased tryptophan consumption, likely because more tryptophan is available for conversion to NAD once tryptophan demand for protein synthesis is met (Patterson, Brown, Linkswiler, & Harper, 1980). Factors that reduce this conversion efficiency increase the

Table 1 Recommended Dietary Allowance (RDA) of Niacin Intake for Humans at Different Point in the Life Span

Group	Age (Years)	Recommended Intake (NE/Day)
Infants (AI)	0–0.5	2
	0.5–1	4
Children (RDA)	1–3	6
	4–8	8
	9–13	12
Adolescents and adults (RDA)		
Males	>14	16
Females	>14	14
Pregnant		18
Lactating		17

AI, adequate intake, recommendation when there is not enough evidence to develop an RDA; *NE*, niacin equivalent, defined as mg niacin + 1/60 mg tryptophan; *RDA*, recommended daily allowance, average daily level of intake that is thought enough to meet the nutrient requirements of nearly all (97%–98%) healthy people.

USDA Agricultural Research Services, Dietary Reference Intakes, Information Retrieved on April 7, 2017.

requirement for niacin. Examples in addition to low tryptophan intake are carcinoid syndrome, in which increased serotonin synthesis depletes tryptophan, and Hartnup disease, which is characterized by deficient tryptophan uptake, and long-term treatment with the drug Isoniazid, which interferes indirectly with the tryptophan to NAD pathway.

2.4 Niacin Food Sources

Plants, bacteria, and yeast produce niacin and nicotinamide. From there, the vitamin molecules are maintained in the food chain. Niacin content varies in different food sources (see Table 2), and various food groups are suitable sources for dietary niacin. Fish, meat, milk, peanuts, and enriched flour products are good niacin sources for human consumption.

2.5 Risk Factors Predisposing to Niacin Deficiency

The major risk factor for niacin deficiency is the ongoing consumption of a diet that relies mostly on nonfortified refined grains and grain products, as well as a diet with little variety that is low in animal or dairy products and legumes. Maize- or *Sorghum*-based diets are examples of such diets

Table 2 Preformed Niacin Content in Food

Food	Niacin Content (Values in mg per 100 g or 100 mL)	
Meats and Fish	Fish (tuna, skipjack)	12–19
	Fish (salmon, mahimahi, swordfish)	7–9
	Poultry (chicken, turkey)	9–12
	Pork	6–11
	Beef and veal	7–9
	Game meat, bison	7–11
Dairy and Eggs	Milk (liquid, whole—nonfat)	0.09–0.1
	Milk (dry, nonfat)	0.7–1
	Yoghurt (various kinds)	0.1–0.2
	Ice cream (various kinds)	0.07–1.5
	Cheese	0.03–1.2
	Egg	0.075–0.2
Grains, Cereals	Cereals, ready to eat	0.2–90
	Rice, uncooked	5
	Pasta and noodles, enriched, dry	6–8.4
	Flour (wheat, barley, rice, buckwheat, white, enriched)	6–7.5
	Cornmeal (enriched)	4.9–5.2
	Corn flour, cornmeal, corn grain	2.6–3.6
Vegetables	Mushrooms	3–14
	Tomato products (canned, paste)	0.9–3
	Beans	2–3
	Corn	1.1–1.8
	Potatoes	1.3–1.9
	Mustard greens, spinach, cabbage, cauliflower	0.2–0.3
	Yeast extract spread	128
Legumes	Peanuts (raw and roasted nuts, butter)	12–15
	Soy products (beans, meal, flour, sauce)	1–3.2
	Soymilk	0.2
	Beans (kidney, garbanzo, mung, black, adzuki, lima)	0.4–0.7

Table 2 Preformed Niacin Content in Food—cont'd

Food	Niacin Content (Values in mg per 100g or 100mL)	
Nuts	Almonds (raw, roasted, butter, paste)	1.4–3.7
	Cashew (raw and roasted nuts, butter)	1–1.7
	Coconut (milk, meat, cream)	0.3–0.9
	Sunflower seed kernels (dried, roasted, butter)	4–8
	Sesame seeds (meal, flour, paste, butter)	4.5–13.4
Fruit	Physalis (groundcherries, fresh poha)	2.8
	Avocados (raw)	1.9
	Oranges	0.4
	Strawberries	0.4
	Apples	0.08
Baked goods	Leavening agents (active dried yeast)	40.2
	Bagels	1.8–4.6
	Bread	2–6
	Waffles	2–7

USDA Agricultural Research Services, Food Composition Database (version from March 29, 2017).

low in bioavailable niacin. *Sorghum* (millet, jowar) contains suitable amounts of tryptophan, but excessive amounts of leucine in *Sorghum* interfere with utilization of tryptophan for niacin synthesis.

Certain population groups are at risk for niacin deficiency even in the context of supplemented food products, i.e., secondary niacin deficiency. This is usually due to either increased niacin demand with increased metabolic NAD consumption, e.g., as seen in elderly people and pregnant women, or cancer patients undergoing treatments that induced DNA damage, such as radiation therapy or exposure to DNA-damaging drugs.

3. NAD METABOLISM

3.1 NAD Synthesis

NAD can be synthesized from nicotinic acid (NA), nicotinamide (NAM), nicotinamide riboside (NR) (all members of the vitamin B3 molecule group), and tryptophan.

NAD is synthesized from NA by the *Preiss–Handler pathway*. It comprises three individual metabolic conversions (Fig. 2). In a first step, NA is

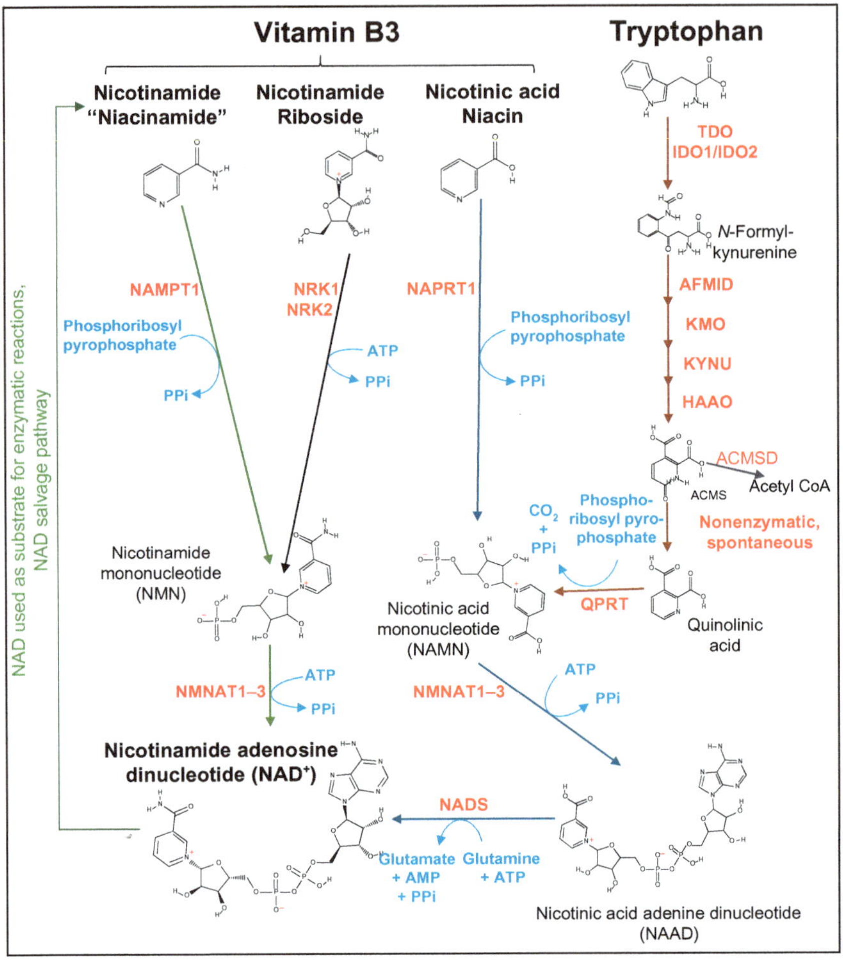

Fig. 2 NAD synthesis pathways present in mammals. *Blue arrows* illustrate the Preiss–Handler pathway of NAD synthesis from the nicotinic acid conversion, and the *NAD salvage pathway* comprises NAD synthesis from nicotinamide (*green arrows*). NAD from tryptophan occurs via the de novo *synthesis* pathway (*brown arrows*). *Abbreviations*: *ACMSD*, ACMS decarboxylase [EC:4.1.1.45]; *ACMS*, alpha-amino-beta-carboxy-muconate-semialdehyde; *AFMID*, *N*-formylkynurenine formamidase [EC:3.5.1.9]; *HAAO*, 3-hydroxyanthranilate 3,4-dioxygenase [EC:1.13.11.6]; *IDO1*, indoleamine 2,3-dioxygenase 1 [EC:1.13.11.52]; *IDO2*, indoleamine 2,3-dioxygenase 2 [EC:1.13.11.52]; *KMO*, kynurenine 3-monooxygenase [EC:1.14.13.9]; *KYNU*, kynureninase [EC:3.7.1.3]; *NAAD*, nicotinic acid adenine dinucleotide; *NAD*, nicotinamide adenine dinucleotide; *NADS*, NAD synthetase [EC:6.3.5.1]; *NAMN*, nicotinic acid mononucleotide; *NAMPT*, nicotinamide phosphoribosyl transferase [EC:2.4.2.12]; *NAPRT*, nicotinic acid phosphoribosyl transferase [EC:6.3.4.21]; *NMN*, nicotinamide mononucleotide; *NMNAT*, nicotinamide mononucleotide adenylyltransferase [EC:2.7.7.1 2.7.7.18]; *NR*, nicotinamide riboside; *NRK1*, NR kinase 1 [EC:2.7.1.22 2.7.1.173]; *NRK2*, NR kinase 2 [EC:2.7.1.22 2.7.1.173]; *QPRT*, quinolinate phosphoribosyl transferase [EC:2.4.2.19]; *TDO*, tryptophan 2,3-dioxygenase [EC:1.13.11.11].

converted to nicotinic acid mononucleotide (NAMN) by the enzyme nicotinic acid phosphoribosyl transferase (NAPT, encoded for by the gene *NPT1*) and use of 5-phosphate-α-D-ribose 1-diphosphate as cosubstrate. NAMN is then converted to nicotinic acid adenine dinucleotide (NAAD) by the enzyme NMNAT (nicotinamide mononucleotide adenylyltransferase) using ATP as cosubstrate. Three isoforms of NMNAT enzymes with distinct subcellular localizations have been described in humans, which permits differentially regulated subcellular NAD concentrations (Emanuelli et al., 2001; Raffaelli et al., 2002; Zhang et al., 2003). As a final step toward NAD synthesis, the NA moiety is amidated by the enzyme NADS (NAD synthetase), which transfers an amino group from glutamine and simultaneously forms glutamate in an ATP-consuming step.

The synthesis of NAD from nicotinamide follows the *salvage pathway*. The name reflects the role of this pathway in the reutilization of nicotinamide released by ADP-ribosylases and NAD glycohydrolases in the body. The enzyme nicotinamide phosphoribosyl transferase (NAMPT) performs the initial step from nicotinamide to nicotinamide mononucleotide (NMN). This step is rate limiting for local NAD synthesis, and NAMPT activity is considered an important regulator of several downstream NAD-consuming enzymes, e.g., sirtuins (Revollo, Grimm, & Imai, 2004, 2007). Making more NAD available is presumed to be the mechanism by which NAMPT overexpression delays senescence in human cells (van der Veer et al., 2007). Additional relevance for NAMPT comes from observations that the expression of NAMPT is regulated by the circadian rhythm machinery. This makes it a direct mediator between circadian rhythms and NAD salvage pathway (Nakahata, Sahar, Astarita, Kaluzova, & Sassone-Corsi, 2009; Ramsey et al., 2009). The final step toward NAD synthesis is performed by the NMNAT enzymes that condense the adenylyl moiety to NMN in a manner comparable to the NAMN to NAD step discussed earlier.

The third molecule of the vitamin B3 group, nicotinamide riboside, comprises the phosphorylation of NR by the enzyme NR kinase (NRK) to form NMN, which subsequently serves as a substrate for NAD synthesis via NMNAT adenylyl transfer, as discussed. Two isoforms for NRK have been described (NRK1 and NRK2), and this pathway appears to be highly conserved between species, e.g., it was found in yeast and humans (Belenky et al., 2007; Bieganowski & Brenner, 2004).

An additional important pathway of NAD synthesis involves the conversion of tryptophan to NAD via the de novo *NAD synthesis* pathway (right section in Fig. 2). This pathway involves seven biochemical reactions in the *kynurenine pathway,* which transforms tryptophan to NAMN. NAMN is transformed to NAAD and finally to NAD by NMNAT/NADS activities, using the identical metabolic transformation that synthesizes NAD from the NAMN generated by the NA metabolisms in the Preiss–Handler pathway.

The first, and rate-limiting step, of the de novo synthesis pathway involves the conversions of tryptophan to *n*-formylkynurenine. This step can be catalyzed by each of three enzymes: indoleamine 2,3-dioxygenase 1 (IDO1), *indoleamine 2,3-dioxygenase 2* (IDO2), or tryptophan 2,3-dioxygenase (TDO). These three enzymes have partially overlapping expression patterns and potentially distinct functions (Ball, Jusof, Bakmiwewa, Hunt, & Yuasa, 2014). TDO is the major enzyme relevant for de novo synthesis in liver, while IDOs have highest activity in lung, small intestine, and placenta (Kudo & Boyd, 2000; Yamazaki, Kuroiwa, Takikawa, & Kido, 1985).

The four subsequent enzymatic reactions result in the formation of alpha-amino-beta-carboxy-muconate-semialdehyde (ACMS). ACMS has two distinct fates that influence niacin status; either it is decarboxylated by the enzyme ACMS decarboxylase to alpha-amino-muconate-semialdehyde (AMS), or it undergoes spontaneous enzyme-independent cyclization and forms quinolinic acid. Quinolinic acid subsequently serves as a precursor for NAD synthesis. AMS on the other hand serves as a substrate for a pathway, leading to the formation of acetyl CoA. The efficiency of NAD synthesis from tryptophan therefore hinges on the accumulation of ACMS. High activity of ACMSD can enzymatically deplete this intermediate from the NAD-directed pathway and prevent efficient tryptophan to NAD conversion, while low levels of ACMSD activity allow enough ACMS accumulation to form quinolinic acid (Fukuwatari, Sugimoto, & Shibata, 2002; Fukuwatari, Ohta, Kimtjra, Sasaki, & Shibata, 2004). Quinolinate phosphoribosyl transferase (QPRT) condenses quinolinic acid with 5-phospho-alpha-D-ribose 1-diphosphate to form NA mononucleotide (NAMN), which is further metabolized to NAAD by NMNAT and subsequently to NAD by NADS, as described for the NA to NAD pathway earlier. QPRT represents the second rate-limiting step in the tryptophan to NAD de novo *synthesis pathway.* It is expressed primarily in liver, kidney, and brain (Magni et al., 2004; Rongvaux, Andris, Van Gool, & Leo, 2003). Deficiencies of QPRT activity are associated with neurological problems, e.g., epilepsy and Huntington's disease, likely reflecting the neurotoxic properties of quinolinic acid (Feldblum et al., 1988; Foster, Whetsell, Bird, & Schwarcz, 1985).

3.2 NAD Recycling

Due to the continuous activity of NAD^+-consuming enzymes, e.g., poly(ADP-ribose)polymerases (PARPs), sirtuins, and CD38, there is an ongoing loss of NAD^+ which is enzymatically cleaved usually with production of nicotinamide as a reaction by-product. Simple dietary niacin intake could not sustainably compensate for this NAD^+ loss. Instead, nicotinamide has to be efficiently recycled to NAD^+. This is achieved by the consecutive enzymatic activity of nicotinamide phosphoribosyltransferase (NAMPT) and nicotinamide mononucleotide adenylyltransferase (NMNAT1–3), as illustrated in Fig. 2 (green arrows, salvage pathway). The importance of this pathway becomes clear when considering that levels of NAMPT in cells had a larger effect on achievable NAD^+ levels than concentrations of nicotinamide, as was demonstrated in cell culture models in which NAD^+ levels correlated with modulated expression of NAMPT (Yang et al., 2007).

3.3 NADPH Synthesis

NADH can be converted to NADPH phosphate (NADP) by the enzymatic activity of NADH 2′-phosphotransferase or NADH kinase, which transfers a phosphate group from ATP to NADH. NADPH is essential for many reductive biosynthetic reactions in cells, e.g., fatty acid, steroid, and nucleotide synthesis, as outlined later.

During cellular metabolism, NADPH is regenerated from NADP via several pathways, including the pentose phosphate pathway (aka hexose monophosphate shunt), reductive reactions that reduce cytoplasmic oxaloacetate to pyruvate in the so-called pyruvate/malate cycle via the malic enzyme, and $NADP^+$-specific forms of isocitrate dehydrogenase (IDH) and aldehyde dehydrogenase (ADLH) (Pollak, Dölle, & Ziegler, 2007).

3.4 NAD and NADP Utilization

Enzymes utilizing NAD and NADP fall into several distinct categories (see Fig. 3). The first category of enzymes utilize NAD[*H*] and the phosphorylated form NADP[*H*] *as cofactors* for cellular reduction and oxidation (redox) reactions, in which NAD and NADP oscillate between the oxidized forms (NAD^+ and $NADP^+$) and the reduced forms (NAD(H) and NADP(H)). While the ratio of oxidized to reduced forms depends on the metabolic reaction involved, the total amount of NAD or NADP is not affected.

The second group of NAD^+-dependent enzymes represents a category of enzymes that utilized only the oxidized form, i.e., NAD^+, *as a substrate* in reactions that require ADP-ribose moieties (see Fig. 3, red arrows). Such

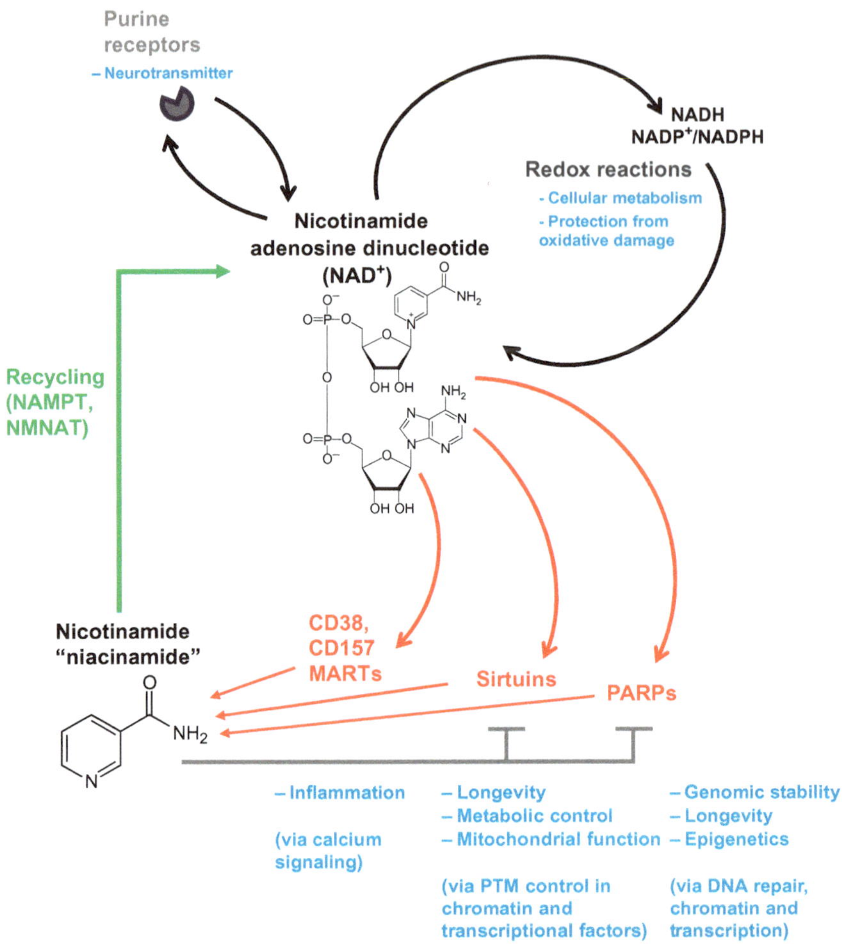

Fig. 3 NAD as essential cofactor for a wide variety of cellular processes. NAD is an essential enzymatic cofactor for most biochemical reactions in cellular metabolism processes and is required to regenerate glutathione necessary to prevent and minimize cellular damage due to naturally occurring reactive oxygen species (*black circles, right upper corner*). Additionally, NAD serves as a neurotransmitter, when it binds to and activates purine receptors (*black arrows, left upper corner*). Three enzyme groups use NAD as an enzymatic cosubstrate. The groups comprise MARTs, PARPs, and sirtuins (*red arrows, bottom part of diagram*). Activity of these enzymes leads to a cleavage of NAD, which causes loss of available NAD and the production of nicotinamide as a reaction by-product. Increasing concentration of nicotinamide has inhibitory effects on PARPs and sirtuins and reduces their activity (indicated by *gray blocking arrows*). Nicotinamide is used as a substrate for NAD regeneration through activity of NAMPT and NMNAT in the NAD salvage pathway (*green arrow*). *Abbreviations*: *MART*, mono(ADP-ribose) transferase; *NAMPT*, nicotinamide phosphoribosyl transferase; *NMNAT*, nicotinamide mononucleotide adenylyltransferase; *PARP*, poly(ADP-ribose)polymerase.

reactions require enzymatic cleavage of NAD^+ and therefore result in a change of total NAD^+ amounts that are available for all cellular reactions. Enzyme families in this category comprise PARP enzymes (PARPs, also known as ADP-ribose transferases (ARTDs)), sirtuins (members of the class III NAD-dependent histone deacetylases, HDACs), NAD^+-dependent ADP cyclases CD38 and CD157, and indoleamine 2,3-dioxygenases (IDOs).

The most recently discovered third function of NAD^+ is *as a ligand* to a small group of purine receptors, e.g., P2Y1 and P2Y11. Regulated NAD^+ release occurs in neurosecretory cells, vascular and visceral smooth muscles, urinary bladder, large intestine, as well as the brain (Breen, Smyth, Yamboliev, & Mutafova-Yambolieva, 2006; Durnin et al., 2012; Mutafova-Yambolieva, 2012; Mutafova-Yambolieva et al., 2007; Smyth, Bobalova, Mendoza, Lew, & Mutafova-Yambolieva, 2004; Yamboliev, Smyth, Durnin, Dai, & Mutafova-Yambolieva, 2009). NAD^+-responsive receptors are present on colonic cells, nucleated blood cells such as monocytes, and endothelial cells of arteries, where NAD^+ neurotransmitter-like actions at purine receptors alter vascular tone.

3.4.1 Redox Reactions

NAD^+ and $NADP^+$ are indispensable cofactors in over 400 enzymatic redox reactions central to most metabolic processes (https://www.cmescribe.com/vitamin-dependent-gene-databases/) (Penberthy & Kirkland, 2012). The following section provides a general and abbreviated overview highlighting some of the most important biochemical processes that rely on sufficient NAD and NADP availability, and hence indirectly on sufficient dietary niacin consumption.

The redox couple NAD^+/NADH [designated NAD(H) from here on] primarily promotes oxidative and catabolic reaction, while $NADP^+$/NADPH [designated NADP(H)] most often drives reductive and anabolic reactions. This difference is due to the wide difference in redox potentials between those two redox couples. Under normal cellular conditions, most of the molecules of the NAD(H) redox couple are present in the form of oxidized NAD^+, while most molecules in the NADP(H) redox couple occur in the reduced form NADPH.

The oxidized NAD^+ is necessary mostly for catabolism, e.g., in the breakdown of glucose to pyruvate via glycolysis and the formation of acetyl coenzyme A and subsequently complete catabolism in the Krebs cycle, which generates NADH reduction equivalents that drive energy production via the respiratory chain in the mitochondria. In addition, NAD^+ reduction to NADH facilitates β-oxidation of fatty acids in the mitochondria. The overview illustrates some select NAD^+-dependent enzymes and their specific enzymatic functions in diverse catabolic processes are shown in Fig. 3. NAD^+-dependent steps are central to energy production from glucose via glycolysis, Krebs cycle, and anaerobic oxidation (steps 1–3 in Fig. 4). Enzymes mediating central oxidative steps, such as glyceraldehyde 3-phosphate dehydrogenase, pyruvate dehydrogenase, IDH, and lactate dehydrogenase, all require NAD^+ as coenzyme. Various enzymes required for β-oxidation of fatty acid processes during lipolysis also require NAD^+ as a cofactor, e.g., L-3-hydroxyacyl coenzyme A dehydrogenase and short-chain hydroxyacyl coenzyme A dehydrogenase (step 4 in Fig. 4). Energy generation via amino acid catabolism similarly is mediated by enzymes that require NAD^+ as a cofactor to catalyze oxidations steps, e.g., glutamate dehydrogenase, branched chain a-ketoacid dehydrogenase, etc. (step 5 in Fig. 4). This overview is far from being complete and is merely intended to illustrate how central NAD^+ is to most biochemical processes in the cells.

Both NAD^+ and $NADH/H^+$ are cofactors for some anabolic processes, as illustrated in Fig. 5. If glucose levels are insufficient, cells can synthesis glucose via gluconeogenesis. In this process, the enzymatic equilibrium of three enzymes in the glycolysis pathway is reverted to allow them to utilize the reductive energy of $NADH/H^+$ to facilitate the synthesis of glucose molecules starting from substrates such as lactate, pyruvate, and acetyl coenzyme A that have been generated by degradation of either amino acids or fatty acids. Common enzymes with such a reversible enzymatic equilibrium are lactate dehydrogenase, malate dehydrogenase, and glyceraldehyde 3-phosphate dehydrogenase (see step 1 in Fig. 5). $NADH/H^+$ cofactors also facilitate the synthesis of fat via the synthesis of glycerol 3-phosphate (step 2 in Fig. 5) and are required for the synthesis of dihydrotestosterone from testosterone as a cofactor for the 5-α-reductase enzyme that mediates this last conversion (step 3 in Fig. 5).

NADPH, the reduced form of $NADP^+$, is an essential cofactor for many anabolic biochemical processes, e.g., the synthesis of fatty acids and cholesterol (see Fig. 6). It also required to support the reductive processes necessary in cytochrome P450-mediated functionalization reactions, glutathione/fatty acid hydroperoxidases, and thioredoxin defense against reactive oxygen stress, and as a substrate for immune response-associated oxidative defense reactions (Pollak et al., 2007).

During fatty acid synthesis, two molecules of NADPH are required for each acetyl group that is added and reduced during lipogenesis. Enzymes that

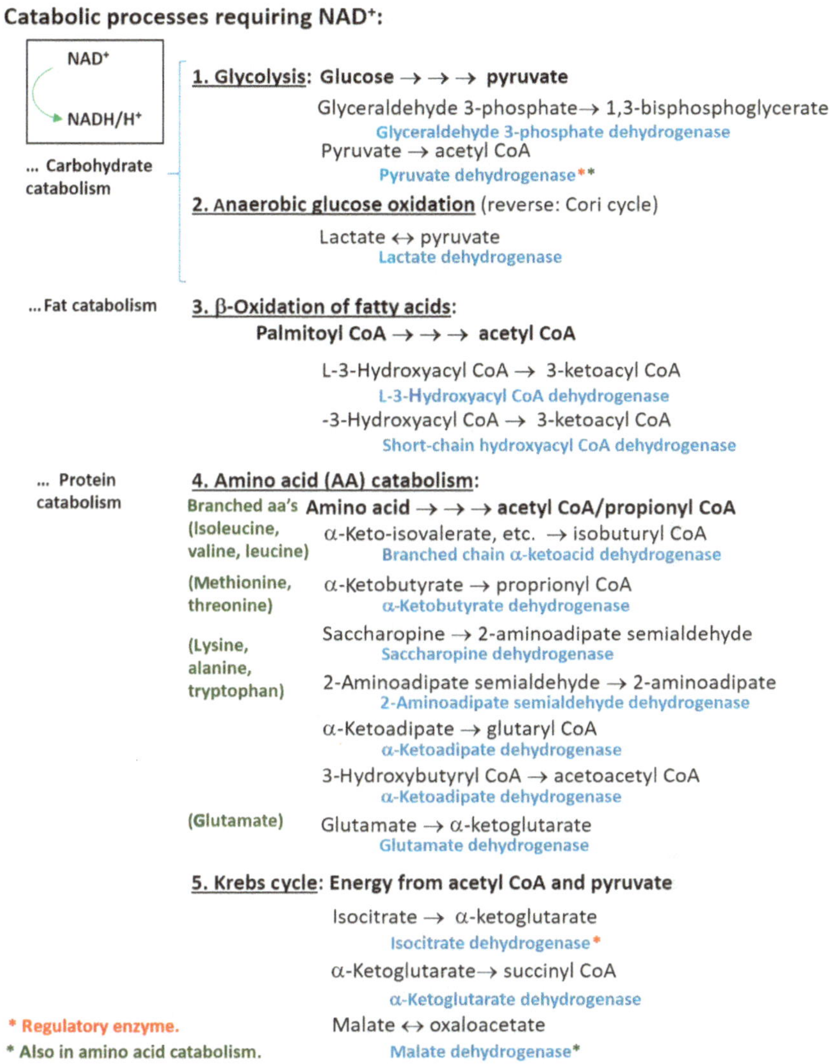

Fig. 4 Overview of biochemical redox reactions in catabolic processes that utilize NAD^+ as a cofactor. This list shows important catabolic processes that lead to reduction of NAD^+ to $NADH/H^+$. The general pathway name is indicated next to the numbers. Specific reactions occurring during the indicated general process are shown by name of reaction intermediated (in *black*); enzymes that catabolize the reaction are indicated in *blue*; relevant amino acid examples are indicated in *green*. *Abbreviations*: *AA*, amino acid; *CoA*, coenzyme A.

are active in the process of fatty acid synthesis and require NADPH as a coenzyme include β-ketoacyl ACP reductase and enoyl ACP reductase (step 1 in Fig. 6). In liver cells, the availability of NADPH therefore depends on the activity of ongoing fatty acid synthesis processes.

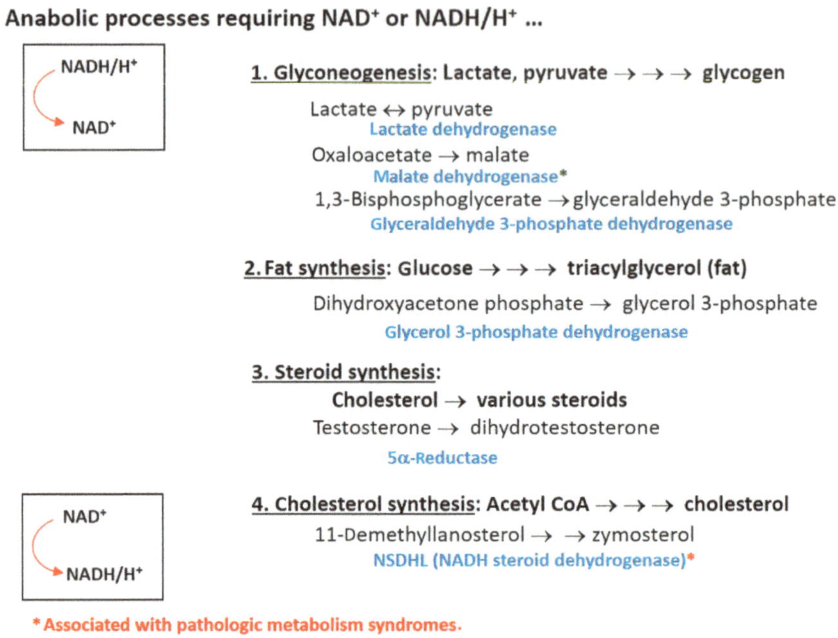

Fig. 5 Overview of biochemical redox reactions in anabolic processes that utilize NADH or NAD^+ as a cofactor. This list shows important anabolic processes that are associated with either the reduction of $NADH/H^+$ to NAD^+ (steps 1–3) or the reduction of NAD^+ to $NADH/H^+$ (step 4). The general pathway name is indicated next to the numbers. Specific reactions occurring during the indicated general process are shown by name of reaction intermediated (in *black*); enzymes that catabolize the reaction are indicated in *blue*.

Cholesterol synthesis is a complex biochemical synthesis process that requires consumption of NAD^+ as well as NADPH. Early in this process, the enzyme 3-hydroxy-3-methylglutaryl CoA reductase requires NADPH as a cofactor to synthesize mevalonate from 3-hydroxy-methylglutaryl CoA. Mevalonate is one of the basic building blocks for cholesterol synthesis. Several pharmacological substances that prevent this enzymatic step are administered to human patients as cholesterol-lowering drugs, e.g., lovastatin, atorvastatin, and simvastatin. Interestingly, combinations of such drugs with niacin are used to treat dyslipidemia. Later sections of this chapter will review this in more detail. In a similar fashion, the enzyme squalene synthase, which catalyzes the step from farnesyl pyrophosphate to squalene, requires NADPH. It is inhibited by the drug squalestatin. Several biochemical pathways can be used to synthesis cholesterol and all require NADPH as a cofactor for the respective enzymes (see step 2 in Fig. 6).

Anabolic processes requiring NADPH/H⁺ ...

NADPH/H⁺ → NADP⁺

1. <u>Fatty acid synthesis</u>: Acetyl CoA → → → palmitate

Acetoacetyl APC* → D-3-Hydroxybuturyl ACP
β-Ketoacyl ACP reductase
Enoyl ACP → acyl ACP
Enoyl ACP reductase

2. <u>Cholesterol synthesis</u>: Acetyl CoA → → → cholesterol

3-Hydroxyl-3-methylglutaryl CoA → mevalonate
3-Hydroxy-3-methylglutaryl CoA reductase*

Farnesyl pyrophosphate → squalene
Squalene synthase*

Lanosterol → 24,25-dihydrolanesterol
Sterol Δ24-reductase*

24,25-Dihydrolanosterol → → → lathosterol
Sterol Δ14-reductase*
Lanosterol 14-α-demethylase (CYP 51A1)*
3β-Ketosterol-reductase,
Sterol- C4-demethyloxidase*

Lathosterol → → → cholesterol
Lathosterol 5-desaturase,
7-Dihydrocholesterol reductase,
Sterol Δ24-reductase

3. <u>Bile acid synthesis</u>:

Cholesterol → → → cholate, chenodeoxycholate
Cholesterol → 7-α-hydroycholesterol
7-α-Hydroxylase

4. <u>Steroid synthesis</u>:

Cholesterol → various steroids

Cholesterol → pregnenolone
Cholesterol monooxygenase

Pregnenolone → progesterone
3β-Hydroxysteroid dehydrogenase

Progesterone → → → **aldosterone, (hydro)cortisol**
21α-Hydroxylase*, 11β-hydroxylase**, 18-hydroxylase

Progesterone → → **testosterone**
17α-Hydroxylase*, 17β-hydroxylase*

Testosterone → estradiol
Aromatase*

* Can be inhibited pharmacologically.
* Deficiency associated with pathologic metabolic abnormalities.

Fig. 6 Overview of biochemical redox reactions in anabolic processes that utilize NADPH as a cofactor. This list shows some important anabolic processes that are associated with either the reduction of NADPH/H^+ to $NADP^+$. The general pathway name is indicated next to the numbers. Specific reactions occurring during the indicated general process are shown by name of reaction intermediated (in *black*); enzymes that catabolize the reaction are indicated in *blue*.

Further important biochemical processes that require NAD^+ and NADPH as redox cofactors are various synthesis reactions that originate from cholesterol: the synthesis of bile acids is initiated by conversion of cholesterol to 7-α-hydroxycholesterol by 7-α-hydroxylase under the consumption of NADPH and O_2 (step 3 in Fig. 6). Cholesterol also is the basic building block for the generation of steroid hormones.

High levels of NADPH are present in red blood cell due to the fact that red blood cells lack mitochondria, and therefore have to rely on the pentose phosphate pathway for their energy metabolism. The resulting abundance of NADPH supplies these cells with the important cofactor that allows ongoing activity of glutathione reductase, which allows glutathione/fatty acid hydroperoxidase enzymes to reduce oxidative damage within the iron and oxygen-rich environment of red blood cells.

NAD(H) and NADP(H) also play a central role in the metabolism of ethanol and other alcohols. Ethanol is popular as a mood-altering compound that is consumed mostly as wine, beer, and spirits. Small quantities of alcohol intake have been associated with health benefits, while consumption of large ethanol quantities causes severe physiological complications, i.e., liver cirrhosis, alcoholic fatty liver disease (AFLD), and hypoglycemia (Gaziano & Manson, 1996). Through a series of consecutive enzymatic oxidation reactions, ethanol is converted to acetaldehyde and finally to acetate. The initial metabolic step, the conversion of ethanol to acetaldehyde, can be catalyzed by either the enzyme alcohol dehydrogenase with production of NADH or enzymes of the microsomal ethanol-oxidizing system, i.e., cytochrome P450 complex enzymes, with consumption of O_2 and $NADPH/H^+$, and leakage of superoxide. Higher ethanol doses, and chronic exposure, will force more ethanol through the P450 route, leading to greater risk of hepatic injury. Also, detoxification of the resulting acetaldehyde to acetate requires the enzymatic activity of ADLH with NAD^+ as the enzymatic cofactor. Chronic ethanol consumption has been associated with a decrease in blood NAD^+ levels, and since the accumulation of acetaldehyde is associated with the unpleasant consequences often experiences with alcohol consumption "hangover," supplemental niacin intake is thought to help with more efficient ethanol metabolism. If ethanol consumption occurs to a degree at which ethanol metabolism leads to markedly decreased NAD^+ and increased $NADH/H^+$ levels, the enzymatic equilibrium in NAD-dependent enzymes is shifted. Such a shift in NAD to NADH ratio results in a shift of the lactate dehydrogenase product equilibrium from pyruvate to lactate, which results in inhibited gluconeogenesis, and consequently can cause hypoglycemia. A similar shift from α-ketoglutarate to isocitrate is observed for IDH, and from oxaloacetate to malate for malate dehydrogenase. Together with a direct NADH concentration-dependent inhibition of the enzyme α-ketoglutarate dehydrogenase, this results in an inhibition of energy production via the Krebs cycle in the liver tissues that are detoxifying the consumed ethanol.

Both NAD(H) and NADP(H) are essential to maintain the cellular respiratory chain within all cells. Within cells, NAD and NADP are present at higher concentrations in the mitochondria than the rest of the cell. This compartmentalization explains why mitochondria-rich tissues, e.g., heart, contain more NAD than tissues with fewer mitochondria, e.g., liver.

Absolute tissue content of NADP(H) and NAD(H) varies widely. Liver contains about double the amounts of NAD(H) than NADP(H) ($\sim$800 nmol/g fresh rat liver tissue weight compared to $\sim$420 nmol/g tissue weight), while skeletal muscle contains closer to 30 times more NAD(H) than NADP(H) (590 nmol/g NAD(H) and 30 nmol/g NADP(H)).

Taken together, the role of NAD in redox metabolism relies heavily on the ratio of oxides to reduced molecules, i.e., how many of the NAD(H) molecules are present as NAD^+ vs NADH/H^+, since this ratio can cause a significant shift in the product equilibrium of enzymatic reaction. This can be seen, e.g., for lactate dehydrogenase, which produces mostly lactate, if there is excessive NADH, while the presence of excessive oxidized NAD^+ allows for the predominant formation of pyruvate. The ratio of NAD^+ and NADH/H^+ is influenced (and influences) not only by the activity of a single or few enzyme(s), but is the result of the sum of cellular metabolic processes (glycolysis, citric acid cycle, fatty acid synthesis, mitochondrial respiratory chain, etc.). The redox state of NAD(H) therefore reflects a complex readout of ongoing biochemical energy metabolism in a given cell type and situation, making NAD(H) an ideal candidate indicator molecule to monitor cellular metabolism, due to its capability to integrate information regarding nutritional quality, overall energy intake, and energy expenditure.

3.5 NAD as Substrate

Supporting the notion that NAD(H) can function as an indicator of the cellular metabolic environment is the fact that NAD^+ is the substrate for enzymatic reactions that permit modulation of cellular epigenetic information. NAD^+, the oxidized form of NAD(H), is the form whose cellular concentration varies most significantly and whose concentration most closely reflects metabolic situations such as insufficient dietary niacin intake, malnutrition, or excessive exercise. NAD^+ levels fluctuate not only with nutritional intake and exercise but also with circadian cues (Nakahata et al., 2009). The close relationship between NAD^+ availability and energy metabolism has resulted in studies testing if increased muscular NAD^+ synthesis could promote oxidative phosphorylation, and prompted thoughts of modulating the NAD^+ pool for therapeutic purposes, e.g., using NAD^+ supplementation to improve energy metabolism in sporting performance (Cantó, Menzies, & Auwerx, 2015; Frederick et al., 2015).

At the same time, activity of the NAD^+-consuming enzymes will directly control NAD^+ concentrations and therefore directly change metabolic processes in response to external exposures and situation that activate these NAD-consuming enzymes. Several families of enzymes continuously consume NAD^+. As mentioned earlier, the NAD^+-consuming enzyme families comprise sirtuins, PARPs, ADP cyclases, and IDO. The degree of NAD^+ consumption is dependent on several factors, e.g., the presence of external DNA-damaging agents. The enzymatic activity of the major PARP enzyme, the nuclear enzyme PARP1, is strongly activated in enzymes that bind to DNA strand breaks. Activation of PARP1 induces enzymatic cleavage of NAD and utilization of the ADP-ribose moiety to form poly(ADP-ribose) (PAR). The degree of enzymatic activation corresponds to the degree of DNA damage. Moderate levels of PAR formation facilitate efficient DNA repair and prevent cell cycle progression to proliferation of abnormal cells that could lead to carcinogenesis (Masutani & Fujimori, 2013; Masutani, Nakagama, & Sugimura, 2003). Exposure to severe DNA damage will lead to PARP1 hyperactivation. This results in dramatically increased cellular NAD^+ consumption, and to a lethal degree of cellular NAD^+ depletion. Taken together, cellular NAD^+ concentration varies directly as a consequence of exposure to any kind of external DNA-damaging conditions, e.g., radiation or exposure to alkylating agents (Burkle & Virag, 2013; Fouquerel & Sobol, 2014; Kupper, van Gool, & Burkle, 1995; Masutani et al., 2000; Meyer, Muller, Beneke, Kupper, & Burkle, 2000; Schreiber et al., 1995; Van Gool et al., 1997). Reduced NAD^+ availability will not only change cellular metabolic processes but can also result in altered gene expression, DNA repair, apoptosis, cell death, or carcinogenesis. This example illustrates the close interdependence of cellular levels of NAD^+ not only on metabolic processes and the nutritional situation but also on additional environmental conditions, e.g., to any kind of exposure, that modulate the activities of individual NAD^+-consuming enzymes.

3.5.1 ADP-Ribosyl Cyclases

The NADase/ADP-ribosyl cyclase enzyme CD38 and its homolog CD157 use NAD^+ and $NADP^+$ to generate cyclic ADP-ribose (Aarhus, Graeff,

Dickey, Walseth, & Lee, 1995; Ferrero, Lo Buono, Horenstein, Funaro, & Malavasi, 2014; Guse & Lee, 2008). Both enzymes are highly expressed in neutrophil cells and endothelial cells where they participate in immune response-associated signal transduction. Thus, they were long regarded as an immune cell "activation marker." This view has evolved significantly in the last years to CD38 as a multifunctional molecule (Quarona et al., 2013). The synthesized cyclic ADP-ribose functions as a signaling molecule that triggers an intracellular signaling cascade and results in Ca^{2+} elevation, likely by activating receptors in the endoplasmic/sarcoplasmic reticulum (Pollak et al., 2007). Immunocytes deficient in CD38 display a defect in migration to inflammatory sites and render mice with such a defect more susceptible to infections (Partida-Sánchez et al., 2001; Partida-Sanchez et al., 2007; Schuber & Lund, 2004). Interestingly, CD38-deficient mice retain higher NAD^+ levels and are protected against obesity and metabolic syndrome (Barbosa et al., 2007), potentially through changes in NAD^+ levels, which then alter other enzymes, like the sirtuins (Escande et al., 2013). CD38 was recently identified as a major factor contributing to the so far enigmatic observation that NAD^+ levels steadily decline with age in a nutrition-independent fashion, and that this decline causes mitochondrial dysfunction (Camacho-Pereira et al., 2016; Schultz & Sinclair, 2016).

CD38/157-induced cyclic ADP-ribose and calcium release are also involved in controlling oxytocin secretion and thus play a regulatory role in social behaviors. In addition to support from animal models, a human survey has shown that expression levels and polymorphisms in these genes have been associated with the Autism Quotient of individuals. These observations provide some mechanistic insight to the appearance of dementia in pellagrins, and the rapid disappearance of psychoses following niacin supplementation (Chong et al., 2017).

3.5.2 Poly(ADP-Ribose) Polymerases (PARPs) or ADP-Ribose Transferases (ATRDs)

PAR formation, using NAD^+ as a substrate, has been known for more than 50 years. Chambon and colleagues first observed NAD^+ depletion and PAR formation upon DNA damage (Chambon, Weill, & Mandel, 1963). Most of the initial studies focused on the most prevalent and most active PAR-forming enzyme, PARP1, whose central role in DNA repair and the maintenance of genomic stability shaped the general and prevailing assessment of poly(ADP-ribosyl)ation as the "guardian angel of the genome" (Chatterjee, Berger, & Berger, 1999; Jeggo, 1998). Subsequent studies revealed the existence of additional proteins with structural homology to PARP

(Johansson, 1999), and eventually of a large protein family of related enzymes (Ame, Spenlehauer, & de Murcia, 2004; Meyer, Meyer-Ficca, Jacobson, & Jacobson, 2004; Meyer-Ficca, Meyer, & Jacobson, 2005). So far, 17 enzymes have been identified as members of the human PARP superfamily, and our knowledge on the evolutionary origins, protein domains, and cellular functions of these proteins grows continually (Aravind, Zhang, de Souza, Anand, & Iyer, 2015; Gupte, Liu, & Kraus, 2017; Kraus, 2015; Perina et al., 2014; Poltronieri & Miwa, 2016).

In general, members of the PARP superfamily generate ADP-ribose posttranslational modification at specific sites in target proteins. Among the members of the PARP family, only some enzymes turned out to be true PARPs. This means that they cleave the glycosylic bond between nicotinamide and ribose in NAD^+, release nicotinamide, attach the liberated ADP-ribose unit to a given target protein, and in subsequent steps keep adding more ADP-ribose units to the initially attached ADP-ribose, thereby generating polymers of (ADP-ribose) (PAR) as posttranslational modifications. Fig. 7A illustrates the chemical structure of PAR molecules (Meyer-Ficca et al., 2005). Detailed structural analyses demonstrated that PAR can comprise hundreds of ADP-ribose units, either as linear or as branched molecules (Alvarez-Gonzalez & Jacobson, 1987; Juarez-Salinas, Levi, Jacobson, & Jacobson, 1982; Kiehlbauch, Aboul-Ela, Jacobson, Ringer, & Jacobson, 1993; Miwa et al., 1981; Miwa, Saikawa, Yamaizumi, Nishimura, & Sugimura, 1979; Miwa & Sugimura, 1984).

In the majority of PARP family members, the enzymatic activity is limited to the initial first step, i.e., cleaving NAD^+, releasing nicotinamide, and attaching one single ADP-ribose unit to the given target protein. Since the subsequent addition of several units does not occur and only mono(ADP-ribose) (MAR) is attached to the target protein, these enzymes have to be classified as functional mono(ADP-ribose)transferases (MARTs). The presence of an HYE amino acid motif in the catalytic center was identified as required, but not sufficient for PAR formation, as it is present in all four enzymes with true PARP activity, i.e., PARP1, PARP2, PARP5a, and PARP5b (better known as tankyrase 1 or TNKS1 and tankyrase 2 or TNKS2). Several in vitro and in vivo studies ascertained that other members of the PARP family either exhibited mono(ADP-ribosyl)transferase activity (PARP 3, 4, 6–12, 14–16) or no activity could yet be demonstrated experimentally (PARP13) (Karlberg et al., 2015; Meyer-Ficca et al., 2015; Vyas et al., 2014; Yang et al., 2017). The discrepancies between common names and enzymatic properties led to an effort aiming to change the nomenclature from PARPs to "ADP-ribosyl transferases" in order to prevent the obvious

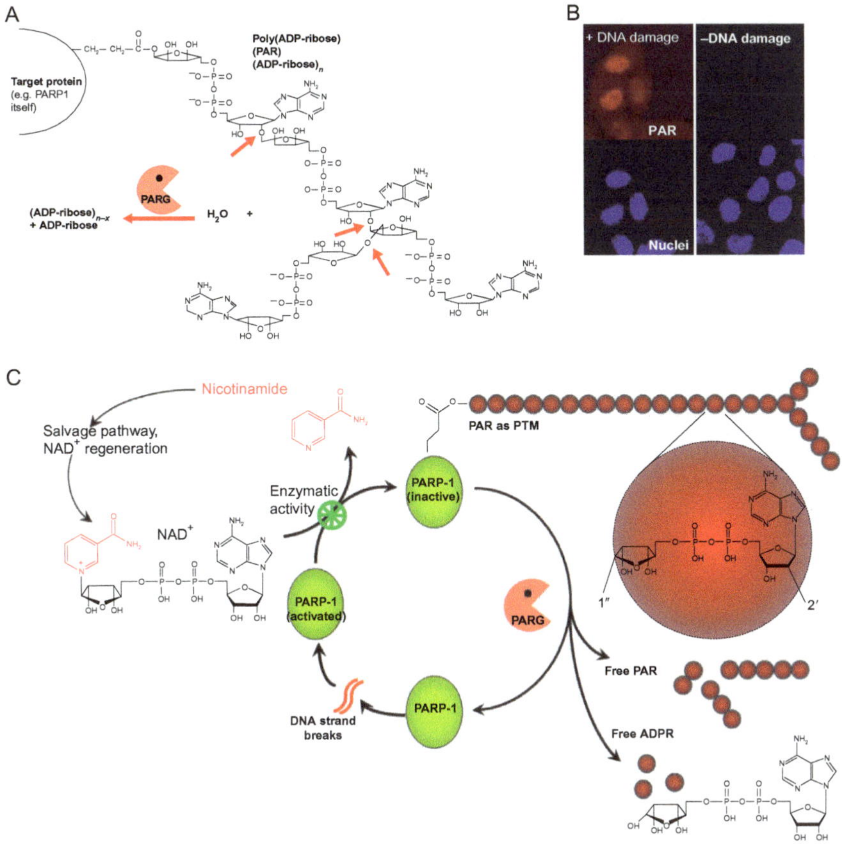

Fig. 7 Poly(ADP-ribosyl)ation by PARP1. (A) The chemical structure of a poly(ADP-ribose) molecule with a single branch point is shown. Glycosylic bonds that can be cleaved by poly(ADP-ribose)glycohydrolase (PARG, indicated in *red*) are indicated by *red arrows*. (B) HeLa cells treated with 500 μM MNNG as DNA-damaging agent accumulation amounts of PAR that can easily be stained with PAR-specific antibodies (visible as red fluorescence signals in the *left top panel*), while PAR levels in untreated cells cannot be visualized clearly (*right top panel*). *Bottom panel*: *Blue*, DAPI-stained nuclei for the cells seen in the *top panels*. (C) Poly(ADP-ribose) is rapidly synthesized upon PARP1 activation. PAR automodification of PARP1 temporarily inactivates the enzyme until PAR is degraded by PARG activity and PARP1 is liberated in the not automodified form. *Abbreviations*: *DAPI*, 2-(4-amidinophenyl)-1*H*-indole-6-carboxamidine; *MNNG*, *N*-methyl-*N*′-nitro-*N*-nitrosoguanidine.

discrepancy between the term "PARP" as the common name for most members of the PARP superfamily that are limited to MART activity (Hottiger, Hassa, Luscher, Schuler, & Koch-Nolte, 2010). A low general acceptance of this novel nomenclature has led to the simultaneous occurrence of several names for the same enzyme in the current scientific literature.

In PARP1, the most active and best-studied PARP, enzymatic activity is rapidly activated upon detection of and binding to DNA strand breaks. Activated PARP1 uses NAD to form large and branched PAR molecules, attached to various target proteins, including PARP1 itself, where it temporarily inactivates enzymatic activity (see Fig. 7C). The effects of PAR are normally short-lived due to rapid degradation by poly(ADP-ribose) glycohydrolase (PARG). Treatment of culture cells with DNA-damaging agents leads to a rapid increase in PAR that can be detected using specific antibodies, e.g., in Fig. 7B, where nuclear PAR is visible as red signals in the top-left panel, while untreated control cells have no visible signals.

The degree of enzymatic activity of PARP1 after activation by DNA breaks directly reflects the degree of DNA damage that is present. The synthesized PAR temporarily blocks cell cycle progression in order to allow sufficient DNA repair and to prevent proliferation of damaged or mutated cells. Extensive DNA damage leads to hyperactivation of PARP1, which causes severe NAD^+ depletion and results in cell death. This process has long been regarded as a cancer prevention mechanism that allows for elimination of severely damaged, potentially malignant cells. In line with this concept are studies that show that *Parp1*-deficient mice are more sensitive to radiation exposure and display increased genomic instability upon gamma radiation and exposure to DNA-damaging drugs and have an increased predisposition to develop tumors (Trucco et al., 1999; Wang et al., 1995). At the same time, preventing PARP1 hyperactivation with its associated NAD^+ loss and spike in PAR formation either in genetic knock out model, or pharmacologically with PARP inhibitors protects from cell death after inflammatory processes or ischemia–reperfusion injuries [reviewed by Herceg and Wang (2001) and Shall and de Murcia (2000)]. These varied functions of PARylation have led to ongoing interest in the development and clinical use of PARP inhibitors for cancer treatment, as well as spiked interest in exploring the benefit of PARP inhibition in nononcological diseases (Berger et al., 2017; Lord & Ashworth, 2017; Rajawat, Shukla, & Mishra, 2017).

Research over the last years has broadly expanded our understanding of cellular functions of ADP-ribosylation beyond these accepted roles in DNA repair.

It has grown from primarily being regarded as mechanism safeguarding genetic information via facilitation of DNA repair to the emerging view of ADP-ribosylation as a versatile posttranslational modification that is associated with many key biological processes. In addition to its undisputed major role in DNA repair, ADP-ribosylation facilitates and regulates DNA replication, cell division, transcription and signal transduction, cellular responses to stress, infections, and aging (Palazzo, Mikoč, & Ahel, 2017). This is achieved through modulation of a wide array of protein/protein interactions and through control of transcription and epigenetic information encoded in the specific DNA/protein interactions in chromatin.

Table 3 briefly illustrates major functions identified to date for individual enzymes in the PARP superfamily.

3.5.3 Sirtuins

One other group of enzymes that consume NAD^+ by enzymatic cleavage are sirtuins. These are members of the class III NAD-dependent HDACs. Sirtuins moved to the limelight of scientific interest by the discovery of their central role mediating life span extension in situations of caloric intake restriction (Anderson, Bitterman, Wood, Medvedik, & Sinclair, 2003; Cohen et al., 2004; Howitz et al., 2003). To date, they have remained at the center of ongoing scientific interest as pharmacological targets with a potential to slow down or ameliorate the aging process (Grabowska, Sikora, & Bielak-Zmijewska, 2017; Pan & Finkel, 2017).

Mammals possess seven ubiquitously expressed sirtuin homologs, Sirt1–7. The common feature of sirtuins as class III HDACs is that they consume NAD^+ to be catalytically active (see Fig. 8). Consequently, changes of NAD^+ availability as well as the NAD^+/NADH ratio directly regulate their activity, which makes them excellent sensors for both cellular energy status and redox status (Michan & Sinclair, 2007).

As HDACs, one of their major sirtuin function is the enzymatic deacetylation of lysine residues on target proteins, and SIRT1–SIRT3, SIRT5, and SIRT6 indeed preferentially act as deacetylases (see Fig. 8, top section). The reaction is a two-step process that consumes NAD^+: sirtuins cleave the nicotinamide moiety of an NAD^+ molecule, which releases the ADP-ribose moiety. In a subsequent step, the acetyl or acyl group of the target protein is then transferred from directly to this ADP-ribose, resulting in a deacetylated target protein and the formation of 2″-*O*-acetyl-ADP-ribose (Tanner, Landry, Sternglanz, & Denu, 2000).

Table 3 The PARP Enzyme Superfamily

PARP	Synonyms	Enzymatic Product	Currently Known Predominant Function
1	ARTD1	PAR	DNA repair (multiple pathways), transcriptional regulation, chromatin structure regulation, cell death control, inflammation
2	ARTD2	PAR	DNA repair, transcriptional regulation, chromatin structure regulation
3	ARTD3	MAR	Transcriptional regulation
4	ARTD4, vault PARP	MAR	
5a	ARTD5, tankyrase 1	PAR	Telomere length regulation, stress signal transduction
5b	ARTD6, tankyrase 2	PAR	Telomere length regulation, protein degradation
6	ARTD17	MAR	
7	ARTD14, tiPARP	MAR	TCDD-mediated toxicity, cell division
8	ARTD16	MAR	Nuclear membrane integrity
9	ARTD9, BAL1	MAR	Cell migration, cytoskeleton
10	ARTD10	MAR	Signal transduction
11	ARTD11	MAR	Nuclear membrane integrity
12	ARTD12, ZC3HDC1	MAR	Cytoplasmic stress response
13	ARTD13, ZC3HAV1, ZAP	Unknown	miRNA-mediated gene silencing
14	ARTD8, BAL2	MAR	Cell migration, cytoskeleton, signal Transduction
15	ARTD7, BAL3	MAR	Cytoplasmic stress response
16	ARTD15	MAR	Unfolded protein response

Overview of current nomenclature, known enzymatic activity, and primary function.

The remaining sirtuins, SIRT4 and SIRT6, preferentially utilize NAD^+ to transfer ADP-ribose units to target proteins, an enzymatic property that SIRT2 and SIRT3 execute in addition to their deacetylase properties. In this reaction, the sirtuin enzyme cleaves NAD, releases the nicotinamide

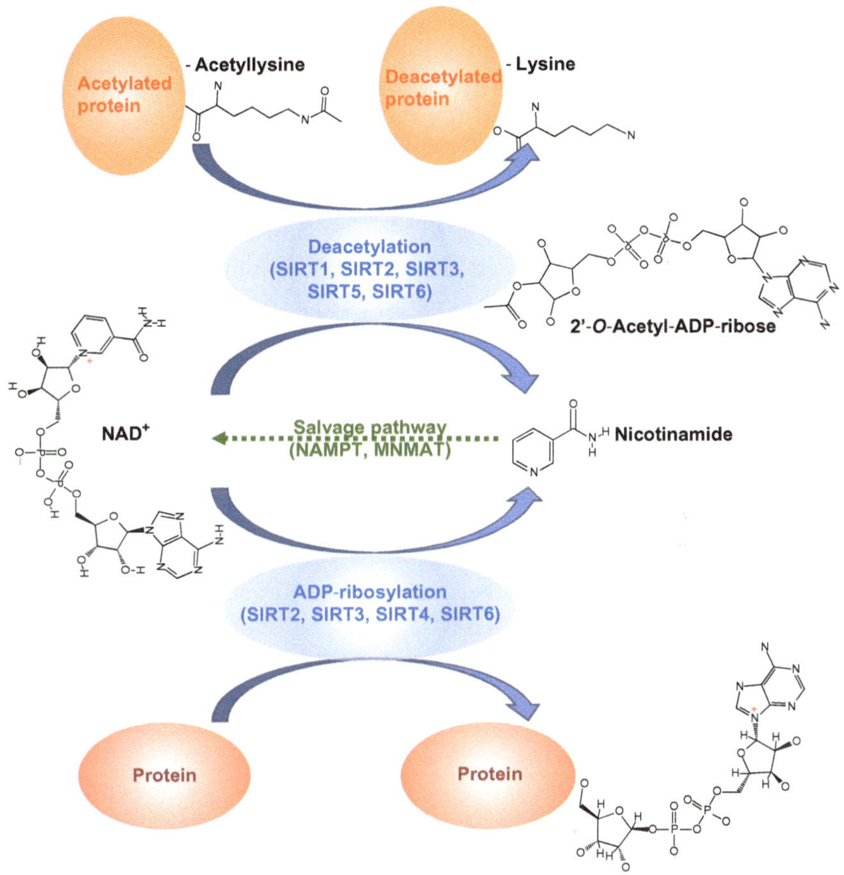

Fig. 8 Sirtuin-mediated activities. Sirtuins SIRT1, 2, 3, 5, and 6 utilize NAD^+ to remove posttranslational acetylation marks from target proteins, producing *O*-acetyl-ADP-ribose and nicotinamide (*top*). Activities of SIRT4 and SIRT5 are limited to transfer of ADP-ribose moieties to target proteins (*bottom*).

moiety, and enzymatically transfers the liberated ADP-ribose moiety to target proteins (see Fig. 8, bottom section). Functionally SIRT2, SIRT3, SIRT4, and SIRT6 thus are MARTs.

As an overall consequence of collective sirtuin activity, epigenetic information is changed, e.g., not only in the form of altered chromatin structure following histone deacetylation but also in the form of direct changes of transcription factor activities.

Among the sirtuin target proteins are transcription factors, e.g., FOXO1, PGC-1α, and HIF-1α, i.e., factors that regulate oxidative metabolism, antioxidant defense, and gene expression in the mitochondria both directly and indirectly. Consequently, reduced sirtuin activity, e.g., as expected in situations of NAD^+ paucity, will reduce the cell's ability to perform oxidative metabolism in the mitochondria (Imai & Guarente, 2016). The presence of multiple sirtuin proteins, each with specific subcellular locations and target protein preferences, allow sirtuin activity to influence a variety of cellular processes in the organisms, e.g., myogenesis, gluconeogenesis and insulin secretion, adipogenesis, DNA repair, development, and senescence (Jiang, Liu, Chen, Yan, & Zheng, 2017; Michan & Sinclair, 2007; van de Ven, Santos, & Haigis, 2017; Vazquez, Thackray, & Serrano, 2017).

Sirtuins are functionally connected to NAD^+ on several levels. Sirtuins directly depend on NAD^+ availability to be enzymatically active. Their activity increases with rising NAD^+ concentrations, while increasing levels of NADH and nicotinamide directly inhibits sirtuin enzymatic activity. This regulation allows sirtuins to directly sense and respond to cellular NAD^+ concentrations and cellular redox status. At the same time, sirtuins compete with other NAD^+-consuming enzymes for available NAD^+. PARP1 is among the enzymes that have a high capacity for NAD^+ utilization and can deplete cellular NAD^+ to levels that change the activity of sirtuins. Generally, tissue concentrations of NAD^+ exceed the levels that sirtuins need for activity. It still is conceivable that subcellular NAD^+ concentrations in the organelles vary and are in ranges that limit sirtuin activity; it is also possible that DNA damage-induced activation of PARP1 depletes NAD^+ to a degree where it actively curbs sirtuin activity. On the other hand, sirtuin activity directly downregulates PARP1 activity via deacetylation, pointing to a complex situation where PARP1 and sirtuins compete for NAD^+ resources.

A third link between sirtuins and NAD^+ is provided by the reported connection between sirtuins, the circadian transcription factors CLOCK/BMAL1, and their regulator effect on expression of NAMPT, the enzyme that is central for mammalian NAD^+ synthesis in the NAD salvage pathway. Together these factors form a circadian feedback loop that allows for the circadian oscillation of NAD^+ directly connected to the activity of sirtuins SIRT1 and SIRT6 (Asher et al., 2008; Belden & Dunlap, 2008; Kritikou, 2008; Nakahata et al., 2008, 2009). An interesting extension of the NAD^+/circadian rhythm connection is the finding that liver PARP1 activity also oscillates in a daily manner and that is regulated by feeding (Asher et al., 2010; Kumar & Takahashi, 2010; Peek et al., 2013; Schuldt, 2010).

Collectively, there is a broad interplay of nutrition, metabolism, and circadian rhythm control, with NAD^+ as a central player connecting dietary intake information with the individual facets of the transcription machinery (Asher & Sassone-Corsi, 2015).

This close connection between NAD^+ and sirtuin activity is seen as a central component of the life span extending properties that sirtuins like, e.g., Sir2 (the yeast homolog to the human SIRT1) and SIRT1, exert. Mutations that result in a loss of a functional NAD^+ biosynthesis pathway cause significantly decreasing life span (Lin et al., 2002; Smith et al., 2000). A decrease in NAD^+ occurs also naturally during aging (Braidy et al., 2011; Massudi et al., 2012), potentially exacerbating age-related health decline. Interestingly, yeast with mutations in SIR2 orthologs that allow higher enzymatic activity in an NAD-depleted environment was recently identified, raising the question if polymorphisms in sirtuin genes that allow efficient enzymatic activity in the presence of low NAD^+ concentration have an evolutionary advantage (Ondracek, Frappier, Ringel, Wolberger, & Guarente, 2017).

In conclusion, the collective activities of NAD^+-dependent enzymes (poly- and mono(ADP-ribosyl)transferases, ADP-ribosyl cyclases, and sirtuins) provide mechanisms that integrate cellular signaling and with the regulation of chromatin structure and epigenetic control, allowing for a spatiotemporal fine-tuning of transcription and gene expression to adjust for changed environmental conditions (see Fig. 9).

3.5.4 Indoleamine 2,3-Dioxygenase

Among the enzymes that potentially control NAD^+ concentrations is IDO, the enzyme mediating the initial and rate-limiting first step of the kynurenine pathway. TDO and IDO activity controls the entry of the normally limiting essential amino acid tryptophan into the kynurenine pathway, which results in decreased plasma tryptophan and increased NAD^+. In contrast to mostly hepatic expression of TDO, IDO is expressed in various tissues, e.g., neurological tissues, where it is subject to extensive regulation.

IDO activity can be activated by proinflammatory cytokines, e.g., interferon gamma, IL-6, and nitric oxide. Immune responses like HIV/AIDS as well as cancers and autoimmune disease can cause IDO overactivation and consequently result in depletion of plasma tryptophan and increased concentrations of neurotoxic metabolites like quinolinic acid (Brown et al., 1991). This was recently demonstrated in HIV/AIDS patient in low-income African populations (Bipath, Levay, & Viljoen, 2015). Similarly, patients with flaky dermatitis displayed metabolic changes, e.g., increased kynurenine/tryptophan ratio and lowered tryptophan, which are consistent with increased IDO activity (Maltos et al., 2015). In astroglia, metabolic, and

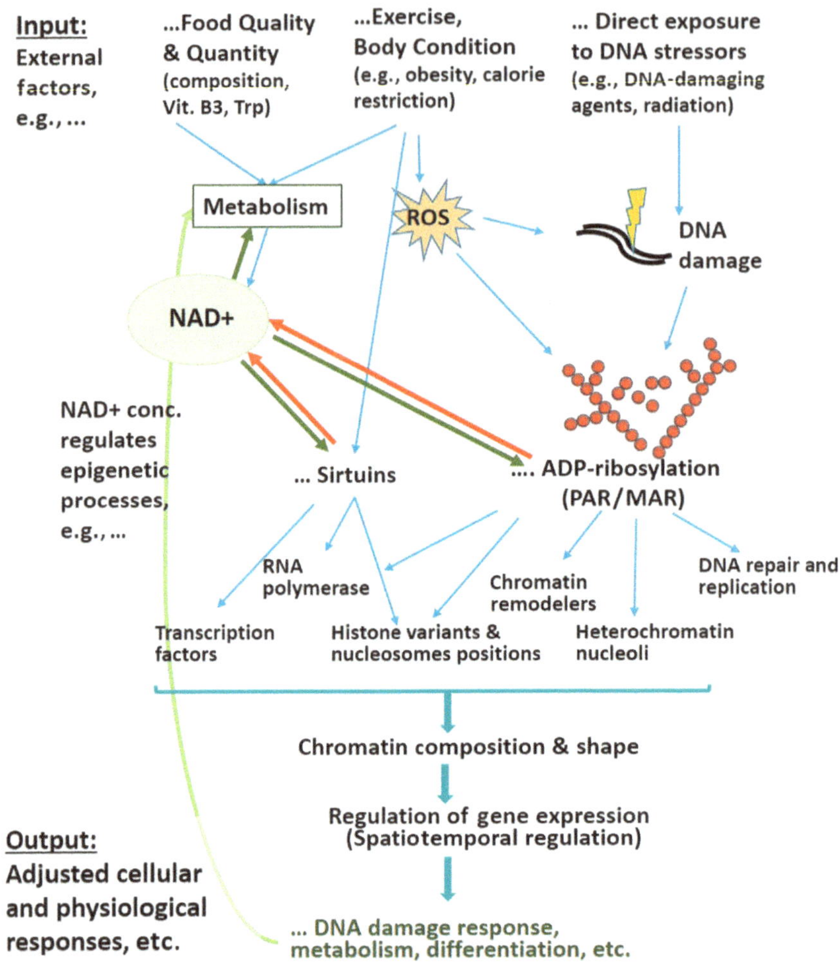

Fig. 9 NAD availability regulates ADP-ribosylation and sirtuin-mediated epigenetic changes. External factors, e.g., composition and quantity of food, metabolic parameters, such as body condition, exercise, and caloric intake, as well as direct exposures to DNA stressors (certain DNA-damaging toxicants, radiation, etc.) control the activity of the sirtuins and ADP-ribose-transferase, two powerful epigenetic regulators. ADP-ribosylation and sirtuin-dependent processes regulate the chromatin structure in a large variety of ways, e.g., histone PTM modifications of histones and transcription factors, heterochromatin condensation, transcriptional regulation, and DNA repair and replication. Change in chromatin structure results in adjusted gene expression, which in turn results in an overall adjustment of the physiological situation by controlling many cellular processes, e.g., metabolism, cellular differentiation, and DNA damage response. The extent of the enzymatic ADP-ribosylation and sirtuin response directly depends on availably NAD^+ levels (*green arrows*), allowing for a direct readout and integration of complex nutritional and external factors into epigenetic gene expression control. At the same time, the consumption of NAD^+ in these processes changes NAD^+ levels (*dark green arrows*). The sum of the adjusted overall cellular responses, e.g., gene expression response, will directly adjust cellular metabolism which in turn influences energy metabolism and NAD^+ oxidation status (*light green arrow*).

neurotropic support cells, activation of IDO by the proinflammatory factor IFN-gamma resulted in increased neurotoxic metabolites, but at the same time maintained NAD levels via de novo synthesis by the kynurenine pathway, which supports the hypothesis that one role of increased IDO activity during inflammation is to maintain NAD^+ levels (Grant & Kapoor, 2003).

Local IDO upregulation causes cellular tryptophan starvation in certain immune cells (e.g., T cells). This is required for the induction of immune tolerance, as it is required to prevent fetal rejection during pregnancy, but occurs similarly in autoimmune diseases and constitutes an immune-suppressive mechanism that regulates the tumor microenvironment (Muller & Prendergast, 2007; Munn & Mellor, 2016; Penberthy, 2007). Induction of IDO and resulting immune suppression is also beneficial for the suppression of graft-vs-host diseases and may have the therapeutic potential in a clinical setting (Jasperson et al., 2009). Inflammation induces IDO activity and results in tryptophan depletion. Inflammation simultaneously increases the production of iNOS-peroxynitrate activation of PARP activity, which results in chronic NAD^+ depletion and persistent TNF-alpha synthesis.

In addition to the connection to immune diseases, abnormal IDO regulation has been associated with cancer and neurological disorders and neurodegenerative diseases, likely also via accumulation of neurotoxic metabolites and depletion of neuroprotective tryptophan and NAD^+ (Badawy, 2017). Increased tryptophan degradation via overactivation of the kynurenine pathways has recently also been associated with cardiovascular disease (Song, Ramprasath, Wang, & Zou, 2017).

In conclusion, the control of IDO activity by the immune system allows for a local metabolic adjustment of innate and adaptive responses to inflammatory and immunological signals on a cellular and systemic level (Munn & Mellor, 2013) and is associated with a control of NAD^+ levels.

3.6 Turnover of PAR and MAR Modifications

Synthesis of PAR by PARPs, e.g., after exposure to DNA damage, leads to a rapid and significant loss of cellular NAD^+. Sirtuin activity, either as deacetylation or as mono(ADP-ribosyl)ation (MAR), is a cause for NAD^+ loss in a similar manner. Both reactions release nicotinamide as a reaction by-product, which will contribute to the NAD^+ regeneration via the NAD^+ salvage pathway as described earlier (see Figs. 2, 3, 7, and 8).

The additional moieties of NAD^+, the ADP-ribose subunit that was transferred to acceptor proteins during the enzymatic NAD^+ cleavage reaction, can undergo a similar regeneration. This regeneration is mediated by enzymatic activities of PARG and ADP-ribosyl hydrolase-3 (ARH3), two enzymes that cleave the linkage between ADP-ribose and the acceptor proteins.

3.6.1 Poly(ADP-Ribose)Glycohydrolase

PARG releases PAR polymers and degrades the polymer into individual ADP-ribose moieties by cleaving the *O*-glycosidic bond of the PAR chains. Through alternative splicing, a single *PARG* gene gives rise to several PARG protein isoforms of different sizes. Importantly, the different subcellular localization allows PARG to degrade PAR molecules present in different parts of the cell (Cortes et al., 2004; Koh et al., 2004; Meyer, Meyer-Ficca, Whatcott, Jacobson, & Jacobson, 2007; Meyer-Ficca, Meyer, Coyle, Jacobson, & Jacobson, 2004; Winstall et al., 1999). The terminal ADP-ribose unit at the receptor glutamate amino acid residue in a PAR-modified target protein can be removed by MACROD1 and MACROD2 proteins, and by TARG1/C6ord130 (Chen et al., 2011; Sharifi et al., 2013). Deficiency of TARG1 is associated with neurological defects comparable to the cytotoxic effects observed with neuronal PAR accumulation (Andrabi et al., 2006; Sharifi et al., 2013; Yu et al., 2006). This emphasizes how important turnover and recycling of PAR and MAR signals are for cellular functions.

3.6.2 ADP-Ribosyl Hydrolases (ARH1 and ARH3)

The ADP-ribosyl hydrolase family of enzymes consists of several members. Of interest in this context are ARH1 and ARH3. ARH1 releases the ADP-ribose subunit from proteins with mono-ADP-ribosylated arginines, while ARH3 releases ADP-ribose moieties from PAR as well as from *O*-acetyl-ADP-ribose, the by-product of sirtuin-mediated protein deacetylation (Kasamatsu et al., 2011; Mashimo, Kato, & Moss, 2013, 2014).

4. NAD: RELEVANCE TO HEALTH AND DISEASE

Impaired niacin status can decrease total NAD(H), decrease NAD^+, shift the cellular NAD^+ to NADH ratio, and change the ratios between NAD(H) and NADP(H) pools. The ratios are likely important to health, recalling that a shift toward reduced NADH occurs during aging.

Furthermore, the shifted ratio also affects enzymatic reactions pertinent to cellular metabolism. It changes the cells' preference for oxidative phosphorylation toward glycolysis, a process often observed in cancer cells. This decreases pyruvate production and results in a net increased lactate formation. This increased glucose uptake and glucose to lactate fermentation is known as the Warburg effect. The Warburg effect occurs even in the presence of completely functional mitochondria and is known to promote cellular growth and rapid proliferation, i.e., promoting tumorigenic properties (Liberti & Locasale, 2016).

Furthermore, the reduced NAD^+ directly affects a large number of NAD^+-dependent signaling pathways discussed earlier (MARTs, PARPs, sirtuins, ADP-ribose cyclases). Those pathways are involved in a large array of biological processes, e.g., control of energy metabolism and mitochondrial functions, calcium homeostasis, control of exposure oxidative stress (antioxidation mechanisms and generation of oxidative stress), control of signaling and gene regulation pathways, immunological functions, aging, and cell death (Ying, 2008). A common feature of the NAD^+-dependent pathways is the release of nicotinamide as a reaction by-product that can be recycled to NAD (Opitz & Heiland, 2015).

In summary, disturbance of the NAD homeostasis and recycling processes may potentiate an array of disorders that range from metabolic syndrome, to cancer, neurological diseases, and age-associated decline. The following section briefly illustrated some recent findings regarding the connection between niacin, NAD, and clinical problems.

4.1 NAD Decline During Aging and Age-Related Disorders

Aging is a complex process caused by the interplay of many pathophysiological processes. Two current theories on aging see either DNA damage accumulation or mitochondrial deficiencies due to DNA damage following nuclear to mitochondrial signaling as major contributors to aging. Both processes are dependent on cellular NAD^+ availability. Availability and redox status of available NAD directly link DNA-damage repair with mitochondrial function and nuclear to mitochondria signaling, indicating the central role the adequate availability of niacin plays as a dietary precursor of NAD for aging-related pathophysiology.

For many years, a connection has been proposed between NAD^+ levels and the link between aging and altered metabolism. This led to the term "NAD world" as a relevant term in aging and with the view of NAD as a metabolic oscillator that regulates aging and metabolism (Imai, 2009a, 2009b, 2010, 2011; Imai & Guarente, 2014, 2016; Rehan, Laszki-Szcząchor, Sobieszczańska, & Polak-Jonkisz, 2014).

A significant and steady decline of NAD^+ levels has been observed with increasing aging in various species (Braidy et al., 2011; Camacho-Pereira et al., 2016; Massudi et al., 2012; Schultz & Sinclair, 2016). In rats, a significant decline in tissue NAD and a shift in NAD^+ to NADH ratio were observed with aging, concomitantly with an increase in lipid peroxidation and DNA damage as well as a decline in antioxidant capacity (Braidy et al., 2011). A similar strong negative correlation between age and tissue NAD content was observed in humans, with similarly increased lipid oxidation and PARP activity (Massudi et al., 2012). Factors that contribute to this age-associated NAD^+ loss include increased activity of NAD^+-consuming enzymes, e.g., PARPs, CD38, and sirtuins (Jasper, 2013; Mouchiroud et al., 2013; Schmeisser et al., 2013). Reducing NAD^+ levels in worms resulted in reduced life span, while pharmacological or genetic restoration to normal NAD^+ levels restored longevity and avoided age-associated metabolic decline (Mouchiroud et al., 2013).

Increased PARP activity in this context is thought to be a response to the age-associated accumulation of DNA damage, and at the same time one of the contributing factors responsible for the increased NAD^+ catabolism seen in aging. Furthermore, a recent study implicated an age-related increase in the activity of the NAD-consuming enzyme CD38 with this age-associated drop in NAD^+ and mitochondrial dysfunction that was at least partially dependent on a pathway regulated by SIRT3 (Camacho-Pereira et al., 2016). This finding provided new clues to the long-known but puzzling observation that NAD^+ declines with age, independent of nutritional niacin intake (Schultz & Sinclair, 2016). Keeping NAD^+ levels up via a knockout of CD38 protected from NAD^+ decline and increased SIRT3 activity. Consequently, improved mitochondrial function and increased glucose tolerance prevented aging-associated defects (Chini, Tarragó, & Chini, 2016).

Reduced NAD^+ availability, in turn, prevents optimal sirtuin activity, which is central to preventing age-associated changes in metabolic processes and mitochondrial dysfunction (van de Ven et al., 2017). A simplified scheme (Fig. 10) illustrates the vicious cycle of aging-related NAD decline, and the exacerbating effects on cellular senescence, DNA damage, and metabolic dysfunction.

In model organisms, supplementation of NAD^+ precursors, e.g., nicotinamide ribose, could increase net NAD^+ synthesis and increase life span

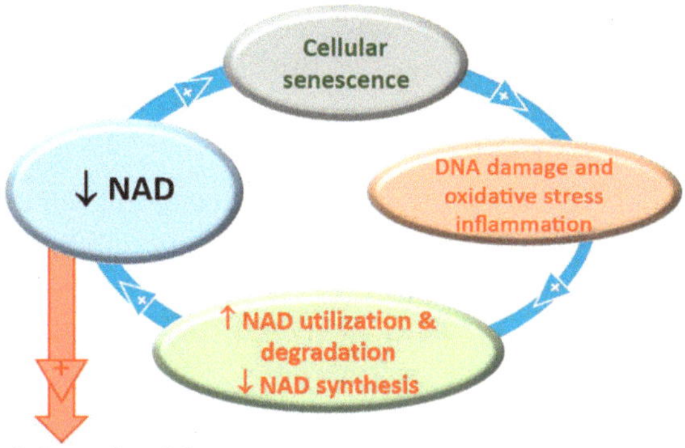

Fig. 10 Vicious cycle of age-dependent NAD^+ loss. Aging (as cellular senescence) is associated with exposure to increasing DNA damage and oxidative stress. Cellular defense mechanisms against such stressors, e.g., PARPs and sirtuins, utilize NAD^+. Increased NAD^+ use decreases NAD^+ availability, which in turn increases aging-related dysfunctions and promotes cellular senescence.

even in the absence of caloric restriction (Belenky et al., 2007). There are current efforts to translate the growing knowledge about the correlation between aging and declining blood NAD^+ into clinical efforts to augment intracellular NAD to prevent aging-associated disorders and decline (Srivastava, 2016).

4.2 DNA Damage Accumulation

Aging is associated with an increase in basal DNA damage and increased rate of mutations and cell death. PARP1 enzyme activation as a first response to external DNA damage can facilitate DNA repair and, in the case of DNA damage overwhelming the cellular repair capacity, induce apoptotic cell death. This process requires utilization and cleavage of significant amounts of NAD^+ and therefore is indirectly dependent on nicotinamide as the dietary precursor for NAD^+ in humans (for a recent review, see Surjana, Halliday, & Damian, 2010). A moderate degree of cell damage can be managed by the PARP1 systems while consuming a tolerable portion of the cellular NAD^+. Exposure to severe genotoxic stress in contrast causes irreversible depletion of the cellular NAD^+ content and results in cell death (Schraufstatter, Hinshaw, Hyslop, Spragg, & Cochrane, 1986). Supplementing NA to cultured human mononuclear blood cells improved their genomic integrity after X-irradiation, as measured by increased viability and reduced micronuclei formation. Supplemented cells also maintained higher cellular NAD^+ levels, produced more PAR, and maintained higher SIRT1 activity (Weidele, Beneke, & Bürkle, 2017; Weidele, Kunzmann, Schmitz, Beneke, & Bürkle, 2010). An increase of intrinsic maximal PARP activity correlates with the maximum life span observed on a species level (Grube & Bürkle, 1992). This indicates that species' higher capability to efficiently recognize and repair DNA damage has a beneficial effect to prevent damage by the increased ROS insults that occur with aging.

4.3 Cell Death/Cancer

Genotoxic stress causes a rapid and oftentimes dramatic depletion of NAD^+ from the nucleus and the cytoplasm, which triggers or at least contributes to cell death. An increased mitochondrial NAD^+ level can rescue cells from cell death mediated by NAD^+ depletion (Yang et al., 2007). This indicates that mitochondrial NAD^+ is a major determinant of apoptosis and presents the intriguing possibility that boosting mitochondrial NAD via adequate diet can determine organ health and disease.

Many of these protective pathways are dysregulated in cancer cells, which makes NAD^+ pathways intriguing targets for cancer therapeutics. NAD^+ depletion reduces energy synthesized by citric acid cycle activity and oxidative phosphorylation and instead promotes Warburg-like metabolism focused on glycolysis. Recently studies have explored the effect of well-known cancer therapeutics in combination with pharmacological depletion of NAD^+ levels. In many cases, there is a synergistic effect on cancer cell cytotoxicity. The combinations predispose cancer cells to oxidative damage by disrupting their antioxidant defense system, increase cell death by preventing DNA repair, and shift cell signaling pathways (e.g., SIRT1 and p53) to a cytotoxic route (Kennedy et al., 2016).

4.4 Mitochondrial Dysfunction

4.4.1 Direct Connection Between Niacin, NAD, and Mitochondrial Function

The correlation between aging and NAD^+ levels and functionality are of interest for mitochondrial health. The ratio of available NAD^+ to NADH regulates the activity of enzymes central to glycolysis, TCA cycle, fatty acid oxidation, and fatty acid synthesis, i.e., pathways essential to cellular metabolism. Aging and insufficient dietary niacin intake cause reduced NAD^+ availability and a shift toward more NADH that in turn leads to reduced oxidative phosphorylation and impaired mitochondrial respiration, while benefitting glucose consumption and Warburg effect.

The reduced availability of NAD^+ prevents optimal activity of the mitochondrial sirtuins (Sirt3–5). Sirt3–5 dependency on NAD^+ allows them to act as metabolic sensors that sense the metabolic state of the cell and translate this info into mitochondrial regulation. Activity of the mitochondrial sirtuins has been implicated in protection against age-related neuropathies, insulin resistance, and cardiopathologies (van de Ven et al., 2017).

PARPs, the other major group of NAD^+-consuming enzymes, also have been identified as modulators of mitochondrial activity (Bai, Nagy, Fodor, Liaudet, & Pacher, 2015). PARP hyperactivation causes a rapid loss of mitochondrial potential and decreases mitochondrial oxygen consumption, in particular upon ischemia–reperfusion injuries (Klaidman et al., 2003; Módis et al., 2012). It can result in uncoupling of the electron transport chain as well as superoxide production (Fang & Bohr, 2017; Fang et al., 2014). Those observations have led to the suggestion of using PARP inhibitors therapeutically to prevent cell death in ischemia–reperfusion scenarios, such as myocardial infarction and stroke.

Examples of mitochondrial-related disease associated include glaucoma. A recent study found an age-related decrease in retinal NAD^+ levels that makes neurons vulnerable to disease-related insults. Increased NAD^+ content via oral administration or through gene therapy driving local NAD^+ production prevented glaucomas in 93% of eyes in the tested rodent model (Williams et al., 2017).

4.4.2 Indirect Connection Between Niacin, NAD, and Mitochondrial Function

Repletion of NAD^+ can increase life span and—most importantly—health span in animal models with defects in DNA repair response associated with genetic aging disorders (such as, ATM/ataxia telangiectasia mutated, XPA/xeroderma pigmentosum group A, CS/Cockayne syndrome). This is thought to occur via the restoration of a defective mitophagy, the selective clearance of defective mitochondria, as well as possible activation of general autophagic proteins and through improved DNA repair (Croteau, Fang, Nilsen, & Bohr, 2017; Fang & Bohr, 2017). One of the cellular problems caused by these genetic defects is a loss of intracellular NAD^+ that results from persistent PARP1 activation due to increased DNA damage in the mitochondria. This decreased NAD^+ availability in turn causes other NAD^+-dependent enzymes, e.g., sirtuins, to suffer a loss of activity (Croteau et al., 2017). SIRT1 is a multifunctional protein deacetylase whose proper activity is central to longevity, metabolism, cellular senescence, genome maintenance, DNA repair, and inflammation. Hampering SIRT1 activity by restricting or making the rate limiting coenzyme NAD^+ unavailable therefore leads to defects in mitochondrial homeostasis, to increased production of reactive oxygen species, decreased DNA repair capability and results in reduced cell survival.

4.5 Neurological Disorders

4.5.1 Age-Related Neurodegeneration

Although pellagrins with dementia display remarkable and rapid improvement with niacin supplementation, suggesting a defect in cell signaling, some brain pathologies are also observed with chronic pellagra. In the aging brain, CNS decline underlies the progressive loss of cognitive, social, and physical abilities that occur with aging. In this context, declining sirtuin activity likely contributes to the aging process, opening up potential opportunities for treating or preventing these age-associated dysfunctions (Satoh, Imai, & Guarente, 2017).

Memory formation and circadian rhythms are closely intertwined, and disruption of circadian rhythms, as it occurs, e.g., during aging, seems to contribute to the age-associated memory loss. The close connection between epigenetic changes, e.g., DNA methylation and sirtuin 1 activity, and both memory formation and circadian rhythm, has given rise to the concept that epigenetic changes in these processes contribute to the age-associated cognitive decline (Deibel, Zelinski, Keeley, Kovalchuk, & McDonald, 2015; Mazucanti et al., 2015).

Neurodegeneration in aging is often associated with mitochondrial dysfunction. Recent studies have been able to link such mitochondrial-associated neurodegeneration in animal models of several progeriatric genetic diseases, e.g., CS and xeroderma pigmentosum group A (XPA) to a decreased availability of mitochondrial NAD^+ (Fang et al., 2014; Scheibye-Knudsen et al., 2014).

4.5.2 Parkinson's Disease

Parkinson's disease (PD) is characterized by degeneration of neurons in the *substantia nigra* of the brain stem, which leads to tremors associated with insufficient dopamine synthesis. The complex etiology of the disease is not completely understood, but includes genetic and environmental factors. Dopamine is synthesized from tyrosine through several enzymatic steps that among others requires NADH as a coenzyme.

A direct link between dietary niacin intake, niacin index, and PD progression was recently described by Wakade and Chong (2014) and Wakade, Chong, Bradley, Thomas, and Morgan (2014). In PD patients, low blood levels of NAD correlated with high expression levels of NA receptor GPR109A PD patients compared to age-matched control individuals. Dietary niacin supplementation normalized NAD levels and GPR109A expression to levels seen in healthy control individuals and ameliorated motor and cognitive functions in the PD patient (Wakade, Chong, Bradley, & Morgan, 2015).

Recent genome-wide association studies have identified several polymorphic gene loci associated with altered risks or age of onset for PD (Bandrés-Ciga et al., 2016; Davis et al., 2016). Among the commonly found genes is ACMSD, an enzyme central in the tryptophan to NAD^+ metabolic pathway that links PD risk to NAD metabolism, albeit though currently unknown mechanistic details.

Familial forms of PD are associated with a gene mutation of *PINK1* that results in mitochondrial impairment, and subsequent neurodegeneration due to mitochondrial deterioration. The mitochondrial disruption is associated with alteration of the NAD^+ redox state and a reduction in NAD^+ metabolites. Dietary supplementation with nicotinamide as well as genetic suppression of PARP activity was able to prevent mitochondrial dysfunction in an animal model of this *pink1* mutation-associated form of PD and could prevent neurodegeneration (Lehmann, Loh, & Martins, 2017).

Similar beneficial effects of exogenous niacin supplementation had previously been observed in a human cellular PD model, in which niacin supplementation protected from exogenously induced mitochondrial defects and oxidative damage (Jia et al., 2008). In a transgenic *Drosophila* PD model, Jia et al. (2008) found that niacin supplementation facilitated improved motor function. Taken together, it appears that mitochondrial dysfunction caused by NAD deficiency is becoming apparent as one of the many factors contributing to the pathogenesis of PD, and that a potential benefit of dietary niacin supplementation warrants further studies of this treatment option.

4.5.3 Multiple Sclerosis

Multiple sclerosis (MS) is a disease characterized by progressive neurodegeneration associated with inflammation. Current treatment options target inflammation, but appear unable to prevent the neurodegeneration that underlies the progressiveness of the disease. Mitochondrial dysfunction, associated with reduced NAD^+ levels and potentially caused by inflammation-associated oxidative stress, was found to contribute to the neurodegeneration seen in MS (reviewed by Nimmagadda et al., 2017).

Neuroprotective effects preventing axon degeneration were observed in neurons with increased activity of NMNAT, i.e., enzymes that increased NAD^+ and preserved ATP production and mitochondrial function in those neurons (Araki, Sasaki, & Milbrandt, 2004; Park et al., 2013; Yahata, Yuasa, & Araki, 2009), and in animal models overexpressing SIRT1 (Nimmagadda et al., 2013). Both pathways point toward a central role of NAD^+ depletion in the pathogenesis of the MS-associated neurodegeneration and indicate that pharmacological interventions designed to increase the intracellular NAD^+ content, e.g., via dietary niacin supplementation, should be studied in the context of clinical applications.

4.5.4 Schizophrenia

While the pathophysiology underlying most schizophrenia cases is not well understood, a recent study observed that chronically ill schizophrenia patients had a significant reduction in NAD^+/NADH ratio compared to a matched healthy control group (Kim et al., 2017). The study provides evidence that schizophrenia is linked to redox imbalance in the brain, pointing toward a role of niacin intake and resulting NAD^+ in schizophrenia etiology. Additionally, full expression of pellagrous dementia is similar to schizophrenia, suggesting that some of the underlying changes in calcium signaling in the niacin-deficient brain may be similar to schizophrenia, although schizophrenia patients, on the whole, do not improve with niacin supplementation (Prousky, Millman, & Kirkland, 2011).

4.5.5 Memory Loss

Ataxia telangiectasia (AT) in humans is associated with progressive neurodegeneration. Treatments that replenished intracellular NAD^+ in AT animal models (mouse and worms) improved neuromuscular function, delayed memory loss, and extended the life span of the test animals. Mechanistically, the intervention prevented DNA damage accumulation and mitochondrial dysfunction, and at the same time stimulated neuronal DNA repair and improved mitochondrial quality, pointing toward a therapeutic potential of NAD-stabilizing treatment to prevent memory loss and neurodegeneration (Fang et al., 2016).

4.6 Skin Health and Skin Cancer Prevention

Severe niacin deficiency, as seen in clinical pellagra, is characterized by erythematous rashes in sun-exposed areas, hyperkeratosis, and dermal fibrosis. NAD^+ restoration via niacin supplementation allows the skin to recover.

NAD deficiency in human keratinocytes causes reduced growth rates, increases rates of apoptotic cell death, and leads to formation of reactive oxygen species and increased rates of DNA damage (Benavente, Schnell, & Jacobson, 2012). In addition, niacin restriction caused photosensitization in skin cell lines, as illustrated by increased amounts of DNA damage, a reduction in DNA damage-induced PAR formation, and alterations of sirtuin expression levels (Benavente et al., 2012).

Skin cells are also prone to exhibit epidermal differentiation defects following chronic UV skin exposure. This effect could largely be averted by administration of pharmacological doses of a topically available NA derivative (Jacobson et al., 2007). A potential underlying mechanism for this protective effect of niacin might be altered cell signal mediated by stimulation of the NA-responsive G-protein-coupled receptors GPR109A and GPR109B. Expression pattern analyses demonstrate the presence of functional receptors in normal human primary and immortalized keratinocytes, while squamous cell carcinoma (SCC) cells only have nearly nonfunctional receptors available (Bermudez et al., 2011).

Using a mouse animal model, Gensler, Williams, Huang, and Jacobson (1999) showed that nutritive administration of nicotinamide could inhibit photocarcinogene. Conversely, mild niacin deficiency in a different mouse study demonstrated an increased incidence of skin cancers after UV-B treatment, indicating that already mild niacin deficiency similar to what is frequently seen in humans might be sufficient to increase the skin cancer risk. In clinical settings and human treatments, benefits of niacin application for skin growth and differentiation have long been demonstrated in the form of improved wound healing, treatment of psoriasis with 6-aminonicotinamide, smoothed skin surface structure, and reduced light damage in skin cells (Gehring, 2004). In line with these findings, another recent large-scale clinical trial on the protective role of oral nicotinamide against skin cancer recurrence found that niacin intake was associated with a reduced risk of SCC (Park et al., 2017). The same study demonstrated a marginally positive association between niacin intake and the risk of basal cell carcinoma and, in men only, with melanoma formation. Overall, Park et al. (2017) concluded that niacin intake might be of potential benefit with respect to development of SCC, but likely not for prevention of BCC or melanoma.

4.7 Cardiovascular Disorders and Niacin in Clinical Trials

In pharmacological doses, niacin intake can reduce myocardial infarction, atherosclerosis, and stroke. The majority of these beneficial effects are thought to be mediated by changes of lipoprotein composition. The ability of pharmacological niacin intake to lower blood lipids, and to increase high-density lipoprotein (HDL) cholesterol, i.e., the "good" cholesterol, has long been recognized. Niacin was the first hypolipidemic drug that significantly reduced cardiovascular clinical events and mortality in patients with cardiovascular disease. For current reviews on function and the use of niacin, see la Paz et al. (2016) and Zeman et al. (2016).

Niacin in cultured human aortic cells increased levels of NADH and reduced glutathione, which caused a reduction of reactive oxygen species synthesis, low-density lipoprotein oxidation, and synthesis of TNF-alpha and TNF-alpha-associated monocyte adhesion. This study thus indicated that therapeutic niacin administration inhibits vascular inflammation, an independent protective effect on the cardiovascular system that functions in addition to niacin's antiatherosclerotic effects on lipoprotein composition (Ganji, Qin, Zhang, Kamanna, & Kashyap, 2009).

Early clinical studies consistently demonstrated that pharmacological doses of niacin, either as monotherapy or in combination with bile acid and cholesterol-sequestering drugs such as colestipol, significantly reduced serum cholesterol and triglyceride levels and improved clinical outcomes: 3-year mortality rate in survivors of myocardial infarction was significantly reduced, the frequency of coronary lesion was reduced, and atherosclerosis regressed and coronary status improved (Blankenhorn et al., 1987; Brown et al., 1990; Carlson, Danielson, Ekberg, Klintemar, & Rosenhamer, 1977; Carlson & Rosenhamer, 1988; Kane, 1990).

The Coronary Drug Project, a niacin monotherapy trial intended to evaluate the frequency of recurrent myocardial infarction over a 6-year interval and the associated 15-year total mortality, found significantly reduced infarct and mortality rates in niacin monotherapy patients (Canner, Furberg, & McGovern, 2006).

Increased levels of low-density lipoprotein cholesterol (LDL-C) and lower-than-average levels of high-density lipoprotein cholesterol (HDL-C) in patients with dyslipidemia were found to be independent risk factors for CVD (Barr, Russ, & Eder, 1951; Gordon, Castelli, Hjortland, Kannel, & Dawber, 1977). This has led to the so-called HDL hypothesis that high HDL protects from arteriosclerosis, and simultaneous correction of the LDL-C to HDL-C ratio together with a general reduction of blood lipids may provide reduced CV morbidity and mortality (reviewed, e.g., in Vergeer, Holleboom, Kastelein, & Kuivenhoven, 2010). Niacin (nicotinic acid) intake increases blood HDL-C levels and reduces LDL-C levels. While the underlying mechanism for increased HDL-C is not yet completely understood, increased production of apolipoprotein A-1 (ApoA-1) and ATP-binding cassette subfamily A member 1 (ABCA1) contributes to the effect (Lamon-Fava et al., 2008; Pang et al., 2014; Rubic, Trottmann, & Lorenz, 2004; Wu & Zhao, 2009).

The unique property of niacin to correct the HDL-C:LDL-C ratio has generated interest in the potential for combined niacin–statin therapies to reduce blood lipid levels by statins and correct HDL-C:LDL-C ratios by niacin in dyslipidemic patients. A study comparing safety and effectiveness of combination formulations of extended release niacin with either a statin or a bile acid sequestrant showed that both, niacin alone and its combination preparations, effectively reduced LDL-C and triglycerides and increased HDL-C (Guyton et al., 1998). The combination of niacin with statin appeared to be most effective. Further trials (including HATS, ARBITER-2, ARBITER-3, ARBITER-6, Oxford Niaspan Study, AVANT, and COMPELL) corroborated that treatment with niacin in combination with statins effectively reduced LDL-C levels and increased HDL-C levels. In these trials, significantly reduced thickening of the carotid intima media, a surrogate marker for atherosclerosis, and reduced stenosis progression were also observed (Brown et al., 2001; deGoma et al., 2015; Lee et al., 2009; McKenney et al., 2007; Taylor, 2004; Taylor, Lee, & Sullenberger, 2006; Taylor et al., 2009). In a clinical trial called the "NIA Plaque study," patients with statin therapy-adjusted LDL-C blood levels were supplemented with niacin or placebo. While the addition of niacin to the statin therapy increased HDL-C levels, the reduction in internal carotid artery wall volume was comparable in both treatment groups. This indicates that it may be sufficient to control LDL-C levels in order to reduce atherosclerosis independent of HDL-C levels (Sibley et al., 2013).

Further clinical trials addressed the growing concern that the predictive value of surrogate measurements, e.g., arterial wall thickness, as indicators for the risk of actual CVD events may be limited. This would mean that reduction of blood lipids and prevention of atherosclerosis by supplementing niacin in a statin treatment regimen might not necessarily translate into improved clinical outcomes (Hochholzer, Berg, & Giugliano, 2011). Two trials (AIM HIGH and HPS2-THRIVE) therefore tested whether supplementing a statin therapy with niacin is able to reduce the risk of CV events in patients with atherogenic dyslipidemia below the levels achieved by statins alone or not (The AIM-HIGH Investigators, 2011b, 2011c). Despite significantly reduced triglyceride and increased HDL-C levels, the incidence of cardiovascular incidents (stroke, nonfatal myocardial infarction, ischemic stroke, acute coronary syndrome, etc.) appeared not to be reduced by adding niacin to the statin therapy during a 36-month follow-up period (The AIM-HIGH Investigators, 2011a; The HPS2-THRIVE Collaborative Group, 2014). Niacin augmentation was associated with significantly increased rates of certain serious adverse events, leading to early termination of one study (Anderson et al., 2014; Lloyd-Jones, 2014; The HPS2-THRIVE Collaborative Group, 2014). Skin-related adverse effects, diabetic complications and new onset diabetes, infections, gastrointestinal and musculoskeletal problems, heart failure, and a 9% increased risk of death were all observed. There is also a growing concern that there is no true causal relationship between low HDL-C and cardiovascular risk, and that low HDL-C levels should be considered a risk marker instead of a risk factor (Voight et al., 2012). If this is the case, additional niacin treatment to correct HDL-C levels cannot be expected to yield clinical benefit (Lloyd-Jones, 2014).

Reports of other clinical trials utilizing such combination therapies conclude that niacin could be safely added to statin or combined statin/cholesterol sequestrator regimens, but recommend monitoring of blood glucose levels as an indicator of intermittent occurrence of transient diabetes in patients on niacin (Fazio et al., 2010; Guyton et al., 2012).

The increased incidence of diabetes in patients in the niacin-supplemented groups raised the question if the benefit of niacin therapy depended on metabolic status of patients. While the niacin-mediated correction of plasma lipid levels occurs regardless of the presence of metabolic syndrome or diabetes mellitus (Ooi et al., 2015), evidence is accumulating from primary studies and meta-analyses that niacin treatment can indeed induce hepatic insulin resistance and increase the risk of developing diabetes (Blond et al., 2014; Goldie et al., 2015).

Vasodilation resulting in skin flushing and itching is a well-known side effect of niacin intake that often leads to increased dropout rates in niacin treatment clinical trial groups and to reduced patient compliance.

A clinical trial focusing on safety and tolerability of prolonged-release niacin found that a daily dose of 2000mg over a period of 15 weeks did not lead to significant hepatotoxicity or muscular adverse effects, but to high incidence of skin flushing (>40% of patients), and a related rate of 10% of patient that withdrew from the study (NAUTILUS study, Vogt, Kassner, Hostalek, Peiter, & Steinhagen-Thiessen, 2006; Vogt, Kassner, Hostalek, Steinhagen-Thiessen, & NAUTILUS Study Group, 2006). Coadministration of a selective prostaglandin D2 receptor subtype 1 antagonist, laropiprant, was shown to reduce the niacin-induced vasodilation (Lai et al., 2007), to safely attenuate skin flushing and to improve patient compliance (Hussein & Nicholls, 2010). In addition, niacin/laropiprant combination therapy improves lipid levels across a range of patient types, irrespective of sex, race, age, statin use, and presence of coronary heart disease, including type 2 diabetes mellitus patients irrespective of glycemic control (Bays et al., 2015, 2012; Bays, Shah, Dong, McCrary Sisk, & Maccubbin, 2011).

The current therapeutic role of niacin for dyslipidemia treatment remains unclear. The perceived negative outcome of the clinical trials that failed to show a potential for niacin to further improve on clinical benefits of statin treatments led to a slight decrease in current niacin therapy for improved cardiovascular health (Bittner et al., 2015). The patients in those trials were already treated to optimal low LDL-C levels with statins, and niacin favorably changed HDL-C and triglyceride levels. Niacin therefore could be used as adjuvant therapy to reduce atherogenic lipoprotein burden in patients that cannot reach their target HDL-C and LDL-C values, or that have contraindications for taking statins or bile-acid sequestrants (Boden, Sidhu, & Toth, 2014; Lloyd-Jones, 2014). There is also growing evidence that the low HDL-C might not be a causal risk factor for cardiovascular disease, but rather a marker of high risk for cardiovascular disease (often associated with metabolic syndrome and insulin resistance) (Voight et al., 2012), in which case pharmacologically increasing HDL-C could not be expected to yield any clinical benefit. Current reevaluations of the AIM-HIGH and the HPS2-THRIVE clinical trial showed that the perceived negative outcome might not be justified due to several methodological flaws in the studies. The reevaluation indicates that some of the beneficial antiatherosclerotic effects of niacin supplementation can be due to niacin supplementation boosting antioxidative and antiinflammatory mechanisms (Zeman et al., 2016). It should be emphasized that these studies are using doses in the range of 1–3g/day of nicotinic acid, and these side effects are unlikely in studies supplementing 100 mg/day of nicotinamide or nicotinamide riboside, for example.

5. SUMMARY

The central role dietary niacin plays for cellular NAD levels and, thus for most metabolic processes, is directly reflected in the variety and large number of clinical problems associated with a suboptimal niacin status and emphasizes how important the maintenance of a good niacin status is for overall health. Niacin has a long and successful history as a safe vitamin supplement, which makes it a promising candidate for therapeutic interventions to ameliorate or treat many of the clinical problems described earlier, and furthermore might prove beneficial in the prevention of health decline associated with aging.

REFERENCES

Aarhus, R., Graeff, R. M., Dickey, D. M., Walseth, T. F., & Lee, H. C. (1995). ADP-ribosyl cyclase and CD38 catalyze the synthesis of a calcium-mobilizing metabolite from NADP. *The Journal of Biological Chemistry*, *270*(51), 30327–30333.

Alvarez-Gonzalez, R., & Jacobson, M. K. (1987). Characterization of polymers of adenosine diphosphate ribose generated in vitro and in vivo. *Biochemistry*, *26*(11), 3218–3224.

Ame, J. C., Spenlehauer, C., & de Murcia, G. (2004). The PARP superfamily. *BioEssays: News and Reviews in Molecular, Cellular and Developmental Biology*, *26*(8), 882–893. https://doi.org/10.1002/bies.20085.

Anderson, R. M., Bitterman, K. J., Wood, J. G., Medvedik, O., & Sinclair, D. A. (2003). Nicotinamide and PNC1 govern lifespan extension by calorie restriction in *Saccharomyces cerevisiae*. *Nature*, *423*(6936), 181–185. https://doi.org/10.1038/nature01578.

Anderson, T. J., Boden, W. E., Desvigne-Nickens, P., Fleg, J. L., Kashyap, M. L., McBride, R., et al. (2014). Safety profile of extended-release niacin in the aim-high trial. *New England Journal of Medicine*, *371*(3), 288–290. https://doi.org/10.1056/NEJMc1311039.

Andrabi, S. A., Kim, N. S., Yu, S. W., Wang, H., Koh, D. W., Sasaki, M., et al. (2006). Poly (ADP-ribose) (PAR) polymer is a death signal. *Proceedings of the National Academy of Sciences of the United States of America*, *103*(48), 18308–18313. https://doi.org/10.1073/pnas.0606526103.

Araki, T., Sasaki, Y., & Milbrandt, J. (2004). Increased nuclear NAD biosynthesis and SIRT1 activation prevent axonal degeneration. *Science (New York, NY)*, *305*(5686), 1010–1013. https://doi.org/10.1126/science.1098014.

Aravind, L., Zhang, D., de Souza, R. F., Anand, S., & Iyer, L. M. (2015). The natural history of ADP-ribosyltransferases and the ADP-ribosylation system. *Current Topics in Microbiology and Immunology*, *384*, 3–32. https://doi.org/10.1007/82_2014_414.

Asher, G., Gatfield, D., Stratmann, M., Reinke, H., Dibner, C., Kreppel, F., et al. (2008). SIRT1 regulates circadian clock gene expression through PER2 deacetylation. *Cell*, *134*(2), 317–328. https://doi.org/10.1016/j.cell.2008.06.050.

Asher, G., Reinke, H., Altmeyer, M., Gutierrez-Arcelus, M., Hottiger, M. O., & Schibler, U. (2010). Poly(ADP-ribose) polymerase 1 participates in the phase entrainment of circadian clocks to feeding. *Cell*, *142*(6), 943–953. https://doi.org/10.1016/j.cell.2010.08.016.

Asher, G., & Sassone-Corsi, P. (2015). Time for food: The intimate interplay between nutrition, metabolism, and the circadian clock. *Cell, 161*(1), 84–92. https://doi.org/10.1016/j.cell.2015.03.015.
Aykroyd, W. R. (1971). Conquest of deficiency diseases achievements and prospects. 3. Pellagra. *Indian Journal of Medical Sciences, 25*(3), 203–207.
Badawy, A. A.-B. (2017). Kynurenine pathway of tryptophan metabolism: Regulatory and functional aspects. *International Journal of Tryptophan Research: IJTR, 10*. 1178646917691938. https://doi.org/10.1177/1178646917691938.
Bai, P., Nagy, L., Fodor, T., Liaudet, L., & Pacher, P. (2015). Poly(ADP-ribose) polymerases as modulators of mitochondrial activity. *Trends in Endocrinology and Metabolism: TEM, 26*(2), 75–83. https://doi.org/10.1016/j.tem.2014.11.003.
Baker, H., DeAngelis, B., Holland, B., Gittens-Williams, L., & Barrett, T. (2002). Vitamin profile of 563 gravidas during trimesters of pregnancy. *Journal of the American College of Nutrition, 21*(1), 33–37.
Ball, H. J., Jusof, F. F., Bakmiwewa, S. M., Hunt, N. H., & Yuasa, H. J. (2014). Tryptophan-catabolizing enzymes—Party of three. *Frontiers in Immunology, 5*(485). https://doi.org/10.3389/fimmu.2014.00485.
Bandrés-Ciga, S., Price, T. R., Barrero, F. J., Escamilla-Sevilla, F., Pelegrina, J., Arepalli, S., et al. (2016). Genome-wide assessment of Parkinson's disease in a Southern Spanish population. *Neurobiology of Aging, 45*, 213.e3–213.e9. https://doi.org/10.1016/j.neurobiolaging.2016.06.001.
Barbosa, M. T. P., Soares, S. M., Novak, C. M., Sinclair, D., Levine, J. A., Aksoy, P., et al. (2007). The enzyme CD38 (a NAD glycohydrolase, EC 3.2.2.5) is necessary for the development of diet-induced obesity. *FASEB Journal: Official Publication of the Federation of American Societies for Experimental Biology, 21*(13), 3629–3639. https://doi.org/10.1096/fj.07-8290com.
Barr, D. P., Russ, E. M., & Eder, H. A. (1951). Protein-lipid relationships in human plasma. II. In atherosclerosis and related conditions. *The American Journal of Medicine, 11*(4), 480–493.
Barrett-Connor, E. (1967). The etiology of pellagra and its significance for modern medicine. *The American Journal of Medicine, 42*(6), 859–867.
Bays, H. E., Brinton, E. A., Triscari, J., Chen, E., Maccubbin, D., MacLean, A. A., et al. (2015). Extended-release niacin/laropiprant significantly improves lipid levels in type 2 diabetes mellitus irrespective of baseline glycemic control. *Vascular Health and Risk Management, 11*, 165–172. https://doi.org/10.2147/VHRM.S70907.
Bays, H., Shah, A., Dong, Q., McCrary Sisk, C., & Maccubbin, D. (2011). Extended-release niacin/laropiprant lipid-altering consistency across patient subgroups. *International Journal of Clinical Practice, 65*(4), 436–445. https://doi.org/10.1111/j.1742-1241.2010.02620.x.
Bays, H. E., Shah, A., Lin, J., Sisk, C. M., Dong, Q., & Maccubbin, D. (2012). Consistency of extended-release niacin/laropiprant effects on Lp(a), ApoB, non-HDL-C, Apo A1, and ApoB/ApoA1 ratio across patient subgroups. *American Journal of Cardiovascular Drugs: Drugs, Devices, and Other Interventions, 12*(3), 197–206. https://doi.org/10.2165/11631530-000000000-00000.
Belden, W. J., & Dunlap, J. C. (2008). SIRT1 is a circadian deacetylase for core clock components. *Cell, 134*(2), 212–214. https://doi.org/10.1016/j.cell.2008.07.010.
Belenky, P., Racette, F. G., Bogan, K. L., McClure, J. M., Smith, J. S., & Brenner, C. (2007). Nicotinamide riboside promotes Sir2 silencing and extends lifespan via Nrk and Urh1/Pnp1/Meu1 pathways to NAD+. *Cell, 129*(3), 473–484. https://doi.org/10.1016/j.cell.2007.03.024.
Benavente, C. A., Schnell, S. A., & Jacobson, E. L. (2012). Effects of niacin restriction on sirtuin and PARP responses to photodamage in human skin. *PLoS One, 7*(7). e42276 https://doi.org/10.1371/journal.pone.0042276.
Beretich, G. R. (2005). Do high leucine/low tryptophan dieting foods (yogurt, gelatin) with niacin supplementation cause neuropsychiatric symptoms (depression) but not

dermatological symptoms of pellagra? *Medical Hypotheses*, *65*(3), 628–629. https://doi.org/10.1016/j.mehy.2005.04.002.

Berger, N. A., Besson, V. C., Boulares, A. H., Bürkle, A., Chiarugi, A., Clark, R. S., et al. (2017). Opportunities for the repurposing of PARP inhibitors for the therapy of non-oncological diseases. *British Journal of Pharmacology*. https://doi.org/10.1111/bph.13748.

Bermudez, Y., Benavente, C. A., Meyer, R. G., Coyle, W. R., Jacobson, M. K., & Jacobson, E. L. (2011). Nicotinic acid receptor abnormalities in human skin cancer: Implications for a role in epidermal differentiation. *PLoS One*, *6*(5). e20487 https://doi.org/10.1371/journal.pone.0020487.

Bieganowski, P., & Brenner, C. (2004). Discoveries of nicotinamide riboside as a nutrient and conserved NRK genes establish a Preiss-Handler independent route to NAD+ in fungi and humans. *Cell*, *117*(4), 495–502.

Bipath, P., Levay, P. F., & Viljoen, M. (2015). The kynurenine pathway activities in a sub-Saharan HIV/AIDS population. *BMC Infectious Diseases*, *15*(346) https://doi.org/10.1186/s12879-015-1087-5.

Bittner, V., Deng, L., Rosenson, R. S., Taylor, B., Glasser, S. P., Kent, S. T., et al. (2015). Trends in the use of nonstatin lipid-lowering therapy among patients with coronary heart disease: A retrospective cohort study in the medicare population 2007 to 2011. *Journal of the American College of Cardiology*, *66*(17), 1864–1872. https://doi.org/10.1016/j.jacc.2015.08.042.

Blankenhorn, D. H., Nessim, S. A., Johnson, R. L., Sanmarco, M. E., Azen, S. P., & Cashin-Hemphill, L. (1987). Beneficial effects of combined colestipol-niacin therapy on coronary atherosclerosis and coronary venous bypass grafts. *JAMA*, *257*(23), 3233–3240.

Blond, E., Rieusset, J., Alligier, M., Lambert-Porcheron, S., Bendridi, N., Gabert, L., et al. (2014). Nicotinic acid effects on insulin sensitivity and hepatic lipid metabolism: An in vivo to in vitro study. *Hormone and Metabolic Research = Hormon- Und Stoffwechselforschung = Hormones Et Métabolisme*, *46*(6), 390–396. https://doi.org/10.1055/s-0034-1372600.

Boden, W. E., Sidhu, M. S., & Toth, P. P. (2014). The therapeutic role of niacin in dyslipidemia management. *Journal of Cardiovascular Pharmacology and Therapeutics*, *19*(2), 141–158. https://doi.org/10.1177/1074248413514481.

Bouma, G., van Faassen, M., Kats-Ugurlu, G., de Vries, E. G. E., Kema, I. P., & Walenkamp, A. M. E. (2016). Niacin (Vitamin B3) supplementation in patients with serotonin-producing neuroendocrine tumor. *Neuroendocrinology*, *103*(5), 489–494. https://doi.org/10.1159/000440621.

Braidy, N., Guillemin, G. J., Mansour, H., Chan-Ling, T., Poljak, A., & Grant, R. (2011). Age related changes in NAD+ metabolism oxidative stress and Sirt1 activity in wistar rats. *PLoS One*, *6*(4). e19194 https://doi.org/10.1371/journal.pone.0019194.

Breen, L. T., Smyth, L. M., Yamboliev, I. A., & Mutafova-Yambolieva, V. N. (2006). Beta-NAD is a novel nucleotide released on stimulation of nerve terminals in human urinary bladder detrusor muscle. *American Journal of Physiology Renal Physiology*, *290*(2), F486–495. https://doi.org/10.1152/ajprenal.00314.2005.

Bröer, S., Cavanaugh, J. A., & Rasko, J. E. J. (2005). Neutral amino acid transport in epithelial cells and its malfunction in Hartnup disorder. *Biochemical Society Transactions*, *33*(Pt. 1), 233–236. https://doi.org/10.1042/BST0330233.

Brown, G., Albers, J. J., Fisher, L. D., Schaefer, S. M., Lin, J.-T., Kaplan, C., et al. (1990). Regression of coronary artery disease as a result of intensive lipid-lowering therapy in men with high levels of apolipoprotein B. *New England Journal of Medicine*, *323*(19), 1289–1298. https://doi.org/10.1056/NEJM199011083231901.

Brown, R. R., Ozaki, Y., Datta, S. P., Borden, E. C., Sondel, P. M., & Malone, D. G. (1991). Implications of interferon-induced tryptophan catabolism in cancer, auto-immune diseases and AIDS. *Advances in Experimental Medicine and Biology*, *294*, 425–435.

Brown, B. G., Zhao, X. Q., Chait, A., Fisher, L. D., Cheung, M. C., Morse, J. S., et al. (2001). Simvastatin and niacin, antioxidant vitamins, or the combination for the prevention of coronary disease. *The New England Journal of Medicine*, *345*(22), 1583–1592. https://doi.org/10.1056/NEJMoa011090.

Burkle, A., & Virag, L. (2013). Poly(ADP-ribose): PARadigms and PARadoxes. *Molecular Aspects of Medicine*, *34*, 1046–1065. https://doi.org/10.1016/j.mam.2012.12.010.

Camacho-Pereira, J., Tarragó, M. G., Chini, C. C. S., Nin, V., Escande, C., Warner, G. M., et al. (2016). CD38 dictates age-related NAD decline and mitochondrial dysfunction through an SIRT3-dependent mechanism. *Cell Metabolism*, *23*(6), 1127–1139. https://doi.org/10.1016/j.cmet.2016.05.006.

Canner, P. L., Furberg, C. D., & McGovern, M. E. (2006). Benefits of niacin in patients with versus without the metabolic syndrome and healed myocardial infarction (from the Coronary Drug Project). *The American Journal of Cardiology*, *97*(4), 477–479. https://doi.org/10.1016/j.amjcard.2005.08.070.

Cantó, C., Menzies, K. J., & Auwerx, J. (2015). NAD+ metabolism and the control of energy homeostasis: A balancing act between mitochondria and the nucleus. *Cell Metabolism*, *22*(1), 31–53. https://doi.org/10.1016/j.cmet.2015.05.023.

Carlson, L. A., Danielson, M., Ekberg, I., Klintemar, B., & Rosenhamer, G. (1977). Reduction of myocardial reinfarction by the combined treatment with clofibrate and nicotinic acid. *Atherosclerosis*, *28*(1), 81–86.

Carlson, L. A., & Rosenhamer, G. (1988). Reduction of mortality in the Stockholm Ischaemic Heart Disease Secondary Prevention Study by combined treatment with clofibrate and nicotinic acid. *Acta Medica Scandinavica*, *223*(5), 405–418.

Chambon, P., Weill, J. D., & Mandel, P. (1963). Nicotinamide mononucleotide activation of new DNA-dependent polyadenylic acid synthesizing nuclear enzyme. *Biochemical and Biophysical Research Communications*, *11*, 39–43.

Chatterjee, S., Berger, S. J., & Berger, N. A. (1999). Poly(ADP-ribose) polymerase: A guardian of the genome that facilitates DNA repair by protecting against DNA recombination. *Molecular and Cellular Biochemistry*, *193*(1–2), 23–30.

Chen, D., Vollmar, M., Rossi, M. N., Phillips, C., Kraehenbuehl, R., Slade, D., et al. (2011). Identification of macrodomain proteins as novel O-acetyl-ADP-ribose deacetylases. *The Journal of Biological Chemistry*, *286*(15), 13261–13271. https://doi.org/10.1074/jbc.M110.206771.

Chini, C. C. S., Tarragó, M. G., & Chini, E. N. (2016). NAD and the aging process: Role in life, death and everything in between. *Molecular and Cellular Endocrinology*, *455*, 62–74. https://doi.org/10.1016/j.mce.2016.11.003.

Chong, A., Malavasi, F., Israel, S., Khor, C. C., Yap, V. B., Monakhov, M., et al. (2017). ADP ribosyl-cyclases (CD38/CD157), social skills and friendship. *Psychoneuroendocrinology*, *78*, 185–192. https://doi.org/10.1016/j.psyneuen.2017.01.011.

Cohen, H. Y., Miller, C., Bitterman, K. J., Wall, N. R., Hekking, B., Kessler, B., et al. (2004). Calorie restriction promotes mammalian cell survival by inducing the SIRT1 deacetylase. *Science (New York, NY)*, 305(5682), 390–392. https://doi.org/10.1126/science.1099196.

Cortes, U., Tong, W. M., Coyle, D. L., Meyer-Ficca, M. L., Meyer, R. G., Petrilli, V., et al. (2004). Depletion of the 110-kilodalton isoform of poly(ADP-ribose) glycohydrolase increases sensitivity to genotoxic and endotoxic stress in mice. *Molecular and Cellular Biology*, *24*(16), 7163–7178. https://doi.org/10.1128/MCB.24.16.7163-7178.2004.

Croteau, D. L., Fang, E. F., Nilsen, H., & Bohr, V. A. (2017). NAD(+) in DNA repair and mitochondrial maintenance. *Cell Cycle (Georgetown, TX)*, *16*, 491–492. https://doi.org/10.1080/15384101.2017.1285631.

Davis, A. A., Andruska, K. M., Benitez, B. A., Racette, B. A., Perlmutter, J. S., & Cruchaga, C. (2016). Variants in GBA, SNCA, and MAPT influence Parkinson disease risk, age at onset, and progression. *Neurobiology of Aging*, 37, 209.e1–209.e7. https://doi.org/10.1016/j.neurobiolaging.2015.09.014.

deGoma, E. M., Salavati, A., Shinohara, R. T., Saboury, B., Pollan, L., Schoen, M., et al. (2015). A pilot trial to examine the effect of high-dose niacin on arterial wall inflammation using fluorodeoxyglucose positron emission tomography. *Academic Radiology*, *22*(5), 600–609. https://doi.org/10.1016/j.acra.2014.12.015.

Deibel, S. H., Zelinski, E. L., Keeley, R. J., Kovalchuk, O., & McDonald, R. J. (2015). Epigenetic alterations in the suprachiasmatic nucleus and hippocampus contribute to age-related cognitive decline. *Oncotarget*, *6*(27), 23181–23203. https://doi.org/10.18632/oncotarget.4036.

Durnin, L., Dai, Y., Aiba, I., Shuttleworth, C. W., Yamboliev, I. A., & Mutafova-Yambolieva, V. N. (2012). Release, neuronal effects and removal of extracellular β-nicotinamide adenine dinucleotide (β-NAD^+) in the rat brain. *The European Journal of Neuroscience*, *35*(3), 423–435. https://doi.org/10.1111/j.1460-9568.2011.07957.x.

Elvehjem, C. A., Madden, R. J., Strong, F. M., & Woolley, D. W. (1937). Relation of nicotinic acid and nicotinic acid amide to canine black tongue. *Journal of the American Chemical Society*, *59*(9), 1767–1768. https://doi.org/10.1021/ja01288a509.

Emanuelli, M., Carnevali, F., Saccucci, F., Pierella, F., Amici, A., Raffaelli, N., et al. (2001). Molecular cloning, chromosomal localization, tissue mRNA levels, bacterial expression, and enzymatic properties of human NMN adenylyltransferase. *The Journal of Biological Chemistry*, *276*(1), 406–412. https://doi.org/10.1074/jbc.M008700200.

Escande, C., Nin, V., Price, N. L., Capellini, V., Gomes, A. P., Barbosa, M. T., et al. (2013). Flavonoid apigenin is an inhibitor of the NAD+ ase CD38: Implications for cellular NAD+ metabolism, protein acetylation, and treatment of metabolic syndrome. *Diabetes*, *62*(4), 1084–1093. https://doi.org/10.2337/db12-1139.

Etheridge, E. W. (1972). *The butterfly caste: A social history of pellagra in the South*. Westport, CT: Greenwood Publishing Co.

Fang, E. F., & Bohr, V. A. (2017). NAD(+): The convergence of DNA repair and mitophagy. *Autophagy*, *13*(2), 442–443. https://doi.org/10.1080/15548627.2016.1257467.

Fang, E. F., Kassahun, H., Croteau, D. L., Scheibye-Knudsen, M., Marosi, K., Lu, H., et al. (2016). NAD(+) replenishment improves lifespan and healthspan in ataxia telangiectasia models via mitophagy and DNA repair. *Cell Metabolism*, *24*(4), 566–581. https://doi.org/10.1016/j.cmet.2016.09.004.

Fang, E. F., Scheibye-Knudsen, M., Brace, L. E., Kassahun, H., SenGupta, T., Nilsen, H., et al. (2014). Defective mitophagy in XPA via PARP-1 hyperactivation and NAD(+)/SIRT1 reduction. *Cell*, *157*(4), 882–896. https://doi.org/10.1016/j.cell.2014.03.026.

Fazio, S., Guyton, J. R., Lin, J., Tomassini, J. E., Shah, A., & Tershakovec, A. M. (2010). Long-term efficacy and safety of ezetimibe/simvastatin coadministered with extended-release niacin in hyperlipidaemic patients with diabetes or metabolic syndrome. *Diabetes, Obesity & Metabolism*, *12*(11), 983–993. https://doi.org/10.1111/j.1463-1326.2010.01289.x.

Feldblum, S., Rougier, A., Loiseau, H., Loiseau, P., Cohadon, F., Morselli, P. L., et al. (1988). Quinolinic-phosphoribosyl transferase activity is decreased in epileptic human brain tissue. *Epilepsia*, *29*(5), 523–529.

Ferrero, E., Lo Buono, N., Horenstein, A. L., Funaro, A., & Malavasi, F. (2014). The ADP-ribosyl cyclases—The current evolutionary state of the ARCs. *Frontiers in Bioscience (Landmark Edition)*, *19*, 986–1002.

Foster, A. C., Whetsell, W. O., Bird, E. D., & Schwarcz, R. (1985). Quinolinic acid phosphoribosyltransferase in human and rat brain: Activity in Huntington's disease and in quinolinate-lesioned rat striatum. *Brain Research*, *336*(2), 207–214.

Fouquerel, E., & Sobol, R. W. (2014). ARTD1 (PARP1) activation and NAD+ in DNA repair and cell death. *DNA Repair*, 23, 27–32. https://doi.org/10.1016/j.dnarep.2014.09.004.

Frederick, D. W., Davis, J. G., Dávila, A., Agarwal, B., Michan, S., Puchowicz, M. A., et al. (2015). Increasing NAD synthesis in muscle via nicotinamide phosphoribosyltransferase is not sufficient to promote oxidative metabolism. *Journal of Biological Chemistry*, *290*(3), 1546–1558. https://doi.org/10.1074/jbc.M114.579565.

Fu, C. S., Swendseid, M. E., Jacob, R. A., & McKee, R. W. (1989). Biochemical markers for assessment of niacin status in young men: Levels of erythrocyte niacin coenzymes and plasma tryptophan. *The Journal of Nutrition*, *119*(12), 1949–1955.

Fukuwatari, T., Ohta, M., Kimtjra, N., Sasaki, R., & Shibata, K. (2004). Conversion ratio of tryptophan to niacin in Japanese women fed a purified diet conforming to the Japanese Dietary Reference Intakes. *Journal of Nutritional Science and Vitaminology*, *50*(6), 385–391.

Fukuwatari, T., Sugimoto, E., & Shibata, K. (2002). Growth-promoting activity of pyrazinoic acid, a putative active compound of antituberculosis drug pyrazinamide, in niacin-deficient rats through the inhibition of ACMSD activity. *Bioscience, Biotechnology, and Biochemistry*, *66*(7), 1435–1441. https://doi.org/10.1271/bbb.66.1435.

Ganji, S. H., Qin, S., Zhang, L., Kamanna, V. S., & Kashyap, M. L. (2009). Niacin inhibits vascular oxidative stress, redox-sensitive genes, and monocyte adhesion to human aortic endothelial cells. *Atherosclerosis*, *202*(1), 68–75. https://doi.org/10.1016/j.atherosclerosis.2008.04.044.

Gaziano, J. M., & Manson, J. E. (1996). Diet and heart disease. The role of fat, alcohol, and antioxidants. *Cardiology Clinics*, *14*(1), 69–83.

Gehring, W. (2004). Nicotinic acid/niacinamide and the skin. *Journal of Cosmetic Dermatology*, *3*(2), 88–93. https://doi.org/10.1111/j.1473-2130.2004.00115.x.

Gensler, H. L., Williams, T., Huang, A. C., & Jacobson, E. L. (1999). Oral niacin prevents photocarcinogenesis and photoimmunosuppression in mice. *Nutrition and Cancer*, *34*(1), 36–41. https://doi.org/10.1207/S15327914NC340105.

Goldie, C., Taylor, A. J., Nguyen, P., McCoy, C., Zhao, X.-Q., & Preiss, D. (2015). Niacin therapy and the risk of new-onset diabetes: A meta-analysis of randomised controlled trials. *Heart (British Cardiac Society)*, *102*, 198–203. https://doi.org/10.1136/heartjnl-2015-308055.

Gopalan, C. (1969). Possible role for dietary leucine in the pathogenesis of pellagra. *The Lancet (London, England)*, *1*(7587), 197–199.

Gopalan, C. (2009). Leucine and pellagra. *Nutrition Reviews*, *26*(11), 323–326. https://doi.org/10.1111/j.1753-4887.1968.tb00835.x.

Gordon, T., Castelli, W. P., Hjortland, M. C., Kannel, W. B., & Dawber, T. R. (1977). High density lipoprotein as a protective factor against coronary heart disease. The Framingham Study. *The American Journal of Medicine*, *62*(5), 707–714.

Goudarzi, A., Zhang, D., Huang, H., Barral, S., Kwon, O. K., Qi, S., et al. (2016). Dynamic competing histone H4 K5K8 acetylation and butyrylation are hallmarks of highly active gene promoters. *Molecular Cell*, *62*(2), 169–180. https://doi.org/10.1016/j.molcel.2016.03.014.

Grabowska, W., Sikora, E., & Bielak-Zmijewska, A. (2017). Sirtuins, a promising target in slowing down the ageing process. *Biogerontology*, *18*, 447–476. https://doi.org/10.1007/s10522-017-9685-9.

Grant, R., & Kapoor, V. (2003). Inhibition of indoleamine 2,3-dioxygenase activity in IFN-gamma stimulated astroglioma cells decreases intracellular NAD levels. *Biochemical Pharmacology*, *66*(6), 1033–1036.

Grube, K., & Bürkle, A. (1992). Poly(ADP-ribose) polymerase activity in mononuclear leukocytes of 13 mammalian species correlates with species-specific life span. *Proceedings of the National Academy of Sciences of the United States of America*, *89*(24), 11759–11763.

Gupte, R., Liu, Z., & Kraus, W. L. (2017). PARPs and ADP-ribosylation: Recent advances linking molecular functions to biological outcomes. *Genes & Development*, *31*(2), 101–126. https://doi.org/10.1101/gad.291518.116.

Guse, A. H., & Lee, H. C. (2008). NAADP: A universal Ca2+ trigger. *Science Signaling*, *1*(44), re10. https://doi.org/10.1126/scisignal.144re10.

Guyton, J. R., Fazio, S., Adewale, A. J., Jensen, E., Tomassini, J. E., Shah, A., et al. (2012). Effect of extended-release niacin on new-onset diabetes among hyperlipidemic patients treated with ezetimibe/simvastatin in a randomized controlled trial. *Diabetes Care*, *35*(4), 857–860. https://doi.org/10.2337/dc11-1369.

Guyton, J. R., Goldberg, A. C., Kreisberg, R. A., Sprecher, D. L., Superko, H. R., & O'Connor, C. M. (1998). Effectiveness of once-nightly dosing of extended-release niacin alone and in combination for hypercholesterolemia. *The American Journal of Cardiology*, *82*(6), 737–743.

Hegyi, J., Schwartz, R. A., & Hegyi, V. (2004). Pellagra: Dermatitis, dementia, and diarrhea. *International Journal of Dermatology*, *43*(1), 1–5.

Herceg, Z., & Wang, Z.-Q. (2001). Functions of poly(ADP-ribose) polymerase (PARP) in DNA repair, genomic integrity and cell death. *Mutation Research/Fundamental and Molecular Mechanisms of Mutagenesis*, *477*(1–2), 97–110. https://doi.org/10.1016/S0027-5107(01)00111-7.

Hochholzer, W., Berg, D. D., & Giugliano, R. P. (2011). The facts behind niacin. *Therapeutic Advances in Cardiovascular Disease*, *5*(5), 227–240. https://doi.org/10.1177/1753944711419197.

Horwitt, M. K., Harper, A. E., & Henderson, L. M. (1981). Niacin-tryptophan relationships for evaluating niacin equivalents. *The American Journal of Clinical Nutrition*, *34*(3), 423–427.

Hottiger, M. O., Hassa, P. O., Luscher, B., Schuler, H., & Koch-Nolte, F. (2010). Toward a unified nomenclature for mammalian ADP-ribosyltransferases. *Trends in Biochemical Sciences*, *35*(4), 208–219. https://doi.org/10.1016/j.tibs.2009.12.003.

Howitz, K. T., Bitterman, K. J., Cohen, H. Y., Lamming, D. W., Lavu, S., Wood, J. G., et al. (2003). Small molecule activators of sirtuins extend *Saccharomyces cerevisiae* lifespan. *Nature*, *425*(6954), 191–196. https://doi.org/10.1038/nature01960.

Hussein, A. A., & Nicholls, S. J. (2010). Critical appraisal of laropiprant and extended-release niacin combination in the management of mixed dyslipidemias and primary hypercholesterolemia. *Therapeutics and Clinical Risk Management*, *6*, 183–190.

Imai, S. (2009a). From heterochromatin islands to the NAD World: A hierarchical view of aging through the functions of mammalian Sirt1 and systemic NAD biosynthesis. *Biochimica et Biophysica Acta*, *1790*(10), 997–1004. https://doi.org/10.1016/j.bbagen.2009.03.005.

Imai, S.-I. (2009b). The NAD world: A new systemic regulatory network for metabolism and aging—Sirt1, systemic NAD biosynthesis, and their importance. *Cell Biochemistry and Biophysics*, *53*(2), 65–74. https://doi.org/10.1007/s12013-008-9041-4.

Imai, S.-I. (2010). "Clocks" in the NAD World: NAD as a metabolic oscillator for the regulation of metabolism and aging. *Biochimica et Biophysica Acta*, *1804*(8), 1584–1590. https://doi.org/10.1016/j.bbapap.2009.10.024.

Imai, S. (2011). Dissecting systemic control of metabolism and aging in the NAD World: The importance of SIRT1 and NAMPT-mediated NAD biosynthesis. *FEBS Letters*, *585*(11), 1657–1662. https://doi.org/10.1016/j.febslet.2011.04.060.

Imai, S., & Guarente, L. (2014). NAD+ and sirtuins in aging and disease. *Trends in Cell Biology*, *24*(8), 464–471. https://doi.org/10.1016/j.tcb.2014.04.002.

Imai, S., & Guarente, L. (2016). It takes two to tango: NAD+ and sirtuins in aging/longevity control. *NPJ Aging and Mechanisms of Disease*, *2*(1), 16017. https://doi.org/10.1038/npjamd.2016.17.

Institute of Medicine. (1998). *Dietary reference intakes for thiamin, riboflavin, niacin, vitamin B_6, folate, vitamin B_{12}, pantothenic acid, biotin, and choline*. Washington, DC: The National Academies Press.

Jackson, T. M., Rawling, J. M., Roebuck, B. D., & Kirkland, J. B. (1995). Large supplements of nicotinic acid and nicotinamide increase tissue NAD+ and poly(ADP-ribose) levels but do not affect diethylnitrosamine-induced altered hepatic foci in Fischer-344 rats. *The Journal of Nutrition, 125*(6), 1455–1461.

Jacobson, E. L. (1993). Niacin deficiency and cancer in women. *Journal of the American College of Nutrition, 12*(4), 412–416.

Jacobson, E. L., & Jacobson, M. K. (1997). Tissue NAD as a biochemical measure of niacin status in humans. *Methods in Enzymology, 280*, 221–230.

Jacobson, E. L., Kim, H., Kim, M., Williams, J. D., Coyle, D. L., Coyle, W. R., et al. (2007). A topical lipophilic niacin derivative increases NAD, epidermal differentiation and barrier function in photodamaged skin. *Experimental Dermatology, 16*(6), 490–499. https://doi.org/10.1111/j.1600-0625.2007.00553.x.

Jacobson, E. L., Shieh, W. M., & Huang, A. C. (1999). Mapping the role of NAD metabolism in prevention and treatment of carcinogenesis. *Molecular and Cellular Biochemistry, 193*(1–2), 69–74.

Jagielska, G., Tomaszewicz-Libudzic, E. C., & Brzozowska, A. (2007). Pellagra: A rare complication of anorexia nervosa. *European Child & Adolescent Psychiatry, 16*(7), 417–420. https://doi.org/10.1007/s00787-007-0613-4.

Jasper, H. (2013). Sirtuins: Longevity focuses on NAD+. *Nature Chemical Biology, 9*(11), 666–667. https://doi.org/10.1038/nchembio.1369.

Jasperson, L. K., Bucher, C., Panoskaltsis-Mortari, A., Mellor, A. L., Munn, D. H., & Blazar, B. R. (2009). Inducing the tryptophan catabolic pathway, indoleamine 2,3-dioxygenase (IDO), for suppression of graft-versus-host disease (GVHD) lethality. *Blood, 114*(24), 5062–5070. https://doi.org/10.1182/blood-2009-06-227587.

Jeggo, P. A. (1998). DNA repair: PARP—Another guardian angel? *Current Biology: CB, 8*(2), R49–51.

Jia, H., Li, X., Gao, H., Feng, Z., Li, X., Zhao, L., et al. (2008). High doses of nicotinamide prevent oxidative mitochondrial dysfunction in a cellular model and improve motor deficit in a Drosophila model of Parkinson's disease. *Journal of Neuroscience Research, 86*(9), 2083–2090. https://doi.org/10.1002/jnr.21650.

Jiang, Y., Liu, J., Chen, D., Yan, L., & Zheng, W. (2017). Sirtuin inhibition: Strategies, inhibitors, and therapeutic potential. *Trends in Pharmacological Sciences, 38*(5), 459–472. https://doi.org/10.1016/j.tips.2017.01.009.

Johansson, M. (1999). A human poly(ADP-ribose) polymerase gene family (ADPRTL): cDNA cloning of two novel poly(ADP-ribose) polymerase homologues. *Genomics, 57*(3), 442–445. https://doi.org/10.1006/geno.1999.5799.

Juarez-Salinas, H., Levi, V., Jacobson, E. L., & Jacobson, M. K. (1982). Poly(ADP-ribose) has a branched structure in vivo. *The Journal of Biological Chemistry, 257*(2), 607–609.

Kane, J. P. (1990). Regression of coronary atherosclerosis during treatment of familial hypercholesterolemia with combined drug regimens. *JAMA: The Journal of the American Medical Association, 264*(23), 3007. https://doi.org/10.1001/jama.1990.03450230043027.

Karlberg, T., Klepsch, M., Thorsell, A.-G., Andersson, C. D., Linusson, A., & Schüler, H. (2015). Structural basis for lack of ADP-ribosyltransferase activity in poly(ADP-ribose) polymerase-13/zinc finger antiviral protein. *The Journal of Biological Chemistry, 290*(12), 7336–7344. https://doi.org/10.1074/jbc.M114.630160.

Kasamatsu, A., Nakao, M., Smith, B. C., Comstock, L. R., Ono, T., Kato, J., et al. (2011). Hydrolysis of O-acetyl-ADP-ribose isomers by ADP-ribosylhydrolase 3. *The Journal of Biological Chemistry, 286*(24), 21110–21117. https://doi.org/10.1074/jbc.M111.237636.

Kennedy, B. E., Sharif, T., Martell, E., Dai, C., Kim, Y., Lee, P. W. K., et al. (2016). NAD (+) salvage pathway in cancer metabolism and therapy. *Pharmacological Research, 114*, 274–283. https://doi.org/10.1016/j.phrs.2016.10.027.

Kiehlbauch, C. C., Aboul-Ela, N., Jacobson, E. L., Ringer, D. P., & Jacobson, M. K. (1993). High resolution fractionation and characterization of ADP-ribose polymers. *Analytical Biochemistry, 208*(1), 26–34.

Kim, S.-Y., Cohen, B. M., Chen, X., Lukas, S. E., Shinn, A. K., Yuksel, A. C., et al. (2017). Redox dysregulation in schizophrenia revealed by in vivo NAD+/NADH measurement. *Schizophrenia Bulletin, 43*(1), 197–204. https://doi.org/10.1093/schbul/sbw129.

Klaidman, L., Morales, M., Kem, S., Yang, J., Chang, M.-L., & Adams, J. D. (2003). Nicotinamide offers multiple protective mechanisms in stroke as a precursor for NAD+, as a PARP inhibitor and by partial restoration of mitochondrial function. *Pharmacology, 69*(3), 150–157. https://doi.org/72668.

Koehn, C. J., Jr., & Elvehjem, C. A. (1937). Further studies on the concentration of the anti-pellagra factor. *Journal of Biological Chemistry, 118*, 693–699.

Koh, D. W., Lawler, A. M., Poitras, M. F., Sasaki, M., Wattler, S., Nehls, M. C., et al. (2004). Failure to degrade poly(ADP-ribose) causes increased sensitivity to cytotoxicity and early embryonic lethality. *Proceedings of the National Academy of Sciences of the United States of America, 101*(51), 17699–17704. https://doi.org/10.1073/pnas.0406182101.

Kraus, W. L. (2015). PARPs and ADP-ribosylation: 50 years ... and counting. *Molecular Cell, 58*(6), 902–910. https://doi.org/10.1016/j.molcel.2015.06.006.

Kraut, J. A., & Sachs, G. (2005). Hartnup disorder: Unraveling the mystery. *Trends in Pharmacological Sciences, 26*(2), 53–55. https://doi.org/10.1016/j.tips.2004.12.003.

Kritikou, E. (2008). Circadian genetics: An enzymatic rheostat. *Nature Reviews Genetics, 9*(9), 653.

Kudo, Y., & Boyd, C. A. (2000). Human placental indoleamine 2,3-dioxygenase: Cellular localization and characterization of an enzyme preventing fetal rejection. *Biochimica et Biophysica Acta, 1500*(1), 119–124.

Kumar, V., & Takahashi, J. S. (2010). PARP around the clock. *Cell, 142*(6), 841–843. https://doi.org/10.1016/j.cell.2010.08.037.

Kupper, J. H., van Gool, L., & Burkle, A. (1995). Molecular genetic systems to study the role of poly(ADP-ribosyl)ation in the cellular response to DNA damage. *Biochimie*, 77(6), 450–455.

Lai, E., De Lepeleire, I., Crumley, T. M., Liu, F., Wenning, L. A., Michiels, N., et al. (2007). Suppression of niacin-induced vasodilation with an antagonist to prostaglandin d2 receptor subtype 1. *Clinical Pharmacology and Therapeutics*, 81(6), 849–857. https://doi.org/10.1038/sj.clpt.6100180.

Lamon-Fava, S., Diffenderfer, M. R., Barrett, P. H. R., Buchsbaum, A., Nyaku, M., Horvath, K. V., et al. (2008). Extended-release niacin alters the metabolism of plasma apolipoprotein (Apo) A-I and ApoB-containing lipoproteins. *Arteriosclerosis, Thrombosis, and Vascular Biology, 28*(9), 1672–1678. https://doi.org/10.1161/ATVBAHA.108.164541.

la Paz, S. M., Bermudez, B., Naranjo, M. C., Lopez, S., Abia, R., & Muriana, F. J. G. (2016). Pharmacological effects of niacin on acute hyperlipemia. *Current Medicinal Chemistry, 23*(25), 2826–2835.

Lee, J. M. S., Robson, M. D., Yu, L.-M., Shirodaria, C. C., Cunnington, C., Kylintireas, I., et al. (2009). Effects of high-dose modified-release nicotinic acid on atherosclerosis and vascular function: A randomized, placebo-controlled, magnetic resonance imaging study. *Journal of the American College of Cardiology, 54*(19), 1787–1794. https://doi.org/10.1016/j.jacc.2009.06.036.

Lehmann, S., Loh, S. H. Y., & Martins, L. M. (2017). Enhancing NAD(+) salvage metabolism is neuroprotective in a PINK1 model of Parkinson's disease. *Biology Open, 6*(2), 141–147. https://doi.org/10.1242/bio.022186.

Li, F., Chong, Z. Z., & Maiese, K. (2006). Cell life versus cell longevity: The mysteries surrounding the NAD+ precursor nicotinamide. *Current Medicinal Chemistry*, *13*(8), 883–895.

Li, R., Yu, K., Wang, Q., Wang, L., Mao, J., & Qian, J. (2016). Pellagra secondary to medication and alcoholism: A case report and review of the literature. *Nutrition in Clinical Practice: Official Publication of the American Society for Parenteral and Enteral Nutrition*, *31*(6), 785–789. https://doi.org/10.1177/0884533616660991.

Liberti, M. V., & Locasale, J. W. (2016). The Warburg effect: How does it benefit cancer cells? *Trends in Biochemical Sciences*, *41*(3), 211–218. https://doi.org/10.1016/j.tibs.2015.12.001.

Lin, S.-J., Kaeberlein, M., Andalis, A. A., Sturtz, L. A., Defossez, P.-A., Culotta, V. C., et al. (2002). Calorie restriction extends Saccharomyces cerevisiae lifespan by increasing respiration. *Nature*, *418*(6895), 344–348. https://doi.org/10.1038/nature00829.

Lloyd-Jones, D. M. (2014). Niacin and HDL cholesterol—Time to face facts. *New England Journal of Medicine*, *371*(3), 271–273. https://doi.org/10.1056/NEJMe1406410.

Lord, C. J., & Ashworth, A. (2017). PARP inhibitors: Synthetic lethality in the clinic. *Science (New York, NY)*, *355*(6330), 1152–1158. https//doi.org/10.1126/science.aam7344.

Magni, G., Amici, A., Emanuelli, M., Orsomando, G., Raffaelli, N., & Ruggieri, S. (2004). Enzymology of NAD+ homeostasis in man. *Cellular and Molecular Life Sciences: CMLS*, *61*(1), 19–34. https://doi.org/10.1007/s00018-003-3161-1.

Maltos, A. L., Portari, G. V., Moraes, G. V., Monteiro, M. C. R., Vannucchi, H., & da Cunha, D. F. (2015). Niacin metabolism and indoleamine 2,3-dioxygenase activation in malnourished patients with flaky paint dermatosis. *Nutrition (Burbank, Los Angeles County, CA)*, *31*(6), 890–892. https://doi.org/10.1016/j.nut.2014.12.023.

Mashimo, M., Kato, J., & Moss, J. (2013). ADP-ribosyl-acceptor hydrolase 3 regulates poly (ADP-ribose) degradation and cell death during oxidative stress. *Proceedings of the National Academy of Sciences of the United States of America*, *110*(47), 18964–18969. https://doi.org/10.1073/pnas.1312783110.

Mashimo, M., Kato, J., & Moss, J. (2014). Structure and function of the ARH family of ADP-ribosyl-acceptor hydrolases. *DNA Repair*, *23*, 88–94. https://doi.org/10.1016/j.dnarep.2014.03.005.

Massudi, H., Grant, R., Braidy, N., Guest, J., Farnsworth, B., & Guillemin, G. J. (2012). Age-associated changes in oxidative stress and NAD+ metabolism in human tissue. *PLoS One*, 7(7). e42357 https://doi.org/10.1371/journal.pone.0042357.

Masutani, M., & Fujimori, H. (2013). Poly(ADP-ribosyl)ation in carcinogenesis. *Molecular Aspects of Medicine*, *34*(6), 1202–1216. https://doi.org/10.1016/j.mam.2013.05.003.

Masutani, M., Nakagama, H., & Sugimura, T. (2003). Poly(ADP-ribose) and carcinogenesis. *Genes, Chromosomes & Cancer*, *38*(4), 339–348. https://doi.org/10.1002/gcc.10250.

Masutani, M., Nozaki, T., Nakamoto, K., Nakagama, H., Suzuki, H., Kusuoka, O., et al. (2000). The response of Parp knockout mice against DNA damaging agents. *Mutation Research*, *462*(2–3), 159–166.

Mazucanti, C. H., Cabral-Costa, J. V., Vasconcelos, A. R., Andreotti, D. Z., Scavone, C., & Kawamoto, E. M. (2015). Longevity pathways (mTOR, SIRT, insulin/IGF-1) as key modulatory targets on aging and neurodegeneration. *Current Topics in Medicinal Chemistry*, *15*(21), 2116–2138.

McKenney, J. M., Jones, P. H., Bays, H. E., Knopp, R. H., Kashyap, M. L., Ruoff, G. E., et al. (2007). Comparative effects on lipid levels of combination therapy with a statin and extended-release niacin or ezetimibe versus a statin alone (the COMPELL study). *Atherosclerosis*, *192*(2), 432–437. https://doi.org/10.1016/j.atherosclerosis.2006.11.037.

Meyer, R. G., Meyer-Ficca, M. L., Jacobson, E. L., & Jacobson, M. K. (2004). Enzymes in poly(ADP-ribose) metabolism. In A. Burkle (Ed.), *Book section*: *Vol. 1. Poly(ADP-ribosyl) ation:* Landes Bioscience.

Meyer, R. G., Meyer-Ficca, M. L., Whatcott, C. J., Jacobson, E. L., & Jacobson, M. K. (2007). Two small enzyme isoforms mediate mammalian mitochondrial poly(ADP-ribose) glycohydrolase (PARG) activity. *Experimental Cell Research, 313*(13), 2920–2936https://doi.org/10.1016/j.yexcr.2007.03.043.

Meyer, R., Muller, M., Beneke, S., Kupper, J. H., & Burkle, A. (2000). Negative regulation of alkylation-induced sister-chromatid exchange by poly(ADP-ribose) polymerase-1 activity. *International Journal of Cancer, 88*(3), 351–355. https://doi.org/10.1002/1097-0215(20001101)88:3<351::AID-IJC5>3.0.CO;2-H.

Meyer-Ficca, M. L., Ihara, M., Bader, J. J., Leu, N. A., Beneke, S., & Meyer, R. G. (2015). Spermatid head elongation with normal nuclear shaping requires ADP-ribosyltransferase PARP11 (ARTD11) in mice. *Biology of Reproduction, 92*(3), 80. https://doi.org/10.1095/biolreprod.114.123661.

Meyer-Ficca, M. L., Meyer, R. G., Coyle, D. L., Jacobson, E. L., & Jacobson, M. K. (2004). Human poly(ADP-ribose) glycohydrolase is expressed in alternative splice variants yielding isoforms that localize to different cell compartments. *Experimental Cell Research, 297*(2), 521–532. https://doi.org/10.1016/j.yexcr.2004.03.050.

Meyer-Ficca, M. L., Meyer, R. G., Jacobson, E. L., & Jacobson, M. K. (2005). Poly(ADP-ribose) polymerases: Managing genome stability. *The International Journal of Biochemistry & Cell Biology, 37*(5), 920–926. https://doi.org/10.1016/j.biocel.2004.09.011.

Michan, S., & Sinclair, D. (2007). Sirtuins in mammals: Insights into their biological function. *The Biochemical Journal, 404*(1), 1–13. https://doi.org/10.1042/BJ20070140.

Miller, D. F. (1978). Pellagra deaths in the United States. *The American Journal of Clinical Nutrition, 31*(4), 558–559.

Miwa, M., Ishihara, M., Takishima, S., Takasuka, N., Maeda, M., Yamaizumi, Z., et al. (1981). The branching and linear portions of poly(adenosine diphosphate ribose) have the same alpha(1 leads to 2) ribose-ribose linkage. *Journal of Biological Chemistry, 256*(6), 2916–2921.

Miwa, M., Saikawa, N., Yamaizumi, Z., Nishimura, S., & Sugimura, T. (1979). Structure of poly(adenosine diphosphate ribose): Identification of 2′-[1″-ribosyl-2″-(or 3″-)(1″-ribosyl)]adenosine-5′,5″,5″-tris(phosphate) as a branch linkage. *Proceedings of the National Academy of Sciences of the United States of America, 76*(2), 595–599.

Miwa, M., & Sugimura, T. (1984). Structure of poly(ADP-ribose). *Methods in Enzymology, 106*, 441–450.

Módis, K., Gero, D., Erdélyi, K., Szoleczky, P., DeWitt, D., & Szabo, C. (2012). Cellular bioenergetics is regulated by PARP1 under resting conditions and during oxidative stress. *Biochemical Pharmacology, 83*(5), 633–643. https://doi.org/10.1016/j.bcp.2011.12.014.

Mouchiroud, L., Houtkooper, R. H., Moullan, N., Katsyuba, E., Ryu, D., Cantó, C., et al. (2013). The NAD(+)/sirtuin pathway modulates longevity through activation of mitochondrial UPR and FOXO signaling. *Cell, 154*(2), 430–441. https://doi.org/10.1016/j.cell.2013.06.016.

Muller, A. J., & Prendergast, G. C. (2007). Indoleamine 2,3-dioxygenase in immune suppression and cancer. *Current Cancer Drug Targets*, 7(1), 31–40.

Munn, D. H., & Mellor, A. L. (2013). Indoleamine 2,3 dioxygenase and metabolic control of immune responses. *Trends in Immunology, 34*(3), 137–143. https://doi.org/10.1016/j.it.2012.10.001.

Munn, D. H., & Mellor, A. L. (2016). IDO in the tumor microenvironment: Inflammation, counter-regulation, and tolerance. *Trends in Immunology, 37*(3), 193–207. https://doi.org/10.1016/j.it.2016.01.002.

Murray, M. F., Langan, M., & MacGregor, R. R. (2001). Increased plasma tryptophan in HIV-infected patients treated with pharmacologic doses of nicotinamide. *Nutrition (Burbank, Los Angeles County, CA), 17*(7–8), 654–656.

Mutafova-Yambolieva, V. N. (2012). Neuronal and extraneuronal release of ATP and NAD (+) in smooth muscle. *IUBMB Life*, *64*(10), 817–824. https://doi.org/10.1002/iub.1076.

Mutafova-Yambolieva, V. N., Hwang, S. J., Hao, X., Chen, H., Zhu, M. X., Wood, J. D., et al. (2007). Beta-nicotinamide adenine dinucleotide is an inhibitory neurotransmitter in visceral smooth muscle. *Proceedings of the National Academy of Sciences of the United States of America*, *104*(41), 16359–16364. https://doi.org/10.1073/pnas.0705510104.

Nakahata, Y., Kaluzova, M., Grimaldi, B., Sahar, S., Hirayama, J., Chen, D., et al. (2008). The NAD+-dependent deacetylase SIRT1 modulates CLOCK-mediated chromatin remodeling and circadian control. *Cell*, *134*(2), 329–340. https://doi.org/10.1016/j.cell.2008.07.002.

Nakahata, Y., Sahar, S., Astarita, G., Kaluzova, M., & Sassone-Corsi, P. (2009). Circadian control of the NAD+ salvage pathway by CLOCK-SIRT1. *Science (New York, NY)*, *324*(5927), 654–657. https://doi.org/10.1126/science.1170803.

Nimmagadda, V. K., Bever, C. T., Vattikunta, N. R., Talat, S., Ahmad, V., Nagalla, N. K., et al. (2013). Overexpression of SIRT1 protein in neurons protects against experimental autoimmune encephalomyelitis through activation of multiple SIRT1 targets. *Journal of Immunology (Baltimore, Md: 1950)*, *190*(9), 4595–4607. https://doi.org/10.4049/jimmunol.1202584.

Nimmagadda, V. K. C., Makar, T. K., Chandrasekaran, K., Sagi, A. R., Ray, J., Russell, J. W., et al. (2017). SIRT1 and NAD+ precursors: Therapeutic targets in multiple sclerosis a review. *Journal of Neuroimmunology*, *304*, 29–34. https://doi.org/10.1016/j.jneuroim.2016.07.007.

Nogueira, A., Duarte, A. F., Magina, S., & Azevedo, F. (2009). Pellagra associated with esophageal carcinoma and alcoholism. *Dermatology Online Journal*, *15*(5), 8.

Nozaki, J., Dakeishi, M., Ohura, T., Inoue, K., Manabe, M., Wada, Y., et al. (2001). Homozygosity mapping to chromosome 5p15 of a gene responsible for Hartnup disorder. *Biochemical and Biophysical Research Communications*, *284*(2), 255–260. https://doi.org/10.1006/bbrc.2001.4961.

Ondracek, C. R., Frappier, V., Ringel, A. E., Wolberger, C., & Guarente, L. (2017). Mutations that allow SIR2 orthologs to function in a NAD(+)-depleted environment. *Cell Reports*, *18*(10), 2310–2319. https://doi.org/10.1016/j.celrep.2017.02.031.

Ooi, E. M., Watts, G. F., Chan, D. C., Pang, J., Tenneti, V. S., Hamilton, S. J., et al. (2015). Effects of extended-release niacin on the postprandial metabolism of Lp(a) and ApoB-100-containing lipoproteins in statin-treated men with type 2 diabetes mellitus. *Arteriosclerosis, Thrombosis, and Vascular Biology*, *35*(12), 2686–2693. https://doi.org/10.1161/ATVBAHA.115.306136.

Opitz, C. A., & Heiland, I. (2015). Dynamics of NAD-metabolism: Everything but constant. *Biochemical Society Transactions*, *43*(6), 1127–1132. https://doi.org/10.1042/BST20150133.

Palazzo, L., Mikoč, A., & Ahel, I. (2017). ADP-ribosylation: New facets of an ancient modification. *The FEBS Journal*, *284*, 2932–2946. https://doi.org/10.1111/febs.14078.

Pan, H., & Finkel, T. (2017). Key proteins and pathways that regulate lifespan. *The Journal of Biological Chemistry*, *292*(16), 6452–6460. https://doi.org/10.1074/jbc.R116.771915.

Pang, J., Chan, D. C., Hamilton, S. J., Tenneti, V. S., Watts, G. F., & Barrett, P. H. R. (2014). Effect of niacin on high-density lipoprotein apolipoprotein A-I kinetics in statin-treated patients with type 2 diabetes mellitus. *Arteriosclerosis, Thrombosis, and Vascular Biology*, *34*(2), 427–432. https://doi.org/10.1161/ATVBAHA.113.302019.

Park, J. Y., Jang, S. Y., Shin, Y. K., Koh, H., Suh, D. J., Shinji, T., et al. (2013). Mitochondrial swelling and microtubule depolymerization are associated with energy depletion in axon degeneration. *Neuroscience*, *238*, 258–269. https://doi.org/10.1016/j.neuroscience.2013.02.033.

Park, S. M., Li, T., Wu, S., Li, W.-Q., Weinstock, M., Qureshi, A. A., et al. (2017). Niacin intake and risk of skin cancer in US women and men. *International Journal of Cancer, 140*(9), 2023–2031. https://doi.org/10.1002/ijc.30630.

Partida-Sánchez, S., Cockayne, D. A., Monard, S., Jacobson, E. L., Oppenheimer, N., Garvy, B., et al. (2001). Cyclic ADP-ribose production by CD38 regulates intracellular calcium release, extracellular calcium influx and chemotaxis in neutrophils and is required for bacterial clearance in vivo. *Nature Medicine*, 7(11), 1209–1216. https://doi.org/10.1038/nm1101-1209.

Partida-Sanchez, S., Gasser, A., Fliegert, R., Siebrands, C. C., Dammermann, W., Shi, G., et al. (2007). Chemotaxis of mouse bone marrow neutrophils and dendritic cells is controlled by ADP-ribose, the major product generated by the CD38 enzyme reaction. *Journal of Immunology (Baltimore, Md: 1950)*, 179(11), 7827–7839.

Patterson, J. I., Brown, R. R., Linkswiler, H., & Harper, A. E. (1980). Excretion of tryptophan-niacin metabolites by young men: Effects of tryptophan, leucine, and vitamin B6 intakes. *The American Journal of Clinical Nutrition, 33*(10), 2157–2167.

Peek, C. B., Affinati, A. H., Ramsey, K. M., Kuo, H.-Y., Yu, W., Sena, L. A., et al. (2013). Circadian clock NAD+ cycle drives mitochondrial oxidative metabolism in mice. *Science (New York, NY), 342*(6158), 1243417. https://doi.org/10.1126/science.1243417.

Penberthy, W. T. (2007). Pharmacological targeting of IDO-mediated tolerance for treating autoimmune disease. *Current Drug Metabolism, 8*(3), 245–266.

Penberthy, W. T., & Kirkland, J. B. (2012). *Niacin. In J. W. Erdman, I. A. Macdonald, & S. H. Zeisel (Eds.), Present knowledge in nutrition* (pp. 293–306). Oxford, UK: Wiley-Blackwell. *https://doi.org/10.1002/9781119946045.ch19.*

Perina, D., Mikoč, A., Ahel, J., Ćetković, H., Žaja, R., & Ahel, I. (2014). Distribution of protein poly(ADP-ribosyl)ation systems across all domains of life. *DNA Repair, 23*, 4–16. https://doi.org/10.1016/j.dnarep.2014.05.003.

Pollak, N., Dölle, C., & Ziegler, M. (2007). The power to reduce: Pyridine nucleotides—Small molecules with a multitude of functions. *Biochemical Journal, 402*(2), 205–218. https://doi.org/10.1042/BJ20061638.

Poltronieri, P., & Miwa, M. (2016). Editorial (thematic issue: Overview on ADP ribosylation and PARP superfamily of proteins). *Current Protein & Peptide Science, 17*(7), 630–632. https://doi.org/10.2174/138920371707160908172601.

Prousky, J., Millman, C. G., & Kirkland, J. B. (2011). Pharmacologic use of niacin. *Journal of Evidence-Based Complementary & Alternative Medicine, 16*(2), 91–101. https://doi.org/10.1177/2156587211399579.

Quarona, V., Zaccarello, G., Chillemi, A., Brunetti, E., Singh, V. K., Ferrero, E., et al. (2013). CD38 and CD157: A long journey from activation markers to multifunctional molecules. *Cytometry Part B, Clinical Cytometry, 84*(4), 207–217. https://doi.org/10.1002/cyto.b.21092.

Raffaelli, N., Sorci, L., Amici, A., Emanuelli, M., Mazzola, F., & Magni, G. (2002). Identification of a novel human nicotinamide mononucleotide adenylyltransferase. *Biochemical and Biophysical Research Communications, 297*(4), 835–840.

Rajakumar, K. (2000). Pellagra in the United States: A historical perspective. *Southern Medical Journal, 93*(3), 272–277.

Rajawat, J., Shukla, N., & Mishra, D. P. (2017). Therapeutic targeting of poly(ADP-ribose) polymerase-1 in cancer: Current developments, therapeutic strategies, and future opportunities. *Medicinal Research Reviews, 37*, 1461–1491. https://doi.org/10.1002/med.21442.

Ramsey, K. M., Yoshino, J., Brace, C. S., Abrassart, D., Kobayashi, Y., Marcheva, B., et al. (2009). Circadian clock feedback cycle through NAMPT-mediated NAD+ biosynthesis. *Science (New York, NY), 324*(5927), 651–654. https://doi.org/10.1126/science.1171641.

Rehan, L., Laszki-Szcząchor, K., Sobieszczańska, M., & Polak-Jonkisz, D. (2014). SIRT1 and NAD as regulators of ageing. *Life Sciences*, *105*(1–2), 1–6. https://doi.org/10.1016/j.lfs.2014.03.015.

Revollo, J. R., Grimm, A. A., & Imai, S. (2004). The NAD biosynthesis pathway mediated by nicotinamide phosphoribosyltransferase regulates Sir2 activity in mammalian cells. *The Journal of Biological Chemistry*, *279*(49), 50754–50763. https://doi.org/10.1074/jbc.M408388200.

Revollo, J. R., Grimm, A. A., & Imai, S. (2007). The regulation of nicotinamide adenine dinucleotide biosynthesis by Nampt/PBEF/visfatin in mammals. *Current Opinion in Gastroenterology*, *23*(2), 164–170. https://doi.org/10.1097/MOG.0b013e32801b3c8f.

Rongvaux, A., Andris, F., Van Gool, F., & Leo, O. (2003). Reconstructing eukaryotic NAD metabolism. *BioEssays: News and Reviews in Molecular, Cellular and Developmental Biology*, *25*(7), 683–690. https://doi.org/10.1002/bies.10297.

Rubic, T., Trottmann, M., & Lorenz, R. L. (2004). Stimulation of CD36 and the key effector of reverse cholesterol transport ATP-binding cassette A1 in monocytoid cells by niacin. *Biochemical Pharmacology*, *67*(3), 411–419. https://doi.org/10.1016/j.bcp.2003.09.014.

Satoh, A., Imai, S., & Guarente, L. (2017). The brain, sirtuins, and ageing. *Nature Reviews Neuroscience*, *18*(6), 362–374.

Scheibye-Knudsen, M., Mitchell, S. J., Fang, E. F., Iyama, T., Ward, T., Wang, J., et al. (2014). A high-fat diet and NAD(+) activate Sirt1 to rescue premature aging in cockayne syndrome. *Cell Metabolism*, *20*(5), 840–855. https://doi.org/10.1016/j.cmet.2014.10.005.

Schmeisser, K., Mansfeld, J., Kuhlow, D., Weimer, S., Priebe, S., Heiland, I., et al. (2013). Role of sirtuins in lifespan regulation is linked to methylation of nicotinamide. *Nature Chemical Biology*, *9*(11), 693–700. https://doi.org/10.1038/nchembio.1352.

Schraufstatter, I. U., Hinshaw, D. B., Hyslop, P. A., Spragg, R. G., & Cochrane, C. G. (1986). Oxidant injury of cells. DNA strand-breaks activate polyadenosine diphosphate-ribose polymerase and lead to depletion of nicotinamide adenine dinucleotide. *The Journal of Clinical Investigation*, 77(4), 1312–1320. https://doi.org/10.1172/JCI112436.

Schreiber, V., Hunting, D., Trucco, C., Gowans, B., Grunwald, D., De Murcia, G., et al. (1995). A dominant-negative mutant of human poly(ADP-ribose) polymerase affects cell recovery, apoptosis, and sister chromatid exchange following DNA damage. *Proceedings of the National Academy of Sciences of the United States of America*, *92*(11), 4753–4757.

Schuber, F., & Lund, F. E. (2004). Structure and enzymology of ADP-ribosyl cyclases: Conserved enzymes that produce multiple calcium mobilizing metabolites. *Current Molecular Medicine*, *4*(3), 249–261.

Schuldt, A. (2010). Circadian rhythms: PARP1 feeds into clocks. *Nature Reviews Molecular Cell Biology*, *11*(11), 754–755. https://doi.org/10.1038/nrm2998.

Schultz, M. B., & Sinclair, D. A. (2016). Why NAD(+) declines during aging: It's destroyed. *Cell Metabolism*, *23*(6), 965–966. https://doi.org/10.1016/j.cmet.2016.05.022.

Shah, G. M., Shah, R. G., Veillette, H., Kirkland, J. B., Pasieka, J. L., & Warner, R. R. P. (2005). Biochemical assessment of niacin deficiency among carcinoid cancer patients. *The American Journal of Gastroenterology*, *100*(10), 2307–2314. https://doi.org/10.1111/j.1572-0241.2005.00268.x.

Shall, S., & de Murcia, G. (2000). Poly(ADP-ribose) polymerase-1: What have we learned from the deficient mouse model? *Mutation Research*, *460*(1), 1–15.

Sharifi, R., Morra, R., Appel, C. D., Tallis, M., Chioza, B., Jankevicius, G., et al. (2013). Deficiency of terminal ADP-ribose protein glycohydrolase TARG1/C6orf130 in neurodegenerative disease. *The EMBO Journal*, *32*(9), 1225–1237. https://doi.org/10.1038/emboj.2013.51.

Shibata, K., & Toda, S. (1997). Effects of sex hormones on the metabolism of tryptophan to niacin and to serotonin in male rats. *Bioscience, Biotechnology, and Biochemistry*, *61*(7), 1200–1202. https://doi.org/10.1271/bbb.61.1200.

Sibley, C. T., Vavere, A. L., Gottlieb, I., Cox, C., Matheson, M., Spooner, A., et al. (2013). MRI-measured regression of carotid atherosclerosis induced by statins with and without niacin in a randomised controlled trial: the NIA plaque study. *Heart*, *99*(22), 1675–1680. https://doi.org/10.1136/heartjnl-2013-303926.

Smith, J. S., Brachmann, C. B., Celic, I., Kenna, M. A., Muhammad, S., Starai, V. J., et al. (2000). A phylogenetically conserved NAD+-dependent protein deacetylase activity in the Sir2 protein family. *Proceedings of the National Academy of Sciences of the United States of America*, *97*(12), 6658–6663.

Smyth, L. M., Bobalova, J., Mendoza, M. G., Lew, C., & Mutafova-Yambolieva, V. N. (2004). Release of beta-nicotinamide adenine dinucleotide upon stimulation of postganglionic nerve terminals in blood vessels and urinary bladder. *The Journal of Biological Chemistry*, *279*(47), 48893–48903. https://doi.org/10.1074/jbc.M407266200.

Song, P., Ramprasath, T., Wang, H., & Zou, M.-H. (2017). Abnormal kynurenine pathway of tryptophan catabolism in cardiovascular diseases. *Cellular and Molecular Life Sciences: CMLS*, *74*, 2899–2916. https://doi.org/10.1007/s00018-017-2504-2.

Srivastava, S. (2016). Emerging therapeutic roles for NAD(+) metabolism in mitochondrial and age-related disorders. *Clinical and Translational Medicine*, *5*(1), 25. https://doi.org/10.1186/s40169-016-0104-7.

Surjana, D., Halliday, G. M., & Damian, D. L. (2010). Role of nicotinamide in DNA damage, mutagenesis, and DNA repair. *Journal of Nucleic Acids*, *2010*, 1–13. https://doi.org/10.4061/2010/157591.

Tang, K., Sham, H., Hui, E., & Kirkland, J. B. (2008). Niacin deficiency causes oxidative stress in rat bone marrow cells but not through decreased NADPH or glutathione status. *The Journal of Nutritional Biochemistry*, *19*(11), 746–753. https://doi.org/10.1016/j.jnutbio.2007.10.003.

Tanner, K. G., Landry, J., Sternglanz, R., & Denu, J. M. (2000). Silent information regulator 2 family of NAD-dependent histone/protein deacetylases generates a unique product, 1-O-acetyl-ADP-ribose. *Proceedings of the National Academy of Sciences of the United States of America*, *97*(26), 14178–14182. https://doi.org/10.1073/pnas.250422697.

Taylor, A. J. (2004). Arterial biology for the investigation of the treatment effects of reducing cholesterol (ARBITER) 2: A double-blind, placebo-controlled study of extended-release niacin on atherosclerosis progression in secondary prevention patients treated with statins. *Circulation*, *110*(23), 3512–3517. https://doi.org/10.1161/01.CIR.0000148955.19792.8D.

Taylor, A. J., Lee, H. J., & Sullenberger, L. E. (2006). The effect of 24 months of combination statin and extended-release niacin on carotid intima-media thickness: ARBITER 3. *Current Medical Research and Opinion*, *22*(11), 2243–2250. https://doi.org/10.1185/030079906X148508.

Taylor, A. J., Villines, T. C., Stanek, E. J., Devine, P. J., Griffen, L., Miller, M., et al. (2009). Extended-release niacin or ezetimibe and carotid intima-media thickness. *The New England Journal of Medicine*, *361*(22), 2113–2122. https://doi.org/10.1056/NEJMoa0907569.

Terada, N., Kinoshita, K., Taguchi, S., & Tokuda, Y. (2015). Wernicke encephalopathy and pellagra in an alcoholic and malnourished patient. *BMJ Case Reports*, *2015*, 1–3. https://doi.org/10.1136/bcr-2015-209412.

The AIM-HIGH Investigators. (2011a). Niacin in patients with low HDL cholesterol levels receiving intensive statin therapy. *New England Journal of Medicine*, *365*(24), 2255–2267. https://doi.org/10.1056/NEJMoa1107579.

The AIM-HIGH Investigators. (2011b). The role of niacin in raising high-density lipoprotein cholesterol to reduce cardiovascular events in patients with atherosclerotic cardiovascular disease and optimally treated low-density lipoprotein cholesterol. *American Heart Journal*, *161*(3), 471–477.e2. https://doi.org/10.1016/j.ahj.2010.11.017.

The AIM-HIGH Investigators. (2011c). The role of niacin in raising high-density lipoprotein cholesterol to reduce cardiovascular events in patients with atherosclerotic cardiovascular disease and optimally treated low-density lipoprotein cholesterol: Baseline characteristics of study participants. *The atherothrombosis intervention in metabolic syndrome with low HDL/high triglycerides: Impact on global health outcomes (AIM-HIGH) trial, American Heart Journal*, *161*(3), 538–543. https://doi.org/10.1016/j.ahj.2010.12.007.

The HPS2-THRIVE Collaborative Group. (2014). Effects of extended-release niacin with Laropiprant in high-risk patients. *New England Journal of Medicine*, *371*(3), 203–212. https://doi.org/10.1056/NEJMoa1300955.

Trucco, C., Rolli, V., Oliver, F. J., Flatter, E., Masson, M., Dantzer, F., et al. (1999). A dual approach in the study of poly (ADP-ribose) polymerase: In vitro random mutagenesis and generation of deficient mice. *Molecular and Cellular Biochemistry*, *193*(1–2), 53–60.

van de Ven, R. A. H., Santos, D., & Haigis, M. C. (2017). Mitochondrial sirtuins and molecular mechanisms of aging. *Trends in Molecular Medicine*, *23*(4), 320–331. https://doi.org/10.1016/j.molmed.2017.02.005.

van der Veer, E., Ho, C., O'Neil, C., Barbosa, N., Scott, R., Cregan, S. P., et al. (2007). Extension of human cell lifespan by nicotinamide phosphoribosyltransferase. *The Journal of Biological Chemistry*, *282*(15), 10841–10845. https://doi.org/10.1074/jbc.C700018200.

Van Gool, L., Meyer, R., Tobiasch, E., Cziepluch, C., Jauniaux, J. C., Mincheva, A., et al. (1997). Overexpression of human poly(ADP-ribose) polymerase in transfected hamster cells leads to increased poly(ADP-ribosyl)ation and cellular sensitization to gamma irradiation. *European Journal of Biochemistry/FEBS*, *244*(1), 15–20.

Vazquez, B. N., Thackray, J. K., & Serrano, L. (2017). Sirtuins and DNA damage repair: SIRT7 comes to play. *Nucleus (Austin, TX)*, *8*(2), 107–115. https://doi.org/10.1080/19491034.2016.1264552.

Verdin, E. (2015). NAD+ in aging, metabolism, and neurodegeneration. *Science*, *350*(6265), 1208–1213. https://doi.org/10.1126/science.aac4854.

Vergeer, M., Holleboom, A. G., Kastelein, J. J. P., & Kuivenhoven, J. A. (2010). The HDL hypothesis: Does high-density lipoprotein protect from atherosclerosis? *The Journal of Lipid Research*, *51*(8), 2058–2073. https://doi.org/10.1194/jlr.R001610.

Vogt, A., Kassner, U., Hostalek, U., Peiter, A., & Steinhagen-Thiessen, E. (2006). Safety and tolerability of nicotinic acid. Results of the multicenter, open, prospective NAUTILUS study. *MMW Fortschritte der Medizin*, *148*(18), 41.

Vogt, A., Kassner, U., Hostalek, U., Steinhagen-Thiessen, E., & NAUTILUS Study Group (2006). Evaluation of the safety and tolerability of prolonged-release nicotinic acid in a usual care setting: The NAUTILUS study. *Current Medical Research and Opinion*, *22*(2), 417–425. https://doi.org/10.1185/030079906X89766.

Voight, B. F., Peloso, G. M., Orho-Melander, M., Frikke-Schmidt, R., Barbalic, M., Jensen, M. K., et al. (2012). Plasma HDL cholesterol and risk of myocardial infarction: A Mendelian randomisation study. *The Lancet (London, England)*, *380*(9841), 572–580. https://doi.org/10.1016/S0140-6736(12)60312-2.

Vyas, S., Matic, I., Uchima, L., Rood, J., Zaja, R., Hay, R. T., et al. (2014). Family-wide analysis of poly(ADP-ribose) polymerase activity. *Nature Communications*, *5*, 1–13. https://doi.org/10.1038/ncomms5426.

Wakade, C., & Chong, R. (2014). A novel treatment target for Parkinson's disease. *Journal of the Neurological Sciences*, *347*(1–2), 34–38. https://doi.org/10.1016/j.jns.2014.10.024.

Wakade, C., Chong, R., Bradley, E., & Morgan, J. C. (2015). Low-dose niacin supplementation modulates GPR109A, niacin index and ameliorates Parkinson's disease symptoms without side effects. *Clinical Case Reports*, *3*(7), 635–637. https://doi.org/10.1002/ccr3.232.

Wakade, C., Chong, R., Bradley, E., Thomas, B., & Morgan, J. (2014). Upregulation of GPR109A in Parkinson's disease. *PLoS One*, *9*(10) . e109818 https://doi.org/10.1371/journal.pone.0109818.

Wang, Z. Q., Auer, B., Stingl, L., Berghammer, H., Haidacher, D., Schweiger, M., et al. (1995). Mice lacking ADPRT and poly(ADP-ribosyl)ation develop normally but are susceptible to skin disease. *Genes & Development*, *9*(5), 509–520.

Weidele, K., Beneke, S., & Bürkle, A. (2017). The NAD(+) precursor nicotinic acid improves genomic integrity in human peripheral blood mononuclear cells after X-irradiation. *DNA Repair*, *52*, 12–23. https://doi.org/10.1016/j.dnarep.2017.02.001.

Weidele, K., Kunzmann, A., Schmitz, M., Beneke, S., & Bürkle, A. (2010). Ex vivo supplementation with nicotinic acid enhances cellular poly(ADP-ribosyl)ation and improves cell viability in human peripheral blood mononuclear cells. *Biochemical Pharmacology*, *80*(7), 1103–1112. https://doi.org/10.1016/j.bcp.2010.06.010.

Williams, P. A., Harder, J. M., Foxworth, N. E., Cochran, K. E., Philip, V. M., Porciatti, V., et al. (2017). Vitamin B3 modulates mitochondrial vulnerability and prevents glaucoma in aged mice. *Science (New York, NY)*, 355(6326), 756–760. https://doi.org/10.1126/science.aal0092.

Winstall, E., Affar, E. B., Shah, R., Bourassa, S., Scovassi, I. A., & Poirier, G. G. (1999). Preferential perinuclear localization of poly(ADP-ribose) glycohydrolase. *Experimental Cell Research*, *251*(2), 372–378. https://doi.org/10.1006/excr.1999.4594.

Wu, Z.-H., & Zhao, S.-P. (2009). Niacin promotes cholesterol efflux through stimulation of the PPARgamma-LXRalpha-ABCA1 pathway in 3T3-L1 adipocytes. *Pharmacology*, *84*(5), 282–287. https://doi.org/10.1159/000242999.

Yahata, N., Yuasa, S., & Araki, T. (2009). Nicotinamide mononucleotide adenylyltransferase expression in mitochondrial matrix delays Wallerian degeneration. *The Journal of Neuroscience: The Official Journal of the Society for Neuroscience*, *29*(19), 6276–6284. https://doi.org/10.1523/JNEUROSCI.4304-08.2009.

Yamazaki, F., Kuroiwa, T., Takikawa, O., & Kido, R. (1985). Human indolylamine 2,3-dioxygenase. Its tissue distribution, and characterization of the placental enzyme. *The Biochemical Journal*, *230*(3), 635–638.

Yamboliev, I. A., Smyth, L. M., Durnin, L., Dai, Y., & Mutafova-Yambolieva, V. N. (2009). Storage and secretion of beta-NAD, ATP and dopamine in NGF-differentiated rat pheochromocytoma PC12 cells. *The European Journal of Neuroscience*, *30*(5), 756–768. https://doi.org/10.1111/j.1460-9568.2009.06869.x.

Yang, C.-S., Jividen, K., Spencer, A., Dworak, N., Ni, L., Oostdyk, L. T., et al. (2017). Ubiquitin modification by the E3 ligase/ADP-ribosyltransferase Dtx3L/Parp9. *Molecular Cell*, 66(4), 503–516.e5. https://doi.org/10.1016/j.molcel.2017.04.028.

Yang, H., Yang, T., Baur, J. A., Perez, E., Matsui, T., Carmona, J. J., et al. (2007). Nutrient-sensitive mitochondrial NAD+ levels dictate cell survival. *Cell*, *130*(6), 1095–1107. https://doi.org/10.1016/j.cell.2007.07.035.

Ying, W. (2008). NAD+/NADH and NADP+/NADPH in cellular functions and cell death: Regulation and biological consequences. *Antioxidants & Redox Signaling*, *10*(2), 179–206. https://doi.org/10.1089/ars.2007.1672.

Yu, S. W., Andrabi, S. A., Wang, H., Kim, N. S., Poirier, G. G., Dawson, T. M., et al. (2006). Apoptosis-inducing factor mediates poly(ADP-ribose) (PAR) polymer-induced cell death. *Proceedings of the National Academy of Sciences of the United States of America*, *103*(48), 18314–18319. https://doi.org/10.1073/pnas.0606528103.

Zeman, M., Vecka, M., Perlík, F., Staňková, B., Hromádka, R., Tvrzická, E., et al. (2016). Pleiotropic effects of niacin: Current possibilities for its clinical use. *Acta Pharmaceutica (Zagreb, Croatia)*, *66*(4), 449–469. https://doi.org/10.1515/acph-2016-0043.
Zhang, X., Kurnasov, O. V., Karthikeyan, S., Grishin, N. V., Osterman, A. L., & Zhang, H. (2003). Structural characterization of a human cytosolic NMN/NaMN adenylyltransferase and implication in human NAD biosynthesis. *The Journal of Biological Chemistry*, *278*(15), 13503–13511. https://doi.org/10.1074/jbc.M300073200.

CHAPTER FOUR

The Emerging Role of Vitamin B6 in Inflammation and Carcinogenesis

Ranjana P. Bird[1]
School of Health Sciences, University of Northern British Columbia, Prince George, BC, Canada
[1]Corresponding author: e-mail address: ranjana.bird@unbc.ca

Contents

Abstract

Vitamin B6 serves as a coenzyme catalyzing more than 150 enzymes regulating metabolism and synthesis of proteins, carbohydrates, lipids, heme, and important bioactive metabolites. For several years vitamin B6 and its vitamers (B6) were recognized as antioxidant and antiinflammatory and in modulating immunity and gene expression. During the last 10 years, there were growing reports implicating B6 in inflammation and inflammation-related chronic illnesses including cancer. It is unclear if the deficiency of B6 or additional intake of B6, above the current requirement, should be the focus. Whether the current recommended daily intake for B6 is adequate should be revisited,

Advances in Food and Nutrition Research, Volume 83
ISSN 1043-4526
https://doi.org/10.1016/bs.afnr.2017.11.004

since B6 is important to human health beyond its role as a coenzyme and its status is affected by many factors including but not limited to age, obesity, and inflammation associated with chronic illnesses. A link between inflammation B6 status and carcinogenesis is not yet completely understood. B6-mediated synthesis of H_2S, a gasotransmitter, and taurine in health and disease, especially in maintaining mitochondrial integrity and biogenesis and inflammation, remains an important area to be explored. Recent developments in the molecular role of B6 and its direct interaction with inflammasomes, and nuclear receptor corepressor and coactivator, receptor-interacting protein 140, provide a strong impetus to further explore the multifaceted role of B6 in carcinogenesis and human health.

1. INTRODUCTION

Vitamin B6 is a molecule of complexity and an essential nutrient for maintaining human health. During the last few decades greater attention is being paid to the antiinflammatory role of vitamin B6. Moreover, several studies alluded to a protective role for B6 in inflammation-related chronic illnesses including cancer. The main thrust of this chapter is to discuss the role of B6 in inflammation and carcinogenesis. Recent studies have provided further evidence supporting a role for B6 as a modifier of molecular events by directly influencing the posttranslational modification of key proteins, involved in inflammation, metabolism, and immunomodulation. An overview on B6 and its emerging role in inflammation and carcinogenesis follows.

1.1 Discovery of Vitamin B6

During the 1930s the structure of vitamin B6 and its role as the "rat pellagra prevention factor" were identified and Paul György named the vitamin pyridoxine due to its structural homology to pyridine (György, 1938; György & Eckardt, 1939). That same year pyridoxine was chemically synthesized (Harris & Folkers, 1939). The historical aspects of vitamin B6 discovery are detailed by Rosenberg (2012).

Later studies showed that vitamin B6 could exist in other chemical forms that differ from pyridoxine by a variable group present at the 4′-position. Pyridoxine (PN) carries a hydroxyl, pyridoxal (PL), an aldehyde, and pyridoxamine (PM), an amino group (Merrill, Henderson, Wang, McDonald, & Millikan, 1984). While all three species can be phosphorylated, it is pyridoxal 5′-phosphate (PLP) that is the biologically most active form and used as a cofactor for many important enzymatic reactions (Ink & Henderson, 1984; Jang et al., 2003; Merrill et al., 1984). In this chapter, vitamin B6 and its vitamers are referred to as B6 unless specified otherwise.

1.2 Requirement, Food Sources, Absorption, and Metabolism of Vitamin B6

The requirement for vitamin B6 depends on the age and physiological state of an individual (Morris, Sakakeeny, Jacques, Picciano, & Selhub, 2010; Stover & Field, 2015) and ranges between 0.1 mg/day (infant) and close to 2 mg/day (pregnancy). Many chronic illnesses may increase the requirement for B6 (Hellmann & Mooney, 2010; Lotto, Choi, & Friso, 2011; Mooney, Leuendorf, Hendrickson, & Hellmann, 2009; Paul, Ueland, & Selhub, 2013; Shen, Lai, Mattei, et al., 2010; Ulvik, Midttun, Pedersen, Nygård, & Ueland, 2012; Ulvik et al., 2014). Meat, fish, and poultry are good sources of B6. PL and PN are the most absorbable forms by the small intestinal tract by facilitated transport with the main circulating forms, PLP and PL, being bound to serum albumin. Liver and intestinal tract are the major sites for B6 metabolism (Albersen et al., 2013; Merrill et al., 1984; Rose, 1980). B6 forms present in mammalian tissue are interconvertible involving kinases, phosphatase, and oxidases (Wilson & Davis, 1983). Catabolism of B6 is also an important aspect for cellular homeostasis of the compound (Schaeffer, Sampson, Skala, Gietzen, & Grier, 1989). A critical step is represented by the dephosphorylating of PLP/PMP/PNP because this step represents a major control for the pool of the available active B6 cofactor (PLP). Human pyridoxal phosphate phosphatase is expressed in various tissues but predominantly in the brain, liver, and testis (Jang et al., 2003). The B6-specific phosphatase has its highest affinity for PLP, followed by PNP and then PMP. Pyridoxal is oxidized to 4-pyridoxic acid, the major excretory form of B6. A significant proportion of B6 (ranging from 5% to 80% of the total B6 content) in many fruits and vegetables is glycosylated. Glycosylated B6 appears to be abundant in plants and is present in soybean, rice, and Ginkgo. Furthermore, in fungi, β-fructosyl and β-galactosyl compounds of pyridoxine have been found. It is anticipated that these derivatives of B6 are not substrates of a pyridoxine kinase; however, specific β-glucosidases are present in humans capable of removing the sugar moiety, making the vitamin accessible for salvage pathway enzymes (Mackey, McMahon, Townsend, & Gregory, 2004). There are factors present in our diet which can affect B6 availability (Chang, 2000; Klosterman, 1974; Mayengbam, Raposo, Aliani, & House, 2016; Reynolds, 1988).

1.3 Indicators of B6 Status

Due to the emerging novel role of B6 in chronic diseases, measurement of B6 status in population is an important and continuously evolving field. A detailed and extensive review on measuring B6 nutritional status has been discussed recently (Ueland, Arve Ulvik, Luisa Rios-Avila, Midttun, & Gregory, 2015). There are direct and functional biomarkers of B6 nutritional status. Nutritional status normally measures intake through food, supplemental use, bioavailability, metabolism, and excretion. There are several factors that could affect B6 status and may confound the measures of B6 status. These factors include age, sex, certain drugs, use of contraceptives, inhibitors, or antagonists of B6 in foods, pregnancy, low serum albumin, alkaline phosphatase in plasma, alcohol intake, coffee, smoking, aging, and number of diseases (Chiang et al., 2005; Ueland et al., 2015; van den Berg, Bode, Mocking, & Löwik, 1990).

Direct measures of B6 status include quantification of plasma, PL, PLP, and PA, or PLP level in erythrocytes and excretion of PA in urine. There has been some concern, however, about the reliability of these markers (Ueland et al., 2015). Functional biomarkers of B6 status measure metabolites or enzymes requiring B6 as coenzyme for their production or their respective activities *are* discussed in detail by Ueland et al. (2015). Novel markers of B6 status have been proposed, which utilizes B6-dependent substrate to product ratio or B6 metabolites (Ueland et al., 2015). Several steps in tryptophan catabolism via the kynurenine pathway are vitamin B6 dependent. The ratio of two downstream metabolites, 3-hydroxykynurenine (HK) and xanthurenic acid (XA), i.e., the HK/XA, was recently proposed as an indicator of functional vitamin B6 status. Two other measures include markers of inflammation or B6 catabolism such as substrate to product ratio in plasma kynurenine/tryptophan ratio and the PA/(PL + PLP) ratio (PAr). It is suggested that PAr is a strong indicator of oxidative stress and inflammation.

1.4 Metabolic Role of B6

Nutrition literature and textbooks are replete with information on the metabolism, function, and requirement of vitamin B6. Briefly B6 is an essential nutrient for mammals. For humans, the major sources of B6 are foods and gut microflora. The coenzyme PLP plays a primary role catalyzing more than 160 distinct enzymatic reactions belonging to 5 enzyme classes (oxidoreductases, transferases, hydrolases, lyases, and isomerases).

Many of the PLP-dependent enzymes catalyze important steps in the metabolism of amino acids, carbohydrates, and fatty acids. PLP is also involved in hemoglobin synthesis (Mooney et al., 2009). An important role for PLP-dependent enzyme is in fatty acid metabolism. The enzyme δ-6-desaturase catalyzes the synthesis of essential long-chain polyunsaturated fatty acids by desaturation of linoleic and γ-linolenic acids, the two essential fatty acids (Horrobin, 1993; Nakamura & Nara, 2004; Saareks, Ylitalo, Mucha, & Riutta, 2002; Tapiero, Ba, Couvreur, & Tew, 2010; Tsuge, Hotta, & Hayakawa, 2000). Physiological importance of essential fatty acids has evolved with time and are critical to maintaining cellular integrity, signal transduction, production of important lipid metabolites involved in immune function and inflammation, are critical to human development (Dennis & Norris, 2015; Fullerton, O'Brien, & Gilroy, 2014; Janssen & Kiliaan, 2014; Zhao et al., 2012).

2. ROLE OF B6 IN OXIDATIVE STRESS AND INFLAMMATION

In addition to its coenzyme role as PLP, B6 vitamers are recognized as having biological activity of their own by having a hydroxyl group on 3-position in their pyridine structure serving as potent antioxidants and antiinflammatory molecules (Bierhaus et al., 2000; Lotto et al., 2011; Wondrak & Jacobson, 2012). Recent research on B6 has focused on its role in inflammation-related chronic illnesses (Morris et al., 2010; Sakakeeny, Roubenoff, Obin, et al., 2012; Selhub, Byun, Liu, Mason, & Bronson, 2013). It is recognized that oxidative damage and inflammation are interrelated. Therefore, it is difficult to separate the role of B6 as an antioxidant from its role as an antiinflammatory agent. To provide a context a brief overview of the process which links oxidative stress and inflammation is provided.

2.1 Interrelationship Between Oxidative Stress and Inflammation

Free radicals are defined as molecules with one or more unpaired electrons and normally involved in controlled chain reaction in vivo. Free radicals are highly reactive and if not processed in a controlled manner could have a devastating effect on tissue integrity (Beckman & Koppenol, 1996; Biswas, 2016; Biswas & Halliwell, 1997; Dröge, 2002; Duracková, 2010).

The incomplete reduction of oxygen by one electron producing superoxide anion ($O_2^{\cdot-}$) is the first step for the formation of most other reactive oxygen species (ROS). Many compounds capable of donating or receiving unpaired electron could serve as antioxidants. Normally antioxidants quench uncontrolled free radicals; however, in the absence of such protection, aberrant cellular responses are encountered. There are oxygen-, nitrogen-, and chlorine-free radicals which in turn could generate highly reactive non-radicals. Collectively the free radical and nonradical species are known as ROS and/or reactive nitrogen species (RNS) or species containing active oxygen and nitrogen (RONS) may involve nitrogen, oxygen, and halide in their moiety. The biology and significance of free radicals are complex and have been researched for many years; several excellent reviews are available (Biswas, 2016; Görlach et al., 2015). Oxidative stress has been implicated to many chronic illnesses as well as normal aging. Organs and subcellular compartments differ from each other for their susceptibility to oxidative damage and free radical production; for example, neuronal health depends a great deal on its ability to neutralize free radicals because of their high oxygen demand and metabolic rate (Shefa et al., 2017).

Free radicals and radical species can be generated by various endogenous sources influenced by exogenous factors (Table 1). In mammalian cells ROS are generated in different cellular components such as membranes, cytoplasm, mitochondria, lysosomes, and peroxisomes. Several enzymes are involved in the generation of oxygen-, nitrogen-, and halide-free radicals and could lead to inflammation and oxidant injury a cyclic event, unless

Table 1 Endogenous and Exogenous Inducers of Free Radicals in Human Body

Endogenous	**Exogenous**
Mitochondrial electron	Radiation
Transport system	Drugs
NADPH oxidase	Chemicals
Xanthine oxidase	Heavy metals
Lipoxygenase	Pollutants
Cyclooxygenase	Diet
Cytochrome P450	Infection
Nitric oxide synthase	Smoking

the cycle is interrupted by antiinflammatory agent, antioxidant, or enzymes evolved in mammalian cells to catabolize free radicals (Biswas, 2016; Görlach et al., 2015).

Immune cells involved in cytocidal activities to eliminate foreign cells or foreign agents use oxidative burst and generate free radicals (Scheibye-Knudsen, Fang, Croteau, Wilson, & Bohr, 2015). Consumption of a high caloric diet on a consistent basis increases mitochondrial activities, the major producers of the ROS (Görlach et al., 2015; Traba & Sack, 2017). Oxidative stressors damage macromolecules and induce genomic instability and aberrant induction or interruption in signal transduction pathway (Biswas, 2016) eventually causing tissue injury. Damaged cells attract proinflammatory cells at the site of injury and these cells in turn further increase oxidative stress leading to a chronic state of inflammation that could result in the pathogenesis of many diseases including but not limited to obesity, diabetes, diabetic complications, hypertension, neurodegenerative diseases, rheumatoid arthritis, inflammatory bowel diseases, atherosclerosis, aging, and cancer (Dröge, 2002; Duracková, 2010; Giorgio, Trinei, Migliaccio, & Pelicci, 2007; Nicolussi, D'Inzeo, Capalbo, Giannini, & Coppa, 2017; Valko et al., 2007). B6 is favorably implicated in several chronic diseases/states and have been referred to as antioxidant and/or as antiinflammatory nutrient.

Several clinical trials involving antioxidant nutrients in the prevention of chronic illnesses especially in cancer have yielded conflicting results, giving rise to the concept "Antioxidant Paradox" (Davies, 2016; Draper & Bird, 1987; Galadari, Rahman, Pallichankandy, & Thayyullathi, 2017; Halliwell, 2000). This concept is critical considering that many nutrients could be toxic in excess, and most importantly, free radicals are involved in elimination of pathogens and cancer cells (Schumacker, 2006) mediated via innate immunity or therapeutic drugs.

2.2 Direct Role of B6 in Neutralizing Oxidative Stress and Inflammation

2.2.1 B6 Vitamers and Their Antioxidant Properties

In the B6 moiety the presence of a phenolic hydroxyl group in the 3-position of the pyridine ring imparts it a strong redox reactivity. The reactions between pyridoxine and several oxygen radicals such as hydroxyl ($^{\cdot}OH$), hydroperoxyl ($^{\cdot}OOH$), and superoxide ($O_2^{\cdot-}$)] have been investigated in great detail (Matxain, Ristilä, Strid, & Eriksson, 2006). Based on addition reactions to the aromatic ring and H-atom abstraction by radical mechanisms, B6 reacts with OH, $^{\cdot}$OOH, and $O_2^{\cdot -}$ forming intermediate B6-free radicals producing H_2O, H_2O_2, and HOO, respectively. Pyridoxine is a highly effective antioxidant in chemical and cellular assays. Pyridoxine and PM protected red blood cells from hyperglycemia induced free radical damage and reduced generation of superoxide radical formation (Baynes, 2001; Baynes & Thorpe, 2000; Jain & Lim, 2001; Jain, Lim, Langford, & Jain, 2002). Pyridoxamine also protects cells from advanced glycoxidation and lipoxidation end products (Ahmad et al., 2014; Amarnath, Amarnath, Amarnath, Davies, & Roberts, 2004; Onorato, Jenkins, Thorpe, & Baynes, 2000). One of the mechanisms could be due to the ability of PM to chelate metal ions.

Pyridoxamine has been known to be a metal chelator that preferentially forms complexes with Cu^{2+} and Fe^{3+} ions. Free and inadequately chelated transition metal ions induce lipid peroxidation, and other redox processes including damage to DNA and other macromolecules are involved in human pathologies (Brewer, 2007; Jomova & Valko, 2011; Tabner, El-Agnaf, German, Fullwood, & Allsop, 2005; Valko, Morris, & Cronin, 2005).

Pyridoxamine is an inhibitor of cellular carbonyl stress and therefore could retard progression of human diseases. ROS lead to carbonylations of amino acid residues present in protein, leading to unrepairable damage and accumulation of other toxic products. PM is known to scavenge various toxic carbonyl species derived from either glucose or lipid degradation through nucleophilic addition (Amarnath et al., 2004; Giannoukakis, 2005; Jain & Lim, 2001; Khalifah, Chen, & Wassenberg, 2005; Voziyan & Hudson, 2005a, 2005b). In contrast, Dainin, Ide, Maeda, Suama, and Akagawa (2017) reported that hepatoma cells exposed to PM and H_2O_2 had significantly reduced levels of carbonylated and aggregated proteins, and improved cell survival rates (Dainin et al., 2017), suggesting that PM in vivo could provide protection to cancer cells against oxidative damage induced by therapeutic agents. The relevance of PM in this context remains elusive.

The photosensitizing role of B6 has been reported. B6 vitamers have 3-hydroxypyridine as core moiety with photodynamic activity (Wondrak, Cervantes-Laurean, et al., 2002; Wondrak, Roberts, Jacobson, & Jacobson, 2002, 2004). Human skin contains B6 in a significant amount which may act as endogenous skin photosensitizers and thus could enhance the phototoxicity of UV radiation (Coburn et al., 2003; Sasaki et al., 1993; Wondrak et al., 2004). Indeed, increased photosensitivity is reported in humans ingesting high dosages of vitamin B6 (Maeda et al., 2000; Murata et al., 1998).

2.2.2 B6 and Inflammasomes

Inflammasomes are nucleotide-binding oligomerization domain leucine-rich repeat and pyrin domain-containing proteins (NLRPs). Regulation of NLRPs in the mammalian system is complex and is involved in a number of inflammation-associated human diseases (Chen & Núñez, 2011; Jha, Brickey, & Ting, 2017; Kolb, Liu, Janowski, Sutterwala, & Zhang, 2011; Menu & Vince, 2013; Rauch et al., 2017; Vandanmagsar et al., 2011). The role of nutrients in modulating inflammasomes remains an important area considering their role in inflammation and in several chronic illnesses. Inflammasomes are multiprotein complexes which are formed mainly to activate inflammatory caspases which in turn leads to maturation of proinflammatory cytokines such as IL-1beta and IL-18. Each inflammasome normally consists of three proteins: the platform, the adapter, and the effector. The platform protein determines the structure of inflammasomes with respect to the type of adapter or effector protein that will be recruited and subsequently determining its function. Therefore, a number of inflammasomes have been identified of which NLRP3 is most studied with respect to its role in inflammation and carcinogenesis (Karki, Man, & Kanneganti, 2017; Lin & Zhang, 2017; Pellegrini, Antonioli, Lopez-Castejon, Blandizzi, & Fornai, 2017; Petrilli, 2017; Zmora, Levy, Pevsner-Fischer, & Elinav, 2017). The inflammasome, NLRP3, is involved in the activation of a pleiotropic inflammatory cytokine IL-1beta, which is primarily produced by myeloid cells such as monocytes and macrophages. IL-1beta induces the proliferation and/or production of other inflammatory cytokines, leukocyte adhesion molecules, or acute-phase proteins in leukocytes, myeloid cells, endothelial cells, hepatocytes, and adipocytes. IL-1beta is synthesized as a precursor form (pro-IL-1beta) that is proteolytically processed into a mature form to gain biological activity (Man & Kanneganti, 2015). The synthesis of pro-IL-1beta is mainly regulated by activation of NF-κB, triggered by several factors including ROS. The conversion of pro-IL-1beta to active form is catalyzed by several proteases. Caspase-1 is the most important among the proteases, and it is synthesized as an inactive precursor form which is activated by autoprocessing.

It is demonstrated that depending on the activators and the tissue type, NLRP3 also inhibits tumor growth by activating caspase-8 and affects intestinal cell homeostasis (Khare, Luc, Dorfleutner, & Stehlik, 2010; Rauch et al., 2017). Inflammasomes could possess different structures and functions in hematopoietic and nonhematopoietic cells (Bakele et al., 2014) and play diverse roles in the development of inflammation-related pathologies.

A recent study (Zhang et al., 2016) reported that vitamin B6 inhibited NLRP3 activation and reduced the maturation of caspase-1 and proteolytic maturation of IL-1beta. Peritoneal macrophages were incubated with lipopolysaccharide to create a proinflammatory environment which in turn induced the synthesis and intracellular accumulation of pro-IL-1beta. These macrophages were treated with vitamin B6 vitamers and then treated with ATP to activate inflammasome (Zhang et al., 2016). Under the experimental conditions, production of mature caspase-1 and the secretion of the mature form of IL-1 were reduced by PL and PLP but not by PN and PM. In the same study (Zhang et al., 2016), it was shown that IL-1beta secretion, induced by other NLRP3 activators, was also inhibited by PL and PLP. In addition to IL-1beta, secretion of IL-18, a proinflammatory cytokine, which, like IL-1beta, undergoes proteolytic maturation catalyzed by caspase-1, was also suppressed by PL and PLP but not by PN or PM. This is an exciting observation and provides a strong impetus to investigate the expanding role of B6 beyond its role as a coenzyme, in human physiology and metabolism.

These findings add to a new molecular role of B6 in alleviating inflammation by interfering the processing of inflammatory cytokines and possibly with the assembly and function of inflammasomes. The mechanism by which PL or PLP affects inflammasome remains an important area to be explored (Fig. 1). Whether PL and PLP are effective in inflammasome inactivation in immune as well as in nonimmune cells and in cells which reside in tumor microenvironment will be of importance in future studies.

2.2.3 B6 a Modulator of Nuclear Receptors, Corepressor, and Coactivators of RIP140

Cellular responses originate from a coordinated interaction among cell surface receptors, nuclear receptors, and signaling molecules. Nuclear receptors regulate reproduction, development, and metabolism. Receptor-interacting protein 140 (RIP140), also known as nuclear receptor-interacting protein 1 (NRIP1), was first discovered in human breast cancer cells and found to interact with estrogen receptor and later with several nuclear receptors (Christian, White, & Parker, 2006; Fritah, Christian, & Parker, 2010; Lee & Wei, 1999; Leonardsson et al., 2004). It is known that RIP140 could function as a corepressor as well as as a coactivator of nuclear receptors and transcription factors. RIP140 is a multifunctional coregulator involved as a global regulator of energy homeostasis with a central role in metabolically active tissues such as adipose, muscle, and liver. It is implicated in metabolic

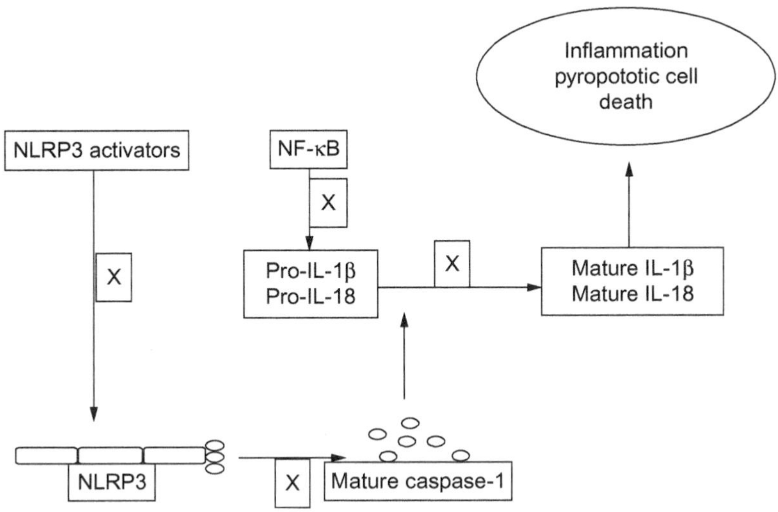

Fig. 1 Inflammasome such as NLRP3 is activated by a number of factors. Once NLRP3 is activated it releases the mature caspase-1 which in turn is involved in the activation of pro-interleukin-1 beta (pro-IL-1beta) and pro-interleukin-18 (pro-IL-18) to mature IL-1beta and IL-18. These cytokines are highly proinflammatory and have been implicated in inflammation-related disorders, proliferation, and pyroptotic cell death. Pyridoxal phosphate (PLP) and pyridoxal (PL) have been shown to inactivate NLRP3 and reduce production of caspase-1, IL-1beta, and IL-18. The symbol X shows where B6 could intervene the process possibly by modifying the assembly of inflammasome by posttranslational modification or by interfering with the access of NLRP3 activators to inflammasome.

syndrome, inflammation, and cancer (Kumar & Lombard, 2015). It binds with nuclear receptors and represses genes implicated in glucose oxidation, glycolysis, fatty acid oxidation, TCA cycle, and mitochondrial biogenesis (Christian et al., 2006). A great deal of work has been conducted in adipose tissue, liver, and muscle. RIP140-null mice do not gain weight on a high-fat diet and are protected from hepatic steatosis. Moreover, RIP140 depletion leads to increase glucose uptake, beta oxidation, mitochondrial biogenesis, O_2 consumption, and breakdown of glucose into CO_2. RIP140 has four repressive domains. As shown in Fig. 2, RIP140 corepresses or coactivates nuclear receptors and E_2F transcription factors, involved in cell cycle and energy metabolism and disease development (Crosby & Almasan, 2004; Hsu & Sage, 2016). NF-κB, a proinflammatory transcription factor, could suppress or enhance tumorigenesis (Hu et al., 2017; Lawrence, Gilroy, Colville-Nash, & Willoughby, 2001) and Wnt signaling, a molecule associated with several disease states including cancer, especially in colorectal cancer with

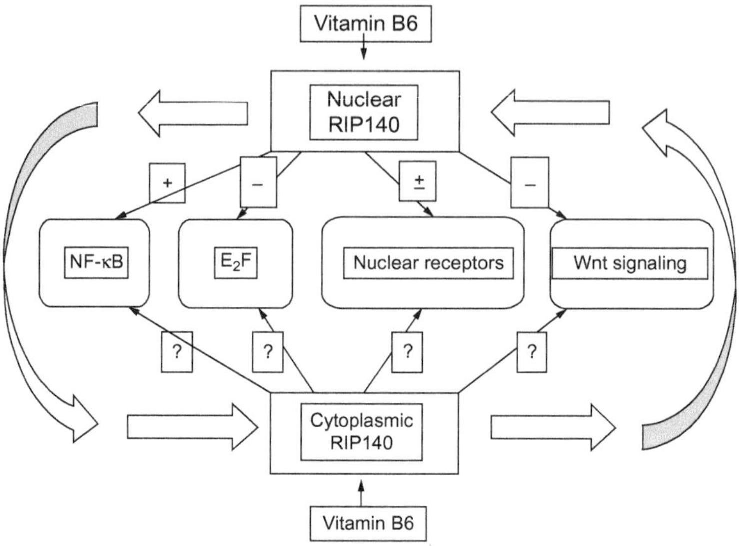

Fig. 2 The nuclear interactive protein 140 is a corepressor as well as a coactivator depending on its posttranslational modification and cellular localization. It corepresses or coactivates nuclear receptors (NR), coactivates nuclear factor κB (NF-κB), corepresses E_2F, cell cycle-related transcriptional factors, and Wnt signaling involved in a number of pathologies including cancer. B6 such as PLP conjugates with corepressor nuclear interactive protein 140 (RIP140). The symbol X denotes the steps where the role of B6 is not known fully. It is postulated that pyridoxilated RIP140 is retained in the nucleus, has enhanced corepressive activity, and thus could further repress E_2F transcription factor which is a positive regulator of cell proliferation and glycolytic pathway. RIP140 is also known to interfere with mitochondrial function and cell viability in cardiac cells. In cardiac cells, it binds with peroxisome proliferator-activated receptor gamma coactivator-1 alpha (PGC-1 alpha) and represses the mitochondrial bioenergetic. The effect of pyridoxilated RIP140 on its translocation to cytoplasm or on the function of already acetylated or phosphorylated RIP140 remains unanswered.

adenomatous polyposis coli (APC) mutation. These molecules have been studied extensively in the pathogenesis of various diseases especially in inflammation and cancer (Augereau et al., 2006; Blanchet et al., 2011; Cavaillès & Lapierre, 2014; Docquier et al., 2010; Görlach et al., 2015; Hallstrom, Mori, & Nevins, 2008; Lawrence et al., 2001; Lapierre et al., 2014; Major et al., 2017; Palomer et al., 2011; Vazquez-Santillan, Melendez-Zajgla, Jimenez-Hernandez, Martínez-Ruiz, & Maldonado, 2015).

RIP140 affects mitochondrial dysfunction by repressing sirtuin 3 (deacetylase) mitochondrial dysfunction in cardiomyocytes. By gain and loss of functional studies, it was demonstrated that RIP140 increased oxidative stress, decreased mitochondrial membrane potential and thus decreased ATP production and decrease in mitochondrial DNA-encoded genes

(You et al., 2017). Therefore, RIP140 could mitigate a negative effect on cellular metabolism. Not surprisingly, there is substantial evidence documenting involvement of RIP140 in tumorigenesis; it has been implicated in tumor suppression as well as rendering tumors a more aggressive phenotype. In the case of colorectal cancer, RIP140 expression is reported to correlate with the expression of tumor suppressor APC protein in 400 cancer biopsies. Most importantly, it was noted that the increased level of RIP140 in tumor tissue increased patients' survival. Role of RIP140 in maintain gut homeostasis its ability to upregulate adenomatous coli gene and in gut inflammation has been reported (Cavaillès & Lapierre, 2014). In breast cancer RIP140 is significantly associated with the malignant breast cancer and was highly expressed in the nucleus. In the benign mammary tumors, RIP140 was highly expressed in the cytoplasm and in the stromal component of the tumors. In breast cancer cell lines, downregulation of RIP140 resulted in increased apoptosis (Docquier et al., 2010). In leukemic cells, increased RIP140 is associated with favorable prognosis and it also inhibits the growth of hepatoma (Hu et al., 2017; Lapierre, Castet-Nicolas, et al., 2015; Lapierre, Docquier, et al., 2015). Posttranslational modification of RIP140 is an important mechanism by which RIP140 is retained in the nucleus and its corepressor or coactivator function can be augmented (Huq & Wei, 2005; Rytinki & Palvimo, 2008).

Huq, Tsai, Lin, Higgins, and Wei (2007) investigated whether PLP could conjugate with RIP140 and alter its activity by posttranslational modification. They demonstrated that PLP conjugated with RIP140 and identified the modification site to be Lysine 613 of the protein. They further demonstrated that this modification enhanced RIP140's transcriptional corepressive activity and its physiological function in adipocyte differentiation. The authors proposed that this effect is due to an increased interaction of RIP140 with histone deacetylases and nuclear retention of RIP140.

Posttranslational modification of RIP140 modulates its corepressor or coactivator activities as well as its translocation to cytoplasm from nucleus. Posttranslational modification of RIP140 could make it more accessible to interact with other proteins and exert its biological effects. It is known that posttranslational modification of RIP140 is an important mechanism by which it is exported to cytoplasm from nucleus and exerted a protective effect against the neuronal death by apoptosis (Feng et al., 2014).

The possibility that B6 could modulate transcriptional activities of nuclear receptors by posttranslational modification of RIP140 remains an important area for exploration. Whether pyridoxilated RIP140 is more prone to be retained in the nucleus and whether it exerts a regulatory role in modulating corepressive or coactivating role of RIP140 and its ability to translocate from nucleus to cytoplasm during inflammation and carcinogenesis remain to be determined and this concept is presented in Fig. 2. It is also possible that pyridoxilation of RIP140 affects further posttranslational modification of rest of the RIP140 moiety. Clearly availability of PLP in a specific tissue to bind with RIP140 and its translocation across nuclear or mitochondrial membrane will be important for its biological activity. One other aspect of RIP140 is that it interacts with sirtuins for exerting its diverse metabolic responses. Sirtuins are family of NAD-dependent deacetylases, involved in posttranslational modification of proteins and histones and thus involved in the development of various diseases (Huang et al., 2016). In this regard sirtuin 3 has received a great deal of attention as it affects nutrient metabolism, ATP balance, antioxidant machinery, and other mechanisms fundamental to mitochondria and is closely associated with the pathogenesis of several disorders and could play an opposite role in carcinogenesis. It promotes and maintains or suppresses the malignant phenotype depending on the type of cancer (Miyo et al., 2015; Xiong, Wang, Zhao, Han, & Jia, 2016; You et al., 2017). The underlying mechanisms involve depletion of intracellular ROS, modulation of metabolic reprogramming, and regulation of intracellular signaling responsible for cellular growth and death. It is anticipated that sirtuin family members will be affected by PLP either directly or via the ability of PLP to pyridoxylate RIP140 and enhance its repressive activity (Wei, Hu, Chandra, Seto, & Farooqui, 2000).

There are synthetic inhibitors of deacetylases with pyridine ring (Cho et al., 2013), alluding to the possibility that PN or PLP could serve to inhibit deacetylases in vivo. We have observed that in Zucker obese rats, a high PN diet reduced liver hypertrophy and lipid accumulation compared to those rats fed a normal level of pyridoxine and the high B6 group fed five times the normal B6 required by rodent (35 mg vs 7 mg/kg diet) had increased nuclear levels of acetylated histones 3 and 4 (H3 and H4) compared to normal B6-fed animals (Naahidi, 2007).

3. INDIRECT ROLE OF B6, ACTING AS A COENZYME AND PRODUCTION OF BIOLOGICALLY ACTIVE H_2S AND TAURINE

Vitamin B6 is involved in the production of important metabolites required by cells to perform vital function. The newly discovered gasotransmitters and taurine are derived from sulfur amino acids utilizing

PLP-dependent enzymes and are involved in inflammation and chronic illnesses. Both metabolites have overlapping functions. They both protect mitochondria from toxic insults. The role of mitochondria in human health and disease is a promising field for future research (Berridge, Dong, & Neuzil, 2015; Berridge & Neuzil, 2017; Cagin & Enriquez, 2015; Kabekkodu et al., 2015; Tokarz & Blasiak, 2014).

Hydrogen sulfide is a gasotransmitter produced from sulfur amino acids requiring mainly two PLP-dependent enzymes cystathionine beta synthase (CBS) and cystathionine γ lyase as well as another enzyme 3-mercaptopyruvate sulfurtransferase (Kimura, 2010). Synthesis of H_2S is regulated by PLP-dependent enzymes and significantly affected by B6 status (Gregory, DeRatt, Rios-Avila, Ralat, & Stacpoole, 2016). Methionine is the main amino acid which is converted into homocysteine and then cysteine; both homocysteine and cysteine serve as the precursor amino acid giving rise to H_2S. Biogenesis and physiological role of H_2S have been reviewed previously and the readers are referred to a number of publications for details (Kabil & Banerjee, 2010; Li, Rose, & Moore, 2011; Rose, Moore, & Zhun Zhu, 2017; Singh & Banerjee, 2011; Wang, 2012). The diverse molecular role of H_2S is complex and beyond the scope of this chapter. However, a summary, relevant to the focus of this chapter, follows.

H_2S is known for its toxic effects for several centuries; it is recently recognized that endogenously produced hydrogen sulfide (H_2S) mediates diverse biological responses at concentrations ranging between ten and several hundred micromolar levels (Singh & Banerjee, 2011). H_2S is anti- as well as proinflammatory, involved in vasodilation, angiogenesis, neurotransmission, apoptosis, cardioprotection, and insulin release. Its biological activity depends on its production and eventually the concentration at the target site. It is generally agreed upon that the level of this gas is very low ~100 pM in blood to ~15 nM in tissues other than aorta, where the concentration is 20- to 100-fold higher than in other tissues. Once H_2S is produced in the cell, it can affect cellular signaling or its sulfur group can be stored in the bound form as acid-labile sulfur or in the form of sulfane sulfur and could be released later as S^{2-}, HS^-, or H_2S. It is observed that at the physiological pH of 7.4 hydrosulfide ion is threefold higher than H_2S, suggesting that hydrosulfide ion could mediate a majority of biological responses.

H_2S has a growth regulatory effect (Baskar & Bian, 2011); it interacts with other two gasotransmitters, NO and CO, in exerting its responses and transmitting redox signals and affects, mitochondrial bioenergetics, cancer, cardiovascular system, inflammation-related pathologies, and glucose homeostasis (Flannigan et al., 2014; Gai et al., 2016; Hellmich & Szabo, 2015; Kabil, Motl, & Banerjee, 2014; Kabil, Vitvitsky, & Banerjee, 2014; Pichette & Gagnon, 2016; Szabo et al., 2014; Vicente et al., 2016). H_2S is hydrophobic and could diffuse through lipid membrane and exerts biological responses in surrounding cells via a paracrine mechanism (Cuevasanta, Möller, & Alvarez, 2017; Kimura, 2010).

The role of H_2S in maintaining mitochondrial health deserves a special attention. H_2S is produced in mitochondria and is associated with mitochondrial biogenesis and gene expression. The steady-state concentration of H_2S is governed by its rate of production, storage, and utilization and by its rate of removal by the mitochondrial sulfide oxidation pathway coupled to oxidative phosphorylation. H_2S can also be removed or sequestered via reaction with disulfides or with heme proteins (Rose et al., 2017).

Studies have reported the increased level of H_2S-producing enzymes in cancer cells. Selected cancer types such as ovarian and pancreatic cancer exhibit overexpression of CBS and produce increased amounts of H_2S. H_2S enhances tumor growth and spread by stimulating cellular bioenergetics, proliferation, angiogenesis, migration, and invasion by affecting underlying signaling pathways. In preclinical models of cancers, pharmacological inhibition or genetic silencing of CBS suppressed cancer cell bioenergetics in vitro and inhibited tumor growth and metastasis in vivo. However, the observation that inhibition of H_2S biosynthesis exerts anticancer effects is contradicted by other studies, showing that increasing H_2S with exogenous donors exerted antitumor actions. Increased expression of CBS is considered a prognostic marker associated with cancer. An increased expression of CBS, in ovarian cancer cells, has been linked to lipogenesis and activation of transcription factor, steroid receptor element-binding protein 1 (Hellmich & Szabo, 2015). It is postulated that H_2S may be involved in post-translational modification of such proteins by glutathionylation or sulfhydration suggested previously (Paul & Snyder, 2012). H_2S could also enhance tumorigenesis by protecting mitochondria and by enhancing mitochondrial biogenesis in tumors. Whether B6 affects H_2S-mediated changes in tumor biology or CBS expression remains to be determined. H_2S is implicated in promoting tissue repair and attenuation of inflammation. In experimental model of colitis, H_2S is produced throughout the gastrointestinal tract, and its synthesis is markedly increased following mucosal injury (Flannigan et al., 2014).

Hydrogen sulfide (H_2S) has also emerged as an important regulator of glucose homeostasis. This is achieved through its production and action in several metabolic and hormone-producing organs including the pancreas, liver, and adipose (Pichette & Gagnon, 2016). Many studies exploring the role of H_2S in modulating physiological responses have used exogenous sources of H_2S or have targeted H_2S-producing enzymes. Recently, it was reported that H_2S production depended on the availability of PLP and cysteine (Gregory et al., 2016). What remains to be determined is how PLP affects production of H_2S in an organ-specific manner including in subcellular organelles especially in mitochondria, which in turn instigate diverse biological responses. Whether physiologically produced H_2S from sulfur amino acids affects inflammation and tumorigenesis and if B6 status has an important role needs further clarification.

The second metabolite that has received considerable attention as an antioxidant is taurine. Taurine is a semiessential amino acid that is abundant in inflamed tissue. It is synthesized from methionine involving B6 as coenzyme in adult liver. Neonates are unable to synthesize taurine, whereas adult liver's ability to synthesize taurine declines with age (Jacobsen & Smith, 1968; Kendler, 1989; Wright, Tallan, Lin, & Gaull, 1986). Taurine, an antioxidant and antiinflammatory amino acid, is implicated in obesity, energy, and lipid metabolism affecting central nervous system, adipose tissue, liver, and skeletal muscles (Kim & Cha, 2014; Marcinkiewicz & Kontny, 2014; Murakami, 2015; Ueki & Stipanuk, 2009). The abundance of taurine at the site of inflammation is purported to neutralize ROS produced by phagocytes and neutrophils using myeloperoxidase system to combat infection and to eliminate foreign cells. The myeloperoxidase system uses chloride and bromide to produce highly toxic HOCl and HOBr, which eliminate foreign objects and cause tissue damage and inflammation.

Experimental and clinical evidence supports the concept that taurine could exert multiprong effects to protect tissue from damage and neutralize toxic HOCl and HOBr. Taurine forms TaurineCl (TauCl) and TaurineBr (TauBr). Taurine also induces superoxide dismutase, glutathione peroxidase, and catalase, the enzymes required for detoxifying ROS. TauCl and TauBr could act as antimicrobial as well as antiinflammatory. TauCl and TauBr, by being antimicrobial, decrease the activities of phagocytic cells and production of ROS, cytokines, nitric oxide, and PGE2. These compounds also induce apoptosis in neutrophils, thus reducing further increase in ROS. TauCl and TauBr reduce production of COX2, iNOS, and heme proteins by increasing heme oxygenase activity and increasing the degradation of heme (Marcinkiewicz & Kontny, 2014).

Taurine affects mitochondrial integrity; it can be transported to mitochondrial compartment or synthesized from cysteine within mitochondria by two thiol dioxygenases, cysteine dioxygenase and 2-aminoethanethiol dioxygenase. Taurine is a constituent of mitochondrial tRNA implicated in a number of human diseases (Stipanuk, Dominy, Lee, & Coloso, 2006; Suzuki, Suzuki, Wada, Saigo, & Watanabe, 2002). It is well known that mitochondria are the cellular powerhouse and their integrity is critical for human health (Hansen, Birkedal, Wibrand, & Grunnet, 2015; Scheibye-Knudsen et al., 2015; Suzuki, Nagao, & Suzuki, 2011a, 2011b; Tokarz & Blasiak, 2014). The role of B6 in sulfur amino acid metabolism, regulation, and production of taurine and H_2S in various tissues and organs and in maintaining mitochondrial metabolism and integrity remains important future research directions. Significant advances have been made in understanding the molecular processes underpinning many pathological states. The emerging molecular role of B6 in modulating the composition and activity of immune cells, RIP140, and inflammasome asserts its importance beyond its role as a vitamin. It is becoming apparent that B6 performs various roles by locating itself in subcellular compartments such as mitochondria and nucleus. This *emphasizes* the importance of understanding the transport mechanism(s) by which B6 enters various subcellular compartments as well as the process which allows the intracellular translocation of B6-conjugated proteins (Whittaker, 2016).

4. B6 AND CARCINOGENESIS

The healing and medicinal properties of food have been recognized for centuries and well captured and recognized by Hippocrates in the quote "*Let food be thy medicine and medicine be thy food*," Hippocrates, 460 BC. The concept that several cancer/chronic diseases are preventable is not new and widely accepted. Foods and nutrient intake constitute an important human environment. The search for causative and preventive novel compounds in our foods has occupied many scientists for decades.

It has been difficult to discern cancer-preventive properties of foods or nutrients because of the complex composition of foods containing nutrients as well as nonnutrient bioactive components. Furthermore, nutrients function in a connected and interconnected manner. Marginal deficiency or excessive intake of one nutrient could tip the metabolic balance and disorganize other nutrients in performing their roles. Moreover, chronic diseases are complex and multifactorial and multistep in nature making it difficult to establish a clear link between dietary factors and the development and prevention of chronic diseases.

Epidemiological studies have linked vitamin B6 with chronic illnesses including cancer. Oxidative stress and inflammation have been purported to be involved in the genesis of obesity, neurodegenerative diseases, diabetes, metabolic syndrome, and cancer. There is evidence that obesity, diabetes, and metabolic syndrome increase the risk of cancer. B6 is known for decades as an antioxidant and antiinflammatory agent and B6 or B6-dependent metabolites have been repeatedly shown to reduce inflammation. During the last few decades, there appears to be heightened interest in understanding the role of B6 in carcinogenesis. A number of epidemiological studies have linked B6 with cancer risk. However, it is unclear if a low B6 level noted among the individuals with cancer is due to the disease and not a causative factor. Cancer as a disease develops through a multistep process involving initiation and postinitiation stages. The question that remains unanswered is: Does low B6 enhance tumorigenesis or a high B6 intake prevents tumorigenesis? A brief overview of the concept of multistep process of carcinogenesis is provided. The overview is simplistic and does not intend to cover the extensive and continuously expanding complex field of cancer biology and carcinogenesis.

4.1 Multistep Process of Carcinogenesis and Involvement of Nutrients

Carcinogenesis is a multistep process which consists of the emergence of transformed cells, some of which are sequentially selected to expand clonally, acquiring new genotypic and phenotypic features compatible with autonomous growth and malignant phenotype (Barcellos-Hoff, Lyden, & Wang, 2013; Loeb & Harris, 2008). It is believed that many solid tumors are preventable and dietary constituents play a pivotal role in the prevention. The concept of cancer prevention is to eliminate or retard the growth of preneoplastic/precancerous lesions. It is also recognized that tumor macro- and microenvironment are critical to its growth. Tumor macroennvironment constitutes host's physiological and nutritional state and microenvironment constitutes cells present in tumor mass which are blood cells, blood vessels, fibroblasts, adipose tissues, and immune cells of various types. These cells profoundly affect tumor growth and are responsible for cancer-associated inflammation (Brock et al., 2015; Quail & Joyce, 2013). Tumor cells as well as host's nutritional state modulate the microenvironment by influencing the composition and behavior of the immune cells.

A limited knowledge exists about the nutritional requirement of tumors. Nevertheless, it is logical to assume that growing tumors will require essential nutrients like developing tissue or organism for their growth; moreover, it is known that tumors reprogram their metabolic machinery to ensure their survival in an environment which may or may not be conducive to their growth (Egeblad et al., 2010).

There is sufficient evidence supporting the contention that diet and nutrients play an important role in the causation and prevention of various types of cancer (Ames, 2001; Ames & Wakimoto, 2002; Ashkenazi et al., 2008; Ross & Milner, 2007; Yang, 2015). Nutrients, caloric intake, and food constituents influence several processes at the genomic and epigenomic levels (Görlach et al., 2015) including but not limited to, carcinogen metabolism, DNA repair, cell proliferation, apoptosis, inflammation, immunity, differentiation, angiogenesis, hormonal regulation, and cellular energetic within an organ and in developing tumor.

Vitamin B6 serves as a coenzyme in several metabolic reactions producing important metabolites and energy vital to growing tissue. Therefore, one would expect that a marginal deficiency of B6 could adversely affect tumor growth so that additional B6 will be needed by *the* growing tumors to sustain its growth. At the same time, if inflammation is the underlying facilitator of carcinogenesis, then B6 could be important in mediating an anticarcinogenic effect.

4.2 Epidemiological Studies Linking B6 to Human Cancer

The concept that vitamin B6 may have a role in carcinogenesis emerged from studies in 1960 and the 1970s when it was reported that plasma PLP levels were lower in cancer patients and that the level further decreased when the disease progressed. Choi and Friso (2012) reviewed the epidemiological, experimental, and mechanistic evidence linking B6 with cancer. A recent publication reported findings of the meta-analyses of a number of epidemiological studies to assess a link between vitamin B6 and cancer outcome (Mocellin et al., 2017). In total, 121 observational and 9 randomized controlled trials were assessed for 19 tumor sites. A higher intake of dietary vitamin B6 was linked to significantly lower the risk of five cancer sites which included esophageal, pancreatic, gastric, colorectal, and breast cancer.

The meta-analyses also revealed that the total intake of B6, which consisted of dietary and supplemental vitamin B6 intake, was weakly associated with cancer risk and limited to three cancer sites—colorectal, gastrointestinal, and urinary tract. The trend from meta-analysis showed a statistically significant inverse dose–response relationship between vitamin B6 intake and all cancer types (6% risk reduction per milligram of vitamin daily intake). Subgroup analysis by tumor site led to the identification of a statistically significant association with colorectal, pancreatic, and lung cancers. The most definitive inverse relationship between B6 and risk of cancer was observed for colorectal cancer. A recent evaluation of four prospective studies along with the roles of three different markers of B6 status on CRC outcome (Gylling et al., 2017) indicated a significant inverse relationship between plasma PLP and risk of CRC.

4.3 B6 and Cancer: Evidence From Studies in Preclinical Models

A handful of studies conducted on animal models explored the role of vitamin B6 in carcinogenesis and alluded to the involvement of vitamin B6 in this complex disease.

An early study reported that experimentally induced hepatomas in a rat model were unable to synthesize coenzymatically active PLP from inactive precursor forms, indicating that the metabolism of vitamin B6 was markedly different in the hepatomas than control livers (Thanassi and Meisler, 1990; Thanassi et al., 1981). In the same animal model, it was demonstrated that the duration of feeding a B6-deficient diet was important. The growth of inoculated Morris hepatomas was significantly lower when inoculated at day 31, but not at 0, 7, and 15 days, after initiation of vitamin B6-deficient diet, suggesting that B6 deficiency impaired either general animal metabolism or the ability of the transformed cells to proliferate (Tryfiates, 1977; Tryfiates & Morris, 1976; Tryfiates et al., 1991).

In another study, mice were fed for 4 weeks with PN at 0.2, 1.2, 7.7, or 74.3 mg/kg of diet (Gridley et al., 1987). The tumor volumes were found to be lower in mice fed 7.7 and 74.3 mg pyridoxine compared with those fed 1.2 mg PN. However, mice fed a diet with 0.2 mg pyridoxine had the lowest tumor volume supporting the contention that B6 deficiency may lead to impaired ability of cancer cells to grow in vivo.

A substantially different approach was taken in a series of animal studies to investigate varying the level of PN in chemically induced model of colon cancer in mice (Komatsu et al., 2005). Mice were fed diets containing 1, 7 (the level required by rodents), 14, or 35 mg pyridoxine HCl/kg diet. The mice maintained on these diets received multiple injections of azoxymethane, a colon carcinogen. It was reported that mice on the

7, 14, and 35 mg pyridoxine diets had significantly lower tumor outcome and suppressed colonic cell proliferation and expression of c-myc and c-fos proteins in a dose-dependent manner. Using the same animal model, Komatsu et al. (2002) suggested that the tumor inhibitory effect of vitamin B6 in part was mediated by reduced oxidative stress and nitric oxide production.

A study by Kular (2007) showed that a low or a high B6 diet when introduced during the postinitiation stages, 1 week after carcinogen injection, did not change the growth of aberrant crypt foci (ACF), a biomarker of early stages of colon carcinogenesis (Bird, 1987; Bird & Good, 2000; Krupnik et al., 1999; Magnuson & Bird, 1994) in Sprague Dawley rats. Whereas a Zucker obese rat model fed a high B6 diet (35 mg/kg of diet) before, during, and after two injections of azoxymethane, exhibited reduced sensitivity to azoxymethane, since the number and growth of ACF were lower in high (35 mg/kg diet) than in normal B6 (7 mg/kg diet) PN-fed animals (Table 2). In the same study a high PN diet alleviated a number of obesity-associated metabolic abnormalities such as enlarged liver, increase in platelets, monocytes, and lymphocytes and increased serum cholesterol (Tables 3 and 4). Compared to the normal B6-fed animals, the liver of high B6 had increased histone H3 and H4 acetylation, suggesting that a high B6 diet was affecting the posttranslational modification of histone (Naahidi, 2007). Physiologically, Zucker obese rats exhibit a proinflammatory environment and are highly sensitive to chemically induced colon carcinogenesis (Jain & Bird, 2010; Raju & Bird, 2006). Colonic tumors induced in these animals also possess a higher proinflammatory environment by having an

Table 2 Effect of a High B6 Diet on the Total Number of ACF With Different Crypt Multiplicities

Group	Total ACF	ACF (1–3 Crypts)	ACF (4–6 Crypts)
Zk-OBN	209.6 ± 19.5*x*	194.9 ± 17.7*x*	14.4 ± 2.5*x*
Zk-OBH	151.3 ± 18.9*y*	141.6 ± 17.7*y*	9.4 ± 1.4*y*
Zk-LN	135.0 ± 16.9*y*	128.6 ± 16.4*y*	6.3 ± 1.4*y*

Zucker obese female rats were fed a normal (7 mg pyridoxine HCl/kg diet) or a high B6 (35 mg pyridoxine HCl/kg diet) for 2 weeks prior to injection with azoxymethane, a colon carcinogen (10 mg/kg body weight) weekly for 2 weeks, and then continued to receive a normal or a high B6 diet for an additional 6 weeks. Zucker lean female rats were fed a normal B6 diet and injected with azoxymethane (10 mg/kg diet) weekly for 2 weeks and continued to receive a normal B6 diet for an additional 6 weeks. Crypt multiplicities represent growth features of ACF. Values are expressed as mean ± SE. $N=8$ per Zucker Obese group and $N=7$ for Zk-LN. Values in the same column without a common symbol are significantly different ($P<0.05$, ANOVA analysis with LSD post hoc tests). Abbreviations: *ACF*, aberrant crypt foci; *Zk-OBH*, Zucker obese high B6 group; *Zk-OBN*, Zucker obese normal B6 group; *Zk-LN*, AOM-injected Zucker Lean group receiving normal B6.

Table 3 A Selected Hematological Parameters of AOM-Injected Female Zucker Obese Rats Before and After Receiving a Normal or High B6 Diet[a,b]

Parameters	Zk-OBN	Zk-OBH	Zk-LN
Platelets (10^9/L)	975.4 ± 43.3*x*	833.5 ± 46.8*y*	800.3 ± 59.4*y*
Lymphocyte (10^9/L)	4.7 ± 0.7*x*	3.0 ± 0.6*x*	3.9 ± 0.9*x*
Monocyte (10^9/L)	0.4 ± 0.1*x*	0.3 ± 0.0*y*	0.1 ± 0.0*y*

[a]Obese rats were fed a normal (7 mg pyridoxine HCl/kg of diet) or a high vitamin B6 (35 mg pyridoxine HCl/kg diet) for 2 weeks prior to injection with azoxymethane (10 mg/kg per week for 2 weeks) and after cessation of injection for an additional 6 weeks (see Kular, 2007 for details) before clinical assessment was carried out. Lean rats received azoxymethane injections employing the same protocol as obese rats, however, received a diet containing normal level of B6 throughout the study. Values are expressed as mean ± SE. $N=8$ for Zk-OBN and Zk-OBH dietary groups and $N=6$ for Zk-LN.
[b]Values in the same row without a common symbol are significantly different ($P<0.05$, ANOVA analyses with LSD post hoc tests).
Abbreviations: *Zk-LN*, AOM-injected Zucker lean rats receiving normal B6; *Zk-OBH*, Zucker obese high B6 group; *Zk-OBN*, Zucker obese normal B6 group.
Modified from Kular, A. J. (2007). An assessment of pyridoxine as a biological response modifier during colon carcinogenesis. M.Sc. Thesis. Waterloo, Ontario, Canada: University of Waterloo.

Table 4 Blood Cholesterol, Glucose, and Triglyceride Levels in Zucker Obese Rats Injected With Azoxymethane and Fed a Normal or a High B6 Diet for 8 Weeks[a,b]

Biochemical Parameters	Zk-OBN	Zk-OBH	Zk-LN
Cholesterol (mmol/L)	11.9 ± 0.4*x*	9.5 ± 1.1*y*	2.5 ± 0.2*z*
Glucose (mmol/L)	14.8 ± 0.8*x*	14.9 ± 1.8*x*	7.6 ± 0.5*y*
Triglycerides (mmol/L)	3.5 ± 0.5*xy*	5.1 ± 1.4*y*	1.2 ± 0.3*x*

[a]Obese rats were fed a normal (pyridoxine HCl/kg diet) or a high vitamin B6 (35 mg pyridoxine HCl/kg diet) for 2 weeks prior to injection with azoxymethane (10 mg/kg per week for 2 weeks) and after cessation of injection for an additional 6 weeks (see Kular, 2007 for details) before clinical assessment was carried out. Lean rats received azoxymethane injections employing the same protocol as obese rats; however, they received a diet containing normal level of B6 throughout the study. Values are expressed as mean ± SE. $N=8$ for Zk-OBN and Zk-OBH dietary groups and $N=6$ for Zk-LN.
[b]Values in the same row without a common symbol are significantly different ($P<0.05$, ANOVA analysis with LSD post hoc tests).
Abbreviations: *Zk-LN*, AOM-injected Zucker lean rats receiving normal B6, ALT, alanine aminotransferase; *Zk-OBH*, Zucker obese high B6 group; *Zk-OBN*, Zucker obese normal B6 group.

increased NF-κB pathway compared to those tumors induced in Zucker lean rats (Jain & Bird, 2010). We have shown previously that cholic acid induces inflammation in the colon and significantly increases azoxymethane-initiated tumor outcome soon after carcinogen treatment (Magnuson et al., 1993). In a later study, we showed that the delayed intervention by cholic acid, 2 weeks after the cessation of carcinogen injection, did not increase tumor outcome (Baijal et al., 1998). This suggested that the effect of cholic acid-mediated inflammation is cocarcinogenic in nature rather than tumor promoting.

The findings from preclinical animal models alluded to the concept that the effect of inflammation on carcinogenesis could be variable depending on the severity and duration of inflammation, and the experimental protocol used (Bird et al., 1985), as well as the level of B6.

In vitro studies involving a variety of cancer cell lines have shown that a high concentration of B6 in the medium exerts growth inhibitory effect (Choi & Friso, 2012). It has been recognized for several decades that tumor cells may differ from normal cells in their ability to metabolize B6 and that cancer-associated B6-conjugated compounds are present in vivo (Gregory, 1992; Tryfiates et al., 1991). Recently, it was reported that pyridoxal kinase, the enzyme which forms PLP, the active coenzyme form of B6 from PL, enhances the sensitivity of cancer cells to the chemotherapeutic drug cisplatin in vivo and in vitro (Aranda et al., 2013; Galluzzi et al., 2013, 2012). Non-small cell lung cancers (NSCLCs) showed resistance to the chemotherapeutic effect of cisplatin. Through a genome-wide siRNA-based screening, PLK was identified as a central regulator of responsiveness of cancer cells to cisplatin. Moreover, it was demonstrated that the higher expression of PLK in cancer tissue was associated with better patient survival, suggesting that PLK could serve as a prognostic marker in non-small cell lung carcinoma.

In an orthotopic mouse model of NSCLCs, treatment with cisplatin and PN produced synergistic therapeutic effects. However, this synergy was only observed in the context of an intact immune system. The authors proposed that the combined treatment of cisplatin plus pyridoxine mediates immune-dependent antineoplastic effects against NSCLCs (Aranda et al., 2013; Galluzzi et al., 2013, 2012). Systemic injections of PN improved the antitumor effects of cisplatin-based chemotherapy, in murine Lewis lung carcinoma cells transplanted into syngeneic C57BL/6 mice.

The same authors (Aranda et al., 2013; Galluzzi et al., 2013, 2012) also reported that PN enhanced cisplatin cytotoxicity in eight additional NSCLC cell lines, four ovarian carcinoma cell lines, cervical carcinoma, osteosarcoma, mesothelioma and wild-type colorectal cancer cells, and murine lung carcinoma cells, as well as in multiple cisplatin-resistant cell lines. It was noted that PN sensitized the cancer cells to cytotoxicity of other chemotherapeutic agents. Cancer cells treated with a subapoptotic concentration of cisplatin exhibited increased phosphorylation of TP53 and increased activation of caspase-3 and -9. Moreover, PN significantly increased cisplatin-DNA adduct formation in treated cancer cell lines. Cells exhibiting resistance to cisplatin had decreased PLK or increased B6-specific phosphatase. These studies conducted in intact animals and in isolated cancer cells provided the insight that PLK enhances the cytotoxic responses on cancer with or without involvement of immune cells. A recent study reported that B6 treatment interfered with the cross talk between colon cancer cells and monocytes and reduced the ability of monocytes to produce cancer-promoting cytokines (Zhang et al., 2016).

Therefore, experimental findings suggest that B6 has a role in carcinogenesis, which probably involves multiple pathways. It is important to consider that the effect of B6 on cancer cells may differ from its effect on precancerous cells. Studies on isolated cells differ substantially from studies conducted in intact animals. The later approach leads to a specific physiological response following a complex cross talk among diverse metabolic pathways involving interactions among different organ systems. Evidence from epidemiological studies demonstrating an inverse relationship between plasma PLP and cancer risk is strong, especially for colorectal cancer. Whether low plasma PLP in individuals with cancer is due to the presence of the disease or low plasma PLP is enhancing the disease remains speculative. Low plasma PLP has been noted in individuals with metabolic syndrome, diabetes, cancer, and obesity. Additional research is needed to establish a clear link between B6, inflammation, and carcinogenesis. Considering the expanding role of vitamin B6 as a biological response modifier, discussed in the preceding sections, makes B6 an important nutrient to be considered for its role in cancer development.

4.4 Potential Mechanisms Underlying Modulation of Carcinogenesis by B6

Vitamin B6 is an essential nutrient influencing several physiological responses critical to carcinogenesis. Some of these responses include one carbon metabolism and methylation of DNA and nucleotide synthesis (Myte et al., 2017; Selhub et al., 2013), neutralization of oxidative stress and inflammation, providing genomic stability, modulation of immune cells and hormone receptors, and posttranslational modification of biologically active proteins (Galluzzi et al., 2013; Myte et al., 2017; Paul et al., 2013; Qian et al., 2017; Rall & Meydani, 1993; Tully et al., 1994; Ueland et al., 2013, 2017). As a matter of fact all these mechanisms are interrelated and converge to the idea that B6 has a multifaceted physiological role and inflammation seems to be the central event. Moreover, pyridoxal 5′-phosphate, the biologically active form of B6, is a coenzyme that takes an essential part in many enzymatic activities associated with synthesis and degradation of amino acids, biogenic amines, bioactive lipids, and carbohydrates. Protein, carbohydrates, and lipid metabolism are critical for the survival of normal and transformed cells. Figs. 3 and 4 both outline the concept implicating the effect of B6 on oxidative stress directly or indirectly by acting as an antioxidant or as a coenzyme producing important metabolites. B6 can also affect oxidative stress, directly by posttranslational modification of specific proteins that are involved in oxidative stress or inflammation (Fig. 3). In Fig. 4, the interrelationships among B6, its effect on metabolism of macromolecules, oxidative stress, and inflammation-associated chronic illnesses are summarized. It shows that inflammation-associated diseases can deplete the biologically available pool.

4.4.1 Role of B6 in Inflammation-Associated Carcinogenesis: A Proposed Concept

The link between chronic inflammation and tumorigenesis has been recognized for centuries. It is estimated that approximately 18% of human tumor burden is infection related and many solid tumors show the increased level of proinflammatory cytokines. In this context, it is important to consider how inflammation could be involved in the multistep process of carcinogenesis.

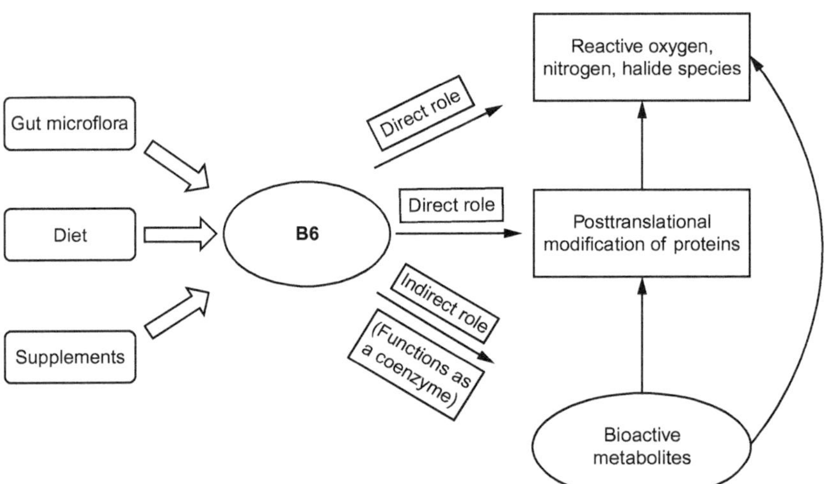

Fig. 3 B6 (pyridoxal, pyridoxine, and pyridoxamine) is derived from diet, gut microflora, and as supplement. B6 directly or as the active vitamin B6 (pyridoxal phosphate) function as antioxidant. PLP serving as the coenzyme is involved in the production of several amino acid and lipid metabolites. B6 especially PLP directly or through taurine and H_2S form various derivatives and conjugates with lipids, proteins, and nucleic acids. These derivatives are involved in modulating oxidative stress and mitochondrial integrity.

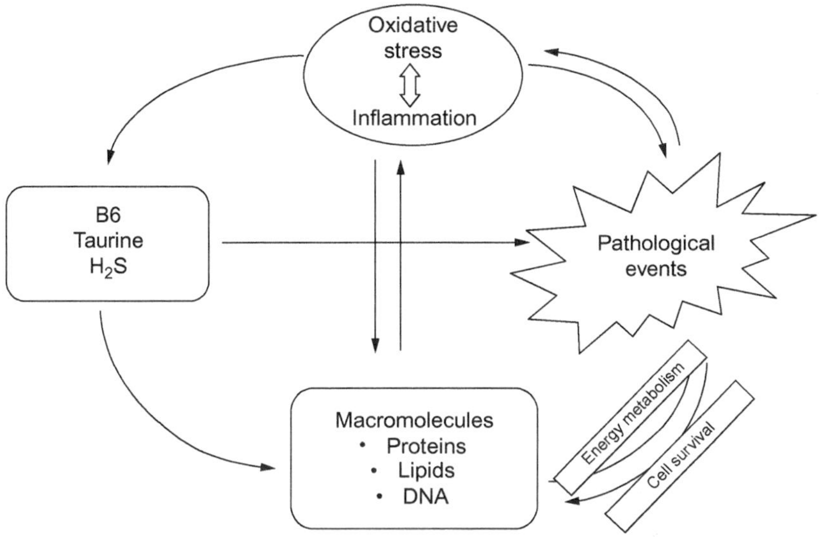

Fig. 4 The interrelationship among B6, metabolic and antioxidant role of B6, oxidative stress, and chronic illnesses. Oxidative stress and inflammation are interrelated unless intervened by antioxidants or antiinflammatory agents. B6 (pyridoxal, pyridoxine, and pyridoxamine) or metabolites such as H_2S and taurine could reduce reactive oxygen nitrogen species. Oxidative stress and inflammation can be induced by metabolic abnormalities such as excessive caloric intake and mitochondrial dysfunction or exogenous factors such as infection, pollution, and foreign cells. B6, H_2S, or taurine can alleviate pathological events by modulating immune responses or metabolic stress. Involvement of pyridoxal phosphate (PLP) in the metabolism of macromolecules could alter the severity of disease, fueling the cells for survival or enhancing the process of immunoediting and eliminating foreign cells such as transformed cells, and reducing the damaging effect of inflammation and oxidative stress on the macromolecules such as chromosomal instability, DNA damage, and carbonylation of proteins. Therefore, oxidative stress, inflammation, and pathological events such as cancer, obesity, diabetes, and rheumatoid arthritis affect the available pool of B6 and could reduce the plasma PLP unless additional B6 becomes available to compensate for increased requirement.

This concept is presented in Fig. 5. Oxidative stress and inflammation induced by exogenous or endogenous factors could damage DNA and inhibit DNA repair resulting in tissue damage and mutation and genomic instability in cells. Genomic instability and mutation in the stem or normal cells generate transformed cells. Damaged tissue initiates inflammatory responses by attracting neutrophils which in turn generate additional oxidative stress and produce cytokines. Several cytokines generated during inflammation are proangiogenic and promitogenic and further stimulate the transformed cells to grow. A chronic state of inflammation develops

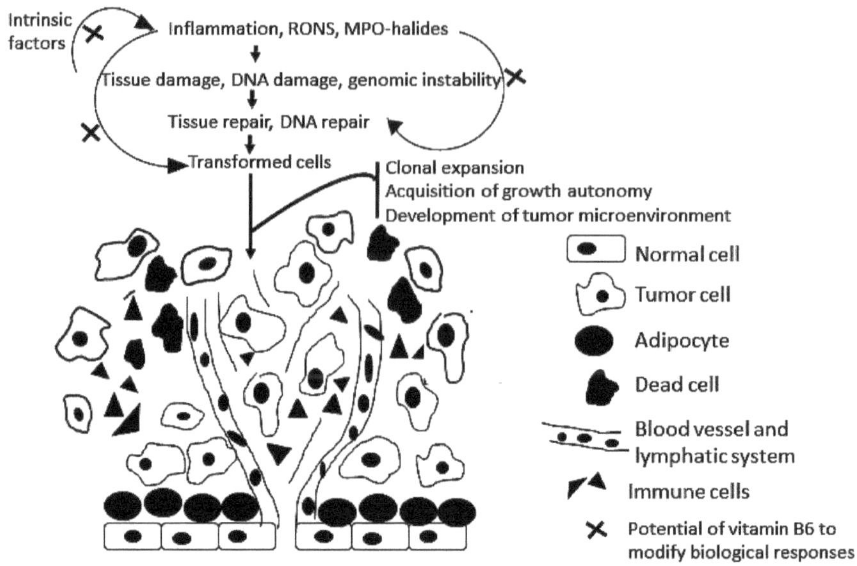

Fig. 5 A simplified scheme showing the role of oxidative stress and inflammation on carcinogenesis. Transformed cells arise due to genomic instability and damage to DNA by reactive oxygen and nitrogen species. In the absence of neutralizing effect of B6 (pyridoxine, pyridoxal, and pyridoxamine) or taurine transformed cells appear. X denotes that the step could be interrupted by B6 or metabolites such as taurine. Presence of altered cells could attract additional proinflammatory cells and a chronic state of inflammation around the clonally expanding transformed cells occurs. Tumor-associated macrophases are the prime regulators of tumor-associated inflammation. Type 1 macrophases enhance the population of T-helper cells 1 (TH-1) which is cytotoxic to cancer cells, whereas type 2 macrophases increase the population of T-helper cells 2 (TH-2), and TH-2 are immunosuppressive. Tumor-associated inflammation enhances epithelial to mesenchymal transition, angiogenesis, and metastasis. Growing tumor establishes a microenvironment and receives ample supply of nutrients, growth factors, and cytokines as needed. In addition to immune cells, vascular endothelial cells, platelets, red blood cells, and adipocytes could also help orchestrate a procarcinogenic environment. Emerging role of B6 in modifying immune responses, inflammasome NLRP3, and RIP140 (see Section 2 for details) provides an attractive platform to further explore its role in altering tumor microenvironment and sensitivity of tumor cells to death signals.

around the clonally expanding transformed cells, which continue to grow and acquire additional genotypic and phenotypic features. The expanding clones of transformed cells slowly develop growth autonomy and their own vascular system (angiogenesis) and microenvironment. Tumor microenvironment mainly consists of blood vessels/lymphatic system, infiltrating immune cells, fibroblasts, red blood cells, plasma proteins, platelets, and adipocytes (Quail & Joyce, 2013; Tape et al., 2016). Tumor microenvironment is responsible for cancer-associated inflammation and for epithelial to mesenchymal transition, angiogenesis, and metastasis by interacting with the tumor cells (Collins et al., 2012). However, tumor microenvironment can be pro- or anticarcinogenic depending on the composition of immune cells and involvement of other cell types such as endothelial cells of blood vessels, fibroblasts, and adipose tissues. Inflammation plays a multifaceted role in cancer (Sethi et al., 2012). Mitochondrial ROS play a significant role in inflammation and chronic diseases and for orchestrating inflammatory responses. Uncoupled mitochondrial ROS in vascular endothelium appear to be involved in immunosurveillance and/or in the recruitment of inflammatory cells (Li et al., 2017). Therefore, tumorigenesis is influenced by inflammation extrinsic and intrinsic to tumor. Whether B6 affects tumorigenesis by reducing inflammatory processes during preinitiation or postinitiation stages and/or reducing tumor-associated proinflammatory responses, a hallmark feature of cancer (Colotta et al., 2009), remain to be evaluated. In this scheme (Fig. 5), it is proposed that B6 will be an effective inhibitor of inflammation-related carcinogenesis. An adequate B6 intake will, directly or through metabolites such as taurine, retard the production of RONS (reactive oxygen nitrogen species) and reduce DNA damage and defective repair system resulting in a decrease in the number of transformed cells. B6 will alleviate chronic state of inflammation which produces growth signals to transformed cells by inactivating inflammasomes in a cell-specific manner. B6 could also affect tumor cells or cells present in the tumor microenvironment by conjugating with RIP140 and enhancing its corepressive or coactivating activity (see Section 2.2.3). We do not know if B6 mediates its effect in vivo, in part, by interacting with RIP140 or inflammasome (see Sections 2.2.2 and 2.2.3). It is acknowledged that the scheme shown in Fig. 5 is simplistic as it does not include the contribution of other nutrient and nonnutrient mediators in the outlined steps. While this scheme outlines a preventive role of B6, we cannot omit the fact that specific molecules or nutrients could exert opposite effects depending on the context and biological make up of a cells or tissue. For example, inflammasome NLRP3 could be involved as a negative or positive effecter of tumor growth (Allen et al., 2010; Bauer et al., 2010; Zivoget et al., 2012). Most importantly, sufficient evidence exists supporting the concept of an *Antioxidant paradox* (Biswas, 2016; Halliwell, 2000). Therefore, one must consider a pro- or anticarcinogenic effect for vitamin B6.

The existing literature presents a persuasive case for B6 to be recognized as an antiinflammatory/antioxidant nutrient. Epidemiological studies have provided a strong inverse association between plasma PLP and risk of cancer, especially colorectal cancer. Furthermore, a molecular role of B6 is emerging as a posttranslation modifier of key signaling molecules including hormone receptors, inflammasomes, and RIP140. Whether B6 affects mitochondrial biogenesis via generation of taurine and H_2S and its importance in tumorigenesis remains to be explored further.

5. FUTURE PERSPECTIVE

The preceding sections on B6 have provided a complex and diverse role for vitamin B6 in inflammation and carcinogenesis. During the last few years, the field of B6 research is dominated by its role in inflammation and related diseases. The clinical and scientific evidence are strong, supporting a role for B6 in inflammation and cancer. Inflammation itself is a complex process varying in length and severity and cellular composition and is associated with immune surveillance and escape, signal transduction, cell senescence, and growth. The nutritional significance of B6 is being recognized at the organ, cellular, subcellular, and molecular levels. The science behind the multifaceted role of B6 in human physiology is an expanding and emerging field. It is apparent that much more work is needed to further understand the metabolism and physiological function of B6. Research areas needing future discussion and attention are highlighted later.

- Whether or not the recommended intake of B6 is sufficient should be reconsidered. In addition to its role as a coenzyme catalyzing several enzymes, B6 plays multiple roles in cellular metabolism and its status is affected by aging, inflammation, obesity, and chronic illnesses.
- How and what form of B6, especially PLP, are transported to subcellular compartments such as mitochondria and nucleus? Is there tissue specificity in this regard? Organ-specific requirement for B6, metabolism, and in maintaining mitochondrial health warrants further considerations.
- Role of inflammation in the development of stepwise progression of solid tumors and whether a higher or lower intake of B6 will be preventive in different stages remains elusive.
- Whether B6 could enhance tumorigenesis by serving as an essential nutrient to tumor cells and/or by providing protection to tumor-associated mitochondrial biogenesis and genomic stability remains to be determined.

- It would be important to determine the effects of nutrients, in this case B6, on the expanding role of tumor microenvironment. Tumor microenvironment is integral to tumor development and cancer cell metastasis and thus its modulation could be critical to secondary prevention of cancer.
- The chemosensitizing role of B6 and the underlying molecular mechanism(s) require further attention.
- Further investigative approaches are required to illustrate the role of B6 in posttranslational modification of key proteins such as inflammasomes, RIP140, NF-κB, and histones and its importance in health and disease.

Briefly, the above research areas are relevant to many chronic illnesses. There is an enormous opportunity for nutritional scientists and molecular biologists to unravel the enigmatic molecular and metabolic role of B6 in human health.

ACKNOWLEDGMENTS

The author is indebted to many researchers who have contributed to the field of vitamin B6. Due to the broad nature of this chapter, the author was unable to cite each worthy article and has opted mainly to refer to excellent reviews. The author acknowledges the assistance of Drs. Jayadev Raju and Vaishali Basu during the preparation of Figs. 1–5.

REFERENCES

Ahmad, S., Khan, M. S., Akhter, F., Khan, M. S., Khan, A., & Ashraf, J. M. (2014). Glycoxidation of biological macromolecules: A critical approach to halt the menace of glycation. *Glycobiology*, *24*(11), 979–990.

Albersen, M., Bosma, M., Knoers, N. V., de Ruiter, B. H., Diekman, E. F., de Ruijter, J., et al. (2013). The intestine plays a substantial role in human vitamin B6 metabolism: A Caco-2 cell model. *PLoS One*, *8*, e54113.

Allen, I. C., Tekipple, E. M., Woodford, R. M., Uronis, J. M., Holl, E. K., Rogers, A. B., et al. (2010). The NLRP3 functions as a negative regulator of tumorigenesis during colitis associated tumorigenesis. *The Journal of Experimental Medicine*, *207*, 1045–1056.

Amarnath, V., Amarnath, K., Amarnath, K., Davies, S., & Roberts, L. J. (2004). Pyridoxamine: An extremely potent scavenger of 1,4-dicarbonyls. *Chemical Research in Toxicology*, *17*, 410–415.

Ames, B. N. (2001). DNA damage from micronutrient deficiencies is likely to be a major cause of cancer. *Mutation Research*, *475*(1–2), 7–20.

Ames, B. N., & Wakimoto, P. (2002). Are vitamin and mineral deficiencies a major cancer risk? *Nature Reviews. Cancer*, *2*(9), 694–704.

Aranda, F., Bloy, N., Lorenzo Galluzzi, L., Kroemer, G., & Laura Senovilla, L. (2013). Vitamin B6 improves the immunogenicity of cisplatin-induced cell death. *Cell Cycle*, *12*(3), 417–421.

Ashkenazi, R., Gentry, S. N., & Jackson, T. L. (2008). Pathways to tumorigenesis—Modeling mutation acquisition in stem cells and their progeny. *Neoplasia*, *10*, 1170–1182.

Augereau, P., Badia, E., Balaguer, P., Carascossa, S., Castet, A., Jalaguier, S., et al. (2006). Negative regulation of hormone signaling by RIP140. *Journal of Steroid Biochemistry and Molecular Biology*, *102*(1–5), 51–59.

Baijal, P. K., Clow, E. P., Fitzpatrick, D., & Bird, R. P. (1998). Tumor-enhancing effects of cholic acid are exerted on the early stages of colon carcinogenesis via induction of aberrant crypt foci with an enhanced growth phenotype. *Canadian Journal of Physiology and Pharmacology*, *76*, 1095–1102.

Bakele, M., Joos, M., Burdi, S., Allgaier, N., Pöschel, S., Fehrenbacher, B., et al. (2014). Localization and functionality of the inflammasome in neutrophils. *The Journal of Biological Chemistry*, *289*(8), 5320–5329.

Barcellos-Hoff, M. H., Lyden, D., & Wang, T. C. (2013). The evolution of the cancer niche during multistage carcinogenesis. *Nature Reviews. Cancer*, *13*, 511–518.

Baskar, R., & Bian, J. (2011). Hydrogen sulfide gas has cell growth regulatory role. *European Journal of Pharmacology*, *656*(1–3), 5–9.

Bauer, C., Duewell, P., Mayer, C., Lehr, H. A., Fitzerald, K. A., Daur, M., et al. (2010). Colitis induced in mice with dextran sulfate sodium (DSS) is mediated by the NLRP3 inflammasome. *Gut*, *59*(9), 1192–1199.

Baynes, J. W. (2001). The role of AGEs in aging: Causation or correlation. *Experimental Gerontology*, *36*, 1527–1537.

Baynes, J. W., & Thorpe, S. R. (2000). Glycoxidation and lipoxidation in atherogenesis. *Free Radical Biology & Medicine*, *28*, 1708–1716.

Beckman, J. S., & Koppenol, W. H. (1996). Nitric oxide, superoxide, and peroxynitrite: The good, the bad, and ugly. *The American Journal of Physiology*, *271*(5), C1424–C1437.

Berridge, M. V., Dong, L., & Neuzil, J. (2015). Mitochondrial DNA in tumor initiation, progression, and metastasis: Role of horizontal mtDNA transfer. *Cancer Research*, *75*(16), 3203–3208.

Berridge, M. V., & Neuzil, J. (2017). The mobility of mitochondria: Intercellular trafficking in health and disease. *Clinical and Experimental Pharmacology & Physiology*. https://doi.org/10.1111/1440-1681.12764. [Epub ahead of print].

Bierhaus, A., Humpert, P. M., Morcos, M., Wendt, T., Chavakis, T., Arnold, B., et al. (2000). Vitamin B6 (pyridoxine) and its derivatives are efficient singlet oxygen quenchers and potential fungal antioxidants. *Photochemistry and Photobiology*, *71*, 129–134.

Bird, R. P. (1987). Observation of aberrant crypts in the murine colons induced by a colon carcinogen: Preliminary findings. *Cancer Letters*, *37*, 147–151.

Bird, R. P., & Good, C. K. (2000). The significance of aberrant crypt foci in understanding the pathogenesis of colon cancer. *Toxicology Letters*, *112–113*, 395–402.

Bird, R. P., Mercer, N. J. H., & Draper, H. H. (1985). Animal models for the study of nutrition and human disease: Colon cancer, atherosclerosis and osteoporosis. In H. H. Draper (Ed.), *Advances in nutritional research* (pp. 155–186): Plenum Publishing Corporation.

Biswas, S. K. (2016). Does the interdependence between oxidative stress and inflammation explain the antioxidant paradox? *Oxidative Medicine and Cellular Longevity*, *2016*, 5698931. https://doi.org/10.1155/2016/5698931.

Biswas, S. K., & Halliwell, B. (1997). What nitrates tyrosine? Is nitrotyrosine specific as a biomarker of peroxynitrite formation in vivo? *FEBS Letters*, *411*(2–3), 157–160.

Blanchet, E., Annicotte, J. S., Lagarrigue, S., Aguilar, V., Clapé, C., Chavey, C., et al. (2011). E2F transcription factor-1 regulates oxidative metabolism. *Nature Cell Biology*, *13*(9), 1146–1152.

Brewer, G. J. (2007). Iron and copper toxicity in diseases of aging, particularly atherosclerosis and Alzheimer's disease. *Experimental Biology and Medicine*, *232*, 323–335.

Brock, A., Krause, S., & Ingber, D. E. (2015). Control of cancer formation by intrinsic genetic noise and microenvironmental cues. *Nature Reviews. Cancer*, *15*, 499–509.

Cagin, U., & Enriquez, J. A. (2015). The complex crosstalk between mitochondria and the nucleus: What goes in between? *International Journal of Biochemistry & Cell Biology*, *63*, 10–15.
Cavaillès, V., & Lapierre, M. (2014). Regulation of intestinal homeostasis and tumorigenesis by the transcriptional coregulator RIP140. *Molecular & Cellular Oncology*, *1*(2), e960761.
Chang, S. J. (2000). Vitamin B6 antagonists alter the function and ultrastructure of mice endothelial cells. *Journal of Nutritional Science and Vitaminology*, *46*, 149–153.
Chen, G. Y., & Núñez, G. (2011). Inflammasomes in intestinal inflammation and cancer. *Gastroenterology*, *141*(6), 1986 1999.
Chiang, E. P., Smith, D. E., Selhub, J., Dallal, G., Wang, Y. C., & Roubenoff, R. (2005). Inflammation causes tissue-specific depletion of vitamin B6. *Arthritis Research & Therapy*, *7*(6), 1254–1262.
Cho, M., Choi, E., Yang, J. S., Lee, C., Seo, J. J., & Kim, B. S. (2013). Discovery of pyridone-based histone deacetylase inhibitors: Approaches for metabolic stability. *ChemMedChem*, *8*, 272–279.
Choi, S. W., & Friso, S. (2012). Vitamins B6 and cancer. *Subcellular Biochemistry*, *56*, 247–264.
Christian, M., White, R., & Parker, M. G. (2006). Metabolic regulation by the nuclear receptor corepressor RIP140. *Trends in Endocrinology and Metabolism*, *17*(6), 243–250.
Coburn, S. P., Slominski, A., Mahuren, J. D., Wortsman, J., Hessle, L., & Millan, J. L. (2003). Cutaneous metabolism of vitamin B-6. *Journal of Investigative Dermatology*, *120*, 292–300.
Collins, M. A., Bednar, F., Zhang, Y., Brisset, J. C., Galba, N. S., et al. (2012). Oncogenic Kras is required for both the initiation and maintenance of pancreatic cancer in mice. *Journal of Clinical Investigation*, *122*, 639–653.
Colotta, F., Allavena, P., Sica, A., Garlanda, C., & Mantovani, A. (2009). Cancer-related inflammation, the seventh hallmark of cancer: Links to genetic instability. *Carcinogenesis*, *30*, 1073–1081.
Crosby, M. E., & Almasan, A. (2004). Opposing roles of E2Fs in cell proliferation and death. *Cancer Biology and Therapy*, *3*(12), 1208–1211.
Cuevasanta, E., Möller, M. N., & Alvarez, B. (2017). Biological chemistry of hydrogen sulfide and persulfides. *Archives of Biochemistry and Biophysics*, *617*, 9–25.
Dainin, K., Ide, R., Maeda, A., Suama, K., & Akagawa, M. (2017). Pyridoxamine scavenges protein carbonyls and inhibits protein aggregation in oxidative stress-induced human HepG2 hepatocytes. *Biochemical and Biophysical Research Communications*, *486*(3), 845–851.
Davies, K. J. (2016). The oxygen paradox, oxidative stress, and ageing. *Archives of Biochemistry and Biophysics*, *595*, 28–32.
Dennis, E. A., & Norris, P. C. (2015). Eicosanoid storm in infection and inflammation. *Nature Reviews. Immunology*, *15*(8), 511–523.
Docquier, A., Harmand, P. O., Fritsch, S., Chanrion, M., Darbon, J. M., & Cavailles, V. (2010). The transcriptional coregulator RIP140 represses E2F1 activity and discriminates breast cancer subtypes. *Clinical Cancer Research*, *16*, 2959–2970.
Draper, H. H., & Bird, R. P. (1987). Micronutrients and cancer prevention: Are the RDAs adequate? *Free Radical Biology and Medicine*, *13*(203–207), 1987.
Dröge, W. (2002). Free radicals in the physiological control of cell function. *Physiological Reviews*, *82*, 47–95.
Duracková, Z. (2010). Some current insights into oxidative stress. *Physiological Research*, *59*, 459–469.
Egeblad, M., Nakasone, E. S., & Werb, Z. (2010). Tumors as organs: Complex tissues that interface with the entire organism. *Developmental Cell*, *18*, 884–901.

Feng, X., Krogh, K. A., Wu, C. Y., Lin, Y. W., Tsai, H. C., Thayer, S. A., et al. (2014). Receptor-interacting protein 140 attenuates endoplasmic reticulum stress in neurons and protects against cell death. *Nature Communications, 28,* 54487. https://doi.org/10.1038/ncomms5487.

Flannigan, K. L., Agbor, T. A., Blackler, R. W., Kim, J. J., Khan, W. I., Verdu, E. F., et al. (2014). Impaired hydrogen sulfide synthesis and IL-10 signaling underlie hyperhomocysteinemia associated exacerbation of colitis. *Proceedings of the National Academy of Sciences of the United States of America, 111*(37), 13559–13564.

Fritah, A., Christian, M., & Parker, M. G. (2010). The metabolic coregulator RIP140: An update. *American Journal of Physiology. Endocrinology and Metabolism, 299,* E335–E340.

Fullerton, J. N., O'Brien, A. J., & Gilroy, D. W. (2014). Lipid mediators in immune dysfunction after severe inflammation. *Trends in Immunology, 35*(1), 12–21.

Gai, J. W., Qin, W., Liu, M., Wang, H. F., Zhang, M., Li, M., et al. (2016). Expression profile of hydrogen sulfide and its synthases correlates with tumor stage and grade in urothelial cell carcinoma of bladder. *Urologic Oncology, 34*(4), 166.e15–20. https://doi.org/10.1016/j.urolonc.2015.06.020 [Epub 2016 Feb 1].

Galadari, S., Rahman, A., Pallichankandy, S., & Thayyullathi, l. F. (2017). Reactive oxygen species and cancer paradox: To promote or to suppress? *Free Radical Biology & Medicine, 104,* 144–164.

Galluzzi, L., Vacchelli, E., Michels, J., Garcia, P., Kepp, O., Senovilla, L., et al. (2013). Effects of vitamin B6 metabolism on oncogenesis, tumor progression and therapeutic responses. *Oncogene, 32,* 4995–5004.

Galluzzi, L., Vitale, I., Senovilla, L., Olaussen, K. A., Pinna, G., Eisenberg, T., et al. (2012). Prognostic impact of vitamin B6 metabolism in lung cancer. *Cell Reports, 2,* 257–269.

Giannoukakis, N. (2005). Pyridoxamine. *Current Opinion in Investigational Drugs, 6,* 410–418.

Giorgio, M., Trinei, M., Migliaccio, E., & Pelicci, P. G. (2007). Hydrogen peroxide: A metabolic by-product or a common mediator of ageing signals? *Nature Reviews. Molecular Cell Biology, 8,* 722–728.

Görlach, A., Dimova, E. Y., Petry, A., Martínez-Ruiz, A., Hernansanz-Agustín, P., Rolo, A. P., et al. (2015). Reactive oxygen species, nutrition, hypoxia and diseases: Problems solved? *Redox Biology, 6,* 372–385.

Gregory, J. F. (1992). A novel vitamin B6 metabolite may be a circulating marker of cancer. *Nutrition Reviews, 50*(10), 295–297.

Gregory, J. F., DeRatt, B. N., Rios-Avila, L., Ralat, M., & Stacpoole, P. W. (2016). Vitamin B6 nutritional status and cellular availability of pyridoxal 5′-phosphate govern the function of the transsulfuration pathway's canonical reactions and hydrogen sulfide production via side reactions. *Biochimie, 126,* 21–26.

Gridley, D. S., Stickney, D. R., Nutter, R. L., Slater, J. M., & Shultz, T. D. (1987). Suppression of tumor growth and enhancement of immune status with high levels of dietary vitamin B6 in BALB/c mice. *Journal of the National Cancer Institute, 78,* 951–959.

Gylling, B., Myte, R., Schneede, J., Hallmans, G., Häggström, J., Johansson, I., et al. (2017). Vitamin B-6 and colorectal cancer risk: A prospective population-based study using 3 distinct plasma markers of vitamin B-6 status. *American Journal of Clinical Nutrition, 105*(4), 897–904.

György, P. (1938). Crystalline vitamin B6. *Journal of the American Chemical Society, 60,* 983–984.

György, P., & Eckardt, R. E. (1939). Vitamin B6 and skin lesions in rats. *Nature, 144,* 512.

Halliwell, B. (2000). The antioxidant paradox. *The Lancet, 355*(9210), 1179–1180.

Hallstrom, T. C., Mori, S., & Nevins, J. R. (2008). An E2F1-dependent gene expression program that determines the balance between proliferation and cell death. *Cancer Cell, 13,* 11–22.

Hansen, S. H., Birkedal, H., Wibrand, F., & Grunnet, N. (2015). Taurine and regulation of mitochondrial metabolism. *Advances in Experimental Medicine and Biology*, *803*, 397–405.

Harris, S. A., & Folkers, K. (1939). Synthetic vitamin B6. *Science*, *9*(89), 347.

Hellmann, H., & Mooney, S. (2010). Vitamin B6: A molecule for human health? *Molecules*, *15*(1), 442–459.

Hellmich, M. R., & Szabo, C. (2015). Hydrogen sulfide and cancer. *Handbook of Experimental Pharmacology*, *230*, 233–241.

Horrobin, D. F. (1993). Fatty acid metabolism in health and disease: The role of delta-6-desaturase. *American Journal of Clinical Nutrition*, *57*, 732S–736S [discussion 736S–737S].

Hsu, J., & Sage, J. (2016). Novel functions for the transcription factor E2F4 in development and disease. *Cell Cycle*, *15*(23), 3183–3190.

Hu, Y. C., Yi, Z. J., Zhou, Y., Li, P. Z., Liu, Z. J., Duan, S. G., et al. (2017). Overexpression of RIP140 suppresses the malignant potential of hepatocellular carcinoma by inhibiting NF-κB-mediated alternative polarization of macrophages. *Oncology Reports*, *37*(5), 2971–2979.

Huang, G., Cheng, J., Yu, F., Liu, X., Yuan, C., Liu, C., et al. (2016). Clinical and therapeutic significance of sirtuin-4 expression in colorectal cancer. *Oncology Reports*, *35*(5), 2801–2810. https://doi.org/10.3892/or.2016.4685 [Epub 2016 Mar 17].

Huq, M. D., Tsai, N. P., Lin, Y. P., Higgins, L., & Wei, L. N. (2007). Vitamin B6 conjugation to nuclear corepressor RIP140 and its role in gene regulation. *Nature Chemical Biology*, *3*(3), 161–165.

Huq, M. D., & Wei, L. N. (2005). Post-translational modification of nuclear co-repressor receptor-interacting protein 140 by acetylation. *Molecular & Cellular Proteomics*, *4*(7), 975–983.

Ink, S. L., & Henderson, L. M. (1984). Vitamin B6 metabolism. *Annual Review of Nutrition*, *4*, 455–470.

Jacobsen, J. G., & Smith, L. H. (1968). Biochemistry and physiology of taurine and taurine derivatives. *Physiological Reviews*, *48*, 424–511.

Jain, S. S., & Bird, R. P. (2010). Elevated expression of tumor necrosis factor-alpha signaling molecules in colonic tumors of Zucker obese (fa/fa) rats. *International Journal of Cancer*, *127*(9), 2042–2050.

Jain, S. K., & Lim, G. (2001). Pyridoxine and pyridoxamine inhibits superoxide radicals and prevents lipid peroxidation, protein glycosylation, and (Na^+ + K^+)-ATPase activity reduction in high glucose-treated human erythrocytes. *Free Radical Biology & Medicine*, *30*, 232–237.

Jain, A. K., Lim, G., Langford, M., & Jain, S. K. (2002). Effect of high-glucose levels on protein oxidation in cultured lens cells, and in crystalline and albumin solution and its inhibition by vitamin B6 and N-acetylcysteine: Its possible relevance to cataract formation in diabetes. *Free Radical Biology & Medicine*, *33*, 1615–1621.

Jang, Y. M., Kim, D. W., Kang, T. C., Won, M. H., Baek, N. I., Moon, B. J., et al. (2003). Human pyridoxal phosphatase. Molecular cloning, functional expression, and tissue distribution. *Journal of Biological Chemistry*, *278*, 50040–50046.

Janssen, C., & Kiliaan, A. J. (2014). Long-chain polyunsaturated fatty acids (LCPUFA) from genesis to senescence: The influence of LCPUFA on neural development, aging, and neurodegeneration. *Progress in Lipid Research*, *53*, 1–17.

Jha, S., June Brickey, W., & Ting, J. P.-Y. (2017). Inflammasomes in myeloid cells: Warriors within. *Microbiology Spectrum*, *5*(1). https://doi.org/10.1128/microbiolspec.MCHD-0049-2016.

Jomova, K., & Valko, M. (2011). Advances in metal-induced oxidative stress and human disease. *Toxicology*, *283*(2–3), 65–87.

Kabekkodu, S. P., Chakrabarty, S., Shukla, V., Varghese, V. K., Singh, K. K., Thangaraj, K., et al. (2015). Mitochondrial biology: From molecules to diseases. *Mitochondrion*, *24*, 93–98.

Kabil, O., & Banerjee, R. (2010). Redox biochemistry of hydrogen sulfide. *The Journal of Biological Chemistry*, *285*, 21903–21907.

Kabil, O., Motl, N., & Banerjee, R. (2014). H_2S and its role in redox signaling. *Biochimica et Biophysica Acta*, *1844*(8), 1355–1366.

Kabil, O., Vitvitsky, V., & Banerjee, R. (2014). Sulfur as a signaling nutrient through hydrogen sulfide. *Annual Review of Nutrition*, *34*, 171–205.

Karki, R., Man, S. M., & Kanneganti, T. D. (2017). Inflammasomes and cancer. *Cancer Immunology Research*, *5*(2), 94–99.

Kendler, B. S. (1989). Taurine: An overview of its role in preventive medicine. *Preventive Medicine*, *18*, 79–100.

Khalifah, R. G., Chen, Y., & Wassenberg, J. (2005). Post-Amadori AGE inhibition as a therapeutic target for diabetic complications: A rational approach to second-generation Amadorin design. *Annals of the New York Academy of Sciences*, *43*, 793–806.

Khare, S., Luc, N., Dorfleutner, A., & Stehlik, C. (2010). Inflammasomes and their activation. *Critical Reviews in Immunology*, *30*(5), 463–487.

Kim, C., & Cha, Y. N. (2014). Taurine chloramine produced from taurine under inflammation provides anti-inflammatory and cytoprotective effects. *Amino Acids*, *46*(1), 89–100.

Kimura, H. (2010). "Hydrogen sulfide" from brain to gut. *Antioxidants & Redox Signaling*, *12*(9), 1111–1123.

Klosterman, H. J. (1974). Vitamin B6 antagonists of natural origin. *Journal of Agricultural and Food Chemistry*, *22*(1), 13–16.

Kolb, R., Liu, G. H., Janowski, A. M., Sutterwala, F. S., & Zhang, W. (2011). Inflammasomes in cancer: A double-edged sword. *Gastroenterology*, *141*(6), 1986–1999.

Komatsu, S., Isobe, M., Yanaka, N., & Kato, N. (2005). A high-fat diet enhances the inhibitory effect of dietary vitamin B6 on colon cell proliferation in mice. *Oncology Reports*, *14*, 265–269.

Komatsu, S., Watanabe, H., Oka, T., Tsuge, H., & Kat, N. (2002). Dietary vitamin B6 suppresses colon tumorigenesis, 8-hydroxyguanosine, 4-hydroxynonenal, and inducible nitric oxide synthase protein in azoxymethane-treated mice. *Journal of Nutritional Science and Vitaminology*, *48*, 65–68.

Krupnik, E., Briere, K. M., Bird, R. P., Littman, C., & Smith, I. C. (1999). 1H magnetic resonance spectroscopy evidence that aberrant crypt foci are preneoplastic lesions in the colon. *Anticancer Research*, *19*(3A), 1699–1703.

Kular, A. j. (2007). *An assessment of pyridoxine as a biological response modifier during colon carcinogenesis*. M.Sc. ThesisWaterloo, Ontario, Canada: University of Waterloo.

Kumar, S., & Lombard, D. B. (2015). Mitochondrial sirtuins and their relationships with metabolic disease and cancer. *Antioxidants & Redox Signaling*, *22*(12), 1060–1077.

Lapierre, M., Bonnet, S., Bascoul-Mollevi, C., Ait-Arsa, I., Jalaguier, S., Del Rio, M., et al. (2014). RIP140 increases APC expression and controls intestinal homeostasis and tumorigenesis. *Journal of Clinical Investigation*, *124*(5), 1899–1913.

Lapierre, M., Castet-Nicolas, A., Gitenay, D., Jalaguier, S., Teyssier, C., Bret, C., et al. (2015). Expression and role of RIP140/NRIP1 in chronic lymphocytic leukemia. *Journal of Hematology & Oncology*, *8*, 20. https://doi.org/10.1186/s13045-015-0116-6.

Lapierre, M., Docquier, A., Castet-Nicolas, A., Gitenay, D., Jalaguier, S., Teyssier, C., et al. (2015). The emerging role of the transcriptional coregulator RIP140 in solid tumors. *Biochimica et Biophysica Acta*, *1856*(1), 144–150.

Lawrence, T., Gilroy, D. W., Colville-Nash, P. R., & Willoughby, D. A. (2001). Possible new role for NF-κB in the resolution of inflammation. *Nature Medicine*, *7*(12), 1291–1297.

Lee, C. H., & Wei, L. N. (1999). Characterization of receptor-interacting protein 140 in retinoid receptor activities. *The Journal of Biological Chemistry*, *274*, 31320–31326.

Leonardsson, G., Steel, J. H., Christian, M., Pocock, V., Milligan, S., Bell, J., et al. (2004). Nuclear receptor corepressor RIP140 regulates fat accumulation. *Proceedings of the National Academy of Sciences of the United States of America*, *101*, 8437–8442.

Li, X., Fang, P., Yang, W. Y., Chan, K., Lavallee, M., Xu, K., et al. (2017). Mitochondrial ROS, uncoupled from ATP synthesis, determine endothelial activation for both physiological recruitment of patrolling cells and pathological recruitment of inflammatory cells. *Canadian Journal of Physiology and Pharmacology*, *95*(3), 247–252.

Li, L., Rose, P., & Moore, P. K. (2011). Hydrogen sulfide and cell signaling. *Annual Review of Pharmacology and Toxicology*, *51*, 169–187.

Lin, C., & Zhang, J. (2017). Inflammasomes in inflammation-induced cancer. *Frontiers in Immunology*, *8*, 271. https://doi.org/10.3389/fimmu.2017.00271.

Loeb, L. A., & Harris, C. C. (2008). Advances in chemical carcinogenesis: A historical review and prospective. *Cancer Research*, *68*, 6863–6872.

Lotto, V., Choi, S. W., & Friso, S. (2011). Vitamin B6: A challenging link between nutrition and inflammation in CVD. *British Journal of Nutrition*, *106*, 183–195.

Mackey, A. D., McMahon, R. J., Townsend, J. H., & Gregory, J. F. (2004). Uptake, hydrolysis, and metabolism of pyridoxine-5′-beta-D-glucoside in Caco-2 cells. *Journal of Nutrition*, *134*(4), 842–846.

Maeda, T., Taguchi, H., Minami, H., Sato, K., Shiga, T., Kosaka, H., et al. (2000). Vitamin B6 phototoxicity induced by UVA radiation. *Archives of Dermatological Research*, *292*, 562–567.

Magnuson, B. A., & Bird, R. P. (1994). Resistance of aberrant crypt foci to apoptosis. *Carcinogenesis*, *15*, 1459–1462.

Magnuson, B. A., Carr, I., & Bird, R. P. (1993). Ability of aberrant crypt foci characteristics to predict colonic tumor incidence in rats fed cholic acid. *Cancer Research*, *53*, 4499–4504.

Major, J. L., Dewan, A., Salih, M., Leddy, J. J., Balwant, S., & Tuana, B. S. (2017). E2F6 impairs glycolysis and activates BDH1 expression prior to dilated cardiomyopathy. *PLoS One*, *12*(1), e0170066.

Man, S. M., & Kanneganti, T. D. (2015). Regulation of inflammasome activation. *Immunology Reviews*, *265*(1), 6–21.

Marcinkiewicz, J., & Kontny, E. (2014). Taurine and inflammatory diseases. *Amino Acids*, *46*, 7–20.

Matxain, J. M., Ristilä, M., Strid, A., & Eriksson, L. A. (2006). Theoretical study of the antioxidant properties of pyridoxine. *The Journal of Physical Chemistry A*, *110*(48), 13068–13072.

Mayengbam, S., Raposo, S., Aliani, M., & House, J. D. (2016). A vitamin B-6 antagonist from flaxseed perturbs amino acid metabolism in moderately vitamin B-6-Deficient male rats. *Journal of Nutrition*, *146*, 14–20.

Menu, P., & Vince, J. E. (2013). The NLRP3 inflammasome in health and disease: The good, the bad and the ugly. *Clinical and Experimental Immunology*, *166*(1), 1–15.

Merrill, A. H., Jr., Henderson, J. M., Wang, E., McDonald, B. W., & Millikan, W. J. (1984). Metabolism of vitamin B-6 by human liver. *Journal of Nutrition*, *114*, 1664–1674.

Miyo, M., Yamamoto, H., Konno, M., Colvin, H., Nishida, N., Koseki, J., et al. (2015). Tumour-suppressive function of SIRT4 in human colorectal cancer. *British Journal of Cancer*, *113*(3), 492–499.

Mocellin, S., Briarava, M., & Pilati, P. (2017). Vitamin B6 and cancer risk: A field synopsis and meta-analysis. *Journal of the National Cancer Institute*, *109*(3), 1–9.

Mooney, S., Leuendorf, J. E., Hendrickson, C., & Hellmann, H. (2009). Vitamin B6: A long known compound of surprising complexity. *Molecules*, *14*, 329–351.

Morris, M. S., Sakakeeny, L., Jacques, P. F., Picciano, M. F., & Selhub, J. (2010). Vitamin B-6 intake is inversely related to, and the requirement is affected by, inflammation status. *Journal of Nutrition*, *140*, 103–110.
Murakami, C. (2015). Role of taurine in the pathogenesis of obesity. *Molecular Nutrition & Food Research*, *59*, 1353–1363.
Murata, Y., Kumano, K., Ueda, T., Araki, N., Nakamura, T., & Tani, M. (1998). Photosensitive dermatitis caused by pyridoxine hydrochloride. *Journal of the American Academy of Dermatology*, *39*, 314–317.
Myte, R., Gylling, B., Häggström, J., Schneede, J., Ueland, P., Hallmans, G., et al. (2017). Untangling the role of one-carbon metabolism in colorectal cancer risk: A comprehensive Bayesian network analysis. *Scientific Reports*, *7*, 43434. Published online 2017 Feb 24. https://doi.org/10.1038/srep43434.
Naahidi, S. (2007). *In vivo epigenetic study of histone acetylation associated with obesity*. M.Sc. Thesis Ontario, Canada: University of Waterloo.
Nakamura, M. T., & Nara, T. Y. (2004). Structure, function., and dietary regulation of delta-6, delta-5, and delta-9 desaturases. *Annual Review of Nutrition*, *24*, 345–376.
Nicolussi, A., D'Inzeo, S., Capalbo, C., Giannini, G., & Coppa, A. (2017). The role of peroxiredoxins in cancer. *Molecular and Clinical Oncology*, *6*(2), 139–153.
Onorato, J. M., Jenkins, A. J., Thorpe, S. R., & Baynes, J. W. (2000). Pyridoxamine, an inhibitor of advanced glycation reactions, also inhibits advanced lipoxidation reactions. *Journal of Biological Chemistry*, *275*, 21177–21184.
Palomer, X., Álvarez-Guardia, D., Davidson, M. M., Chan, T. O., Feldman, A. M., & Vázquez-Carrera, M. (2011). The interplay between NF-kappaB and E2F1 coordinately regulates inflammation and metabolism in human cardiac cells. *PLoS One*, *6*(5), e19724. https://doi.org/10.1371/journal.pone.0019724. [Epub 2011 May 23].
Paul, B. D., & Snyder, S. H. (2012). H2S signalling through protein sulfhydration and beyond. *Nature Reviews. Molecular Cell Biology*, *13*, 499–507.
Paul, L., Ueland, P. M., & Selhub, J. (2013). Mechanistic perspective on the relationship between pyridoxal 5′-phosphate and inflammation. *Nutrition Reviews*, *71*(4), 239–244.
Pellegrini, C., Antonioli, L., Lopez-Castejon, G., Blandizzi, C., & Fornai, M. (2017). Canonical and non-canonical activation of NLRP3 inflammasome at the crossroad between immune tolerance and intestinal inflammation. *Frontiers in Immunology*. https://doi.org/10.3389/fimmu.2017.00036.
Petrilli, V. (2017). The multifaceted roles of inflammasome proteins in cancer. *Current Opinion in Oncology*, *29*(1), 35–40.
Pichette, J., & Gagnon, J. (2016). Implications of hydrogen sulfide in glucose regulation: How H2S can alter glucose homeostasis through metabolic hormones. *Oxidative Medicine and Cellular Longevity*, *2016*. Article ID 3285074, 5 pages. https://doi.org/10.1155/2016/3285074.
Qian, B., Shen, S., Zhang, J., & Jing, P. (2017). Effects of vitamin B6 deficiency on the composition and functional potential of T cell populations. *Journal of Immunology Research*, *2017*, 2197975. https://doi.org/10.1155/2017/2197975.
Quail, D. F., & Joyce, J. A. (2013). Microenvironmental regulation of tumor progression and metastasis. *Nature Medicine*, *19*, 1423–1437.
Raju, J., & Bird, R. P. (2006). Alleviation of hepatic steatosis accompanied by modulation of plasma and liver TNF-alpha levels by Trigonella foenum graecum (fenugreek) seeds in Zucker obese (fa/fa) rats. *International Journal of Obesity*, *30*(8), 1298–1307.
Rall, L. C., & Meydani, S. N. (1993). Vitamin B6 and immune competence. *Nutrition Reviews*, *51*(8), 217–225.
Rauch, I., Deets, K. A., Ji, D. X., von Moltke, J., Tenthorey, J. L., Lee, A. Y., et al. (2017). NAIP-NLRC4 inflammasomes coordinate intestinal epithelial cell expulsion with eicosanoid and IL-18 release via activation of caspase-1 and -8. *Immunity*, *46*(4), 649–659.

Reynolds, R. D. (1988). Bioavailability of vitamin B-6 from plant foods. *American Journal of Clinical Nutrition*, *48*(3 Suppl), 863–867.
Rose, R. C. (1980). Water-soluble vitamin absorption in intestine. *Annual Review of Physiology*, *42*, 157–171.
Rose, P., Moore, P. K., & Zhun Zhu, Y. Z. (2017). H2S biosynthesis and catabolism: New insights from molecular studies. *Cellular and Molecular Life Sciences*, *74*, 1391–1412.
Rosenberg, I. H. (2012). A history of the isolation and identification of vitamin B(6). *Annals of Nutrition and Metabolism*, *61*(3), 236–238.
Ross, S. A., & Milner, J. A. (2007). Epigenetic modulation and cancer: Effect of metabolic syndrome? *American Journal of Clinical Nutrition*, *86*(3), s872–7.
Rytinki, M. M., & Palvimo, J. J. (2008). SUMOylation modulates the transcription repressor function of RIP140. *Journal of Biological Chemistry*, *283*(17), 11586–11595.
Saareks, V., Ylitalo, P., Mucha, I., & Riutta, A. (2002). Opposite effects of nicotinic acid and pyridoxine on systemic prostacyclin, thromboxane and leukotriene production in man. *Pharmacology and Toxicology*, *90*(6), 338–342.
Sakakeeny, L., Roubenoff, R., Obin, M., et al. (2012). Plasma pyridoxal-5-phosphate is inversely associated with systemic markers of inflammation in a population of US adults. *Journal of Nutrition*, *142*, 1280–1285.
Sasaki, N., Fukatsu, R., Tsuzuki, K., Hayashi, Y., Yoshida, T., Fuji, N., et al. (1993). Pyridoxine toxicity to cultured fibroblasts caused by near-ultraviolet light. *Journal of Investigative Dermatology*, *100*, 266–270.
Schaeffer, M. C., Sampson, D. A., Skala, J. H., Gietzen, D. W., & Grier, R. E. (1989). Clinical chemistry of vitamin B6. *Journal of Nutrition*, *119*(10), 1392–1398.
Scheibye-Knudsen, M., Fang, E. F., Croteau, D. L., Wilson, D. M., & Bohr, V. A. (2015). Protecting the mitochondrial powerhouse. *Trends in Cell Biology*, *25*, 158–170.
Schumacker, P. T. (2006). Reactive oxygen species in cancer cells: Live by the sword, die by the sword. *Cancer Cell*, *10*, 175–176.
Selhub, J., Byun, A., Liu, Z., Mason, J. B., Bronson, R. T., & Crott, J. W. (2013). Dietary vitamin B6 intake modulates colonic inflammation in the IL10 −/− model of inflammatory bowel disease. *Journal of Nutritional Biochemistry*, *24*(12), 2138–2143.
Sethi, G., Shanmugam, M. K., Ramachamdran, L., Kumar, A. P., & Tergaonkar, V. (2012). Multifaceted link between cancer and inflammation. *Bioscience Reports*, *32*, 1–15 [Printed in Great Britain]. https://doi.org/10.1042/BSR20100136.
Shefa, U., Yeo, S. G., Kim, M. S., Song, I. O., Jung, J., Jeong, Y., et al. (2017). Role of gasotransmitters in oxidative stresses, neuroinflammation, and neuronal repair. *BioMed Research International*, *2017*, 1689341. https://doi.org/10.1155/2017/1689341. Epub 2017 Mar 12.
Shen, J., Lai, C. Q., Mattei, J., et al. (2010). Association of vitamin B-6 status with inflammation, oxidative stress, and chronic inflammatory conditions: The Boston Puerto Rican health study. *American Journal of Nutrition*, *91*, 337–342.
Singh, S., & Banerjee, R. (2011). PLP-dependent H_2S biogenesis. *Biochimica et Biophysica Acta*, *1814*(11), 1518–1527.
Stipanuk, M. H., Dominy, J. E., Jr., Lee, J. I., & Coloso, R. M. (2006). Mammalian cysteine metabolism: New insights into regulation of cysteine metabolism. *Journal of Nutrition*, *136*, 1652S–1659S.
Stover, P. J., & Field, M. S. (2015). Vitamin B-6. *Advances in Nutrition*, *6*, 132–133.
Suzuki, T., Nagao, A., & Suzuki, T. (2011a). Human mitochondrial diseases caused by lack of taurine modification in mitochondrial tRNAs. *Wiley Interdisciplinary Reviews. RNA*, *2*(3), 376–386.
Suzuki, T., Nagao, A., & Suzuki, T. (2011b). Human mitochondrial tRNAs: Biogenesis, function, structural aspects, and diseases. *Annual Review of Genetics*, *45*, 299–329.
Suzuki, T., Suzuki, T., Wada, T., Saigo, K., & Watanabe, K. (2002). Taurine as a constituent of mitochondrial tRNAs: New insights into the functions of taurine and human mitochondrial diseases. *EMBO Journal*, *21*(23), 6581–6589.

Szabo, C., Ransy, C., Módis, K., Andriamihaja, M., Murghes, B., Coletta, C., et al. (2014). Regulation of mitochondrial bioenergetic function by hydrogen sulfide. Part I. Biochemical and physiological mechanisms. *British Journal of Pharmacology*, *171*(8), 2099–2122.

Tabner, B. J., El-Agnaf, O. M., German, M. J., Fullwood, N. J., & Allsop, D. (2005). Protein aggregation, metals and oxidative stress in neurodegenerative diseases. *Biochemical Society Transactions*, *33*, 1082–1086.

Tape, C. J., Ling, S., Dimitriadi, M., McMahon, K. M., Worboys, J. D., Leong, H. S., et al. (2016). Oncogenic KRAS regulates tumor cell signaling via stromal reciprocation. *Cell*, *165*, 910–920.

Tapiero, H., Ba, G. N., Couvreur, P., & Tew, K. D. (2010). Polyunsaturated fatty acids (PUFA) and eicosanoids in human health and pathologies. *Current Pharmaceutical Biotechnology*, *11*(8), 819–836.

Thanassi, J. W., & Meisler, N. T. (1990). Metabolism of pyridoxine in hepatomas. *Annals of the New York Academy of Sciences*, *85*, 424–437.

Thanassi, J. W., Nutter, L. M., Meisler, N. T., Commers, P., & Chiu, J. F. (1981). Vitamin B6 metabolism in Morris hepatomas. *Journal of Biological Chemistry*, *256*, 3370–3375.

Tokarz, P., & Blasiak, J. (2014). Role of mitochondria in carcinogenesis. *Acta Biochimica Polonica*, *61*, 671–678.

Traba, J., & Sack, M. N. (2017). The role of caloric load and mitochondrial homeostasis in the regulation of the NLRP3 inflammasome. *Cellular and Molecular Life Sciences*, *74*(10), 1777–1791.

Tryfiates, G. P. (1977). Vitamin B6 effects on the growth of Morris hepatomas and the development of enzymatic activity. *Advances in Experimental Medicine and Biology*, *22–24*(92), 607–642.

Tryfiates, G. P., Bishop, R. E., & Murgo, A. J. (1991). Vitamin B6 and cancer: A novel pyridoxal 5-phosphate conjugate in tumor cells and blood of cancer patients. *Anticancer Research*, *11*(3), 1281–1284.

Tryfiates, G. P., & Morris, H. P. (1976). Growth of Morris hepatoma no. 7794A with and without vitamin B6. Effect of inoculation time. *European Journal of Cancer*, *12*, 9–12.

Tsuge, H., Hotta, N., & Hayakawa, T. (2000). Effects of vitamin B-6 on (n-3) polyunsaturated fatty acid metabolism. *Journal of Nutrition*, *130*(2S Suppl), 333S–334S.

Tully, D. B., Allgood, V. E., & Cidlowski, J. A. (1994). Modulation of steroid receptor-mediated gene expression by vitamin B6. *FASEB Journal*, *8*(3), 343–349.

Ueki, I., & Stipanuk, M. H. (2009). 3T3-L1 adipocytes and rat adipose tissue have a high capacity for taurine synthesis by the cysteine dioxygenase/cysteinesulfinate decarboxylase and cysteamine dioxygenase pathways. *Journal of Nutrition*, *139*, 207–214.

Ueland, P. M., Arve Ulvik, A., Luisa Rios-Avila, L., Midttun, Ø., & Gregory, J. F. (2015). Biomarkers of vitamin B6 status. *Annual Review of Nutrition*, *35*, 33–70.

Ueland, P. M., McCann, A., Midttun, Ø., & Ulvik, A. (2013). Inflammation, vitamin B6 and related pathways. *Journal of Nutritional Biochemistry*, *24*(12), 2138–2143.

Ueland, P. M., McCann, A., Midttun, Ø., & Ulvik, A. (2017). Inflammation, vitamin B6 and related pathways. *Molecular Aspects of Medicine*, *53*, 10–27.

Ulvik, A., Midttun, Ø., Pedersen, E. R., Eussen, S. J., Nygård, O., & Ueland, P. M. (2014). Evidence for increased catabolism of vitamin B-6 during systemic inflammation. *American Journal of Clinical Nutrition*, *100*, 250–255.

Ulvik, A., Midttun, Ø., Pedersen, E. R., Nygård, O., & Ueland, P. M. (2012). Association of plasma B-6 vitamers with systemic markers of inflammation before and after pyridoxine treatment in patients with stable angina pectoris. *American Journal of Clinical Nutrition*, *95*, 1072–1078.

Valko, M., Leibfritz, D., Moncol, J., Cronin, M. T., Mazur, M., & Telser, J. (2007). Free radicals and antioxidants in normal physiological functions and human disease. *International Journal of Biochemistry and Cell Biology*, *39*, 44–84.

Valko, M., Morris, H., & Cronin, M. T. (2005). Metals, toxicity and oxidative stress. *Current Medicinal Chemistry*, *12*(10), 1161–1208.

Vandanmagsar, B., Youm, Y.-H., Ravussin, A., Galgani, J. E., Stadler, K., Randall, L., et al. (2011). The NALP3/NLRP3 Inflammasome instigates obesity-induced autoinflammation and insulin resistance. *Nature Medicine*, *17*(2), 179–188.

van den Berg, H., Bode, W., Mocking, J. A., & Löwik, M. R. (1990). Effect of aging on vitamin B6 status and metabolism. *Annals of the New York Academy of Sciences*, *585*, 96–105.

Vazquez-Santillan, K., Melendez-Zajgla, J., Jimenez-Hernandez, L., Martínez-Ruiz, G., & Maldonado, V. (2015). NF-κB signaling in cancer stem cells: A promising therapeutic target? *Cellular Oncology (Dordrecht)*, *38*(5), 327–339.

Vicente, J. B., Malagrinò, F., Arese, M., Forte, E., Sarti, P., & Giuffrè, A. (2016). Bioenergetic relevance of hydrogen sulfide and the interplay between gasotransmitters at human cystathionine β-synthase. *Biochimica et Biophysica Acta*, *1857*(8), 1127–1138.

Voziyan, P. A., & Hudson, B. G. (2005a). Pyridoxamine as a multifunctional pharmaceutical: Targeting pathogenic glycation and oxidative damage. *Cellular and Molecular Life Sciences*, *62*, 1671–1681.

Voziyan, P. A., & Hudson, B. G. (2005b). Pyridoxamine: The many virtues of a maillard reaction inhibitor. *Annals of the New York Academy of Sciences*, *1043*, 807–816.

Wang, R. (2012). Physiological implication of hydrogen sulfide: A whiff exploration that blossomed. *Physiological Reviews*, *92*(2), 791–796.

Wei, L. N., Hu, X., Chandra, D., Seto, E., & Farooqui, M. (2000). Receptor interacting protein 140 directly recruits histone deacetylases for gene silencing. *Journal of Biological Chemistry*, *275*, 40782–40787.

Whittaker, J. W. (2016). Intracellular trafficking of the pyridoxal cofactor. Implications for health and metabolic disease. *Archives of Biochemistry and Biophysics*, *592*, 20–26.

Wilson, R. G., & Davis, R. E. (1983). Clinical chemistry of vitamin B6. *Advances in Clinical Chemistry*, *23*, 1–68.

Wondrak, G. T., Cervantes-Laurean, D., Roberts, M. J., Qasem, J. G., Kim, M., Jacobson, E. L., et al. (2002). Identification of alpha-dicarbonyl scavengers for cellular protection against carbonyl stress. *Biochemical Pharmacology*, *63*, 361–373.

Wondrak, G. T., & Jacobson, E. L. (2012). Vitamin B6: Beyond coenzyme functions. *Subcellular Biochemistry*, *56*, 291–300.

Wondrak, G. T., Roberts, M. J., Jacobson, M. K., & Jacobson, E. L. (2002). Photosensitized growth inhibition of cultured human skin cells: Mechanism and suppression of oxidative stress from solar irradiation of glycated proteins. *Journal of Investigative Dermatology*, *119*, 489–498.

Wondrak, G. T., Roberts, M. J., Jacobson, M. K., & Jacobson, E. L. (2004). 3-Hydroxypyridine chromophores are endogenous sensitizers of photooxidative stress in human skin cells. *Journal of Biological Chemistry*, *279*, 30009–30020.

Wright, C. E., Tallan, H. H., Lin, Y. Y., & Gaull, G. E. (1986). Taurine: Biological update. *Annual Review of Biochemistry*, *55*, 427–453.

Xiong, Y., Wang, M., Zhao, J., Han, Y., & Jia, L. (2016). Sirtuin 3: A Janus face in cancer (review). *International Journal of Oncology*, *49*(6), 2227–2235.

Yang, C. S. (2015). Influences of dietary and other factors on xenobiotic metabolism and carcinogenesis—A review article in memory of Dr. Allan H. Conney (1930-2013). *Nutrition and Cancer*, *67*(8), 1207–1213.

You, J., Yue, Z., Chen, S., Chen, Y., Lu, X., Zhang, X., et al. (2017). Receptor-interacting protein 140 represses sirtuin 3 to facilitate hypertrophy, mitochondrial dysfunction and energy metabolic dysfunction in cardiomyocytes. *Acta Physiologica (Oxford, England)*, *220*(1), 58–71.

Zhang, P., Tsuchiya, K., Kinoshita, T., Kushiyama, H., Suidasari, S., Hatakeyama, M., et al. (2016). Vitamin B6 prevents IL-1 protein production by inhibiting NLRP3 inflammasome activation. *Journal of Biological Chemistry*, *291*(47), 24517–24527.

Zhao, M., Lamers, Y., Ralat, M. A., Coats, B. S., Chi, Y. Y., Muller, K. E., et al. (2012). Marginal vitamin B-6 deficiency decreases plasma (n-3) and (n-6) PUFA concentrations in healthy men and women. *Journal of Nutrition, 42*(10), 1791–1797.

Zivoget, I., Kepp, O., Galluzzi, I., & Kroemer, G. (2012). Inflammasomes in carcinogenesis and anticancer immune responses. *Nature Immunology, 13*(4), 343–351.

Zmora, N., Levy, M., Pevsner-Fischer, M., & Elinav, E. (2017). Inflammasomes and intestinal inflammation. *Mucosal Immunology*. https://doi.org/10.1038/mi.2017.19. [Epub ahead of print] Review.

FURTHER READING

Bierhaus, A., Humpert, P. M., Morcos, M., Wendt, T., Chavakis, T., Arnold, B., et al. (2005). Understanding RAGE, the receptor for advanced glycation end products. *Journal of Molecular Medicine, 83*, 876–886.

Calle, E. E., Rodriguez, C., Walker-Thurmond, K., & Thun, M. J. (2003). Overweight, obesity, and mortality from cancer in a prospectively studied cohort of U.S. adults. *The New England Journal of Medicine, 348*, 1625–1638.

Cellini, B., Montioli, R., Oppici, E., Astegno, A., & Voltattorni, C. B. (2014). The chaperone role of the pyridoxal 5-phosphate and its implications for rare diseases involving B6-dependent enzymes. *Clinical Biochemistry, 47*, 158–165.

Chen, F., Castranova, V., Shi, X., & Demers, L. M. (1999). New insights into the role of nuclear factor-κB, a ubiquitous transcription factor in the initiation of diseases. *Clinical Chemistry, 45*(1), 7–17.

Christian, M., Kiskinis, E., Debevec, D., Leonardsson, G., White, R., & Parker, M. G. (2005). RIP140-targeted repression of gene expression in adipocytes. *Molecular and Cellular Biology, 25*, 9383–9391.

di Salvo, M. L., Contestabile, R., & Safo, M. K. (2011). Vitamin B(6) salvage enzymes: Mechanism, structure and regulation. *Biochimica et Biophysica Acta, 14*(11), 1597–1608.

Draper, H. H., & Bird, R. P. (1984). Antioxidants and cancer. *Journal of Agricultural and Food Chemistry, 32*, 433–435.

Eberhard, Y., Gronda, M., Hurren, R., Datti, A., MacLean, N., Ketela, T., et al. (2014). H_2S and its role in redox signaling. *Biochimica et Biophysica Acta, 1844*, 1355–1366.

Escoté, X., & Fajas, L. (2015). Metabolic adaptation to cancer growth: From the cell to the organism. *Cancer Letters, 356*(2 Pt. A), 171–175.

Galluzzi, L., Marsili, S., Vitale, I., Senovilla, L., Michels, J., Garcia, P., et al. (2014). Vitamin B6 metabolism influences the intracellular accumulation of cisplatin. *Cell Cycle, 12*(3), 417–421.

García-Giménez, J. L., Òlaso, G., Hake, S. B., Bönisch, C., Wiedemann, S. M., & Markovic, J. (2013). Histone H3 glutathionylation in proliferating mammalian cells destabilizes nucleosomal structure. *Antioxidants & Redox Signaling, 19*(12), 1305–1320.

Gonzalez, C. A., & Riboli, E. (2006). Diet and cancer prevention: Where we are, where we are going. *Nutrition and Cancer, 56*(2), 225–231.

Gregory, J. F., & Ink, S. L. (1987). Identification and quantification of pyridoxine-β-glucoside as a major form of vitamin B6 in plant-derived foods. *Journal of Agricultural and Food Chemistry, 35*(1), 76–82.

Gregory, J. F., 3rd, & Kirk, J. R. (1981). The bioavailability of vitamin B6 in foods. *Nutrition Reviews, 39*(1), 1–8.

Hatzoglou, M., Snider, M. D., & Maruvada, P. (2014). It's all about balance: Cellular responses to nutrients and development of disease. *Advances in Nutrition, 5*(5), 558–560.

Herzog, B., Hallberg, M., Seth, A., Woods, A., White, R., & Parker, M. G. (2007). The nuclear receptor cofactor, receptor-interacting protein 140, is required for the regulation of hepatic lipid and glucose metabolism by liver X receptor. *Molecular Endocrinology*, *21*, 2687–2697.

Ho, P. C., & Wei, L. N. (2012). Biological activities of receptor-interacting protein 140 in adipocytes and metabolic diseases. *Current Diabetes Reviews*, *8*, 452–457.

Huntington, J. T., Tang, X., Kent, L. N., Schmidt, C. R., & Leone, G. (2016). The spectrum of E2F in liver disease—Mediated regulation in biology and cancer. *Journal of Cellular Physiology*, *231*(7), 1438–1449.

Kannan, K., & Jain, S. K. (2004). Effect of vitamin B6 on oxygen radicals, mitochondrial membrane potential, and lipid peroxidation in H_2O_2-treated U937 monocytes. *Free Radical Biology & Medicine*, *6*(4), 423–428.

Kim, R., Emi, M., & Tanabe, K. (2007). Cancer immunoediting from immune surveillance to immune escape. *Immunology*, *121*, 1–14.

Larsson, S. C., Orsini, N., & Wolk, A. (2010). Vitamin B6 and risk of colorectal cancer: A meta-analysis of prospective studies. *JAMA*, *303*(11), 1077–1083.

Lin, Y. W., Lee, B., Liu, P. S., & Wei, L. N. (2016). Receptor-interacting protein 140 orchestrates the dynamics of macrophage M1/M2 polarization. *Journal of Innate Immunity*, *8*(1), 97–107.

Lin, Y. W., Liu, P. S., Adhikari, N., Hall, J. L., & Wei, L. N. (2015 Feb). RIP140 contributes to foam cell formation and atherosclerosis by regulating cholesterol homeostasis in macrophages. *Journal of Molecular and Cellular Cardiology*, *79*, 287–294. https://doi.org/10.1016/j.yjmcc.2014.12.009 [Epub 2014 Dec 18].

Liu, T. F., Brown, C. M., El Gazzar, M., McPhail, L., Millet, P., Rao, A., et al. (2012). Fueling the flame: Bioenergy couples metabolism and inflammation. *Journal of Leukocyte Biology*, *2*(3), 499–507.

Marzio, A., Merigliano, C., Gatti, M., et al. (2014). Sugar and chromosome stability: Clastogenic effects of sugars in vitamin B6-deficient cells. *PLoS Genetics*, *10*(3), e1004199.

Maurizi, G., Della Guardia, L., Maurizi, A., & Poloni, A. (2002). Adipocytes properties and crosstalk with immune system in obesity-related inflammation. *Biomedical Pharmacotherapy*, *56*(5), 215–222.

Mayne, S. T., Playdon, M. C., & Rock, C. L. (2016). Diet, nutrition, and cancer: Past, present and future. *Nature Reviews. Clinical Oncology*, *13*(8), 504–515.

McCormick, D. B., & Chen, H. (1999). Update on interconversions of vitamin B-6 with its coenzyme. *Journal of Nutrition*, *129*, 325–327.

Meritxell, R., Jones, M. C., & Parker, M. G. (2011). Role of nuclear receptor corepressor RIP140 in metabolic syndrome. *Biochimica et Biophysica Acta*, *1812*(2011), 919–928.

Merrill, A. H., Jr., & Henderson, J. M. (1990). Vitamin B6 metabolism by human liver. *Annals of the New York Academy of Sciences*, *585*, 110–117.

Metz, T. O., Alderson, N. L., Chachich, M. E., Thorpe, S. R., & Baynes, J. W. (2003). Pyridoxamine traps intermediates in lipid peroxidation reactions in vivo: Evidence on the role of lipids in chemical modification of protein and development of diabetic complications. *Journal of Biological Chemistry*, *8*, 42012–42019.

Metzler, D. E., Ikawa, M., & Snell, E. E. (1954). A general mechanism for vitamin B6-catalyzed reactions. *Journal of the American Chemical Society*, *76*, 648–652.

Monlun, M., Hyernard, C., Blanco, P., Lartigue, L., & Faustin, B. (2017). Mitochondria as molecular platforms integrating multiple innate immune signalings. *Journal of Molecular Biology*, *429*(1), 1–13.

Nautiyal, J., Christian, M., & Parker, M. G. (2013). Distinct functions for RIP140 in development, inflammation, and metabolism. *Trends in Endocrinology and Metabolism*, *24*(9), 451–459. https://doi.org/10.1016/j.tem.2013.05.001 [Epub 2013 Jun 3. Review].

Oka, T. (2001). Modulation of gene expression by vitamin B6. *Nutrition Research Reviews, 14*, 57–265.
Parker, M. G., Christian, M., & White, R. (2006). The nuclear receptor co-repressor RIP140 controls the expression of metabolic gene networks. *Biochemical Society Transactions, 34*(Pt 6), 1103–1106.
Polhemus, D. J., & Lefer, D. J. (2014). Emergence of hydrogen sulfide as an endogenous gaseous signaling molecule in cardiovascular disease. *Circulation Research, 114*(4), 730–737.
Ross, S. A. (2007). Nutritional genomic approaches to cancer prevention research. *Experimental Oncology, 29*(4), 250–256.
Ross, S. A. (2008). Evidence for the relationship between diet and cancer. *Cancer Letters, 269*, 189–198.
Ross, S. A., & Davis, C. D. (2011). MicroRNA, nutrition, and cancer prevention. *Advances in Nutrition*, (6), 472–485.
Ross, S. A., Dwyer, J., Umar, A., Kagan, J., Verma, M., Van Bemmel, D. M., et al. (2008). Introduction: Diet, epigenetic events and cancer prevention. *Nutrition Reviews, 66*(Suppl. 1), S1–6.
Schulz, E., Wenzel, P., Münzel, T., & Daiber, A. (2014). Mitochondrial redox signaling: Interaction of mitochondrial reactive oxygen species with other sources of oxidative stress. *Antioxidants & Redox Signaling, 20*(2), 308–324.
Ulrich, P., & Cerami, A. (2001). Protein glycation, diabetes, and aging. *Recent Progress in Hormone Research, 56*, 1–21.
Wang, J., & Boyd, R. J. (1996). Tautomeric equilibria of hydroxypyridines in different solvents: Ab initio study. *The Journal of Physical Chemistry, 100*, 16141–16146.
Wen, C., Junxia, G., Yanzhen, Z., & Jing, Z. (2016). The beneficial effects of taurine in preventing metabolic syndrome. *Food & Function*, 7, 1849. https://doi.org/10.1039/c5fo01295c. www.rsc.org/foodfunction.
Williams, S. C. P. (2013). Link between obesity and cancer. *Proceedings of the National Academy of Sciences of the United States of America, 110*, 8753–8754.
Yu, X. H., Zheng, X. L., & Tang, C. K. (2015). Nuclear factor-κB activation as a pathological mechanism of lipid metabolism and atherosclerosis. *Advances in Clinical Chemistry, 70*, 1–30.
Zschiedrich, U., Hardeland, A., Krones-Herzig, M., Berriel Diaz, A., Vegiopoulos, J., Muggenburg, D., et al. (2008). Coactivator function of RIP140 for NFkappaB/RelA-dependent cytokine gene expression. *Blood, 112*, 264–276.

CHAPTER FIVE

Recent Developments in Folate Nutrition

Nassim Naderi*,†, James D. House*,1

*Department of Food and Human Nutritional Sciences, University of Manitoba, Winnipeg, MB, Canada
†Institute of Nutrition and Functional Foods (INAF), Université Laval, Quebec City, QC, Canada
1Corresponding author: e-mail address: james.house@umanitoba.ca

Contents

Abstract

The term folate (vitamin B9) refers to a group of water-soluble compounds that are nutritionally essential for the support of optimal human health and development. Folates participate in numerous one-carbon transfer reactions, including the methylation of important biomolecules (lipids, amino acids, DNA). A deficiency of folate leads to pathological outcomes including anemia and impairments in reproductive health and fetal development. Due to the linkage of impaired folate status with an increased prevalence of neural tube defects (NTDs) in babies, several jurisdictions required the fortification of the food supply with folic acid, a synthetic and stable form of folate. Data from the postfortification era have provided strong evidence for the reduction of NTDs due to folic acid fortification. However, concern is now growing with respect to the amount of synthetic folic acid within the human food supply. Excess folic acid intake has been linked to a masking of vitamin B12 deficiency, and concerns regarding the promotion of folate-sensitive cancers, including colorectal cancer. New strategies to ensure the supply of optimal folate to at-risk populations may be needed, including the use of biofortification approaches, in order to address recent concerns.

Advances in Food and Nutrition Research, Volume 83
ISSN 1043-4526
https://doi.org/10.1016/bs.afnr.2017.12.006

1. INTRODUCTION

Folate (vitamin B9) represents the collective term for a class of compounds (vitamers) belonging to the water-soluble family of vitamins. Collectively, the folate vitamers play important roles in one-carbon metabolism, including their involvement in the biosynthetic pathways for key amino acids and nucleic acids (Institute of Medicine, 1998). As such, folate is a key factor in cell replication and intermediary metabolism. A deficiency of folate has been linked to a variety of disorders (Iyer & Tomar, 2009), including anemia and neural tube defects (NTDs). While animal cells can interconvert the various vitamin forms, they do not possess the ability to synthesize the foundational pteroylglutamic acid structure, and thus rely on exogenous sources of this important nutrient (Molloy, 2005). Exogenous supply can come in the form of naturally occurring folates in the food supply and via folate synthesized by commensal microorganisms within the large bowel. Additionally, synthetic crystalline folic acid is routinely used to fortify foods and as a constituent of dietary supplements. While supplemental folic acid has proven effective in addressing critical public health issues, including the reduction of the incidence of NTDs, there is growing evidence that the human food supply is now overfortified with folic acid, and this might pose health risks, including the increased risk of developing colorectal cancer and the masking of vitamin B12 deficiency (Kim, 2004). The risk appears linked to crystalline folic acid, and not the naturally occurring forms of folate. As such, there is interest in positioning alternative sources of folate for use in supplements and to fortify the food supply. This chapter will further discuss recent advances in our understanding of folate nutrition, particularly with respect to addressing alternative approaches to ensure adequate folate intakes in target populations.

2. FOLATE AND FOLIC ACID

2.1 Terminology and Chemical Structures

Folate, also known as folacin or vitamin B9, refers to a family of chemically and functionally related compounds that exhibit a common vitamin activity based on the parent structure of folic acid (Selhub & Rosenberg, 1996). The synthetic folic acid (pteroylglutamate) is the most oxidized and stable form of folate. Folic acid (molecular weight = 441.4 (g/mol)) is odorless, slightly soluble in cold water, with more solubility in aqueous alkali, acetic acid, phenol, pyridine, and other basic solvents. Folic acid is fully oxidized and, in neutral solution, has a maximum absorption at 298 nm. Synthetic folic acid lacks coenzyme activity and must be reduced to the metabolically active tetrahydrofolate (THF) form within the cell. Due to its chemical stability and greater commercial availability compared to natural form of folate, folic acid is the most commonly used form in supplements and food fortification. Both natural folate and synthetic folic acid after intakes are converted to 5-methyltetrahydrofolate (5-MTHF), and this is the predominant folate form usually found in blood plasma (Wien et al., 2012).

The basic structure of folate comprises a pteridine ring attached to *p*-aminobenzoic acid through a methylene group, an L-glutamic acid residue is linked to *p*-aminobenzoic acid (Fig. 1). The terms folate and folic acid are often used interchangeably for this vitamin. The pteridine ring of folic acid is fully oxidized, while natural folates occur as dihydro- or tetrahydrofolates (H_4folate) with different one-carbon substituents linked at the N5- and/or N10-position (Fig. 2; Lavoisier, 2008).

The folates in food and feedstuffs are typically in a reduced form as polyglutamyl derivatives of tetrahydrofolic acid. The polyglutamated form of 5-MTHF is recognized as the predominant form of dietary folate and is naturally found in fresh food sources (Lucock, 2000). The 5-MTHF vitamer has the chemical formula $C_{20}H_{25}N_7O_6$ (molecular weight of 459.5 g/mol) and possesses intermediate oxidative stability relative to folic acid. The

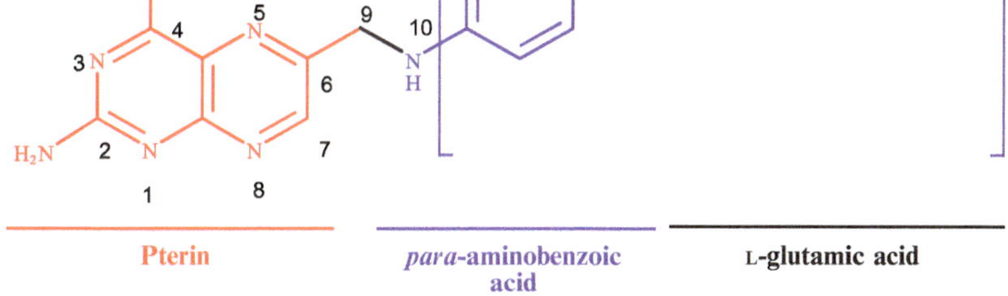

Fig. 1 Chemical structure of crystalline folic acid.

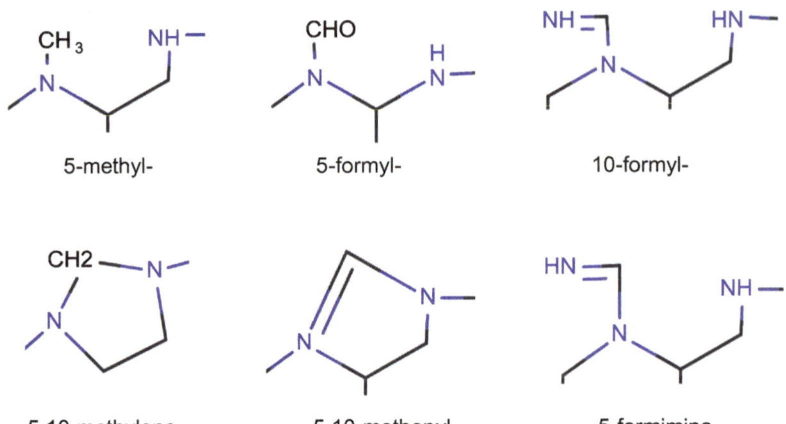

Fig. 2 One-carbon moieties linked at either the N5- and/or N10-position of the folate structure.

compound can be isolated as a white powder; however, in solution it requires ascorbate or another reducing substance to achieve medium-term stability. At neutral pH, it has an absorption maxima of 290 (Scott & Weir, 1993). A calcium salt of L-5-methyltetrahydrofolic acid ($C_{20}H_{23}CaN_7O_6$; MW = 497.5 g/mol) has been synthesized (Ginting & Arcot, 2004) and shown to possess comparative physiological activity, bioavailability, and absorption at equimolar doses to the naturally occurring form (Pietrzik, Bailey, & Shane, 2010).

2.2 Historical Perspective

The natural form of folate was first discovered by Lucy Wills over 80 years ago as a component of yeast with documented abilities to both prevent and cure macrocytic anemia. The structure of polyglutamate folate was determined in the mid-1940s. The chronology of folate discovery is summarized in Table 1. Later in 1943, the monoglutamate form of folate was synthesized in pure crystalline form, and this discovery proved that folic acid is composed of a pteridine ring, *p*-aminobenzoic acid, and glutamic acid. Afterward, it became evident that natural folates usually differed from pteroylglutamic acid. Today, folic acid refers to the fully oxidized chemical compound which does not exist in natural foods. The term "folate" is designated to the large group of compounds having the same vitamin activity and includes natural folates and folic acid (Hoffbrand & Weir, 2001).

Table 1 The Chronology of Folic Acid Discovery From Ulrich and Potter (2006)

1930	Wills and Mehta	Yeast extract prevented the dietary anemia in rats
1931	Wills	Yeast or marmite prevents macrocytic anemia of pregnancy
1934	Vaughan and Turnbull	Marmite corrects anemia of coeliac disease
1938	Wills and Evans	Purified liver extracts do not correct nutritional, pregnancy, or macrocytic anemia
1940	Snell and Peterson	Norit eluate factor—factor absorbed from yeast or liver is growth factor for *Lactobacillus casei*
1941	Mitchell, Snell, and Williams	The term folic acid coined and shown to be a growth factor for *Streptococcus lactis R* (*Streptococcus faecalis*)
1943; 1946	Fullerton; Watson and Castle	Idiopathic steatorrhoea megaloblastic anemia responds to crude liver extracts or yeast extract
1943	Wright and Welch	Enzyme hydrolyzing folate polyglutamates to monoglutamates folate conjugase
1944	Binkley et al.	Yeast extracts effective as a source of vitamin BC only 2%–5% being active for *L. casei*. Required enzymatic digestion to balance activity
1945	Angier et al.	Synthesis of folic acid and using the term pteroylglutamic acid
1945	Day et al.	Purified *L. casei* factor is vitamin M
1946	Pfiffner, Calkins, Bloom, and O'Dell	Naturally occurring folate in liver is a heptaglutamate

2.3 Functional Bioefficacy

The functional bioefficacy of a nutrient refers to the proportion of it that, once ingested, possesses the ability to perform specific metabolic function (Brouwer, van Dusseldorp, West, & Steegers-Theunissen, 2001). Since plasma total homocysteine is a functional index of folate status, by nature of the role that folate plays in sulfur amino acid metabolism (Fig. 3), changes in plasma total homocysteine concentration in response to a given intake of folate or folic acid can be used as a measure of functional bioefficacy. Factors influencing bioavailability and bioefficacy of folate include: (a) species of folate, (b) amount of folate and folic acid consumed, (c) food matrix effects

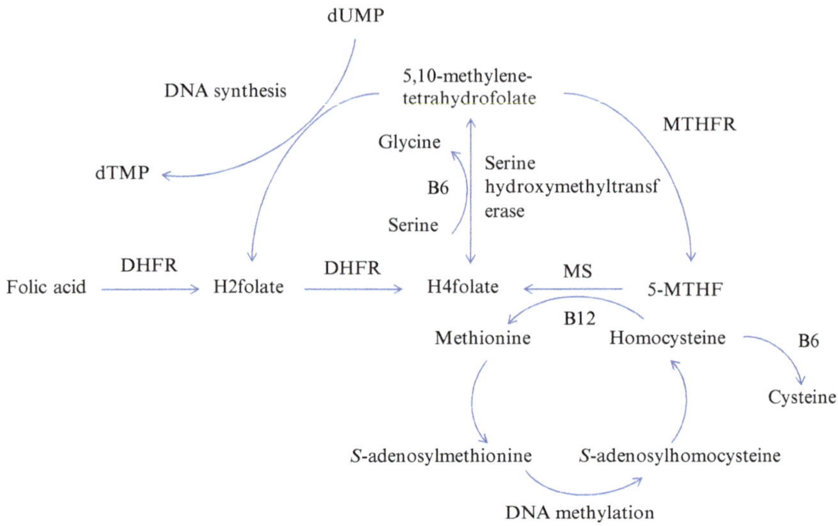

Fig. 3 Schematic presentation of the main metabolic pathways of folate.

(d) overall nutrient status of the host (e) genetic factors, and (f) host-related factors (Gregory, 2001).

Folates and their derivatives occur in nature in forms ranging from mono- to polyglutamates. Synthetic folic acid exists in the monoglutamate form and it is fully oxidized at the N5- and N10-positions. The polyglutamated forms of natural folates, a form needed for cellular folate retention, must be hydrolyzed to monoglutamate forms prior to absorption in the small intestine (see below). The monoglutamate forms of folate, including folic acid, are transported across the proximal small intestine. During the past decades, several attempts have been made to assess the bioavailability of folate polyglutamate compared to monoglutamate form, but the results were not consistent. It is suggested that bioavailability of ingested monoglutamates is significantly greater than that of folate polyglutamates apparently because of the requirement for hydrolysis of the latter (Gregory, 2001; Patanwala et al., 2014). It has been stated that the bioavailability of dietary folate is hampered by the polyglutamate chain to which most of the natural folate is attached. The polyglutamate chain must be removed by the enzyme conjugase present in the brush border of the small intestine prior to absorption and transport as a monoglutamate into the portal vein (Melse-Boonstra et al., 2002). Bioavailability of folate or folic acid is likely to be influenced by the amount ingested. Uncertainty exists as to the extent of folate bioavailability from natural food sources, as this is dependent on both host- and food-related factors (McNulty & Pentieva, 2004). Limitations related to the host organism include folate status, health, age, gender, gastrointestinal function, and the use of medication and alcohol (Melse-Boonstra et al., 2002; Witthöft, 2011).

Food processing can also influence bioavailability by affecting the food matrix and folate stability, resulting in variable folate losses prior ingestion. Bioavailability of folates in food stuff has been difficult to assess quantitatively, but in general folates from animal products for human appear to be more bioavailable compared to plant-derived food (McKillop et al., 2003). By comparing homocysteine and glycine, both sensitive measures of folate status in the folate depleted–repleted rodent model, it was shown that the folate present in chicken egg in the form of 5-MTHF monoglutamate had the same or slightly greater relative bioavailability as crystalline folic acid (House, O'Connor, & Guenter, 2003).

2.4 Stability of Folates

Folates are susceptible to oxidative degradation during food processing which is enhanced by oxygen, light, and heat. Oxidation results in a splitting of the molecule into biologically inactive forms, of which *p*-aminobenzoyl-glutamate is one major form (Strandler, Patring, Jägerstad, & Jastrebova, 2015).The susceptibility of folate is largely influenced by the pH of the medium, reducing agents in the buffer, type of buffer, folate derivatives, and the food system. In most cases, folic acid shows considerably better stability than the reduced folates. The use of antioxidant agents such as ascorbic acid can improve considerably the stability of 5-MTHF (House, Braun, Ballance, O'Connor, & Guenter, 2002). Folate compounds are susceptible to light which is reported to cause cleavage at C^9–N^{10} position. Thereby, folate analysis should be often carried out in low light and glassware that prevents light penetration (i.e., wrapped with aluminum foil) (Indrawati et al., 2004).

2.5 Absorption and Folate Metabolism

After consumption of polyglutamate folates and synthetic monoglutamate, they undergo absorption in the intestinal tract. While folic acid can be absorbed directly, the polyglutamated forms of folates must be converted to their monoglutamate form. Within the intestinal tract of humans, this is principally achieved through the action of glutamate carboxypeptidase II (folate conjugase) located on the brush surface of enterocytes

(Visentin, Diop-Bove, Zhao, & Goldman, 2014). The monoglutamate forms of folate are then transported across the enterocyte via the action of specific transporters: (1) the reduced folate carrier or (2) the proton-coupled folate transporter (PCFT) (Said, 2011). The latter PCFT carrier is postulated to be the primary mechanism of folate transport in the upper intestinal tract (Visentin et al., 2014).

Following transport, monoglutamates are reduced to di- and tetrahydrofolate (THF) by reductase enzymes. THF is the parent compound of all biologically active form of folate. In next step, the THF is converted to 5,10-methylenetetrahydrofolate and 5-MTHF. From there, they are transferred to the hepatic portal vein which leads to the liver and systemic blood circulation and body tissues (Czeizel, Dudás, Paput, & Bánhidy, 2011; Visentin et al., 2014; Witthöft & Jägerstad, 2002).

After reduction to tetrahydrofolate, synthetic folic acid gains metabolic activity similar to other folate species. The enzyme dihydrofolate reductase is responsible in reducing folic acid to dihydrofolate (H_2folate) and next to THF (Fig. 3). The main function of folate in the human body system is the transport of one-carbon moieties, such as methyl and formyl groups. The 5,10-methylenetetrahydrofolate reductase enzyme is responsible for converting the THFs to 5-MTHF via 5,10-methylenetetrahydrofolate. 5-MTHF is the methyl donor in the remethylation of homocysteine to methionine through the action of methionine synthase (MS). Methionine (essential amino acid) will be converted to *S*-adenosylmethionine which is an important intracellular methyl group donor (Fox & Stover, 2008).

Folates act as a cofactor to transport one-carbon units at different oxidation levels (methyl-, formyl-, methylene-) for the biosynthesis of DNA and for hundreds of methylation reactions. Therefore, folates are important for all cell replication activities and restoring DNA and RNA. Folates play a role in regeneration of methionine from homocysteine. This reaction is catalyzed by MS and requires vitamin B12 as coenzyme. During this reaction, the methyl group in folate is transferred first from 5-MTHF to the cofactor to form methylcobalamin, and then the methyl group is ultimately transferred to homocysteine. Deficiency of vitamin B12 in cells results in the accumulation of 5-methylated folate, thus forming a block in folate metabolism, the "methyl trap" (Scott & Weir, 1993). Thus, vitamin B12 deficiency can create a conditional folate deficiency. Folate deficiency causes abnormal nuclear maturation in the blood cells which in turn leads to megaloblastic anemia (Aslinia, Mazza, & Yale, 2006).

2.6 Folate and Health

Beyond its role in normal metabolism, folate status has been linked to the development of chronic diseases, including cardiovascular disease, and also in reproductive health. Given the involvement of folate with other water-soluble vitamins in the methionine cycle, principally vitamin B6 and B12, specific roles for folate are often difficult to delineate. Folate deficiency has traditionally been recognized as a problem in countries struggling with poverty and malnutrition, with the main risk being macrocytic megaloblastic anemia (Gough, Read, McCarthy, & Waters, 1963). Low intake of folate associated with megaloblastic anemia led to a higher risk of giving birth to infants with NTDs and possibly other birth defects (Iyer & Tomar, 2012). With respect to chronic diseases, folate inadequacy has been linked to cardiovascular diseases (Graham et al., 1997), depression, Alzheimer's disease (Kruman et al., 2002), and cancer (Duthie, 1999). Folate plays a part in many metabolic pathways, including DNA replication, repair, and methylation, and in the synthesis of nucleotides, amino acids, and some vitamins. A dietary supply of this vitamin is therefore required to prevent nutritional deficiency in humans (LeBlanc, de Giori, Smid, Hugenholtz, & Sesma, 2007). Traditionally, folate deficiency was more common in populations that had a high intake of refined cereals and a low intake of green leafy vegetables and fruit. Other risk factors for folate deficiency include the malabsorption of folic acid, which can be caused by diseases affecting either intestinal pH or the jejunal mucosa, or due to genetic defects. Prior to the introduction of mandatory wheat flour fortification with folic acid in 1998, about 15% of adult women in the United States had low serum folate levels. Serum folate level is an indicator of recent dietary folate intake, and the most widely used method of assessing folate status (Pfeiffer, Caudill, Gunter, Osterloh, & Sampson, 2005).

The main dietary sources of folate are green leafy vegetables, pulses, liver, egg, and fortified grain products. As such, the requirement for this vitamin is supplied mainly via the dietary intake of folates and the use of synthetic folic acid. Vegetables and pulses are considered as rich sources of folate having 200–600 μg $100\,g^{-1}$. Citrus fruits and cereals are good sources of folate with concentration of 50–100 μg $100\,g^{-1}$. Egg yolk has high content of folate for about 90 μg $100\,g^{-1}$. In addition to dietary sources of folate, either from natural or supplemental/synthetic forms, the human digestive tract harbors microorganisms capable of synthesizing folate. Evidence is accumulating that this source of folate is metabolically available to the host (Aufreiter et al., 2009). However, the extent of this contribution has not been sufficiently quantified to date.

2.6.1 Neural Tube Defects

NTDs (spina bifida, encephalocele, and anencephaly) represent a major group of birth defects. During fetal development, the neural tube closes by the third week of postconception in humans. Failure of the neural tube to close completely results in a neural tube defect or NTD (Czeizel & Dudas, 1992). Maternal nutrition factors contribute significantly to the various etiologies of NTDs. Previous investigation (Smithells, Sheppard, & Schorah, 1976) reported that the diets and postnatal blood of women who had given birth to babies with an NTD were lacking in several micronutrients, principally folic acid. Following two decades of research, it was established that folic acid supplementation reduces 50%–75% of NTD-affected pregnancies (Laurence, James, Miller, Tennant, & Campbell, 1981; Scott & Weir, 1993; Seller & Nevin, 1984; Vergel, Sanchez, Heredero, Rodriguez, & Martinez, 1990). Dietary supplementation with 400 mg of folic acid per day showed 70% prevention of NTD in women with a past history of carrying a fetus with NTDs and 50% of NTDs in the general population. It was found that 30%–50% of the population is nonresponsive to maternal folate level, demonstrating folate-responsive and nonresponsive individuals in the human population (Berry & Li, 2002; Czeizel et al., 2011). The latter observations provided strong evidence that additional folate might provide a public health benefit, particularly with respect to healthy birth outcomes.

2.7 Folate Fortification

Food fortification refers to addition of micronutrients into processed foods. It is a valid technology to improve micronutrient levels, as part of food-based approaches, if access to adequate levels of specific nutrients is limited. Supplementation, on the other hand, refers to the provision of micronutrients in the form of pills, capsules, or syrups. Typically, supplementation is considered to be a rapid approach to achieve enhanced intakes of nutrients of concern, in stable and bioavailable forms, in select population groups. However, with respect to NTDs, given the fact that 3 weeks postconception reflects the maximum time for the protective effect of folate, challenges to the efficacy of folic acid supplementation for NTD risk mitigation exist. These include the fact that many pregnancies are unplanned and the usage of supplements by women of child-bearing age is typically below 40%. In a cohort of young, college-aged women, only 26% of the participants reported taking a supplement containing folic acid (Shuaibi, House, & Sevenhuysen, 2008). Supplement usage by women during pregnancy is, however, considerably higher, with greater than 90% of a cohort of pregnant women in Ontario, Canada, taking supplements with folic acid (Plumptre et al., 2015). However, this level of supplement usage may not be NTD-protective given the need for adequate periconceptional folate status. Fortification of the food supply with folic acid presented an alternative approach to ensuring adequate intakes.

In 1992, to reduce the number of cases of NTDs, the US Public Health Service recommended that all women capable of becoming pregnant consume 400 μg of folic acid daily. Three approaches were advocated to increase folate consumption, including the improvement of dietary habits to ensure the consumption of foods naturally rich in folate, fortify foods with folic acid, and promote the use of dietary supplements containing folic acid (CDC, 1991). However, for reasons discussed above, the governments of the United States and Canada, along with several other countries, introduced the mandatory fortification of grain products with folic acid (FDA, 1996). Crystalline folic acid (monoglutamate form) was chosen as the form of folate to use in fortifying foods owing to its availability, cost, and stability.

Mandatory fortification of cereal grain products started in 1998 from United States and followed by Canada and some Latin American countries but not in Norway or in the United Kingdom (Wien et al., 2012). Thereafter, the results of several studies and surveillance show the number of NTD-affected pregnancies declined by as much as half. Given the significant decline in NTD rates, mandatory folic acid fortification became recognized as a major public health triumph (De Wals et al., 2007). However, in recent years, concern has been raised that populations within folic acid-fortified countries might be exposed to excessive levels of this form of folate.

2.8 Adverse Effects of Folic Acid Food Fortification and Supplementation

Despite documented beneficial effects of folic acid fortification on preventing NTDs, concern has been raised about the amount of population exposure to crystalline folic acid. The potential for folic acid, enhanced via food fortification, in masking vitamin B12 deficiency has been raised, as the latter leads to the progression of irreversible neurological symptoms (Molloy, Brody, Mills, Scott, & Kirke, 2009). The latter findings were of particular importance to those individuals who received folic acid from both supplements and a range of fortified foods because in this situation people may exceed the upper limit for folic acid intake (1 mg from folic acid). Furthermore, there are some reports on the relation between folic acid supplementation during pregnancy and risk for miscarriage or an increase in the occurrence of multiple births (Czeizel & Dudas, 1992).

Another concern that has been raised in relation to increased exposure to dietary folic acid relates to the potential for increased risk of gastrointestinal cancer, including colorectal cancer (Kim, 2004). A systematic review found only prostate cancer to be linked to high folic acid intakes (Wien et al., 2012). While the latter finding may not translate to specific concerns related to maternal folate exposure, recent preclinical studies in animals have shown that high folic acid (and other methyl donor) intakes may lead to epigenetic changes in offspring, increasing the risk of development of diabetes and alterations in food intake behavior (Huot et al., 2016). Perhaps it is time to reevaluate the strategy of fortifying foods with folic acid, taking into account the impacts to the entire population and not just a subsection at risk for NTD-affected pregnancies.

2.9 Replacing Synthetic Folic Acid With Natural Folate

The introduction of mandatory folic acid fortification of grain products in certain jurisdictions, including Canada and the United States, has succeeded in decreasing the rates of NTD by 20%–50% (Persad, Van den Hof, Dubé, & Zimmer, 2002; Ray, 2004). However, in some countries like the Netherlands, folic acid fortification is prohibited due to growing debate about the long-term safety of folic acid consumption by general population (Varela-Moreiras, Murphy, & Scott, 2009). The debate addresses two main risks of long-term consumption of high folic acid intake which could promote the formation of colorectal tumors (Kim, 2004; Mason et al., 2007) and masking the appearance of vitamin B12 deficiency anemia (Morris, Jacques, Rosenberg, & Selhub, 2007, 2010). Recent studies regarding folic acid fortification are rediscovering the harms to the nervous system from long-term exposure to doses of folic acid between 0.5 and 1 mg in the presence of vitamin B12 deficiency (Reynolds, 2016). Thereby, it is highly recommended to review the safe upper intake level of folic acid to prevent the health risk factors.

As discussed above, synthetic folic acid is absorbed by carrier-mediated mechanism, and it must be reduced and then metabolized to 5-MTHF in the human mucosal cell and/or liver in order to be metabolically active. Because the ability of conversion is limited, unmetabolized folic acid can appear in the systemic circulation, even after low-dose application (Pietrzik et al., 2010). Due to the suspected adverse effects of folic acid fortification, attention has been directed toward other solutions that might help increase population intake of natural folates, which do not carry with them the same safety concerns as synthetic folic acid, particularly with respect to the masking of vitamin B12 deficiency. As such, other strategies have been investigated in order to increase population intake of natural folates without adverse effect of synthetic folic acid. The development of novel foods enriched with natural folates has been considered as an innovative solution in nutritional strategies.

One strategy for increasing natural folates in the food supply includes the development of novel functional foods based on animal products enriched with natural folate through the addition of high dose of synthetic folic acid to animal feed (Altic, McNulty, Hoey, McAnena, & Pentieva, 2016; House et al., 2002). Research has shown that hen eggs can be enriched with folate by supplementing hen's diet with folic acid and reaching maximum folate content of about 2.5 times that of a normal egg (Altic et al., 2016; House et al., 2002; Sahlin & House, 2006). Natural folates in novel folate-enriched eggs are highly stable with little or no losses during storage at refrigerator or room temperature for 27 days (Altic et al., 2016). Studies revealed that the folate which appears in enriched eggs is in the natural form, mainly as 5-MTHF, and this strategy could offer an efficient solution to increase folate intake of population, especially where long-term consumption of folic acid-fortified foods has been criticized (Hoey et al., 2009). This researcher stated that no polyglutamate forms were present in egg yolk. House et al. (2003) also reported the relative bioavailability of folate in eggs is higher or equal to 100% compared to folic acid. The 5-MTHF in egg yolk does not oxidize during cold (4°C) storage for 4 weeks and folate levels remained constant relative to freshly collected eggs (House et al., 2002).

Compared to green leafy vegetables, folates in animal products including eggs are also usually more stable during thermal treatment (McKillop et al., 2003). However, a significant obstacle in egg consuming is the perception that egg consumption is associated with a rise in blood cholesterol levels. Food processing techniques such as fractionation methods enable scientists to fractionate egg yolk in different components with different compositions. Fractionation of egg yolk by centrifugation techniques can enable us to separate a low-cholesterol content fraction of yolk which is concentrated with

5-MTHF (Naderi, House, & Pouliot, 2014). The consumption of these novel folate-enriched eggs or egg-derived, folate-enriched fractions could represent alternative and cost-efficient ways of increasing folate intake and potentially folate status of population in countries that mandatory folic acid fortification remains nonexistent (Altic et al., 2016). Another way to introduce a stabilized form of 5-MTHF into the food supply is via direct encapsulation (Liu et al., 2012), a strategy shown to lead to effective fortification of cereal grain products, including noodles and bakery products (Liu, Green, & Kitts, 2015; Liu, Green, Wong, & Kitts, 2013).

Another strategy to increasing the supply of natural folates is via encapsulated probiotics (Divya & Nampoothiri, 2015). The use of folate-producing food-grade microorganisms has been positioned as a natural alternative to fortification with the synthetic folic acid (Laiño, Zelaya, del Valle, de Giori, & LeBlanc, 2015). Fermented milk bioenriched with folate was produced from folate-producing lactic acid bacteria, and this product effectively increased plasma folate concentrations (Laiño et al., 2015). The latter method represents an approach toward biofortification of foods with folate. The use of genetically modified biofortified crops offers additional opportunities. Researchers applied transgenic breeding techniques in order to enhance vitamin or mineral concentrations in some crops such as Golden Rice (De Steur, Feng, Xiaoping, & Gellynck, 2013). However, the genetically modified biofortification faces challenges for commercialization, due to regulatory hurdles and consumer acceptance of genetically modified foods. Folate biofortification through breeding or metabolic engineering in rice, tomato, potato, wheat, and cassava was assessed; however, this strategy needs to be developed to enable adequate biofortification of other staples (Strobbe & Van Der Straeten, 2017).

3. SUMMARY

Folate is a water-soluble vitamin critical for optimal health in humans. Recent attention has focused on the supply of folic acid to the population, with particular emphasis on the reduction of the incidence of NTD rates in newborn infants. Recent evidence is challenging current public health strategies with respect to food fortification, necessitating a review of the current approaches. Innovative alternative approaches to supplying folate to the human population include the use of biofortified foods and novel, stabilized forms of folate.

REFERENCES

Altic, L., McNulty, H., Hoey, L., McAnena, L., & Pentieva, K. (2016). Validation of folate-enriched eggs as a functional food for improving folate intake in consumers. *Nutrients*, *8*, 777.

Angier, R. B., Boothe, J. H., Hutchings, B. L., Mowat, J. H., Semb, J., Stokstad, E. L., et al. (1945). Synthesis of a compound identical with the L. casei factor isolated from liver. *Science*, *102*, 227–228.

Aslinia, F., Mazza, J. J., & Yale, S. H. (2006). Megaloblastic anemia and other causes of macrocytosis. *Clinical Medicine & Research*, *4*, 236–241.

Aufreiter, S., Gregory, J. F., III, Pfeiffer, C. M., Fazili, Z., Kim, Y.-I., Marcon, N., et al. (2009). Folate is absorbed across the colon of adults: Evidence from cecal infusion of (13)C-labeled [6S]-5-formyltetrahydrofolic acid. *The American Journal of Clinical Nutrition*, *90*, 116–123.

Berry, R. J., & Li, Z. (2002). Folic acid alone prevents neural tube defects: Evidence from the China study. *Epidemiology*, *13*, 114–116.

Binkley, S. B., Bird, O. D., Bloom, E. S., Brown, R. A., Calkins, D. G., Campbell, C., et al. (1944). On the vitamin Bc conjugate in yeast. *Science*, *100*, 36–37.

Brouwer, I. A., van Dusseldorp, M., West, C. E., & Steegers-Theunissen, R. P. (2001). Bioavailability and bioefficacy of folate and folic acid in man. *Nutrition Research Reviews*, *14*, 267–294.

Centers for Disease Control and Prevention. (1991). Use of folic acid for prevention of spina bifida and other neural tube defects—1983–1991. *MMWR. Morbidity and Mortality Weekly Report*, *40*, 513.

Czeizel, A. E., & Dudas, I. (1992). Prevention of the first occurrence of neural-tube defects by periconceptional vitamin supplementation. *New England Journal of Medicine*, *327*, 1832–1835.

Czeizel, A. E., Dudás, I., Paput, L., & Bánhidy, F. (2011). Prevention of neural-tube defects with periconceptional folic acid, methylfolate, or multivitamins? *Annals of Nutrition & Metabolism*, *58*(4), 263–271.

Day, P. L., Mims, V., Totter, J. R., Stokstad, E. L., Hutchings, B. L., & Sloane, N. H. (1945). The successful treatment of vitamin M deficiency in the monkey with highly purified Lactobacillus casei factor. *Journal of Biological Chemistry*, *157*, 423–424.

De Steur, H., Feng, S., Xiaoping, S., & Gellynck, X. (2013). Are beneficiaries willing to pay for folate biofortified rice? Findings from a high-risk region in China. In *International Congress Hidden Hunger—From assessment to solution*.

De Wals, P., Tairou, F., Van Allen, M. I., Uh, S. H., Lowry, R. B., Sibbald, B., et al. (2007). Reduction in neural-tube defects after folic acid fortification in Canada. *New England Journal of Medicine*, *357*, 135–142.

Divya, J. B., & Nampoothiri, K. M. (2015). Encapsulated Lactococcus lactis with enhanced gastrointestinal survival for the development of folate enriched functional foods. *Bioresource Technology*, *188*, 226–230.

Duthie, S. J. (1999). Folic acid deficiency and cancer: Mechanisms of DNA instability. *British Medical Bulletin*, *55*, 578–592.

FDA. (1996). Food standards: Amendment of standards of identity for enriched grain products to require addition of folic acid. *Federal Register*, *61*, 8781–8797.

Fox, J. T., & Stover, P. J. (2008). Folate-mediated one-carbon metabolism. *Vitamins & Hormones*, *79*, 1–44.

Fullerton, H. W. (1943). Macrocyte anaemia of pregnancy and the puerperium. *British Medical Journal*, *1*, 158–160.

Ginting, E., & Arcot, J. (2004). High-performance liquid chromatographic determination of naturally occurring folates during tempe preparation. *Journal of Agricultural & Food Chemistry*, *52*, 7752–7758.

Gough, K. R., Read, A. E., McCarthy, C. F., & Waters, A. H. (1963). Megaloblastic anaemia due to nutritional deficiency of folic acid. *The Quarterly Journal of Medicine*, *32*, 243–256.

Graham, I. M., Daly, L. E., Refsum, H. M., Robinson, K., Brattström, L. E., Ueland, P. M., et al. (1997). Plasma homocysteine as a risk factor for vascular disease. The European Concerted Action Project. *Journal of the American Medical Association*, *277*, 1775–1781.

Gregory, J. F. (2001). Case study: Folate bioavailability. *Journal of Nutrition*, *131*, 1376S–1382S.

Hoey, L., McNulty, H. E., McCann, E. M. E., McCracken, K. J., Scott, J. M., Marc, B. B., et al. (2009). Laying hens can convert high doses of folic acid added to the feed into natural folates in eggs providing a novel source of food folate. *British Journal of Nutrition*, *101*, 206–212.

Hoffbrand, A. V., & Weir, D. G. (2001). The history of folic acid. *British Journal of Haematology*, *113*, 579–589.

House, J. D., Braun, K., Ballance, D. M., O'Connor, C. P., & Guenter, W. (2002). The enrichment of eggs with folic acid through supplementation of the laying hen diet. *Poultry Science*, *81*, 1332–1337.

House, J. D., O'Connor, C. P., & Guenter, W. (2003). Plasma homocysteine and glycine are sensitive indices of folate status in a rodent model of folate depletion and repletion. *Journal of Agricultural & Food Chemistry*, *51*, 4461–4467.

Huot, P. S., Ly, A., Szeto, I. M., Reza-Lopez, S. A., Cho, D., Kim, Y. I., et al. (2016). Maternal and postweaning folic acid supplementation interact to influence body weight, insulin resistance, and food intake regulatory gene expression in rat offspring in a sex-specific manner. *Applied Physiology, Nutrition & Metabolism*, *41*(4), 411–420.

Indrawati, C., Arroqui, I., Messagie, M., Nguyen, T., Van Loey, A., & Hendrickx, M. (2004). Comparative study on pressure and temperature stability of 5-methyltetrahydrofolic acid in model systems and in food products. *Journal of Agricultural & Food Chemistry*, *52*, 485–492.

Institute of Medicine. (1998). *Dietary reference intakes for thiamin, riboflavin, niacin, vitamin B6, folate, vitamin B12, pantothenic acid, biotin, and choline*. Washington, DC: National Academies Press (US).

Iyer, R., & Tomar, S. K. (2009). Folate: A functional food constituent. *Journal of Food Science*, *74*(9), R114–R122.

Iyer, R., & Tomar, S. K. (2012). Folate and prevention of neural tube disease. chap. 7 In K. L. Narasimhan (Ed.), *Neural tube defects—Role of folate, prevention strategies and genetics* (pp. 117–138).

Kim, Y. (2004). Folate, colorectal carcinogenesis, and DNA methylation: Lessons from animal studies. *Environmental & Molecular Mutagenesis*, *44*, 10–25.

Kruman, I. I., Kumaravel, T. S., Lohani, A., Pedersen, W. A., Cutler, R. G., Kruman, Y., et al. (2002). Folic acid deficiency and homocysteine impair DNA repair in hippocampal neurons and sensitize them to amyloid toxicity in experimental models of Alzheimer's disease. *Journal of Neuroscience*, *22*, 1752–1762.

Laiño, J. E., Zelaya, H., del Valle, M. J., de Giori, G. S., & LeBlanc, J. G. (2015). Milk fermented with selected strains of lactic acid bacteria is able to improve folate status of deficient rodents and also prevent folate deficiency. *Journal of Functional Foods*, *17*, 22–32.

Laurence, K. M., James, N., Miller, M. H., Tennant, G. B., & Campbell, H. (1981). Double-blind randomised controlled trial of folate treatment before conception to prevent recurrence of neural-tube defects. *British Medical Journal (Clinical Research Edition)*, *282*, 1509–1511.

Lavoisier, A. L. (2008). Chemical and physiological properties of vitamins. In G. F. Combs (Ed.), *The vitamins: Fundamental aspects in nutrition and health* (3rd ed., pp. 35–74). London: Academic Press.

LeBlanc, J. G., de Giori, G. S., Smid, E. J., Hugenholtz, J., & Sesma, F. (2007). Folate production by lactic acid bacteria and other food-grade microorganisms. *Communicating Current Research and Educational Topics and Trends in Applied Microbiology*, *1*, 329–339.

Liu, Y., Green, T. J., & Kitts, D. D. (2015). Stability of microencapsulated L-5-methyltetrahydrofolate in fortified noodles. *Food Chemistry, 171*, 206–211.

Liu, Y., Green, T. J., Wong, P., & Kitts, D. D. (2013). Microencapsulation of L-5-methyltetrahydrofolic acid with ascorbate improves stability in baked bread products. *Journal of Agricultural and Food Chemistry, 61*, 249–255.

Liu, Y., Tomiuk, S., Rozoy, E., Soimard, S., Bazinet, L., Green, T., et al. (2012). Thermal oxidation studies on reduced folate, L-5-methyltetrahydrofolic acid (L-5-MTHF) and strategies for stabilization using food matrices. *Journal of Food Science, 77*, C236–243.

Lucock, M. (2000). Folic acid: Nutritional biochemistry, molecular biology, and role in disease processes. *Molecular Genetics and Metabolism, 71*, 121–138.

Mason, J. B., Dickstein, A., Jacques, P. F., Haggarty, P., Selhub, J., Dallal, G., et al. (2007). A temporal association between folic acid fortification and an increase in colorectal cancer rates may be illuminating important biological principles: A hypothesis. *Cancer Epidemiology Biomarkers & Prevention, 16*, 1325–1329.

McKillop, D. J., Pentieva, K. D., Scott, J. M., Strain, J. J., McCreedy, R., Alexander, J., et al. (2003). Protocol for the production of concentrated extracts of food folate for use in human bioavailability studies. *Journal of Agricultural & Food Chemistry, 51*, 4382–4388.

McNulty, H., & Pentieva, K. (2004). Folate bioavailability. *The Proceedings of the Nutrition Society, 63*, 529–536.

Melse-Boonstra, A., Verhoef, P., Konings, E. J. M., van Dusseldorp, M., Matser, A., Hollman, P. C. H., et al. (2002). Influence of processing on total, monoglutamate and polyglutamate folate contents of leeks, cauliflower, and green beans. *Journal of Agricultural & Food Chemistry, 50*, 3473–3478.

Mitchell, H. K., Snell, E. E., & Williams, R. J. (1941). The concentration of "folic acid" *Journal of the American Chemical Society, 63*, 2284.

Molloy, A. M. (2005). The role of folic acid in the prevention of neural tube defects. *Trends in Food Science & Technology, 16*, 241–245.

Molloy, A. M., Brody, L. C., Mills, J. L., Scott, J. M., & Kirke, P. N. (2009). The search for genetic polymorphisms in the homocysteine/folate pathway that contribute to the etiology of human neural tube defects. *Birth Defects Research Part A: Clinical and Molecular Teratology, 85*, 285–294.

Morris, M. S., Jacques, P. F., Rosenberg, I. H., & Selhub, J. (2007). Folate and vitamin B-12 status in relation to anemia, macrocytosis, and cognitive impairment in older Americans in the age of folic acid fortification. *The American Journal of Clinical Nutrition, 85*, 193–200.

Morris, M. S., Jacques, P. F., Rosenberg, I. H., & Selhub, J. (2010). Circulating unmetabolized folic acid and 5-methyltetrahydrofolate in relation to anemia, macrocytosis, and cognitive test performance in American seniors. *The American Journal of Clinical Nutrition, 91*, 1733–1744.

Naderi, N., House, J. D., & Pouliot, Y. (2014). Scaling-up a process for the preparation of folate-enriched protein extracts from hen egg yolks. *Journal of Food Engineering, 141*, 85–92.

Patanwala, I., King, M. J., Barrett, D. A., Rose, J., Jackson, R., Hudson, M., et al. (2014). Folic acid handling by the human gut: Implications for food fortification and supplementation. *The American Journal of Clinical Nutrition, 100*, 593–599.

Persad, V. L., Van den Hof, M. C., Dubé, J. M., & Zimmer, P. (2002). Incidence of open neural tube defects in Nova Scotia after folic acid fortification. *Canadian Medical Association Journal, 167*, 241–245.

Pfeiffer, C. M., Caudill, S. P., Gunter, E. W., Osterloh, J., & Sampson, E. J. (2005). Biochemical indicators of B vitamin status in the US population after folic acid fortification: Results from the National Health and Nutrition Examination Survey 1999–2000. *The American Journal of Clinical Nutrition, 82*, 442–450.

Pfiffner, J. J., Calkins, D. G., Bloom, E. S., & O'Dell, B. L. (1946). On the peptide nature of vitamin Bc conjugate from yeast. *Journal of the American Chemical Society, 68*, 1392.

Pietrzik, K., Bailey, L., & Shane, B. (2010). Folic acid and L-5-methyltetrahydrofolate. *Clinical Pharmacokinetics*, *49*, 535–548.

Plumptre, L., Masih, S. P., Ly, A., Aufreiter, S., Sohn, K. J., Croxford, R., et al. (2015). High concentrations of folate and unmetabolized folic acid in a cohort of pregnant Canadian women and umbilical cord blood. *The American Journal of Clinical Nutrition*, *102*, 848–857.

Ray, J. G. (2004). Folic acid fortification in Canada. *Nutrition Reviews*, *62*, S35–39.

Reynolds, E. H. (2016). What is the safe upper intake level of folic acid for the nervous system? Implications for folic acid fortification policies. *European Journal of Clinical Nutrition*, *70*, 537–541.

Sahlin, A., & House, J. D. (2006). Enhancing the vitamin content of meat and eggs: Implications for the human diet. *Canadian Journal of Animal Science*, *86*, 181–195.

Said, H. M. (2011). Intestinal absorption of water-soluble vitamins in health and disease. *Biochemical Journal*, *437*, 357–372.

Scott, J., & Weir, D. (1993). Folate/vitamin B12 inter-relationships. *Essays in Biochemistry*, *28*, 63–72.

Selhub, J., & Rosenberg, I. H. (1996). Folic acid. In E. E. Ziegler & L. J. Filer, Jr. (Eds.), *Present knowledge in nutrition* (7th ed., pp. 206–219). ILSI Press.

Seller, M. J., & Nevin, N. C. (1984). Periconceptional vitamin supplementation and the prevention of neural tube defects in south-east England and Northern Ireland. *Journal of Medical Genetics*, *21*, 325–330.

Shuaibi, A., House, J. D., & Sevenhuysen, G. P. (2008). Folate status in Canadian women of childbearing age after folic acid fortification of grain products. *Journal of the American Dietetics Association*, *108*, 2090–2094.

Smithells, R. W., Sheppard, S., & Schorah, C. J. (1976). Vitamin deficiencies and neural tube defects. *Archives of Diseases in Childhood*, *51*, 944–950.

Snell, E. E., & Peterson, W. H. (1940). Growth factors for bacteria: X. Additional factors required by certain lactic acid bacteria 1. *Journal of Bacteriology*, *39*(3), 273–285.

Strandler, H. S., Patring, J., Jägerstad, M., & Jastrebova, J. (2015). Challenges in the determination of unsubstituted food folates: Impact of stabilities and conversions on analytical results. *Journal of Agricultural & Food Chemistry*, *63*, 2367–2377.

Strobbe, S., & Van Der Straeten, D. (2017). Folate biofortification in food crops. *Current Opinions in Biotechnology*, *44*, 202–211.

Ulrich, C. M., & Potter, J. D. (2006). Folate supplementation: Too much of a good thing? *Cancer Epidemiology Biomarkers & Prevention*, *15*, 189–193.

Varela-Moreiras, G., Murphy, M. M., & Scott, J. M. (2009). Cobalamin, folic acid, and homocysteine. *Nutrition Reviews*, *67*(1), S69–S72.

Vaughan, J. M., & Turnbull, H. M. (1934). *The anaemias.* Oxford University Press, H. Milford.

Vergel, R. G., Sanchez, L. R., Heredero, B. L., Rodriguez, P. L., & Martinez, A. J. (1990). Primary prevention of neural tube defects with folic acid supplementation: Cuban experience. *Prenatal Diagnosis*, *10*, 149–152.

Visentin, M., Diop-Bove, N., Zhao, R., & Goldman, I. M. (2014). The intestinal absorption of folates. *The Annual Review of Physiology*, *76*, 251–274.

Watson, J., & Castle, W. B. (1946). Nutritional macrocytic anemia, especially in pregnancy: Response to a substance in liver other than that effective in pernicious anemia. *American Journal of Medical Science*, *211*, 513–530.

Wien, T. N., Pike, E., Wisløff, T., Staff, A., Smeland, S., & Klemp, M. (2012). Cancer risk with folic acid supplements: A systematic review and meta-analysis. *BMJ Open*, *2*, e000653.

Wills, L. (1931). Treatment of "pernicious anaemia" of pregnancy and "tropical anaemia". *British Medical Journal*, *1*, 1059–1064. Available at http://www.ncbi.nlm.nih.gov/pmc/articles/PMC2314785/.

Wills, L., & Evans, B. F. (1938). Tropical macrocytic anaemia: Its relation to pernicious anaemia. *Lancet*, *232*, 416–421.
Wills, L., & Mehta, M. M. (1930). Studies in "pernicious anaemia" of pregnancy. Part I. Preliminary report. *Indian Journal of Medical Research*, *17*, 777–792.
Witthöft, C. M. (2011). Analytical methods to assess the bioavailability of water-soluble vitamins in food—Exemplified by folate. In M. Rychlik (Ed.), *Fortified foods with vitamins: Analytical concepts to assure better and safer products* (pp. 21–36). Weinheim, Germany: Wiley-VCH Verlag GmbH & Co. KGaA.
Witthöft, C. M., & Jägerstad, M. (2002). Folates, nutritional significance. In *Encyclopedia of dairy sciences* (pp. 2714–2721). .
Wright, L. D., & Welch, A. D. (1943). The production of folic acid by rat liver in vitro. *Science*, *98*, 179–182.

FURTHER READING

Combs, G. F., Jr., & McClung, J. P. (2016). *The vitamins: Fundamental aspects in nutrition and health*. Academic Press.
Dickson, T. M., Tactacan, G. B., Hebert, K., Guenter, W., & House, J. D. (2010). Optimization of folate deposition in eggs through dietary supplementation of folic acid over the entire production cycle of Hy-Line W36, Hy-Line W98, and CV20 laying hens. *The Journal of Applied Poultry Research*, *19*, 80–91.
Hebert, K., House, J. D., & Guenter, W. (2005). Effect of dietary folic acid supplementation on egg folate content and the performance and folate status of two strains of laying hens. *Poultry Science*, *84*, 1533–1538.
Hertrampf, E., & Cortés, F. (2008). National food-fortification program with folic acid in Chile. *Food & Nutrition Bulletin*, *29*, S231–S237.
Hewitt, S. M., Crowe, C. M. W., Navin, A. W., & Miller, M. E. (1992). Recommendations for the use of folic acid to reduce the number of cases of spina bifida and other neural tube defects. *MMWR. Morbidity and Mortality Weekly Report*, *41*, 980–984.
Lumley, J., Watson, L., Watson, M., & Bower, C. (2001). Periconceptional supplementation with folate and/or multivitamins for preventing neural tube defects. *Cochrane Database Systematic Reviews*, *3*.
Shaw, G. M., Schaffer, D., Velie, E. M., Morland, K., & Harris, J. A. (1995). Periconceptional vitamin use, dietary folate, and the occurrence of neural tube defects. *Epidemiology*, 219–226.
Tactacan, G. (2011). *Characterization of factors influencing the regulation of dietary folic acid deposition in the eggs*. .
Tactacan, G. B., Jing, M., Thiessen, S., Rodriguez-Lecompte, J. C., O'Connor, D. L., Guenter, W., et al. (2010). Characterization of folate-dependent enzymes and indices of folate status in laying hens supplemented with folic acid or 5-methyltetrahydrofolate. *Poultry Science*, *89*, 688–696.
Wright, A. J. A., Finglas, P. M., & Southon, S. (2001). Proposed mandatory fortification of the UK diet with folic acid: Have potential risks been underestimated? *Trends in Food Science & Technology*, *12*, 313–321.

CHAPTER SIX

Vitamin B_{12}

A. David Smith[1,*], Martin J. Warren[2], Helga Refsum[3]
[1]Department of Pharmacology, University of Oxford, Oxford, United Kingdom
[2]School of Biosciences, University of Kent, Canterbury, Kent, United Kingdom
[3]Department of Nutrition, University of Oslo, Oslo, Norway
*Corresponding author: e-mail address: david.smith@pharm.ox.ac.uk

Contents

Advances in Food and Nutrition Research, Volume 83
ISSN 1043-4526
https://doi.org/10.1016/bs.afnr.2017.11.005

Abstract

The biosynthesis of B_{12}, involving up to 30 different enzyme-mediated steps, only occurs in bacteria. Thus, most eukaryotes require an external source of B_{12}, and yet the vitamin appears to have only two functions in eukaryotes: as a cofactor for the enzymes methionine synthase and methylmalonylCoA mutase. These two functions are crucial for normal health in humans, and in particular, the formation of methionine is essential for providing methyl groups for over 100 methylation processes. Interference with the methionine synthase reaction not only depletes the body of methyl groups but also leads to the accumulation of homocysteine, a risk factor for many diseases. The syndrome pernicious anemia, characterized by lack of intrinsic factor, leads to a severe, sometimes fatal form of B_{12} deficiency. However, there is no sharp cutoff for B_{12} deficiency; rather, there is a continuous inverse relationship between serum B_{12} and a variety of undesirable outcomes, including neural tube defects, stroke, and dementia. The brain is particularly vulnerable; in children, inadequate B_{12} stunts brain and intellectual development. Suboptimal B_{12} status (serum $B_{12} < 300$ pmol/L) is very common, occurring in 30%–60% of the population, in particular in pregnant women and in less-developed countries. Thus, many tens of millions of people in the world may suffer harm from having a poor B_{12} status. Public health steps are urgently needed to correct this inadequacy.

ABBREVIATIONS

AD	Alzheimer's disease
AdoCbl	adenosylcobalamin
ApoE	apolipoprotein E
B_{12}	cobalamin
CDR	clinical dementia rating
DMB	dimethylbenzimidazole
DTI	diffusion tensor imaging
HC	haptocorrin
HoloTC	holotranscobalamin (TC-cobalamin)
IF	intrinsic factor
MCI	mild cognitive impairment
MMA	methylmalonic acid
MMSE	mini-mental state examination
RDA	recommended daily allowance
SAH	*S*-adenosylhomocysteine
SAM	*S*-adenosylmethionine
SPECT	single-photon emission computed tomography
TC	transcobalamin
tHcy	plasma total homocysteine

1. INTRODUCTION

This chapter reviews a selection of topics of current interest in the biochemistry, function, and clinical relevance of vitamin B_{12} (B_{12}). We have chosen topics that the authors are familiar with and regret that a more extensive survey was not possible for space reasons. Nevertheless, we hope that the reader will find material of interest and will be provoked by our recommendations for clinical practice. We believe that B_{12} is a neglected area of public health and that more attention needs to be placed upon this water-soluble vitamin.

2. BIOCHEMISTRY OF VITAMIN B_{12}

2.1 Chemistry

A brief history of the discovery of B_{12} helps to explain the nomenclature associated with this remarkable molecule. The compound was first identified as a nutrient, or extrinsic factor, through the research efforts of Minot, Murphy, and Whipple in the 1920s, who demonstrated that the symptoms of pernicious anemia could be overcome through the addition of liver to the diet (Minot & Murphy, 1926; Whipple & Robscheit-Robbins, 1925). The nutrient was eventually isolated from liver samples after addition of cyanide and extraction into organic solvent, from which it was crystallized to reveal the presence of a deep-red pigment (Rickes, Brink, Koniuszy, Wood, & Folkers, 1948; Smith, 1948). The structure of the compound was solved through the pioneering work of Dorothy Hodgkin by the application of X-ray crystallography, revealing that the vitamin was a cyanolated, cobalt-containing, amidated tetrapyrrole (Hodgkin et al., 1956). Consequently, it was called cyanocobalamin and, therefore, technically, B_{12} specifically refers to cyanocobalamin.

The structure of B_{12} is shown in Fig. 1. The cobalt is located in the center of a ring-contracted modified tetrapyrrole macrocycle, coordinated via the four pyrrole nitrogen atoms. This core of the molecule is known as a corrin ring and is similar to, though quite distinct from, the tetrapyrrole-derived ring systems found in hemes and chlorophylls. Attached to the corrin ring, through a side chain linked to ring D of the macrocycle, is a lower nucleotide loop that contains an unusual base called dimethylbenzimidazole (DMB), which also coordinates to the cobalt ion. In B_{12}, therefore, the cobalt ion is ligated not only through the four pyrrole ring nitrogens of the corrin ring

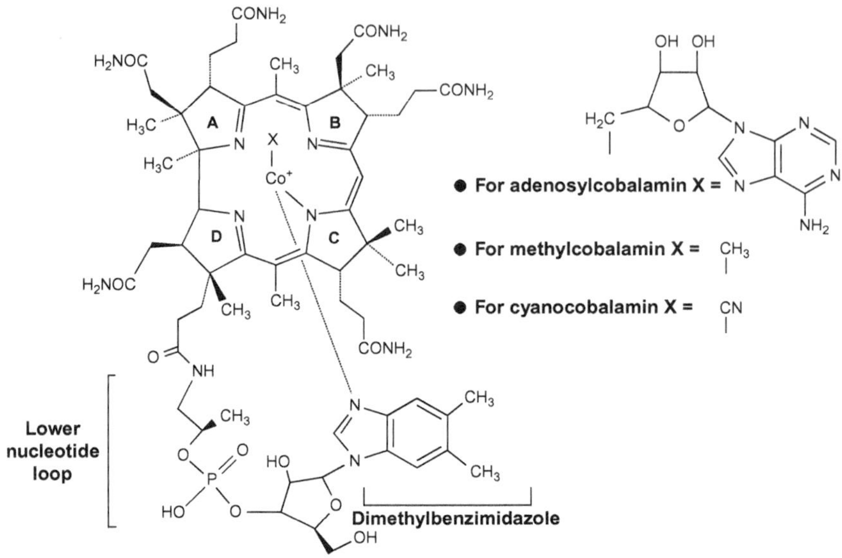

Fig. 1 Structure of vitamin B_{12}. The structure of cyanocobalamin is when X is CN. The main biological forms of cobalamin have the cyano group replaced with an adenosyl, methyl, or hydroxyl group, which are found in adenosylcobalamin, methylcobalamin, and hydroxocobalamin, respectively.

but also through upper (beta) and lower (alpha) ligands. The upper ligand in B_{12} is the cyano group, whereas the lower ligand is the nitrogen from the DMB. In biological systems, the upper cyano ligand is replaced by either an adenosyl group, to give (AdoCbl), a methyl group, to give methylcobalamin, or water, to give hydroxo or aquacobalamin. Similarly, some biological systems utilize a different lower base to which the cobalt is attached (Gruber, Puffer, & Krautler, 2011). In these systems, the DMB is replaced with bases such as adenine, substituted benzimidazoles that include hydroxy or methoxybenzimidazole, or phenolic compounds encompassing phenol or cresol (Hazra, Tran, Crofts, & Taga, 2013).

These variations in the structure of the molecule give rise to a great deal of confusion in terms of nomenclature (Nomenclature of corrinoids, 1976). Technically, B_{12} refers only to cyanocobalamin, but the term B_{12} is used ubiquitously to refer to the different forms of cobalamin, such as methyl or AdoCbl, and they are also referred to as complete corrinoids. The variants that have a different lower ligand are referred to as cobalamin analogues or are called by trivial names. This review relates largely to cobalamin (the complete corrinoids), and hence throughout the text, we will use the term B_{12} loosely to refer to cobalamin.

2.2 B_{12} as a Coenzyme

As a nutrient, B_{12} acts as either a cofactor or a coenzyme in a number of essential biochemical processes (Banerjee & Ragsdale, 2003; Gruber et al., 2011; Krautler, 2005). The biochemistry of B_{12} revolves around the chemistry that is mediated via the central cobalt ion, and especially the ability of the cobalt ion to form direct cobalt–carbon bonds. The chemistry reflects cobalt's ability to switch between different oxidation states, Co(I), Co(II), and Co(III), although as the oxidation state of cobalt changes, so the coordination of the metal changes from four to six, respectively. Of these oxidation states, Co(I) is the most unstable and acts as a supernucleophile. The corrin ring appears to be optimally constructed to help stabilize this form (Widner et al., 2016). Cobalamin-dependent enzymes fall into four broad classes that involve methylation, isomerization, reductive dehalogenation, and radical *S*-adenosylmethionine (SAM) processes (Bridwell-Rabb & Drennan, 2017; Vey & Drennan, 2011). All these B_{12}-mediated reactions are facilitated by the ability of the cobalt ion to change its state of oxidation.

2.2.1 Cobalamin-Dependent Methylation Reactions

These reactions involve the transfer of methyl group from methylcobalamin to an acceptor molecule. This is best exemplified with methionine synthase, where methylcobalamin methylates homocysteine to give methionine. The cobalamin is then recharged through methylation from methyltetrahydrofolate, which results in the formation of tetrahydrofolate (Drennan, Matthews, & Ludwig, 1994). In this respect, methylcobalamin is a cofactor that needs to be reactivated. Mechanistically, the process depends on the ability of the central cobalt ion to form a Co(I) species, which acts as a powerful nucleophile to form the methylcobalamin. The close relationship with the requirement for folate explains why deficiencies in folate and cobalamin share some similar symptoms. It is also interesting to note that nature has evolved a cobalamin-independent methionine synthase, which is found in a range of different organisms, including plants and yeasts, but that the cobalamin-independent forms of the enzyme are less efficient in terms of their turnover number. This means that there is a strong selection pressure for biological systems to retain the cobalamin-dependent methionine synthase.

Other known cobalamin-dependent methyltransferases are deployed by specific bacteria and archaea, especially those involved in methanogenesis, anaerobic methane oxidation, and acetogenesis. Here, cobalamin plays key roles in the formation of methane as well as in the fixation of C1 substrates such as methane or carbon dioxide. These processes are reviewed elsewhere (Ragsdale, 2008).

2.2.2 B_{12}-Dependent Isomerases

Adenosylcobalamin acts as a coenzyme in enzymes that are able to catalyze complex rearrangement reactions of their substrates. Most of these reactions are involved in bacterial fermentation processes, although enzymes such as methylmalonylCoA mutase have application in more general biochemical pathways (Roth, Lawrence, & Bobik, 1996). In the example of methylmalonylCoA mutase, the enzyme is able to interconvert methylmalonylCoA and succinylCoA; it is the only isomerase present in animals. The underpinning process that allows this to happen is the ability of AdoCbl to undergo a homolytic cleavage of the cobalt–carbon bond, which results in the formation of a Co(II) species as well as an adenosyl radical (Dowling, Croft, & Drennan, 2012). The radical promotes hydrogen abstraction from the substrate and initiates the rearrangement chemistry. In bacteria, there are many other AdoCbl enzymes associated with a broad range of isomerization reactions, and all are dependent upon the ability of AdoCbl to undergo homolytic cleavage to generate the adenosyl radical.

2.2.3 Reductive Dehalogenation

It is only recently that molecular insight has been gained on the role of B_{12} in the process of reductive dehalogenation, which takes place in certain bacteria (Bridwell-Rabb & Drennan, 2017; Fincker & Spormann, 2017; Payne et al., 2015). Reductive dehalogenases are able to remove a halogen component from a substrate with the simultaneous addition of electrons. In this case, the enzyme appears to function through the ability to form a direct cobalt–halogen bond. These processes have the potential to be used for the bioremediation of contaminated land and water.

2.2.4 Radical SAM Enzymes

The final class of B_{12}-dependent enzymes outlined earlier are the radical SAM enzymes (Bridwell-Rabb & Drennan, 2017). They are an extremely broad class of enzymes that house at least one Fe–S center and employ the use of SAM to generate an adenosyl radical (Sofia, Chen, Hetzler, Reyes-Spindola, & Miller, 2001). However, within this broad class of radical SAM enzymes, the largest group also have a requirement for B_{12}. Again, we are only just starting to understand how cobalamin chemistry is being coupled with radical SAM chemistry to perform these complex reactions. These enzymes are largely restricted to prokaryotes.

2.2.5 Nonenzymatic Functions

B_{12} is also involved in a number of nonenzymatic processes outside the animal kingdom. B_{12} is able to bind to specific nucleic acid sequences, called riboswitches that control and modulate either transcription or translation (Mandal & Breaker, 2004). The binding of B_{12} is normally associated with the regulation of either the uptake of B_{12} or the biosynthesis of the molecule. A number of transcription factors have also been identified that bind B_{12}. In the best-studied case, the CarH transcription factor has been shown to act as light sensor, employing AdoCbl as its light-sensitive pigment. In its repressor form, CarH is a tetramer that binds AdoCbl. However, when exposed to light, which leads to the scission of the cobalt–adenosyl bond, the tetramer disassociates and is released from the DNA resulting in transcription (Bridwell-Rabb & Drennan, 2017; Jost et al., 2015; Padmanabhan, Jost, Drennan, & Elias-Arnanz, 2017). CarH is a transcriptional repressor for carotenoid biosynthesis and the light deactivation of the repressor therefore results in the production of cellular light-protecting carotenoids.

2.3 Biosynthesis of B_{12}

Enzymes using B_{12} are often very efficient and result in significant rate enhancements when compared to model systems or enzymes that are B_{12}-independent. However, the ability to use B_{12} comes at a cost. It is a structurally intricate molecule whose complexity is reflected in a similarly sophisticated biosynthetic pathway, involving somewhere around 30 enzyme-mediated steps (Warren, Raux, Schubert, & Escalante-Semerena, 2002). The ability to biosynthesize B_{12} is restricted to certain prokaryotes. Some bacteria have the ability to make their own B_{12}, others acquire it from their environment, and some have evolved to live in a B_{12}-less world. Bacteria that make B_{12} generally synthesize it by one of the two distinct pathways, which are referred to as the aerobic and anaerobic routes (Battersby, 1994; Moore et al., 2013) (Supplementary Fig. S1 available on https://doi.org/10.1016/bs.afnr.2017.11.005). Both pathways have now been fully elucidated, and there is also quite a lot of molecular information available concerning the mechanism of the individual steps associated with the pathway. The two pathways for the synthesis of AdoCbl from the common tetrapyrrole primogenitor uroporphyrinogen III diverge at an intermediate called precorrin-2 and converge with the formation of adenosylcobyric acid.

Many bacteria are able to acquire or salvage B_{12} from the environment in which they grow, even if they have the ability to make the nutrient de novo (Escalante-Semerena, 2007). The salvage pathway often involves the acquisition of just the corrin component of the molecule, and in these cases, the bacteria are able to attach the lower loop and add the upper ligand. In these cases, bacteria can take up molecules such as cobinamide or cobyric acid, assemble the lower nucleotide, and attach it to the corrin ring (Supplementary Fig. S1 available on https://doi.org/10.1016/bs.afnr.2017.11.005). However, as some bacteria are able to make cobalamin variants with different bases to DMB, other bacteria have evolved systems to take up these B_{12} variants, remove the lower nucleotide loop, and replace it with another loop containing a more appropriate base. The synthesis and sharing of B_{12} in complex microbial communities are a growing area of interest, as the ability to make, portion, and acquire such a valuable metabolic resource has significant effects in shaping microbial communities (Degnan, Taga, & Goodman, 2014; Escalante-Semerena, 2007).

The structural complexity of B_{12} means that the nutrient cannot be made commercially through chemical synthesis. In this respect, B_{12} is one of the few vitamins that is obtained through bacterial fermentation (Martens, Barg, Warren, & Jahn, 2002). High-producing bacterial strains were developed initially by Rhone-Poulenc in France although similar strains are now also used to produce the vitamin in China. It is estimated that about 35 metric tons of B_{12} are produced by bacterial fermentation annually, some of which is used to supply the human nutrient market but a majority is used as supplements for animal feed.

2.4 B_{12} in Eukaryotes

In contrast to the broad range of enzymes and molecular processes involving B_{12} in prokaryotic systems, the picture is very different in eukaryotes where B_{12} is restricted to just methionine synthase and methylmalonyl-CoA mutase. The eukaryotic systems that use B_{12} include mammals, birds, fish, worms, and protists such as algae. Many eukaryotes do not require B_{12}, including higher plants and yeasts/fungi. The acquisition of B_{12} in eukaryotic systems represents a very interesting challenge. Algae, for instance, gain their B_{12} through association with bacteria in either a symbiotic or mutualistic manner (Croft, Lawrence, Raux-Deery, Warren, & Smith, 2005). Some mammals, especially ruminants, are able to absorb their B_{12} from the enteric bacteria that live in their intestines (Girard,

Santschi, Stabler, & Allen, 2009). Although humans have a rich microbial fauna, present mainly in the large intestine, the human B_{12} uptake system is located in the small intestine. Humans therefore do not acquire B_{12} from their enteric bacteria and have to rely on B_{12} present in their diet (Stabler & Allen, 2004).

As humans require B_{12} for just two enzymes and as they tend to retain and recycle their B_{12}, the daily requirement for B_{12} is very low (Stabler & Allen, 2004). The RDA for B_{12} is around 2.5 μg/day, which is the lowest of all the vitamins. Good dietary sources of B_{12} include meat, in particular liver, fish, shellfish, and dairy products. Vegetarians, and more specifically vegans, are prone to B_{12} deficiency due to the lack of B_{12} in higher plants (Rizzo et al., 2016). However, some macroalgae such as Nori do contain B_{12}. A number of foodstuffs such as breakfast cereals and energy drinks are fortified with B_{12}. Most humans on a healthy varied diet are not prone to frank B_{12} deficiency, but B_{12} insufficiency is very much more common, as we will discuss in Section 3. Sometimes bacteria growing in foods or beverages can provide B_{12}, so, counterintuitively, the greater the hygiene applied to the preparation of foods and drinks, e.g., through washing the food, clean water, and refrigeration, the lower the B_{12} levels. This process may partly explain the finding that the median plasma B_{12} in urban middle-class Indian men was 89 pmol/L, compared with 145 pmol/L for men living in urban slums (Yajnik et al., 2006).

A lack of B_{12} gives rise to a complex series of signs and symptoms, most of which arise from metabolic problems associated with the two human enzymes that require B_{12} (Green et al., 2017). The symptoms of B_{12} deficiency range from megaloblastic anemia, neurological problems, cardiovascular symptoms, methylmalonic aciduria, etc. For instance, impairment of methionine synthase impedes the methionine cycle, whereby homocysteine is converted into methionine and then to SAM. SAM is used in a range of one carbon methylation reactions where it is broken down first to *S*-adenosylhomocysteine (SAH) and then into homocysteine and adenosine. A lack of B_{12} results in a build-up of homocysteine and folate being trapped as methyltetrahydrofolate, which in return leads to depletion of tetrahydrofolates used in thymidylate and purine synthesis. This is how B_{12} deficiency impedes cellular proliferation and protein synthesis and thereby causes development of megaloblastic anemia. A lack of B_{12} has similar detrimental effects on methylmalonylCoA mutase, an enzyme that is associated with the degradation of odd chain fatty acids, certain amino acids, and propionate metabolism. The enzyme allows the interconversion of MMA and succinate, and a lack of B_{12} results in the buildup of MMA causing acidemia in the blood and urine.

2.4.1 Inborn Errors

B_{12} insufficiency in humans is caused not only by a lack of the vitamin in the diet or by B_{12} malabsorption but also by genetic disorders linked with the nutrient uptake and distribution system (Froese & Gravel, 2010; Gherasim, Lofgren, & Banerjee, 2013; Watkins & Rosenblatt, 2013). Our understanding of the processes involved in cellular uptake and subcellular distribution of B_{12} has come from a large volume of work associated with inborn errors of metabolism linked to genetic mutations of specific proteins (Froese & Gravel, 2010; Watkins & Rosenblatt, 2013). Distinct components associated with human B_{12} metabolism were identified, and coupled with biochemical analysis of radioactive metabolites and B_{12}, it was possible to classify a range of different genotypes and phenotypes (Froese & Gravel, 2010). As a result, it was possible to identify a number of complementation groups, termed *cblA-G* and *mut*, that represented genes associated with B_{12} metabolism and transport. The roles of these different loci are shown in Table 1, and their function is outlined in Fig. 2. A number of the genes associated with these complementation groups already had assigned functions (e.g., *cblG* = methionine synthase), whereas in other cases, the identification of the gene led to the discovery of a new functionality (e.g., *cblC* = decyanalase).

2.4.2 B_{12} Uptake and Distribution

In humans, several different B_{12}-binding proteins facilitate the absorption of B_{12} from dietary sources (Fedosov, 2012; Nielsen, Rasmussen, Andersen, Nexo, & Moestrup, 2012). These proteins include the three carrier proteins called haptocorrin (HC), intrinsic factor (IF), and transcobalamin (TC). The structures of all three proteins have been determined (Furger, Frei, Schibli, Fischer, & Prota, 2013; Mathews et al., 2007; Wuerges et al., 2006), including several with the proteins bound to their cognate receptors (Alam et al., 2016; Andersen, Madsen, Storm, Moestrup, & Andersen, 2010). Both HC and IF are glycoproteins, but all three proteins share a degree of sequence similarity, which is reflected in a similar overall topology. Importantly, they have different affinities for B_{12} and B_{12} analogues and play different roles as outlined later.

The first of the B_{12}-binding proteins in the uptake mechanism is HC, which is excreted into the saliva and upper part of the gastrointestinal tract,

Table 1 Genes Associated With B_{12} Metabolism and Transport

Complementation Group	Protein Name	Phenotype	Function
cblA/MMAA		Adenosylcobalamin deficiency in cells	P-loop GTPase—is thought to function in the delivery of the adenosylcobalamin from the adenosyltranferase to the methylmalonylCoA mutase. Acts to protect the activity of the mutase
cblB	Adenosyltransferase	Adenosylcobalamin deficiency in cells	Adenosylates cobalamin in an ATP-dependent manner
cblC/MMACHC		Unable to convert cyanocobalamin into biological forms	Removes upper alkyl or cyano ligand from cobalamin
cblD		Improper targeting of cobalamin to cognate enzymes	Branching of cobalamin within the cell to either the cytosol or the mitochondrion
cblE/MTRR	Methionine synthase reductase	Inactive methionine synthase	Performs the reductive methylation of cobalamin to generate methylcobalamin from cob(II)alamin
cblF/LMBRD1		Cobalamin accumulation within lysosome	Likely helps in the transport of cobalamin out of the lysosome
cblG/MTR	5-Methyl-tetrahydrofolate: homocysteine methyltransferase or methionine synthase	Build-up of homocysteine	Performs the transfer of a methyl group from methyltetrahydrofolate to homocysteine to produce methionine
cblJ/ABCD4		Cobalamin accumulation within lysosome	Transporter
Mut/MCM	MethylmalonylCoA mutase	Accumulation of methylmalonic acid	The enzyme responsible for the reversible interconversion of methylmalonylCoA and succinylCoA

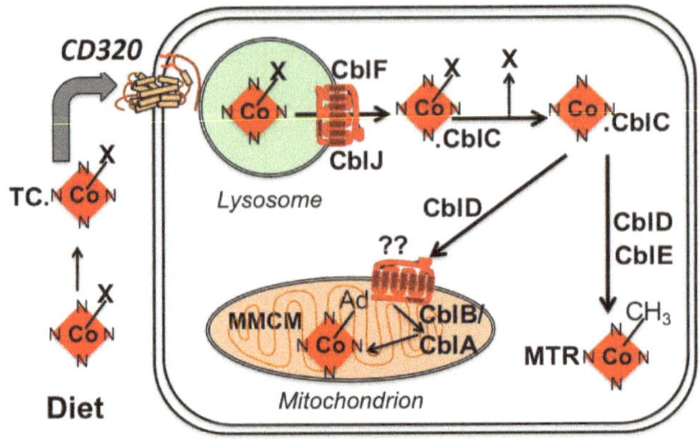

Fig. 2 The trafficking of cobalamin within humans. B_{12} is absorbed from the diet and passed via intestinal cells to transcobalamin (TC) in the bloodstream. The TC–B_{12} complex binds to the TC receptor (CD320), where it is transported into the lysosome, where the TC is degraded and the B_{12} released. The nutrient is then transported via the CblJ and CblF transporters into the cytosol, where the B_{12} is acted upon by CblC, and any upper alkyl or cyano group (X) is removed. For methionine synthase (MTR), the B_{12} is chaperoned by the action of CblD and CblE to ensure correct delivery within the cytoplasm. For methylmalonylCoA mutase activity (MMCM), CblD ensures delivery to the mitochondrion, where the molecule is transported to the inside of the organelle by an unknown transporter. Once inside, the action of CblB and CblA ensure that the molecule is adenosylated and delivered to the enzyme.

and it is also found in blood serum. This protein has a high affinity for B_{12} and for B_{12} analogues and binds them as they are released from the food source through the initial stages of digestion (involving pepsin and acid) in the mouth and stomach. The HC and other B_{12}-bound protein complexes are further digested in the small intestine to release the B_{12}, where it is bound by IF.

IF is produced by the parietal cells in the stomach, and adverse conditions targeting these cells will lead to IF deficiency. Notably, pernicious anemia is a B_{12} deficiency due to lack of IF and is usually caused by an autoimmune condition that targets the parietal cells (Toh, van Driel, & Gleeson, 1997). In contrast to HC, IF is much more specific for binding B_{12} and has limited affinity for cobalamin analogues. In ileum, the IF-cobalamin complex crosses the intestinal membrane by first binding to the apical side of epithelial cells. Here it binds to the cubam receptor, which facilitates endocytosis into the lysosome. In the lysosome, IF is degraded and the liberated cobalamin is released into the cytosol and then transported to the bloodstream via the multidrug resistance protein MRP1. Once in the bloodstream, the majority of B_{12} (80%) and all the cobalamin analogues are bound to the less-specific HC, whereas the remainder is bound to the much more specific TC. Like IF, TC binds only B_{12}. Significantly, only TC is able to facilitate uptake into cells. The role of circulating HC is not fully understood, but it may help remove cobalamin analogues or act as a general store. Interestingly, patients who are unable to produce HC have low serum levels of B_{12} but do not display symptoms of B_{12} deficiency.

The TC-cobalamin complex is taken up into cells (Fig. 2) by receptor-mediated endocytosis after binding to the TC receptor (a transmembrane protein called CD320) (Quadros, 2010). In the lysosome, the TC is digested away and the free cobalamin is exported into the cytosol via the synergistic effort of two membrane proteins called ABCD4 (*cblJ*) and LMBD1 (*cblF*) (Quadros, 2010; Rutsch et al., 2009). In the cytosol, B_{12} is acted upon by CblC (Lerner-Ellis et al., 2006), which is a protein that plays a key role for subsequent compartmentalization of B_{12} within the cell. Up to this stage, the delivery of B_{12} to the cell has been determined by the requirement for an intact lower ligand (i.e., DMB) with little attention paid to the nature of the upper ligand. Thus, the transported cobalamins are likely to have different upper ligands (adenosyl, methyl, hydroxyl, cyano). In the cytosol, CblC has the ability to dealkylate and decyanalate the various forms via distinct processes that have been studied through biochemical and molecular approaches (Hannibal et al., 2009; Kim, Gherasim, & Banerjee, 2008). Alkylated cobalamins (methyl and adenosyl) are cleaved by binding to CblC, which in the presence of glutathione mediates a nucleophilic displacement of the upper ligand to form a glutathione ester and a cob(I)alamin (Kim, Hannibal, Gherasim, Jacobsen, & Banerjee, 2009). The latter is able to react rapidly with water to form hydroxocob(II)alamin. The reaction of CblC with cyanocobalamin is slightly different in that it requires a reduced flavin as a cofactor, which encourages homolytic cleavage of the bond to give cob(II)alamin and cyanide (Kim, Gherasim, et al., 2008). Significantly, this means that people with a defective CblC usually respond well to treatment with hydroxocobalamin, but not to other cobalamins. Interestingly, the structure of CblC reveals that it has a similar structure to the enzyme BluB (Koutmos, Gherasim, Smith, & Banerjee, 2011), which is involved in the biosynthesis of the DMB base (Taga, Larsen, Howard-Jones, Walsh, & Walker, 2007). Structurally, these proteins have a flavoprotein fold. The structure is also similar to the reductive dehalogenases, indicating that this fold has evolved to be involved in a broad range of cobalamin-related reactions (Payne et al., 2015).

CblC also interacts with the cytosolic protein CblD, which acts as a key chaperone protein in the transfer of the nutrient to either methionine synthase (CblG) in the cytoplasm or the mitochondrial transport system

(Coelho et al., 2008). CblD has a similar structure to CblC but does not bind B_{12} (Yamada, Gherasim, Banerjee, & Koutmos, 2015). Generation of functional methionine synthase requires reductive methylation of cob(II)alamin to methylcobalamin using methionine synthase reductase in which SAM acts as a methyl donor (Leclerc et al., 1998).

The import system for B_{12} into the mitochondrion has not yet been identified, but once inside mitochondria, the B_{12} is adenosylated by an adenosyltransferase (*cblB*). The adenosyltransferase binds cob(II)alamin in a 4-coordinate manner, which has the effect of minimizing the reducing power required to convert the cob(II)alamin into cob(I)alamin in the presence of a suitable reductant such as a flavodoxin or ferredoxin system (Stich, Yamanishi, Banerjee, & Brunold, 2005). In the presence of ATP, the cob(I)alamin displaces the triphosphate of the ATP and generates AdoCbl (Mera, St Maurice, Rayment, & Escalante-Semerena, 2007). Significantly, the adenosyltransferase also chaperones AdoCbl to the methylmalonylCoA mutase and thereby allows direct delivery of the coenzyme (Padovani, Labunska, Palfey, Ballou, & Banerjee, 2008). A further complementation group, *cblA*, was initially thought to be caused by mutations in the adenosyltransferase, but, when the gene was isolated, it was found to encode for a protein that was a member of the P-loop GTPases (Dobson et al., 2002). These proteins are linked with the assembly and protection of metal centers. An orthologous protein in bacteria, MeaB, has been shown to form a complex with the methylmalonylCoA mutase and to protect the enzyme from inactivation (Padovani & Banerjee, 2009).

In summary, the identification of a broad range of human inborn errors of metabolism has allowed us to piece together the complex uptake and delivery mechanisms required for cobalamin biochemistry (Froese & Gravel, 2010; Gherasim et al., 2013; Watkins & Rosenblatt, 2013).

3. POOR B_{12} STATUS: DEFICIENCY OR INADEQUACY?

3.1 How Do We Define B_{12} Deficiency?

To answer this question, we need to specify the biological or clinical outcomes of relevance and to decide which marker will be measured. Since megaloblastic anemia was the first B_{12}-deficiency syndrome discovered, it has been traditional to use hematological signs, such as a raised mean corpuscular volume or megaloblastic anemia as markers (Green, 2017). However, in the classic paper by Lindenbaum et al. (1988), it was shown that neurological signs of B_{12} deficiency are often present in the absence of anemia. Even in patients with clinical pernicious anemia, 28% do not have anemia and up to 33% have normal mean corpuscular volume (Carmel, 2000). In fact, anemia is just one possible outcome of deficiency and should not therefore be used as a sole marker. Ideally, a marker should be associated with most, if not all, of the known outcomes. For this reason, biochemical markers, such as serum total B_{12}, holoTC, or the functional markers MMA or tHcy, are usually preferable.

Part of the problem in defining B_{12} deficiency lies in the measurement of serum total B_{12} levels. Serum levels reflect B_{12} that is bound to either HC or TC. Although HC binds the majority of serum B_{12}, up to 60% of bound material comprises incomplete corrinoids or cobalamin analogues (Morkbak, Poulsen, & Nexo, 2007). In this respect, an estimate of total serum B_{12} levels is not very reliable when trying to assess whether patients are B_{12} deficient, since a "normal" reading using some assay methods can mask deficiency if serum contains relatively large amounts of cobalamin analogues. However, since TC only binds biologically active cobalamin (Hardlei & Nexo, 2009), measurement of holoTC, i.e., TC-cobalamin, will give a more accurate diagnosis (Fedosov, 2012). Another advantage in measuring holoTC is in subjects who have markedly elevated total serum B_{12}, found in about 8% of tests; the high B_{12} is usually due to an excessive amount of HC and so could mask any deficiency (Arendt & Nexo, 2013).

But then the question arises, what value of the biochemical marker defines the cutoff for deficiency? Traditionally, the cutoff for serum/plasma B_{12} has been 148 pmol/L (200 pg/mL) or 150 pmol/L (203 pg/mL), according to WHO and the US Institute of Medicine. However, just as with cardiovascular disease risk factors, it is now recognized that there is no sharp cutoff for identifying harmful B_{12} status but that there is a continuous inverse relationship between serum B_{12} and a variety of undesirable outcomes (Smith & Refsum, 2012). Many undesirable outcomes have been associated with serum B_{12} levels that are well above the traditional cutoff value of 150 pmol/L, as will be discussed in Section 4. Distinct biological and biochemical outcomes have different relationships to the serum B_{12} level (see fig. 2 in Smith & Refsum, 2016), which makes it difficult to define "deficiency." For this reason, we prefer the terms "B_{12} inadequacy" or "B_{12} insufficiency."

Insufficiency of B_{12} is also revealed by measurement of metabolic markers such as MMA and tHcy, whose blood levels start to rise as the serum B_{12} level falls below about 400 pmol/L (see Section 3). Recently, a new approach to defining B_{12} inadequacy has been proposed, which incorporates data on serum B_{12}, holoTC, tHcy, and MMA to give a "combined B_{12}

indicator" (Fedosov, Brito, Miller, Green, & Allen, 2015). Although promising, this indicator requires extensive facilities for the assays and needs further validation, especially regarding the cognitive outcomes. For the time being, it remains a research tool. For instance, in a study on cognitive development in children, plasma B_{12} levels in infancy were not related to cognition at 5 years, but the combined indicator was related, as were MMA and tHcy (Kvestad et al., 2017).

As can be seen, the diagnosis of B_{12} deficiency/insufficiency is far from satisfactory, and we agree with the following statement by the British Society for Haematology in their Guidelines: "The clinical picture is the most important factor in assessing the significance of test results assessing cobalamin status because there is no 'gold standard' test to define deficiency" (Devalia, Hamilton, Molloy, & the British Committee for Standards in Haematology, 2014).

3.2 How Prevalent Is B_{12} Deficiency and Inadequacy? (See Supplementary Table 1S available on https://doi.org/10.1016/bs.afnr.2017.11.005)

There are several valuable reviews of the prevalence of B_{12} deficiency (Allen, 2004, 2009; Allen et al., 2018; Baik & Russell, 1999; Green et al., 2017; Hunt, Harrington, & Robinson, 2014; McLean, de Benoist, & Allen,

Table 2 Summary of Prevalence of B_{12} Deficiency

Category	Number of Cohorts	Number of Subjects	B_{12} Deficiency (%, Rounded)
Adults (under 60 years)	18	81,438	6
Women (nonpregnant)	16	18,520	16
Women pregnant	11	11,381	27.5
Children	14	22,331	12.5
Elderly (over 60 years)	25	30,449	19
	23 (not China)	26,249	10
	2 (China)	4,200	75

Defined as serum/plasma $B_{12} < 148$ or 150 pmol/L.
Derived from data in Supplementary Table 1S.

Table 3 Summary of Prevalence of B_{12} Inadequacy/Insufficiency

Category	Number of Cohorts	Number of Subjects	B_{12} Inadequacy (%, Rounded)
Adults (under 60 years)	18	66,529	33
Women (nonpregnant)	11	4,681	51
Women pregnant	4	2,848	60
Children	10	19,704	29
Elderly (over 60 years)	12 (excl. China)	19,023	31

Defined as serum/plasma B_{12} <220, <258, or <300 pmol/L, i.e., the group with deficient or low-normal B_{12} levels associated with signs of metabolic insufficiency with elevated tHcy and/or MMA. Derived from data in Supplementary Table 1S.

2008). In Supplementary Table 1S available on https://doi.org/10.1016/bs.afnr.2017.11.005, we have summarized observational studies, mainly since 2010, on this question. In 62 of these reports, B_{12} deficiency was defined as serum/plasma levels <148 or <150 pmol/L. In 5 studies, the cutoff was lower (ranging from 111 to 130 pmol/L); in 12 studies the cutoff was higher (ranging from 160 to 221 pmol/L). We have amalgamated the data from Supplementary Table 1S available on https://doi.org/10.1016/bs.afnr.2017.11.005 into summary tables (Tables 2 and 3) to make it easier to see the overall picture.

There is a tendency for the prevalence of *B_{12} deficiency* (Table 2) to increase with age, but this is not always so clear, as can be seen in one of the NHANES studies (Bailey et al., 2011). A more marked effect of age was reported in a study from Liechtenstein (Koenig et al., 2014), reproduced in Supplementary Fig. S2 available on https://doi.org/10.1016/bs.afnr.2017.11.005.

It is noteworthy that the prevalence of *B_{12} inadequacy* (Table 3) is high in all subgroups of the population, ranging from 30% to 60%. B_{12} inadequacy is five times more prevalent than deficiency in adults, threefold higher in nonpregnant women, twice as high in pregnant women and in children, and threefold higher in the elderly. Since B_{12} inadequacy is associated with impaired B_{12} function as judged by elevation of MMA and tHcy, and probably also with impaired health outcomes (see Section 4 and Table 4), the

Table 4 Clinical Outcomes Found in the Zone of Metabolic B_{12} Insufficiency Without Deficiency

Outcome	Type of Study	References
Neural tube defect	Prospective (in mothers)	Kirke et al. (1993); Molloy et al. (2009)
Infantile tremor syndrome	Prospective (in mothers)	Goraya, Bansal, Singla, and Kaur (2016)
Cognitive deficit in elderly	Cross-sectional	Hin et al. (2006); Mizrahi, Lubart, and Leibovitz, 2017
Cognitive decline in elderly	Prospective	Clarke et al. (2007); Hooshmand et al. (2012); Tangney, Tang, Evans, and Morris (2009)
Alzheimer's disease	Cross-sectional	Clarke et al. (1998); Refsum and Smith (2003)
White matter damage	Cross-sectional	de Lau, Smith, Refsum, Johnston, and Breteler (2009); Graber, Sherman, Kaufmann, Kolodny, and Sathe (2010); van Overbeek, Staals, and van Oostenbrugge (2013)
Impaired regional brain microstructure and memory impairment	Cross-sectional	Kobe et al. (2016)
Whole-brain atrophy	Prospective, cross-sectional	Hooshmand et al. (2016); Tangney et al. (2011); Vogiatzoglou et al. (2008)
Depression	Prospective	Kim, Stewart, et al. (2008)
Response to treatment of depression	Prospective	Hintikka, Tolmunen, Tanskanen, and Viinamaki (2003)
Stroke	Prospective	Spence (2016); Weikert et al. (2007)
Cognitive and motor development in children	Prospective	Strand et al. (2013)
	Intervention	Torsvik, Ueland, Markestad, and Bjorke-Monsen (2013); Torsvik, Ueland, Markestad, Midttun, and Bjorke Monsen (2015); Kvestad et al. (2017); Kvestad et al. (2015)

Table 4 Clinical Outcomes Found in the Zone of Metabolic B_{12} Insufficiency Without Deficiency—cont'd

Outcome	Type of Study	References
Age-related macular degeneration	Prospective	Gopinath, Flood, Rochtchina, Wang, and Mitchell (2013)
Low bone mineral density in women	Cross-sectional	Dhonukshe-Rutten et al. (2003)
Autonomic dysfunction (orthostatic hypotension)	Clinical (head-up tilt) Treatment intervention	Ganjehei, Massumi, Razavi, and Wilson (2012); Moore et al. (2004); Oner et al. (2014)
DNA damage in lymphocytes	Treatment intervention	Fenech et al. (1998)
Uracil misincorporation into DNA	Cross-sectional	Kapiszewska, Kalemba, Wojciech, and Milewicz (2005)

The table shows clinical outcomes in subjects whose serum/plasma B_{12} level is above the cut-off for deficiency (148 pmol/L) but below the proposed cut-point for insufficiency (300–400 pmol/L). This table is an expanded version of one given in Smith and Refsum (2012).

high prevalence is of concern to public health. Other matters of concern can be deduced from Tables 2, 3, and Supplementary Table 1S available on https://doi.org/10.1016/bs.afnr.2017.11.005 notably that about half of women have a poor B_{12} status, with one-quarter of pregnant women being classed as deficient. Infants, on average, appear to have poorer status than older children and adults. The latter result may be due to low B_{12} status usually observed in breastfed infants. Furthermore, many of these studies are from regions (e.g., parts of South Asia, Africa, Middle, and South America) with a general poor B_{12} status. In parts of China, the prevalence of deficiency in the elderly is very high indeed, with 75% having concentrations below 185 pmol/L (Zhang et al., 2014; Zhang, Liu, et al., 2016). In Hong Kong, 35% of institutionalized elderly had B_{12} below 150 pmol/L (Wong et al., 2015). As expected, high proportions of vegetarians and vegans have poor B_{12} status, with deficiencies ranging from 11% to 90% (Gilsing et al., 2010; Pawlak, Lester, & Babatunde, 2014; Pawlak, Parrott, Raj, Cullum-Dugan, & Lucus, 2013; Rizzo et al., 2016). In countries like India, where lifelong vegetarians form a high proportion of the population, up to 71% of pregnant women are B_{12} deficient (Yajnik et al., 2008), with major implications for the health of the mother and child (see Section 6).

We would like to draw attention to a novel observation from a study in Chile: the prevalence of B_{12} deficiency in the elderly varied from 19.1% in the north of the country to 5.7% in the south (Cabrera, Benavente, Alvo, de Pablo, & Ferro, 2014). The authors found that the prevalence of B_{12} deficiency was inversely associated with geographical latitude and was positively related to solar radiation, and they speculated that solar radiation increases the degradation of B_{12}. In Europe, a much less marked gradient in median serum B_{12} has been observed from 330 pmol/L in the north to 290 pmol/L in the south ($P < 0.001$), and this was attributed to differences in diet (Eussen et al., 2013). In China, the gradient is in the opposite direction, with a much higher prevalence of low B_{12} status in the north (21% deficiency in adults) than in the south (4% deficiency) (Hao et al., 2007); this may also be due to differences in diets.

The likely causes of B_{12} deficiency and inadequacy are reviewed in a book chapter (Green & Miller, 2014), reviews (Green, 2017; Hunt et al., 2014; Stabler, 2013), a recent primer (Green et al., 2017), and a BOND report (Allen et al., 2018) and will not be covered here. However, regarding pernicious anemia, it should be recognized that the diagnosis and treatment of this disease are far from ideal, as revealed in a survey of patients in the United Kingdom (Hooper, Hudson, Porter, & McCaddon, 2014). We recommend the interested reader to the website of the Pernicious Anaemia Society (https://pernicious-anaemia-society.org/). The clinical consequences of frank B_{12} deficiency are well reviewed in the articles cited earlier and will not be covered here, but we will examine the consequences of low-normal B_{12} status.

4. CLINICAL RELEVANCE OF B_{12} INADEQUACY/INSUFFICIENCY

4.1 Concept of "Subclinical Cobalamin Deficiency"

This concept was introduced by Carmel (2011) and is defined as "a state of mild metabolic abnormalities without clinical signs or symptoms." By "mild metabolic abnormalities," Carmel meant that plasma tHcy and MMA are elevated and/or holoTC is low. While such changes very often occur without overt hematological signs, this does not in our view make them necessarily "subclinical." We have pointed out that the zone of "metabolic insufficiency," as defined in Fig. 3, may be associated with one or more clinical signs and symptoms, which are listed in Table 4 (Smith & Refsum,

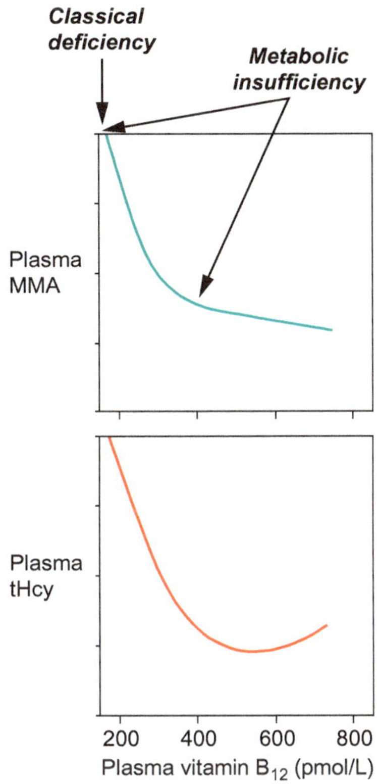

Fig. 3 Defining B_{12} inadequacy by use of markers of metabolic insufficiency. Relationship between plasma vitamin B_{12} and plasma total homocysteine (tHcy) or methylmalonic acid (MMA) in 3262 community-dwelling people aged 71–74 years in Norway. *Based on fig. 1 from Smith, A. D., & Refsum, H. (2012). Do we need to reconsider the desirable blood level of vitamin B12?* Journal of Internal Medicine, 271*(2), 179–182.*

2012). This zone lies well above the typical cutoff for B_{12} deficiency of 148 pmol/L and extends up to about 350–400 pmol/L.

A threshold between 350 and 400 pmol/L separates optimal from suboptimal B_{12} levels, based on metabolic B_{12} markers, and is consistent with several population studies. A detailed statistical analysis on data from 7260 Americans from NHANES (1999–2000) used segmented regression to obtain two B_{12} cut-points for tHcy of 200 pmol/L (CI 187, 214) and 426 pmol/L (CI 367, 485) (Bang, Mazumdar, & Spence, 2006). Vogiatzoglou, Oulhaj, et al. (2009) used segmented regression to obtain B_{12} cut-points of 334 pmol/L (SE 33) and 393 pmol/L (SE 11) for MMA and tHcy, respectively, for 6946 middle-aged and elderly Norwegians. A detailed analysis of data from 12,683 adults (≥19 years) in NHANES (1999–2004) used 5 different modeling methods to obtain cut-points related to raised MMA (Bailey et al., 2013). The derived cut-point varied according to the model used. A linear splines model best fitted the data and gave two cut-points of 126 pmol/L (CI 117, 182) and 287 pmol/L (CI 253, 434), with a marked increase in MMA in the group below 126 pmol/L, and a more modest increase in those with B_{12} in the range 126–287 pmol/L. Accordingly, these authors divided the United States population into three groups: Group 1 ($B_{12} < 126$ pmol/L) represented about 1% of adults, while Group 3 ($B_{12} > 287$ pmol/L) represented almost ~68% of adults. The middle Group 2, about 31% of adults, had B_{12} cutoff values between 126 and 287 pmol/L. This group corresponds most closely to our category of "metabolic insufficiency" and would be likely to show a higher prevalence of clinical outcomes listed in Table 4.

Notably, in a study examining median and upper limits for MMA according to creatinine, age, and B_{12} status, it appears that it is mainly the upper MMA limit that changes according to B_{12} status (see fig. S4 in Vogiatzoglou, Oulhaj, et al., 2009), suggesting that the MMA increase occurs in a subgroup. Thus, we believe that the two NHANES thresholds of 126 and 287 pmol/L may distinguish between a level where a majority of the subjects have metabolic defects (<126 pmol/L), while the MMA increase in the intermediate levels (126–287) is mainly confined to a subgroup of people requiring higher B_{12} level to obtain optimal metabolic function.

A good example of how low-normal B_{12} status can lead to potentially serious harm is that the rate of atrophy of the whole brain in the normal elderly, measured by serial MRI scans, is linearly inversely related to the serum B_{12} level over almost the entire normal range (Hooshmand et al., 2016; Vogiatzoglou et al., 2008). We will discuss this further in Section 7.

4.2 Clinical Implications of Low-Normal B_{12} Status

The range and variety of the clinical outcomes in Table 4 suggest that it can no longer be stated that metabolic insufficiency related to low-normal B_{12} status is without clinical consequence. We therefore do not agree with the recent "Primer" on B_{12} deficiency (Green et al., 2017), which states that B_{12} inadequacy is "of no clinical relevance."

Our conclusion has several important implications and *we make the following recommendations*:

1. Patients with symptoms or signs consistent with B_{12} deficiency and with low-normal B_{12} levels in the range above the cutoff value of 148 pmol/L, extending up to around 350 pmol/L, should not be told that their B_{12} is "normal," but that they have "low-normal" levels.
2. Patients with any one of the outcomes listed in Table 4 should have their B_{12} status measured and possible causes for their low B_{12} status should be identified.
3. If the B_{12} falls in the range of 150 to about 350 pmol/L, then these patients should be offered high-dose oral B_{12} (1 mg daily) for several months, or regular i.m. injections, to see if their symptoms are reduced and metabolic function markers are normalized. One month of treatment is unlikely to be sufficient (Favrat et al., 2011). *The aim of the treatment, whatever form or dose used, should always be the relief of signs and symptoms, especially neurological signs.*
4. Clinicians should look out for other unexpected clinical signs or symptoms in patients who have B_{12} status in this borderline range. Clinicians need to look outside the conventional hematological and neurological signs, and also consider conditions such as infertility, birth defects, orthostatic hypotension, skin changes, and psychiatric and cognitive problems, as listed in Table 4.

4.3 What Oral Dose of B_{12} Is Needed to Obtain Optimal B_{12} Status?

The literature on population dietary intake of B_{12} is surveyed in the recent BOND report (Allen et al., 2018) and so will not be reviewed here. However, we must note that the bioavailability of B_{12} from food sources is far from uniform and that the blood levels of B_{12} first reach a plateau when about 10 μg/day is consumed from food sources (Tucker et al., 2000; Vogiatzoglou, Smith, et al., 2009). Our purpose here is to discuss the supplemental dose of B_{12} that will correct the signs of metabolic insufficiency. First, we must point out that the widespread use of intramuscular B_{12} treatment is often not necessary. Parenteral B_{12} is only required to correct the initial very poor B_{12} status when pernicious anemia is diagnosed. Later, although some patients prefer regular injections, oral high-dose (1–2 mg/day) maintenance treatment is adequate for most patients with pernicious anemia (Berlin, Berlin, & Brante, 1968; Kuzminski, DelGiacco, Allen, Stabler, & Lindenbaum, 1998; Vidal-Alaball et al., 2005). For subjects with low B_{12} status caused by poor dietary intake, or by food malabsorption, it will be simpler and cheaper to use oral B_{12} preparations. In Sweden, a national survey found that in the year 2000, two out of three B_{12} prescriptions were for oral tablets (Nilsson et al., 2005). This statistic does not include the small proportion (<1%) of oral B_{12} tablets freely purchased over the counter.

The normal mechanism of absorption of oral B_{12}, via the IF pathway, is easily saturated: While ~70% of B_{12} is absorbed from doses of 0.1 to 0.5 μg, this falls to 56% for 1 μg, 16% for 10 μg, and 3% for doses of 25–50 μg (Chanarin, 1979). High oral doses will very largely be absorbed passively: the classical study of Berlin et al. (1968) using radiolabeled cyanocobalamin showed that an upper limit of 2 μg can be absorbed via the IF pathway, while at higher doses, from 100 to 100,000 μg, the absorption is passive with 1.2% of the dose being absorbed. The authors recommended that an oral dose of 1000 μg of cyanocobalamin per day should be given to patients with pernicious anemia since they found that this dose was equivalent in every respect to the traditional intramuscular dose regime.

Modern studies on the relation between oral B_{12} dose and blood levels have been conflicting. Notably, these studies have been performed in general populations or subjects with low B_{12}, i.e., only few of them suffered from pernicious anemia. A Danish study on 98 postmenopausal women found that an intake of ~6 μg/day from food and supplements was adequate to bring the plasma B_{12} and holoTC to a plateau and to maximally lower MMA and tHcy. The median plasma B_{12} at this level of intake was 380 pmol/L. The effectiveness of a relatively low dose was also found a randomized trial over 1 year in India (Deshmukh et al., 2010), where B_{12} status is very poor due to the largely vegetarian diet: 300 adults and children (48% were B_{12} deficient, i.e., <148 pmol/L) were given either 2 or 10 μg of B_{12} and the B_{12} level rose, respectively, from 168 to 242 pmol/L and from 159 to 307 pmol/L. There were also significant decreases in tHcy levels. Other reports have suggested that much higher doses might be needed to normalize the biomarkers. Rajan et al. (2002) recruited 23 elderly Americans with B_{12} deficiency due to food malabsorption and administered daily oral doses of 25, 100, and 1000 μg of cobalamin (form not stated) and they found that almost all participants required 1000 μg to normalize the serum MMA level. This result was confirmed by a dose-finding study in the Netherlands (Eussen et al., 2005) in which oral cyanocobalamin in doses of 2.5, 100, 250, 500, and 1000 μg was administered for 16 weeks to elderly people with a baseline serum B_{12} of 208 (IQR 113–362) pmol/L. The median serum B_{12} at the end of the trial was 404 (293) pmol/L for those receiving 500 μg/day and was 574 (418) pmol/L for those receiving 1000 μg/day. The lowest

MMA values were obtained after 500 and 1000 μg, while 1000 μg was required to maximally lower tHcy. The authors concluded that "much higher doses of cyanocobalamin are required to normalize vitamin B_{12} deficiency than were previously believed." A similar conclusion was reached by Hill et al. (2013), who found that in elderly subjects with marginal plasma B_{12} status (median 201, range 107–249 pmol/L), a dose of 10 μg/day for 8 weeks was sufficient to bring plasma B_{12} up to >200 pmol/L in all but 15% of subjects but >40% still had elevated plasma MMA. A dose of 500 μg/day still left 10% with B_{12} below 200 pmol/L and 23% with elevated MMA. The discrepancy between the studies showing that low B_{12} doses (<10 μg/day) and those finding that only high doses (500 μg/day or higher) are effective may appear puzzling, but can probably be explained: First, if the supplement is taken with food, it will have a higher absorption due to the secretion of IF; this would also apply if several small doses were repeated throughout the day, e.g., by enriching food sources (Winkels, Brouwer, Clarke, Katan, & Verhoef, 2008). Second, the form of the supplementary B_{12} might be important, whether it is cyano-, hydroxo-, methyl-, or some other form of cobalamin. Third, the effect of supplementary B_{12} may differ if the cause of low B_{12} status is inadequate diet compared with malabsorption of food-bound B_{12}, or if it is due to frank pernicious anemia.

Two excellent reviews on this topic (Dullemeijer et al., 2013; Hoey, Strain, & McNulty, 2009) have not fully resolved the above discrepancies, although the elegant meta-analysis by Dullemeijer et al. (2013) has provided a deeper understanding of the relationship between oral B_{12} intake and serum/plasma B_{12} levels. These authors examined outcomes from 19 observational studies on 12,570 subjects and 37 randomized controlled trials (RCT) on 3398 subjects and obtained a pooled estimate of the dose–response relationship between B_{12} intake (x) and serum or plasma B_{12} (y) that satisfied the equation:

$$\log_e(y) = 0.15 \times \log_e(x) + 5.31$$

This estimate means that the doubling of B_{12} intake increases the B_{12} concentration in serum or plasma by $2^{0.15}$ fold, which corresponds to 11% (95% CI 9.4%, 12.5%). "This calculation meant that a person with an intake of 2 μg B_{12}/day would have a serum/plasma concentration that was 11% higher than that of a person with and intake of 1 μg/day; or a person with an intake of 8 μg/day would have a B_{12} serum/plasma

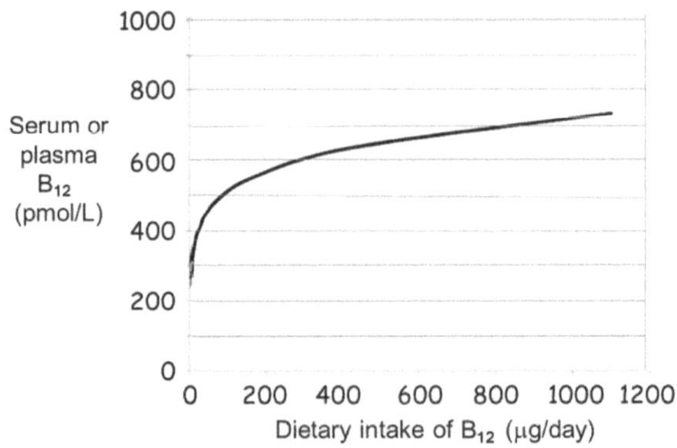

Fig. 4 Blood level of B_{12} in relation to oral intake. Serum or plasma vitamin B_{12} concentration (pmol/L) as a function of dietary vitamin B_{12} intake (μg/day) estimated by using random effects meta-analyses of a combination of observational studies and randomized controlled trials. *Based on fig. 3 from Dullemeijer, C., Souverein, O. W., Doets, E. L., van der Voet, H., van Wijngaarden, J. P., de Boer, W. J., et al. (2013). Systematic review with dose-response meta-analyses between vitamin B_{12} intake and European micronutrient recommendations Aligned's prioritized biomarkers of vitamin B_{12} including randomized controlled trials and observational studies in adults and elderly.* American Journal of Clinical Nutrition, 97*(2), 390–402.*

concentration that was 11% higher than that of a person who had an intake of 4 μg/day" (Dullemeijer et al., 2013).

The authors used the equations derived from the observational and RCT reports to derive a graph showing the relationships; this important outcome is shown in Fig. 4. It is instructive to compare this graph with the results from the VITACOG trial, in which subjects with mild cognitive impairment (MCI) were given a supplement containing 500 μg/day of cyanocobalamin for 2 years: the geometric mean plasma B_{12} before treatment was 330 pmol/L (CI 303, 360) and was 672 pmol/L (95% CI 626, 722) at the end of the study. The value of 672 pmol/L is very close to that predicted for an intake of 500 μg from the plot for RCTs in Fig. 4.

We recommend that an oral dose of 500 μg/day should be the minimum used to treat those with B_{12} in the low-normal range.

5. B_{12} STATUS AND NEUROLOGICAL AND COGNITIVE FUNCTIONING IN ADULTS

5.1 Background

An association between B_{12} deficiency and impaired neurological or cognitive function in adults has been known since Addison's description of pernicious anemia in the 19th century (McCaddon, 2006, 2013; Stabler, 2013). Lindenbaum et al. (1988) described neurological and neuropsychiatric impairments in patients with pernicious anemia who did not display the classical hematological signs; the patients' neurological signs improved upon treatment with B_{12}. In another study of patients with $B_{12} < 148$ pmol/L, it was shown that a beneficial neuropsychiatric response to B_{12} treatment was largely confined to subjects with elevated tHcy or MMA (Stabler, Allen, Savage, & Lindenbaum, 1990). In relation to normal aging, an important concept was introduced in 1992 by Rosenberg and Miller (1992), who stated "For most people, including elderly people, overt vitamin deficiencies are unlikely; it is more likely that mild or 'subclinical' vitamin deficiencies may play a role in the pathogenesis of declining cognitive function in aging." Thus, as discussed in Section 3.1, we need to consider the status of B_{12} over a range of values, not just focus on deficiency, when looking at cognitive outcomes. A good example is that in nondemented subjects, plasma holoTC was correlated ($r = 0.25$, $P = 0.04$) with global cognitive test scores over the entire normal range (35–200 pmol/L), higher concentrations being associated with better test scores (Refsum & Smith, 2003)—illustrated in Smith and Refsum (2009).

The topic has been reviewed extensively (Doets et al., 2013; Health Quality Ontario, 2013; Li et al., 2014; McCaddon, 2013; McCaddon & Miller, 2015; Moore et al., 2012; O'Leary, Allman-Farinelli, & Samman, 2012; Reynolds, 2006; Smith, 2008; Smith & Refsum, 2009, 2016; Venkatramanan, Armata, Strupp, & Finkelstein, 2016; Werder, 2010; Zhang, Ye, Mu, & Cui, 2017). Several of these reviews come to the conclusion that there is no convincing evidence for a link between B_{12} status and cognitive impairment or dementia. In our view, there are flaws in the analyses in some of these reviews, which we have not got space to discuss in detail here. We outlined some of the problems in our earlier review (Smith & Refsum, 2009) and in particular we pointed out that the association of poor B_{12} status with impaired cognition often depended on how B_{12} status was assessed and also upon the context. We will now summarize these aspects.

5.2 Use of Markers of B_{12} Function and the Question of Power

Since only the 20% of the total serum B_{12} that is bound to TC is available to the tissues, the use of total B_{12} may mask an association with cognition. For example, in a case–control study on 51 patients with histopathologically confirmed Alzheimer's disease (AD) and 65 control subjects, the serum total B_{12} levels were not significantly different, whereas the holoTC levels (41.1 and 57.1 pmol/L, respectively) were different ($P < 0.001$) (Refsum & Smith, 2003). It was also found that both MMA and tHcy were significantly elevated in the AD cases compared with the controls, so it was concluded that a poor B_{12} status was associated with AD. Logistic regression confirmed this conclusion, with low holoTC showing a significant odds ratio for AD of 2.82 and raised MMA an odds ratio of 3.83. Total serum B_{12} showed a tendency to be associated with AD, but the confidence intervals for the odds ratio included 1.0. However, this appeared to be a power issue since when the same 51 subjects were combined in an analysis with the whole cohort (76 AD cases, 108 controls), there was a highly significant association between low serum B_{12} and AD with an odds ratio of 4.3 for those in the bottom tertile of serum B_{12} vs those in the top tertile (Clarke et al., 1998). In a cross-sectional study from Chicago, while total serum B_{12} was not related to global cognitive test scores, the markers of B_{12} status (tHcy, MMA) were significantly related to poorer scores (Tangney et al., 2011). These results indicate that it could be an error to conclude that there is no association between B_{12} status and a cognitive outcome, either because there were not enough subjects in the study or because a more sensitive marker was not used. A recent striking finding may also reflect the importance of sufficient power: a data mining study on 5821 patients in the placebo arms of several AD clinical trials found that the strongest biomarker, out of a total of 51 biomarkers, for a high score in the MMSE was serum vitamin B_{12} (Szalkai, Grolmusz, & Grolmusz, 2017).

Further examples of the importance of using markers of B_{12} function and not relying on total B_{12} are two large community studies in Oxfordshire, United Kingdom. In the first, cross-sectional study, the odds ratios for cognitive impairment in 830 elderly people were significant only for those in the bottom quartile of total serum B_{12}, but were significant for the bottom three quartiles of holoTC (Hin et al., 2006). The second, prospective, study on 692 elderly subjects over 10 years found that low baseline concentrations of holoTC and high baseline concentrations of MMA and of tHcy were associated with an increased rate of cognitive decline, whereas there was no relationship to baseline concentrations of total B_{12} (Clarke et al., 2007). The effects were highly significant: a doubling of the baseline concentration of holoTC was associated with a 30% slower rate of cognitive decline, while doubling of the MMA or of tHcy concentrations was each associated with a 50% increased rate of cognitive decline. It is noteworthy that in this study, there was no association of serum folate with cognitive decline. The authors suggested that "the variability in the rates of cognitive decline were primarily explained by the reduced vitamin B_{12} status."

5.3 Importance of the Context in Which B_{12} Is Measured

Some of the discrepancies between different studies might have arisen because the populations studied had different characteristics. There are several reports that the relationship between B_{12} status and cognition in older people is stronger in, or even unique to, those who carry the ε4 allele of *ApoE* (Bunce, Kivipelto, & Wahlin, 2004; Feng, Li, Yap, Kua, & Ng, 2009; Vogiatzoglou et al., 2013). The relationship between B_{12} status and cognition is also more marked in people with higher depression scores (Bell et al., 1990; Garrod et al., 2008; Vogiatzoglou et al., 2013). Thus, the prevalence of *ApoE* ε4 and/or of depression in a population could influence the observed association of B_{12} status with cognition. Another contextual effect has been observed in a study on >2000 Norwegian elderly people, where there was an interaction between plasma choline and B_{12} (Nurk et al., 2012): subjects with either low choline (8.36 μmol/L) or low B_{12} (<257 pmol/L) had no significant risk of cognitive impairment, but subjects with both low choline and low B_{12} had a two- to threefold increased risk in several cognitive domains. Thus, if a population under study has a very good choline status, it may not be possible to demonstrate an association between B_{12} status and cognition. The final contextual effect we should mention is that of high folate status. There are reports that in elderly subjects with high folate status the cognitive impairment due to low B_{12} status is made worse (Castillo-Lancellotti et al., 2015; Morris, Jacques, Rosenberg, & Selhub, 2007), and that this effect is related to circulating levels of unmetabolized folic acid (Morris, Jacques, Rosenberg, & Selhub, 2010). The folate/ B_{12} interaction is controversial and has recently been reviewed (Selhub & Rosenberg, 2016; Smith & Refsum, 2016).

We conclude from cross-sectional and prospective observational studies that low B_{12} status, especially when assessed by functional markers, is often associated with cognitive impairment in elderly subjects, but that this association is context dependent and so may not be detected in some populations.

5.4 Intervention Trials

The crucial question is whether neurological and cognitive impairments in adults due to low B_{12} status can be reversed. The classical case study of

Lindenbaum et al. (1988) showed that the neurological and neuropsychiatric impairments due to B_{12} deficiency in pernicious anemia could largely be reversed by treatment. Relatively few modern studies, mainly on elderly subjects, have tested whether such impairments can be reversed by administering B_{12}. It should be obvious, but nevertheless has to be stated, that randomized trials of this nature are meaningful only if carried out on subjects with low B_{12} status who show clear, and ideally progressive, deficits in neurological and/or cognitive functioning (Smith & Refsum, 2016). If the placebo group does not show impairment, what effect can be expected in the group treated with B_{12}? The published randomized trials are listed in Table 5. It can be concluded from this table that all of the trials listed have quite serious weaknesses that make it difficult to draw any conclusions about the possible benefit of B_{12} supplementation. Eight out of the 10 trials either had no placebo group or there was no decline in scores in the placebo group over time. In one trial (Hvas et al., 2001), there was neurological improvement in the B_{12} group in subjects with poor B_{12} status at baseline, but this was not confirmed in the larger OPEN trial (Miles et al., 2017). The OPEN trial (Dangour et al., 2015) was on asymptomatic older people (aged 80 years) who had low-normal serum B_{12} (median 223 pmol/L) and were given 1 mg oral B_{12} daily for 12 months. In the placebo group, there was no change in B_{12} over 1 year, whereas in the B_{12}-treated group the B_{12} increased to 641 pmol/L. The placebo group showed no significant change in the various neurological parameters tested, and so it is hardly surprising that no effect of B_{12} treatment was found. The OPEN trial has been criticized (Kalita & Misra, 2015) for the choice of neurological endpoints, which are not very sensitive to poor B_{12} status. The primary neurological endpoint was the posterior tibial compound muscle action potential, whereas more sensitive endpoints would have been sensory- and motor-evoked potentials (Kalita & Misra, 2015). Furthermore, there was no decline over time in the placebo group in any of the several cognitive tests used; indeed, in the primary outcome (memory), there was a tendency for the performance to get better in the placebo group (Dangour et al., 2015), probably due to learning effect. Clearly, slowing of cognitive decline by B_{12} treatment cannot be expected if there is no cognitive decline in the placebo group.

An open trial in 51 asymptomatic participants (age 73 years) with B_{12} deficiency (median = 120 pmol/L) in Chile used a combination of B_{12} (10 mg), B_6 (100 mg), and thiamine (100 mg) in a single intramuscular injection given 4 months before the second neurological assessment. At the end of the trial, the sensory latency of both sural nerves improved in parallel with

Table 5 Clinical Trials of B_{12} in Older Adults in Relation to Neurological or Cognitive Outcomes

Trial Subjects	N	Dose of B_{12}	Duration (Months)	Outcome and *Interpretation*	References
Patients with low B_{12} (112 pmol/L):		Parenteral, but no details given		Dementia: no difference in cognition in B_{12}-treated group cf. matched untreated group with normal B_{12}.	Eastley, Wilcock, and Bucks (2000)
Dementia	132		7.5		
Cognitive impairment	22		9	Cognitive impairment: improved score on verbal fluency in B_{12} group cf. matched untreated group with normal B_{12}. *Cannot be interpreted since no information about dose and timing of B_{12} treatment*	
High plasma MMA ($\geq$0.4 µmol/L). Median B_{12}: 254 and 278 pmol/L Randomized	140	1 mg i.m. weekly for 4 weeks	Follow-up at 3	Overall no improvement in neurological scores in B_{12} group, but significant improvements in neurological symptom scores in subjects with baseline MMA $\geq$ 0.6 µmol/L, tHcy $\geq$ 15 µmol/L, and trend for $B_{12} \leq$ 250 pmol/L. *Treatment period very short but possible beneficial effect in subjects with worse baseline B_{12} status*	Hvas, Ellegaard, and Nexo (2001)
High plasma MMA (same subjects as Hvas et al., 2001) Randomized	140	1 mg i.m. weekly for 4 weeks	Follow-up at 3	Cognitive scores improved in both placebo and B_{12} groups. *Improved scores likely to be a learning effect. Treatment period too short*	Hvas, Juul, Lauritzen, Nexo, and Ellegaard (2004)

Continued

Table 5 Clinical Trials of B_{12} in Older Adults in Relation to Neurological or Cognitive Outcomes—cont'd

Trial Subjects	N	Dose of B_{12}	Duration (Months)	Outcome and *Interpretation*	References
Mild B_{12} deficiency (100–200 pmol/L; or MMA ≥ 0.32 μmol/L) Randomized	111	1 mg orally, daily	6	Cognitive scores in several domains "improved slightly" in both placebo and B_{12} groups; improvement in memory in placebo was significantly better than in B_{12} group. *No decline, and often improvement, in placebo group so difficult to conclude anything about treatment*	Eussen et al. (2006)
Dementia (serum $B_{12} < 200$ pmol/L)	21	1 mg i.m., then daily orally	10	No cognitive improvement, slight improvement in delirium. *Open trial, no placebo group*	Kwok, Lee, Lam, and Woo (2008)
Dementia (serum $B_{12} < 184$ pmol/L)	28	1 mg i.m., repeated	4	No significant clinical benefits. *Open trial, no placebo group*	van Dyck, Lyness, Rohrbaugh, and Siegal (2009)
Adults with neurological syndrome ($B_{12} < 155$ pmol/L)	33	1 mg i.m., repeated	6	Functional neurological recovery in the "majority." Cognitive improvement in several domains. P3 improved. *Lack of placebo group and imprecise reporting*	Kalita, Agarwal, Chandra, and Misra (2013)
Asymptomatic elderly (80 years) with low-normal B_{12} (median 228 pmol/L). "OPEN" randomized trial	201	1 mg daily oral	12	Neurological measure: primary outcome was tibial compound muscle action potential. Slight (nonsign.) increase in placebo; no change B_{12}	Dangour et al. (2015)

				group. Cognitive measure: word-list memory. A slight increase in both placebo and B_{12} groups, but no difference. *Lack of decline in placebo group scores so difficult to conclude anything about treatment*	
Asymptomatic elderly (80 years) with low-normal B_{12} (median 228 pmol/L). "OPEN" randomized trial	201	1 mg daily oral	12	Analysis based on subgroups with differing baseline B_{12} status showed no significant effect of B_{12} treatment on primary outcome of tibial compound muscle action potential in those with low B_{12} status at baseline. *Difficult to interpret as placebo data not reported*	Miles et al. (2017)
Elderly diabetic patients with low B_{12} (mean 227 pmol/L)	271	1 mg daily oral methylcobalamin	27	Both groups showed cognitive decline but no difference in CDR or cognitive tests between treated and placebo groups. However, placebo group cognitive scores tended to increase. *Lack of decline in placebo group scores, so difficult to conclude anything about treatment*	Kwok et al. (2016)

Trials were included if B_{12} was the only treatment administered.

improvement in the B_{12} status (Brito et al., 2016). Interpretation of this trial must be seen in light of the combination treatment and the lack of a placebo group.

In view of the paucity of well-designed trials to test the effect of B_{12} treatment on neurological and cognitive deficits, we will briefly summarize the VITACOG trial in which three B vitamins (0.8 mg folic acid; 0.5 mg B_{12}; 20 mg B_6) were given over 2 years to older people with MCI. In this trial (reviewed in Smith & Refsum, 2016, 2017), the placebo group showed worsening performance in scores for global cognition (MMSE), episodic memory, semantic memory, and clinical dementia rating (CDR sum-of-boxes) but only in those with tHcy above the median (11.3 μmol/L) at baseline. Participants with tHcy below the median did not show a significant cognitive decline. B vitamin treatment markedly slowed cognitive decline in those with tHcy above the median and led to an improvement in clinical status (CDR score) in those with baseline tHcy $>$13 μmol/L (de Jager, Oulhaj, Jacoby, Refsum, & Smith, 2012). A similar beneficial effect of B vitamin treatment was found on the rate of whole-brain atrophy, which was slowed by 30% compared with the placebo group. The slowing of brain atrophy rate was greater, the higher the baseline tHcy concentration, reaching 53% for participants with tHcy in the top quartile (Smith et al., 2010). Thus, B vitamin treatment that includes B_{12} can slow the brain atrophy and cognitive decline that occurs in people with MCI, but only in those with relatively poor B vitamin status as revealed by high tHcy (above the median). The results of this trial show that modification of the disease process underlying cognitive decline in MCI (a prodromal stage of AD) is feasible using nutritional interventions in those with low-normal B vitamin status (Smith & Refsum, 2017). No specific neurological tests were done in VITACOG, but it was observed that 13 participants (9.8%) taking the placebo showed a loss of vibration sense over 2 years, whereas only three subjects (2.2%) taking B vitamins showed such a loss ($P = 0.019$). VITACOG will be further discussed in Section 7.2.

6. POOR B_{12} STATUS IN MOTHER AND CHILD: HEALTH OUTCOMES

6.1 Background

This important topic requires a detailed review on its own, and we will only summarize a few findings here. As pointed out in Section 3.2, there is a high prevalence of poor B_{12} status in pregnant women in many parts of the world.

This has major implications for maternal health, for example, increasing the risk of gestational diabetes (Knight et al., 2015; Krishnaveni et al., 2009; Lai et al., 2017; Sukumar, Venkataraman, et al., 2016) and obesity (Sukumar, Venkataraman, et al., 2016). For the embryo and infant, low maternal B_{12} status is associated with slowing of embryonic growth rate (Parisi et al., 2017), increasing the risk of low birth weight (Rogne et al., 2017; Sukumar, Rafnsson, et al., 2016), of preterm birth (Rogne et al., 2017), and of neural tube defects (Molloy et al., 2009; Tang, Li, & Wang, 2015). The mother's B_{12} status may also be an important determinant of the later health of the child, as shown in the Pune Maternal Nutrition Study in which low B_{12} and high folate in the mother were risk factors for the development of obesity and insulin resistance in the child at age 6 years (Yajnik et al., 2008). This topic has recently been reviewed (Krishnaveni & Yajnik, 2017). Several studies have shown that maternal B_{12} status is an important determinant of the B_{12} status of the infant (Bjorke-Monsen & Ueland, 2011; Finkelstein, Kurpad, Thomas, Srinivasan, & Duggan, 2017; Hay et al., 2010; Ueland & Monsen, 2003). Furthermore, infants who are breastfed have a lower B_{12} status than nonbreastfed infants. In Norway, breastfed infants at 6 months had a mean serum B_{12} of 242 pmol/L, while those not breastfed had a B_{12} of 365 pmol/L (Hay, Johnston, Whitelaw, Trygg, & Refsum, 2008). In India, the respective figures were medians of 184 and 334 pmol/L (Taneja et al., 2007). Notably, in the Norwegian study of healthy infants of omnivore mothers, the lower B_{12} status was found in all breastfed infants, not only in those that were exclusively breastfeeding (Hay et al., 2008). Thus, a critical question is whether the low B_{12} status in breastfeeding is solely explained by low B_{12} content in breast milk or whether metabolic effects of breast milk or breastfeeding cause a change in B_{12} homeostasis. At this stage, we do not know whether the low B_{12} in healthy breastfed infants is a major concern in relation to later health and development, or if it just reflects different "normal" B_{12} levels in breastfed vs nonbreastfed infants.

6.2 Severe B_{12} Deficiency in Exclusively Breastfed Infants

To further understand the potential harmful effect of low B_{12} status in infancy, we need to recognize that most data on major health concerns are from case studies of exclusively breastfed infants of mothers with poor B_{12} status (Goraya, Kaur, & Mehra, 2015), or in infants with poor B_{12} status combined with poor health, e.g., low birth weight or stunted growth.

In a recent report of a case series in India, the literature of neonatal B_{12} deficiency in exclusively breastfed infants was also reviewed (Goraya et al.,

2015). Briefly, the infants usually develop normally in the first 4–6 months. Initial symptoms include failure to thrive, lethargy, feeding difficulties, refusal of solid foods, and pallor, followed by involuntary movements in the form of tremors, myoclonus, or choreoathetosis. Upon examination, anemia with macrocytosis and megaloblastic bone marrow is usually present and the infants have a low B_{12} combined with elevated tHcy and MMA. CT or MRI of the brain reveals cerebral atrophy and delayed myelination. If diagnosed early and treated promptly, the infants improve rapidly and usually catch up within some few months. Even reversal of cerebral atrophy has been observed (Casella, Valente, de Navarro, & Kok, 2005; Goraya et al., 2015; Lövblad et al., 1997; see also Section 6.2 and Fig. 6). Nevertheless, long-term neurocognitive symptoms in later childhood may occur (Bhate et al., 2008; Zengin, Sarper, & Caki Kilic, 2009), which is probably due a combination of severity of the conditions and delayed treatment. Hence, it is critically important to discover B_{12} deficiency before severe symptoms arise.

6.3 B_{12} in Children's Motor and Cognitive Development and Intervention Trials

It is notable that inborn errors of remethylation (see Section 2 and Table 1) have revealed several crucial roles for B_{12} in development (Huemer et al., 2017). In the general population, as shown in Supplementary Table 1S available on https://doi.org/10.1016/bs.afnr.2017.11.005, the B_{12} status in infants and children is often poor, especially in less developed countries, with deficiencies of up to 30% (Nepal), 40% (Kenya), 58% (Honduras), and 68% (India). At this critical stage in childhood development, it is crucial to know if low B_{12} status observed in infants has harmful effects and also if there is beneficial effect of treatment. There are several reviews (Bjorke-Monsen & Ueland, 2011; Dror & Allen, 2008; Goraya et al., 2015; Venkatramanan et al., 2016) that suggest that a good B_{12} status may also improve health and development in a more general population of infants. A good example of these findings is a study in New Delhi on 538 infants 12–18 months old in which cognitive performance was tested, using Bayley's Mental Development Index, 4 months after blood sampling (Strand et al., 2013). The mean plasma B_{12} was 221 pmol/L and 27% were B_{12} deficient (<150 pmol/L): there was a positive association between plasma B_{12} and cognitive performance, and an inverse association between tHcy and MMA and the cognitive scores. It is noteworthy that the associations with cognition were approximately linear over the whole normal range of B_{12} and its markers (Supplementary Fig. S3 available on https://doi.org/10.1016/bs.afnr.2017.11.005). An important prospective study in Nepal has shown that B_{12} status in the infant is positively associated with development and performance on social perception tasks and visuospatial abilities at 5 years of age (Kvestad et al., 2017).

Table 6 summarizes some of the associations between poor B_{12} status and childhood development according to observational studies cited in the earlier reviews (see also Demir, Koc, Ustyol, Peker, & Abuhandan, 2013; Goraya et al., 2015; Zengin et al., 2009).

Unfortunately, there are relatively few randomized trials in this area. In Norway, 79 infants (mean age 4 months) referred for feeding difficulties and who had elevated tHcy were given a sham injection or a single intramuscular injection of hydroxocobalamin (0.4 mg): 1 month later, those treated with B_{12} showed a marked fall in tHcy and MMA and clinically a larger improvement than the placebo group in a test for gross motor development and also in regurgitation (Torsvik et al., 2013). Another randomized trial by the same authors showed that 6-month-old infants with low- or low-normal birth weight (2000–3000 g) and modestly elevated tHcy (>6.5 μmol/L) and who had been exclusively breastfed for >1 month had a lower serum B_{12} (median 321 pmol/L) than formula-fed infants (497 pmol/L) and had poorer gross motor development. A single injection of hydroxocobalamin (0.4 mg) in the breastfed infants led to an improvement in motor performance (Torsvik et al., 2015). These studies are important because they raise the idea that exclusive breastfeeding in susceptible groups (e.g., low birth weight,

Table 6 Some Associations of B_{12} Deficiency and Poor B_{12} Status With Childhood Development[a]

Neurological Development	Cognitive Development
Hypotonia	Impairment in social perception
Hypokinesia	Impaired visuospatial abilities
Lethargy	Lower mental development scores
Ataxia	Impaired attention
Insufficient head control	Impaired short-term memory
Involuntary movements	Impaired academic performance at school
Swallowing dysfunction	Absenteeism from school
Seizures	Grade repetition at school
Brain atrophy	
White matter disease	

[a]Signs of neurological development have mainly been seen in infants with severe B_{12} deficiency, while signs of impaired cognitive impairment also include children with low B_{12} status during early childhood.

failure to thrive, elevated tHcy) might lead to impaired motor development through an effect on B_{12} status. Unfortunately, these trials did not further evaluate whether the beneficial effect was confined to certain subgroups, e.g., such as infants with more marked elevation of tHcy. Furthermore, the results cannot be extrapolated to healthy infants of mothers with normal B_{12} status.

An important trial has been carried out in New Delhi, in which 1000 children (aged 6–30 months) were randomized in a factorial design to four groups: a paste containing two RDAs of B_{12}, folic acid, both, or placebo (folic acid 150 μg; B_{12} 1.8 μg; doses were halved for infants <12 months). The vitamins were administered for 6 months after which several outcomes were assessed, including common infections (Taneja et al., 2013), growth (Strand et al., 2015), and hemoglobin (Kumar et al., 2017), and, in 422 children, cognitive and motor skills were also assessed (Kvestad et al., 2015). Notably, this is a study in a population where mothers often have adhered to a lifelong vegetarian diet, have low B_{12} status before, during, and after pregnancy, and where malnutrition is common. Furthermore, the vast majority of mothers were still breastfeeding at the time of inclusion. In the analyses of the effect of treatment on infections and hemoglobin, there was no clear benefit of B_{12} treatment. In terms of growth, the effect of B_{12} was confined to children that were stunted, underweight, or wasted. In tests of gross motor performance and problem-solving skills, children who received both folic acid and B_{12} showed significantly better scores than those receiving placebo. B_{12} on its own improved gross motor scores, but folic acid alone had no effect. The effect was largest in those with high tHcy (>10 μmol/L), with no apparent effect on those with lower tHcy (Kvestad et al., 2015). It should be noted that in these trials in India, the dose of B_{12} was only twice the RDA for infants. Nevertheless, the treatment lasted for 6 months, and there was a marked improvement in B_{12} status with increase in B_{12} of about 180 pmol/L in those receiving B_{12}. Interestingly, the findings are quite similar to the Norwegian studies in that the effect seems to be largely confined to infants with poor B_{12} status and who have signs of growth impairment.

We conclude that the maternal B_{12} status during pregnancy is an essential determinant of B_{12} status at birth and in infancy. Furthermore, there is no doubt that breastfed infants have lower B_{12} status than nonbreastfed infants. However, given the evidence that breastfeeding confers clear health benefits to both mother and child (Bar, Milanaik, & Adesman, 2016; Victora et al., 2016), it would be a serious error to recommend against breastfeeding, even exclusive breastfeeding for the recommended time period of 6 months.

Rather, we recommend that a public health initiative should be taken worldwide to provide B_{12} oral supplements to women throughout pregnancy and during the breastfeeding period to women, in particular to those adhering to vegetarian or vegan diet or in regions with overall low intake of animal-derived foods. Furthermore, in children that are exclusively breastfed for a prolonged time, B_{12} status (including tHcy) should be assessed if they had low birth weight or show failure to thrive.

7. HOW DOES THE BRAIN MEDIATE THE EFFECTS OF B_{12} ON COGNITION?

Deficits in cognition due to B_{12} insufficiency could be due directly to impairment in B_{12}-dependent processes in the brain, or indirectly following an increase in brain homocysteine (Smith & Refsum, 2016), or to a combination of both. It is difficult to distinguish between these alternatives and we will not attempt to do so in what follows. Two main avenues through which low B_{12} or raised tHcy can influence cognition are directly by damage to nerve cells and neuronal networks or less directly by damage to the cerebrovascular system.

7.1 Damage to the Cerebrovascular System

Functioning of the nervous system depends upon a well-functioning cerebrovascular system, as shown by the catastrophic consequences when blood supply to a brain region is compromised in ischemic stroke. Raised tHcy is a well-known risk factor for ischemic stroke (Casas, Bautista, Smeeth, Sharma, & Hingorani, 2005), and low-normal B_{12} is also a risk factor for cerebral ischemia (ischemic stroke or transient ischemic attack) (Spence, 2016; Weikert et al., 2007). Animal studies have thrown light on possible mechanisms: in mice given a B-vitamin-deficient diet for 10 weeks, there was a sevenfold increase in tHcy and the mice showed impaired spatial memory and learning (Troen et al., 2008). The cognitive impairment was highly correlated with a rarefaction of the vasculature in the hippocampus, specifically with a shortening of the length of the blood capillaries. Reduced capillary length was also correlated with higher tHcy. These changes in the vasculature occurred without any obvious signs of neurodegeneration, such as gliosis, leading the authors to suggest that changes in the vasculature may be more important mediators of cognitive impairment than effects on nerve cells. The vascular mediation of cognitive impairment caused by raised tHcy has recently been reviewed (Hainsworth, Yeo, Weekman, & Wilcock,

2016; Smith & Refsum, 2016). The former authors concluded that at physiological concentrations there are important vascular effects of homocysteine on myocyte proliferation, vessel wall fibrosis, impairment of nitric oxide signaling, generation of superoxide, and procoagulant actions. We will focus here on the effects of low B_{12} on the nervous system.

7.2 Damage to the Nervous System

While there may be controversy about the extent to which homocysteine-lowering treatments can prevent vascular disease (Spence, 2016), it is clear that raised homocysteine and low B_{12} have unique effects directly on the nervous system, and that many of these effects are probably independent of vascular disease.

There are currently four distinct hypotheses to account for the harmful effects of low B_{12} status on the nervous system, apart from the view that all the harmful effects of low B_{12} are mediated by increased tHcy: (i) accumulation of unusual fatty acids as a result of inactivation of methylmalonylCoA mutase; (ii) inhibition of methylation reactions; (iii) accumulation of toxic cytokines and decline in protective cytokines (Hathout & El-Saden, 2011; Scalabrino, 2009); and (iv) increase in cellular prion levels (Veber & Scalabrino, 2015). We only have space to deal with the first two hypotheses and refer the reader to the comprehensive reviews cited for the last two hypotheses.

If we consider that B_{12} insufficiency might cause damage to the nervous system independent of homocysteine, then the obvious pathway is that involving AdoCbl as a cofactor in the conversion of methylmalonylCoA to succinylCoA (Section 1). Impairment in this pathway leads to the accumulation of MMA and to increased amounts of atypical fatty acids, such as odd-chain and branched-chain fatty acids. For a long time, it was thought that these unusual fatty acids might cause neuropathy and cognitive impairment by being incorporated into myelin lipids, but in his landmark review, Metz dismissed this idea: "...a large body of evidence can now be marshalled against the hypothesis that decreased function of AdoCbl is the mechanism of the cobalamin neuropathy" (Metz, 1992). As Metz pointed out, neuropathy is not found in those genetic diseases that lead to accumulation of MMA without effect on methionine synthase functions, such as *cblA*, *cblB*, and *Mut/MCM* (Table 1). However, although neurological impairments are not characteristic of these diseases, they do occur in about 10% of such patients and may be due to metabolic intoxication (Nizon et al., 2013).

The alternative hypothesis, proposed by Gandy, Jacobson, and Sidman (1973), that neurological deficits in B_{12} deficiency are due to impaired methylation reactions was strongly supported by the observation of Scott and colleagues in 1981 that administration of methionine could prevent the subacute combined degeneration of the spinal cord produced in monkeys by exposure to N_2O, which inactivates methionine synthase (Scott, Dinn, Wilson, & Weir, 1981).

Low B_{12} status impairs the formation of methionine and so decreases the supply of SAM, the body's main methyl donor. Low B_{12} status also causes increased levels of homocysteine, which, in turn, leads to increased levels of SAH, a potent inhibitor of methylation reactions throughout the body (Deguchi & Barchas, 1971). Thus, we propose the that most of the harmful effects of low B_{12} or high tHcy on the brain are initially caused by impaired methylation due to one or both of these biochemical events. We will distinguish four consequences of deficient methylation in the nervous system: (i) impairment of transmission of signals from one brain region to another, thus disrupting neural networks; (ii) impairment of synaptic mechanisms that mediate cognition; (iii) regional brain atrophy; and (iv) formation of Alzheimer-type neuropathology.

7.2.1 Impairment of Transmission of Signals in the Brain

A cardinal sign of B_{12} deficiency is the disorganization of myelin in the long tracts of the spinal cord. This leads to the phenomenon called "subacute combined degeneration." It presumably also underlies the neuropathy found in sensory and motor nerves. A plausible hypothesis to account for the disruption of myelin is that myelin basic protein, which is essential to maintain the compact structure of myelin, needs to be methylated at one of its 18 arginine residues (residue 107) to facilitate the compaction of myelin (Kim, Lim, Park, & Paik, 1997). The enzyme that methylates this residue is protein methylase-1, and inhibition of this enzyme in embryonic neuronal cultures leads to loss of compact myelin structure (Amur, Shanker, Cochran, Ved, & Pieringer, 1986). Deficient methylation due to B_{12} deficiency or in the presence of elevated tHcy would therefore lead to disruption of the compact structure of myelin that is essential for its normal function. This effect on myelin structure could explain the changes observed in the cerebral white matter (i.e., areas of the brain mainly made up of myelinated axons) that are revealed as regions of hyperintensity in MRI scans. These hyperintensities are often attributed to small vessel disease, but it appears that they can also be caused by high tHcy (reviewed in Smith & Refsum, 2016) or by low B_{12} status (de Lau et al., 2009; Graber et al., 2010; van Overbeek et al., 2013). It was found in the Rotterdam scan study that such white matter damage could be detected over the entire normal range of holoTC (Supplementary Fig. S4 available on https://doi.org/10.1016/bs.afnr.2017.11.005) (de Lau et al., 2009). The white matter lesions in people who are B_{12} deficient can in part be reversed by B_{12} treatment, as shown in Supplementary Fig. S5 available on https://doi.org/10.1016/bs.afnr.2017.11.005 (Vry et al., 2005), so it is unlikely that they are caused only by small vessel disease in these patients. Reversibility of MRI changes in the white matter of the spinal cord after B_{12} treatment has also been reported in 15 out of 29 cases of subacute combined degeneration (Ertan, Tanriverdi, Kiziltan, & Tata, 2002).

A more direct way of revealing the integrity of white matter tracts in the brain is the MRI technique called diffusion tensor imaging (DTI), based on restrictions in the random movements of water molecules by macromolecules and myelin. Application of DTI to patients with B_{12} deficiency has revealed widespread abnormalities in many white matter tracts, notably corpus callosum and fornix; these abnormalities were partly reversed upon treatment with B_{12}. Furthermore, they were correlated with cognitive impairment in several domains (Gupta, Garg, et al., 2014; Gupta, Gupta, et al., 2014). Importantly, one of the abnormalities found in B_{12}-deficient patients was an increase in radial diffusivity in white matter tracts (Supplementary Fig. S6 available on https://doi.org/10.1016/bs.afnr.2017.11.005) (Gupta, Gupta, et al., 2014), which reflects disruption of myelin (Song et al., 2002).

Since white matter tracts link regions of the brain involved in networks that underlie cognition, it can be predicted that the operation of these networks will be compromised in B_{12} insufficiency. This outcome has been demonstrated in a resting-state functional MRI study from India (Gupta et al., 2016). These authors showed that the regional homogeneity was lower overall in the cerebral cortex of patients with B_{12} deficiency (116 pmol/L) than in control subjects. Regional homogeneity correlated with cognitive test scores in several domains, which were lower in the patients with B_{12} deficiency. Regional homogeneity was particularly low in regions that are components of networks important for cognition, such as the default mode network, where the hippocampus and the medial prefrontal cortex showed deficits. It is notable that several of these deficits were reversed upon B_{12} treatment for 6 weeks (Supplementary Fig. S7 available on https://doi.org/10.1016/bs.afnr.2017.11.005), as were the deficits in neuropsychological tests. This elegant study shows that B_{12} deficiency leads to impairment in the functioning of neural networks in the brain, notably the network that includes the hippocampus, a structure that is crucial for memory.

Another fine-structural MRI study (Kobe et al., 2016) has examined the hippocampus in subjects with MCI using DTI to assess the mean diffusivity of water molecules in this region: a high mean diffusivity indicates impaired structural integrity of nerve cells and their processes in gray matter. Two subgroups of MCI patients were compared: those with low-normal B_{12} status (mean 253 pmol/L; range 153–303) and those with high-normal B_{12} status (mean 463 pmol/L; range 306–934). Those with low-normal B_{12} status performed less well on tests of episodic memory (learning ability and recognition memory), and on DTI/MRI, they had an increased mean diffusivity specifically in a small subfield of the hippocampus (CA4 and dentate gyrus). Mediation analysis showed that the microstructural impairment in this specific subfield mediated 48% of the association between low B_{12} status and the deficit in learning ability (Kobe et al., 2016). This landmark study shows, first, that B_{12} deficiency is not essential for impairment of memory since it already occurs at low-normal B_{12} levels and, second, that the memory deficit is in part directly caused by damage to the neurons in the hippocampus; wider implications of these results have been discussed elsewhere (Smith, 2016).

A different way of looking at the functioning of neural networks in the brain is single-photon emission computed tomography (SPECT), which reveals regional blood flow. A SPECT study on 12 patients with low B_{12} status (185 pmol/L), most with cognitive impairment, showed impairments in blood flow bilaterally in frontotemporal regions and in some cases in the prefrontal cortex that correlated with cognitive impairments (Tu, Lo, Chen, & Huang, 2015).

We can conclude that low B_{12} status has profound effects on the functioning of neural networks in the brain and that this can account for many of the associations with cognition.

7.2.2 Impairment of Synaptic Mechanisms That Mediate Cognition

Omega-3 fatty acids play a crucial role in the structure and function of synapses in the brain (Bazinet & Laye, 2014), and they are involved in the synaptic plasticity that underlies learning and memory (Thomazeau, Bosch-Bouju, Manzoni, & Laye, 2017). The omega-3 fatty acids are mainly present in brain in an esterified form in phospholipids such as phosphatidylcholine where they are involved in maintenance of membrane structure.

Importantly, phosphatidylcholine species that contain omega-3 fatty acids can be formed by methylation of phosphatidylethanolamine, using SAM as methyl donor. Hence, a deficit in B_{12} could lead to impaired synthesis of omega-3-containing phosphatidylcholine by decreasing the supply of methionine and SAM, leading to the build-up of homocysteine and so of SAH, which inhibits the methylation reactions (Fig. 5). A seminal paper by Selley (2007) used this hypothesis to account for his findings in patients with AD. Selley found that the raised tHcy and plasma SAH in patients with AD were associated with the following changes: a decrease in erythrocyte membrane content of phosphatidylcholine, a decreased proportion of docosahexaenoic acid in erythrocyte phosphatidylcholine, but an increase in the erythrocyte content of phosphatidylethanolamine. In cognitively normal men, a similar relationship has been reported: high tHcy is associated with lower levels of omega-3 fatty acids in plasma phospholipids (Li, Mann, & Sinclair, 2006). Significantly, the latter authors also found a positive correlation between serum B_{12} and plasma omega-3 fatty acids. A study on chick embryonic brain is consistent with the hypothesis (Miller, Leanza, Phillips, & Blacquire, 2003): exposure of chick embryos to homocysteine led to a decrease in brain phosphatidylcholine; an increase in brain phosphatidylethanolamine; and a 76% fall in brain docosahexaenoic levels. These

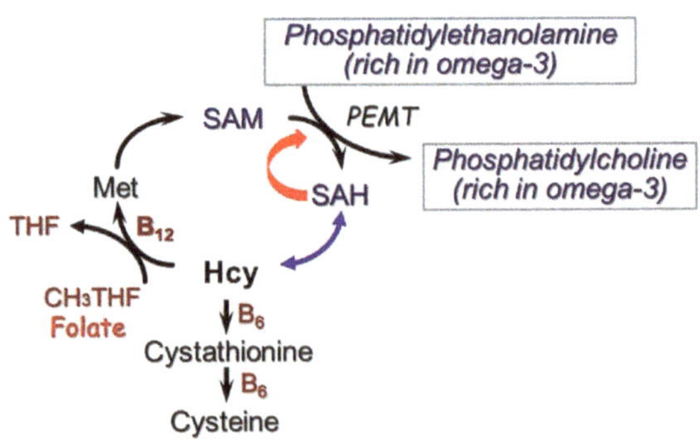

Fig. 5 Interface between homocysteine (Hcy), B vitamins, and omega-3 fatty acids. Elevated total Hcy (tHcy) leads to elevated *S*-adenosylhomocysteine (SAH), which inhibits the methylation of phosphatidylethanolamine to phosphatidylcholine by phosphatidylethanolamine *N*-methyltransferase (PEMT). Lowering tHcy by B vitamins releases this inhibition and allows phosphatidylcholine enriched in omega-3 fatty acids to be synthesized. Abbreviations: *Met*, methionine; *SAM*, *S*-adenosylmethionine; *THF*, tetrahydrofolate.

striking changes would lead to changes in synaptic membrane fluidity and organization (Shaikh & Teague, 2012). Thus, through its role in the methylation of phosphatidylethanolamine, B_{12} is likely to play a key role in the structure and function of synaptic membrane lipids and proteins (Sidhu, Huang, Desai, Kevala, & Kim, 2016). It is therefore of interest that beneficial interactions have been observed between omega-3 fatty acids and B vitamins in relation to brain atrophy (Jernerén et al., 2015) and to cognitive decline (Oulhaj, Jernerén, Refsum, Smith, & de Jager, 2016) in people with MCI.

Another possible role for B_{12} in synaptic mechanisms related to cognition has been revealed by the surprising finding that DNA methylation is required for memory formation and consolidation involving hippocampal synapses. Furthermore, the synaptic plasticity believed to underlie memory is impaired by inhibitors of DNA methylation, although the precise molecular mechanisms linking DNA methylation with memory consolidation are not known (Day & Sweatt, 2011). A similar role for the methylation of lysine residues of histones in the consolidation of hippocampal long-term memory formation has been described (Gupta et al., 2010). These findings provide a novel way in which alteration in B_{12} status might influence cognition via epigenetic mechanisms.

7.2.3 Regional Brain Atrophy

Descriptions of B_{12} deficiency in very young children often mention that the brain shows atrophy. For example, brain atrophy was found in 8 out of 15 infants in one study (Taskesen, Yaramis, Pirinccioglu, & Ekici, 2012), in 7 out of 14 in a second study (Ekici et al., 2016), and in 9 out of 9 in another study (Goraya et al., 2015). In several reports on infants, it was shown that treatment with B_{12} (usually intramuscular) could reverse some or all of the signs of atrophy and some of the associated cognitive deficits (Casella et al., 2005; Glaser, Girschick, Schropp, & Speer, 2015; Lövblad et al., 1997); an example is shown in Fig. 6.

In adults with B_{12} deficiency, reports in which neuroimaging methods have been used to examine the brain are rare, and as a result, there are few descriptions of atrophy. There are, however, many studies on brain atrophy in adults in relation to tHcy. These studies were reviewed recently (Smith & Refsum, 2016) and so will not be discussed here. There are some reports on brain volume in adults with B_{12} in the normal range. In a study on 32 elderly volunteers, it was found that total B_{12} intake (diet plus supplements = mean of 62 μg/day) was related to the volume of the left and right superior parietal lobules, with a greater intake being associated

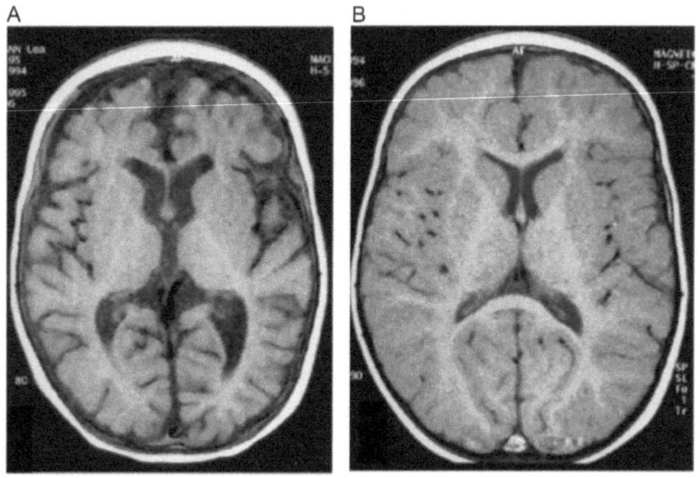

Fig. 6 Brain atrophy in a child with B_{12} deficiency and its reversal by B_{12} treatment. A 6-month-old girl of vegetarian parents had serum B_{12} of 92 pmol/L and serious neurological impairment. MRI of the brain showed marked atrophy, especially in the frontal and temporal lobes (A). After 5 months of daily treatment with B_{12}, there was regression of atrophy and neurological signs were normal (B). *Based on Figs. 2 and 5 from Lövblad, K., Ramelli, G., Remonda, L., Nirkko, A. C., Ozdoba, C., & Schroth, G. (1997). Retardation of myelination due to dietary vitamin B12 deficiency: Cranial MRI findings.* Pediatric Radiology, *27(2), 155–158.*

with a larger volume (Erickson et al., 2008). A cross-sectional study in Chicago on 121 elderly people found that serum B_{12} (mean 331 pmol/L) was not associated with total brain volume, but that higher MMA and tHcy were associated with a reduced total brain volume (Tangney et al., 2011).

Two longitudinal studies, in the United Kingdom and Sweden, have shown that B_{12} status, assessed by either total serum B_{12} or holoTC, is inversely associated with the rate of whole-brain atrophy (Hooshmand et al., 2016; Vogiatzoglou et al., 2008). The important finding was that the association extended over the whole normal range of B_{12} (200–600 pmol/L) or holoTC (Fig. 7). This result is consistent with the extension of some clinical associations with B_{12} over a greater range than the classical cutoff for deficiency, as discussed in Section 4. Despite normal serum levels of B_{12}, it appears that in the brain the concentrations of some forms of B_{12} may be reduced in the elderly. A recent study showed that there was a striking 10-fold decline in methylcobalamin levels in aging (Zhang, Hodgson, et al., 2016). This unexpected finding may partly explain the increased sensitivity of neurological functions to low B_{12} with aging. The study needs replicating.

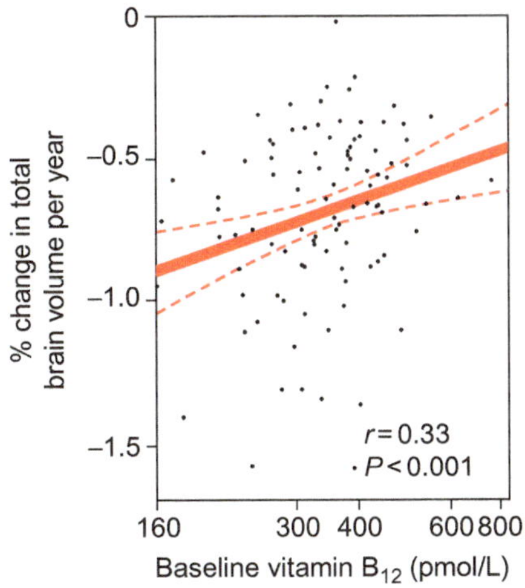

Fig. 7 Rate of brain atrophy in elderly in relation to serum B_{12} concentration. Annual rate of brain atrophy in 107 volunteers (mean age 73.2 years) measured over 5 years. There was no obvious threshold, and the linear association between atrophy rate and B_{12} concentration indicates that up to 13% of the variance in atrophy rate can be explained by B_{12}. *Solid line* mean, *dashed lines* show the 95% confidence intervals. *Based on fig. A from Vogiatzoglou, A., Refsum, H., Johnston, C., Smith, S. M., Bradley, K. M., de Jager, C., et al. (2008). Vitamin B12 status and rate of brain volume loss in community-dwelling elderly.* Neurology, 71, *826–832.*

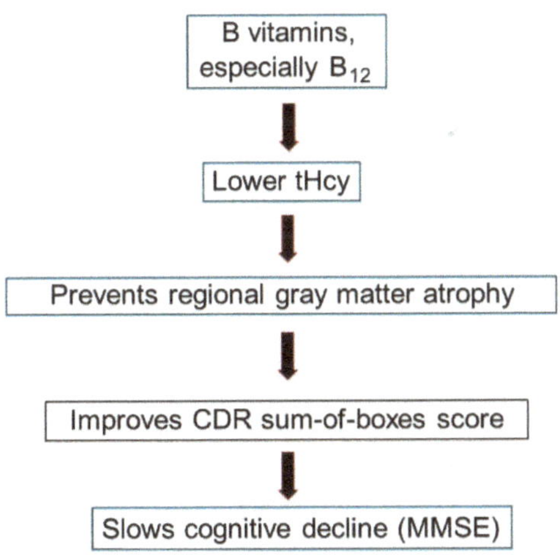

Fig. 8 VITACOG trial: pathway mediating the effect of B vitamin treatment on cognition. A directed acyclic graph analysis of B vitamin treatment and consequential changes in brain structure and function in mild cognitive impairment. The mediating pathway (causal links indicated by *arrows*) shows the optimal Bayesian network that explains the findings from the VITACOG trial. *From data in Douaud, G., Refsum, H., de Jager, C. A., Jacoby, R., Nichols, T. E., Smith, S. M., et al. (2013). Preventing Alzheimer's disease-related gray matter atrophy by B-vitamin treatment.* Proceedings of the National Academy of Sciences of the United States of America, 110*(23), 9523–9528.*

The most detailed study of brain atrophy in relation to B vitamins is the VITACOG trial, summarized in Section 5.4. In this trial in people with MCI, the placebo group showed a rate of whole brain atrophy of 1.08% per year, while those treated with B vitamins for 2 years showed an atrophy rate of 0.76% per year. The slowing of atrophy depended upon baseline tHcy, and for those in the top quartile (>13 μmol/L) the rate was slowed by 53% in the B vitamin group (Smith et al., 2010). In the B vitamin-treated group, the plasma B_{12} increased from a geometric mean of 330 to 672 pmol/L after 2 years of treatment. The greater the improvement in B_{12} status (assessed by plasma B_{12}, holoTC, or TC saturation), the slower the rate of atrophy. More detailed analysis showed that the rate of atrophy of those regions of the brain that shrink in AD (such as the hippocampus) was slowed by almost 90% by the B vitamin treatment (Supplementary Fig. S8 available on https://doi.org/10.1016/bs.afnr.2017.11.005) and that the likely driver of this effect was the supplementation with B_{12} (Douaud et al., 2013). By means of a Bayesian directed acyclic graph analysis, causal links mediating the modifying effect of B_{12} on cognitive decline were revealed (Fig. 8).

Subsequent work showed that the effects of the B vitamins on brain atrophy (Jernerén et al., 2015) and on cognitive decline (Oulhaj et al., 2016) depended upon the omega-3 fatty acid status of the subject; only those participants with a good omega-3 status benefitted from the B vitamin treatment.

7.2.4 Formation of Alzheimer-Type Neuropathology

Cognitive impairment due to B_{12} insufficiency could be an early expression of the role of low B_{12} in the causation of AD (see Section 5.2). So is there any evidence that low B_{12} is related to the formation of the two cardinal histopathological markers of AD, beta-amyloid, and phosphorylated tau? There is little direct evidence on this question, but a great deal of evidence that raised tHcy is indeed related to these markers, as reviewed elsewhere (Fuso, 2013; Smith & Refsum, 2016; Sontag & Sontag, 2014; Zhuo, Wang, & Pratico, 2011). The basic hypotheses are that low B_{12} (or folate) status leads to raised tHcy which, via its effect on SAH, (i) inhibits the methylation of the promotor regions of genes that generate beta-amyloid, so leading to increased expression of these genes; and (ii) inhibits the methylation of protein phosphatase 2A, which inactivates the enzyme that converts phosphorylated-tau into tau, thereby causing a build-up of neurofibrillary tangles.

In animal experiments where beta-amyloid formation is increased in hyperhomocysteinemia, increased deposition of beta-amyloid and the accompanying memory deficits can be reversed by administration of SAM (Fuso et al., 2012). Evidence more directly related to B_{12} is that the greater dietary intake of B_{12} over the range 1–13 μg/day in older humans, the lower the beta-amyloid load is in the parietal cortex, as revealed by PET scans (Mosconi et al., 2014). Plasma B_{12} was inversely related to the CSF concentration of amyloid-beta$_{1-42}$ in another study on normal elderly, with a regression coefficient almost as large as that for tHcy (Oikonomidi et al., 2016).

There is considerable evidence of a link between raised tHcy and the formation of phosphorylated tau and of neurofibrillary tangles (Sontag & Sontag, 2014). In patients with AD, it has been found that raised tHcy up to 10 years before death is associated with an increased density of neurofibrillary tangles in the cerebral cortex (Hooshmand et al., 2013), but not with beta-amyloid deposition. In normal elderly as well as in AD patients, there is a correlation between the CSF concentrations of SAH and phosphorylated tau and an inverse correlation between the ratio of SAM/SAH and phosphorylated tau (Obeid et al., 2007; Popp et al., 2009). Experiments in rats have shown that hyperhomocysteinemia leads to deposition of hyperphosphorylated tau in the hippocampus and to spatial memory deficits, while also activating several protein kinases and inhibiting protein phosphatase 2A. It is striking that all these effects of hyperhomocysteinemia were reduced or abolished by administration of folate and B_{12} to the rats (Wei et al., 2011). In humans, the formation of hyperphosphorylated tau and deposition of neurofibrillary tangles are strongly linked to the degree of cortical atrophy and to cognitive impairment (Xia et al., 2017).

We conclude that there are multiple possible mechanisms in the brain through which B_{12} status could influence cognition. It is significant that several of these mechanisms are reversible. These results are consistent with clinical trials in which treatments that include B_{12} can slow or prevent cognitive decline. More such trials are needed.

8. CONCLUSIONS

There is a considerable interest today in B_{12} even nearly 200 years since the first probable description of pernicious anemia (Combe, 1824) and just over 60 years since the structure of B_{12} was discovered (Hodgkin et al., 1956). One thing is clear: it is not appropriate to tell people that all they need is "a good, well-balanced diet." So far as B_{12} is concerned, tens of millions of people in the world do not get sufficient B_{12} for optimum health due to dietary inadequacy and/or to decreased ability to absorb B_{12}. More research is needed to identify the best way to rectify this situation and to understand better why B_{12}, "Nature's most beautiful cofactor" (Stubbe, 1994), is so important for health. In the meantime, we agree with the consensus statement in the recent Primer on B_{12} deficiency: "Universal improvement of B_{12} status... seems to be a nutritional imperative with possibly profound beneficial effects on the nervous system, particularly at the bookends of life" (Green et al., 2017).

Supplementary data to this article can be found online at https://doi.org/10.1016/bs.afnr.2017.11.005.

REFERENCES

Alam, A., Woo, J. S., Schmitz, J., Prinz, B., Root, K., Chen, F., ... Locher, K. P. (2016). Structural basis of transcobalamin recognition by human CD320 receptor. *Nature Communications*, 7, 12100. https://doi.org/10.1038/ncomms12100.

Allen, L. H. (2004). Folate and vitamin B_{12} status in the Americas. *Nutrition Reviews*, *62*(6 Pt. 2), S29–33.

Allen, L. H. (2009). How common is vitamin B-12 deficiency? *The American Journal of Clinical Nutrition*, *89*(Suppl), 693S–696S.

Allen, L. H., Miller, J. W., de Groot, L., Rosenberg, I. H., Smith, A. D., Refsum, H., & Raiten, D. J. (2018). Biomarkers of nutrition for development (BOND): Vitamin B_{12} review. *Journal of Nutrition*. in press.

Amur, S. G., Shanker, G., Cochran, J. M., Ved, H. S., & Pieringer, R. A. (1986). Correlation between inhibition of myelin basic protein (arginine) methyltransferase by sinefungin and lack of compact myelin formation in cultures of cerebral cells from embryonic mice. *Journal of Neuroscience Research*, *16*(2), 367–376. https://doi.org/10.1002/jnr.490160204.

Andersen, C. B., Madsen, M., Storm, T., Moestrup, S. K., & Andersen, G. R. (2010). Structural basis for receptor recognition of vitamin-B(12)-intrinsic factor complexes. *Nature*, *464*(7287), 445–448. https://doi.org/10.1038/nature08874.

Arendt, J. F., & Nexo, E. (2013). Unexpected high plasma cobalamin: Proposal for a diagnostic strategy. *Clinical Chemistry and Laboratory Medicine*, *51*(3), 489–496. https://doi.org/10.1515/cclm-2012-0545.

Baik, H. W., & Russell, R. M. (1999). Vitamin B-12 deficiency in the elderly. *Annual Review of Nutrition*, *19*, 357–377.

Bailey, R. L., Carmel, R., Green, R., Pfeiffer, C. M., Cogswell, M. E., Osterloh, J. D., ... Yetley, E. A. (2011). Monitoring of vitamin B-12 nutritional status in the United States by using plasma methylmalonic acid and serum vitamin B-12. *American Journal of Clinical Nutrition*, *94*(2), 552–561. https://doi.org/10.3945/ajcn.111.015222.

Bailey, R. L., Durazo-Arvizu, R. A., Carmel, R., Green, R., Pfeiffer, C. M., Sempos, C. T., ... Yetley, E. A. (2013). Modeling a methylmalonic acid-derived change point for serum vitamin B-12 for adults in NHANES. *American Journal of Clinical Nutrition*, *98*, 460–467. https://doi.org/10.3945/ajcn.113.061234.

Banerjee, R., & Ragsdale, S. W. (2003). The many faces of vitamin B_{12}: Catalysis by cobalamin-dependent enzymes. *Annual Review of Biochemistry*, *72*, 209–247.

Bang, H., Mazumdar, M., & Spence, D. (2006). Tutorial in biostatistics: Analyzing associations between total plasma homocysteine and B vitamins using optimal categorization and segmented regression. *Neuroepidemiology*, *27*(4), 188–200.

Bar, S., Milanaik, R., & Adesman, A. (2016). Long-term neurodevelopmental benefits of breastfeeding. *Current Opinion in Pediatrics*, *28*(4), 559–566. https://doi.org/10.1097/mop.0000000000000389.

Battersby, A. R. (1994). How nature builds the pigments of life: The conquest of vitamin B_{12}. *Science*, *264*(5165), 1551–1557.

Bazinet, R. P., & Laye, S. (2014). Polyunsaturated fatty acids and their metabolites in brain function and disease. *Nature Reviews Neuroscience*, *15*(12), 771–785. https://doi.org/10.1038/nrn3820.

Bell, I. R., Edman, J. S., Miller, J., Hebben, N., Linn, R. T., Ray, D., & Kayne, H. L. (1990). Relationship of normal serum vitamin B_{12} and folate levels to cognitive test performance in subtypes of geriatric major depression. *Journal of Geriatric Psychiatry and Neurology*, *3*, 98–105.

Berlin, H., Berlin, R., & Brante, G. (1968). Oral treatment of pernicious anemia with high doses of vitamin B_{12} without intrinsic factor. *Acta Medica Scandinavica*, *184*(4), 247–258.

Bhate, V., Deshpande, S., Bhat, D., Joshi, N., Ladkat, R., Watve, S., … Yajnik, C. (2008). Vitamin B_{12} status of pregnant Indian women and cognitive function in their 9-year-old children. *Food and Nutrition Bulletin*, *29*(4), 249–254.

Bjorke-Monsen, A. L., & Ueland, P. M. (2011). Cobalamin status in children. *Journal of Inherited Metabolic Disease*, *34*(1), 111–119.

Bridwell-Rabb, J., & Drennan, C. L. (2017). Vitamin B_{12} in the spotlight again. *Current Opinion in Chemical Biology*, *37*, 63–70.

Brito, A., Verdugo, R., Hertrampf, E., Miller, J. W., Green, R., Fedosov, S. N., … Allen, L. H. (2016). Vitamin B-12 treatment of asymptomatic, deficient, elderly Chileans improves conductivity in myelinated peripheral nerves, but high serum folate impairs vitamin B-12 status response assessed by the combined indicator of vitamin B-12 status. *The American Journal of Clinical Nutrition*, *103*(1), 250–257. https://doi.org/10.3945/ajcn.115.116509.

Bunce, D., Kivipelto, M., & Wahlin, A. (2004). Utilization of cognitive support in episodic free recall as a function of apolipoprotein E and vitamin B_{12} or folate among adults aged 75 years and older. *Neuropsychology*, *18*(2), 362–370.

Cabrera, S., Benavente, D., Alvo, M., de Pablo, P., & Ferro, C. J. (2014). Vitamin B_{12} deficiency is associated with geographical latitude and solar radiation in the older population. *Journal of Photochemistry and Photobiology. B, Biology*, *140*, 8–13. https://doi.org/10.1016/j.jphotobiol.2014.07.001.

Carmel, R. (2000). Current concepts in cobalamin deficiency. *Annual Review of Medicine*, *51*(1), 357–375.

Carmel, R. (2011). Biomarkers of cobalamin (vitamin B-12) status in the epidemiologic setting: A critical overview of context, applications, and performance characteristics of cobalamin, methylmalonic acid, and holotranscobalamin II. *American Journal of Clinical Nutrition*, *94*(1), 348S–358S.

Casas, J. P., Bautista, L. E., Smeeth, L., Sharma, P., & Hingorani, A. D. (2005). Homocysteine and stroke: Evidence on a causal link from Mendelian randomisation. *Lancet*, *365*(9455), 224–232.

Casella, E. B., Valente, M., de Navarro, J. M., & Kok, F. (2005). Vitamin B_{12} deficiency in infancy as a cause of developmental regression. *Brain & Development*, *27*(8), 592–594.

Castillo-Lancellotti, C., Margozzini, P., Valdivia, G., Padilla, O., Uauy, R., Rozowski, J., & Tur, J. A. (2015). Serum folate, vitamin B_{12} and cognitive impairment in Chilean older adults. *Public Health Nutrition*, *18*(14), 2600–2608. https://doi.org/10.1017/s1368980014003206.

Chanarin, I. (1979). *The megaloblastic anaemias* (2nd ed.). . Oxford, UK: Blackwell Scientific.
Clarke, R., Birks, J., Nexo, E., Ueland, P. M., Schneede, J., Scott, J., … Evans, J. G. (2007). Low vitamin B-12 status and risk of cognitive decline in older adults. *American Journal of Clinical Nutrition*, *86*(5), 1384–1391.
Clarke, R., Smith, A. D., Jobst, K. A., Refsum, H., Sutton, L., & Ueland, P. M. (1998). Folate, vitamin B_{12}, and serum total homocysteine levels in confirmed Alzheimer disease. *Archives of Neurology*, *55*(11), 1449–1455.
Coelho, D., Suormala, T., Stucki, M., Lerner-Ellis, J. P., Rosenblatt, D. S., Newbold, R. F., … Fowler, B. (2008). Gene identification for the cblD defect of vitamin B_{12} metabolism. *The New England Journal of Medicine*, *358*(14), 1454–1464.
Combe, J. S. (1824). History of a case of anaemia. *Transactions Royal Medical & Chirurgical Society Edinburgh*, *1*, 194–203.
Croft, M. T., Lawrence, A. D., Raux-Deery, E., Warren, M. J., & Smith, A. G. (2005). Algae acquire vitamin B_{12} through a symbiotic relationship with bacteria. *Nature*, *438*(7064), 90–93.
Dangour, A. D., Allen, E., Clarke, R., Elbourne, D., Fletcher, A. E., Letley, L., … Mills, K. (2015). Effects of vitamin B-12 supplementation on neurologic and cognitive function in older people: A randomized controlled trial. *American Journal of Clinical Nutrition*, *102*(3), 639–647.
Day, J. J., & Sweatt, J. D. (2011). Cognitive neuroepigenetics: A role for epigenetic mechanisms in learning and memory. *Neurobiology of Learning and Memory*, *96*(1), 2–12.
de Jager, C. A., Oulhaj, A., Jacoby, R., Refsum, H., & Smith, A. D. (2012). Cognitive and clinical outcomes of homocysteine-lowering B-vitamin treatment in mild cognitive impairment: A randomized controlled trial. *International Journal of Geriatric Psychiatry*, *27*(6), 592–600.
de Lau, L., Smith, A. D., Refsum, H., Johnston, C., & Breteler, M. M. (2009). Plasma vitamin B_{12} status and cerebral white matter lesions. *Journal of Neurology, Neurosurgery & Psychiatry*, *80*, 149–157.
Degnan, P. H., Taga, M. E., & Goodman, A. L. (2014). Vitamin B_{12} as a modulator of gut microbial ecology. *Cell Metabolism*, *20*(5), 769–778.
Deguchi, T., & Barchas, J. (1971). Inhibition of transmethylations of biogenic amines by S-adenosylhomocysteine. Enhancement of transmethylation by adenosylhomocysteinase. *Journal of Biological Chemistry*, *246*(10), 3175–3181.
Demir, N., Koc, A., Ustyol, L., Peker, E., & Abuhandan, M. (2013). Clinical and neurological findings of severe vitamin B deficiency in infancy and importance of early diagnosis and treatment. *Journal of Paediatrics and Child Health*, *48*, 820–840.
Deshmukh, U. S., Joglekar, C. V., Lubree, H. G., Ramdas, L. V., Bhat, D. S., Naik, S. S., … Yajnik, C. S. (2010). Effect of physiological doses of oral vitamin B_{12} on plasma homocysteine: A randomized, placebo-controlled, double-blind trial in India. *European Journal of Clinical Nutrition*, *64*(5), 495–502.
Devalia, V., Hamilton, M. S., Molloy, A. M., & the British Committee for Standards in Haematology. (2014). Guidelines for the diagnosis and treatment of cobalamin and folate disorders. *British Journal of Haematology*, *166*(4), 496–513. https://doi.org/10.1111/bjh.12959.
Dhonukshe-Rutten, R. A., Lips, M., de Jong, N., Chin, A. P. M. J., Hiddink, G. J., van Dusseldorp, M., … van Staveren, W. A. (2003). Vitamin B-12 status is associated with bone mineral content and bone mineral density in frail elderly women but not in men. *Journal of Nutrition*, *133*(3), 801–807.
Dobson, C. M., Wai, T., Leclerc, D., Wilson, A., Wu, X., Dore, C., … Gravel, R. A. (2002). Identification of the gene responsible for the cblA complementation group of vitamin B_{12}-responsive methylmalonic acidemia based on analysis of prokaryotic gene arrangements. *Proceedings of the National Academy of Sciences of the United States of America*, *99*(24), 15554–15559.

Doets, E. L., van Wijngaarden, J. P., Szczecinska, A., Dullemeijer, C., Souverein, O. W., Dhonukshe-Rutten, R. A., ... de Groot, L. C. (2013). Vitamin B_{12} intake and status and cognitive function in elderly people. *Epidemiologic Reviews*, *35*(1), 2–21.

Douaud, G., Refsum, H., de Jager, C. A., Jacoby, R., Nichols, T. E., Smith, S. M., & Smith, A. D. (2013). Preventing Alzheimer's disease-related gray matter atrophy by B-vitamin treatment. *Proceedings of the National Academy of Sciences of the United States of America*, *110*(23), 9523–9528.

Dowling, D. P., Croft, A. K., & Drennan, C. L. (2012). Radical use of Rossmann and TIM barrel architectures for controlling coenzyme B_{12} chemistry. *Annual Review of Biophysics*, *41*, 403–427.

Drennan, C. L., Matthews, R. G., & Ludwig, M. L. (1994). Cobalamin-dependent methionine synthase: The structure of a methylcobalamin-binding fragment and implications for other B_{12}-dependent enzymes. *Current Opinion in Structural Biology*, *4*(6), 919–929.

Dror, D. K., & Allen, L. H. (2008). Effect of vitamin B_{12} deficiency on neurodevelopment in infants: Current knowledge and possible mechanisms. *Nutrition Reviews*, *66*(5), 250–255.

Dullemeijer, C., Souverein, O. W., Doets, E. L., van der Voet, H., van Wijngaarden, J. P., de Boer, W. J., ... van 't Veer, P. (2013). Systematic review with dose-response meta-analyses between vitamin B-12 intake and European Micronutrient Recommendations Aligned's prioritized biomarkers of vitamin B-12 including randomized controlled trials and observational studies in adults and elderly. *American Journal of Clinical Nutrition*, *97*(2), 390–402.

Eastley, R., Wilcock, G. K., & Bucks, R. S. (2000). Vitamin B_{12} deficiency in dementia and cognitive impairment: The effects of treatment on neuropsychological function. *International Journal of Geriatric Psychiatry*, *15*(3), 226–233.

Ekici, F., Tekbas, G., Hattapoglu, S., Yaramis, A., Onder, H., & Bilici, A. (2016). Brain MRI and MR spectroscopy findings in children with nutritional vitamin B_{12} deficiency. *Clinical Neuroradiology*, *26*(2), 215–220.

Erickson, K. I., Suever, B. L., Prakash, R. S., Colcombe, S. J., McAuley, E., & Kramer, A. F. (2008). Greater intake of vitamins B_6 and B_{12} spares gray matter in healthy elderly: A voxel-based morphometry study. *Brain Research*, *1199*, 20–26.

Ertan, S., Tanriverdi, T., Kiziltan, G., & Tata, G. (2002). Complete resolution of myelopathy in a patient with vitamin B_{12} deficiency: A case report. *Internet Journal of Neurology*, *2*(1), 1–7.

Escalante-Semerena, J. C. (2007). Conversion of cobinamide into adenosylcobamide in bacteria and archaea. *Journal of Bacteriology*, *189*(13), 4555–4560.

Eussen, S. J., de Groot, L. C., Clarke, R., Schneede, J., Ueland, P. M., Hoefnagels, W. H., & van Staveren, W. A. (2005). Oral cyanocobalamin supplementation in older people with vitamin B_{12} deficiency: A dose-finding trial. *Archives of Internal Medicine*, *165*(10), 1167–1172.

Eussen, S. J., de Groot, L. C., Joosten, L. W., Bloo, R. J., Clarke, R., Ueland, P. M., ... van Staveren, W. A. (2006). Effect of oral vitamin B-12 with or without folic acid on cognitive function in older people with mild vitamin B-12 deficiency: A randomized, placebo-controlled trial. *American Journal of Clinical Nutrition*, *84*(2), 361–370.

Eussen, S. J., Nilsen, R. M., Midttun, O., Hustad, S., IJssennagger, N., Meyer, K., ... Vollset, S. E. (2013). North-south gradients in plasma concentrations of B-vitamins and other components of one-carbon metabolism in Western Europe: Results from the European Prospective Investigation into Cancer and Nutrition (EPIC) Study. *British Journal of Nutrition*, *110*(2), 363–374.

Favrat, B., Vaucher, P., Herzig, L., Burnand, B., Ali, G., Boulat, O., ... Verdon, F. (2011). Oral vitamin B_{12} for patients suspected of subtle cobalamin deficiency: A multicentre pragmatic randomised controlled trial. *BMC Family Practice*, *12*(1), 2.

Fedosov, S. N. (2012). Physiological and molecular aspects of cobalamin transport. *Sub-Cellular Biochemistry, 56*, 347–367.
Fedosov, S. N., Brito, A., Miller, J. W., Green, R., & Allen, L. H. (2015). Combined indicator of vitamin B_{12} status: Modification for missing biomarkers and folate status and recommendations for revised cut-points. *Clinical Chemistry and Laboratory Medicine, 53*(8), 1215–1225.
Fenech, M., Aitken, C., & Rinaldi, J. (1998). Folate, vitamin B_{12}, homocysteine status and DNA damage in young Australian adults. *Carcinogenesis, 19*(7), 1163–1171.
Feng, L., Li, J., Yap, K. B., Kua, E. H., & Ng, T. P. (2009). Vitamin B-12, apolipoprotein E genotype, and cognitive performance in community-living older adults: Evidence of a gene-micronutrient interaction. *American Journal of Clinical Nutrition, 89*(4), 1263–1268.
Fincker, M., & Spormann, A. M. (2017). Biochemistry of catabolic reductive dehalogenation. *Annual Review of Biochemistry, 86*, 357–386.
Finkelstein, J. L., Kurpad, A. V., Thomas, T., Srinivasan, K., & Duggan, C. (2017). Vitamin B_{12} status in pregnant women and their infants in South India. *European Journal of Clinical Nutrition, 71*, 1046–1053. https://doi.org/10.1038/ejcn.2017.29.
Froese, D. S., & Gravel, R. A. (2010). Genetic disorders of vitamin B(12) metabolism: Eight complementation groups—Eight genes. *Expert Reviews in Molecular Medicine, 12*, e37.
Furger, E., Frei, D. C., Schibli, R., Fischer, E., & Prota, A. E. (2013). Structural basis for universal corrinoid recognition by the cobalamin transport protein haptocorrin. *Journal of Biological Chemistry, 288*(35), 25466–25476.
Fuso, A. (2013). The "golden age" of DNA methylation in neurodegenerative diseases. *Clinical Chemistry and Laboratory Medicine, 51*(3), 523–534.
Fuso, A., Nicolia, V., Ricceri, L., Cavallaro, R. A., Isopi, E., Mangia, F., ... Scarpa, S. (2012). S-adenosylmethionine reduces the progress of the Alzheimer-like features induced by B-vitamin deficiency in mice. *Neurobiology of Aging, 33*. 1482.e1481–1482.e1416.
Gandy, G., Jacobson, W., & Sidman, R. (1973). Inhibition of a transmethylation reaction in the central nervous system-an experimental model for subacute combined degeneration of the cord. *Journal of Physiology, 233*(1), 1P–3P.
Ganjehei, L., Massumi, A., Razavi, M., & Wilson, J. M. (2012). Orthostatic hypotension as a manifestation of vitamin B_{12} deficiency. *Texas Heart Institute Journal, 39*(5), 722–723.
Garrod, M. G., Green, R., Allen, L. H., Mungas, D. M., Jagust, W. J., Haan, M. N., & Miller, J. W. (2008). Fraction of total plasma vitamin B_{12} bound to transcobalamin correlates with cognitive function in elderly Latinos with depressive symptoms. *Clinical Chemistry, 54*(7), 1210–1217.
Gherasim, C., Lofgren, M., & Banerjee, R. (2013). Navigating the B_{12} road: Assimilation, delivery, and disorders of cobalamin. *Journal of Biological Chemistry, 288*(19), 13186–13193.
Gilsing, A. M., Crowe, F. L., Lloyd-Wright, Z., Sanders, T. A., Appleby, P. N., Allen, N. E., & Key, T. J. (2010). Serum concentrations of vitamin B_{12} and folate in British male omnivores, vegetarians and vegans: Results from a cross-sectional analysis of the EPIC-Oxford cohort study. *European Journal of Clinical Nutrition, 64*(9), 933–939.
Girard, C. L., Santschi, D. E., Stabler, S. P., & Allen, R. H. (2009). Apparent ruminal synthesis and intestinal disappearance of vitamin B_{12} and its analogs in dairy cows. *Journal of Dairy Science, 92*(9), 4524–4529.
Glaser, K., Girschick, H. J., Schropp, C., & Speer, C. P. (2015). Psychomotor development following early treatment of severe infantile vitamin B_{12} deficiency and West syndrome—Is everything fine? A case report and review of literature. *Brain & Development, 37*(3), 347–351.

Gopinath, B., Flood, V. M., Rochtchina, E., Wang, J. J., & Mitchell, P. (2013). Homocysteine, folate, vitamin B-12, and 10-y incidence of age-related macular degeneration. *American Journal of Clinical Nutrition, 98*(1), 129–135.
Goraya, J. S., Bansal, K., Singla, S., & Kaur, S. (2016). Vitamin B_{12} status of mothers of children with infantile tremor syndrome. *Indian Pediatrics, 53*(3), 256–257.
Goraya, J. S., Kaur, S., & Mehra, B. (2015). Neurology of nutritional vitamin B_{12} deficiency in infants: Case series from India and literature review. *Journal of Child Neurology, 30*(13), 1831–1837.
Graber, J. J., Sherman, F. T., Kaufmann, H., Kolodny, E. H., & Sathe, S. (2010). Vitamin B_{12}-responsive severe leukoencephalopathy and autonomic dysfunction in a patient with "normal" serum B_{12} levels. *Journal of Neurology, Neurosurgery & Psychiatry, 81*(12), 1369–1371.
Green, R. (2017). Vitamin B_{12} deficiency from the perspective of a practicing hematologist. *Blood, 129*, 2603–2611.
Green, R., Allen, L. H., Bjorke-Monsen, A. L., Brito, A., Gueant, J. L., Miller, J. W., ... Yajnik, C. (2017). Vitamin B_{12} deficiency. *Nature Reviews Disease Primers, 3*, 17040.
Green, R., & Miller, J. W. (2014). Vitamin B_{12}. In J. Zempleni, J. W. Suttie, J. F. I. Gregory, & P. J. Stover (Eds.), *Handbook of vitamins* (5th ed., pp. 447–489). Boca Raton, FL: CRC Press.
Gruber, K., Puffer, B., & Krautler, B. (2011). Vitamin B_{12}-derivatives-enzyme cofactors and ligands of proteins and nucleic acids. *Chemical Society Reviews, 40*(8), 4346–4363.
Gupta, P. K., Garg, R. K., Gupta, R. K., Malhotra, H. S., Paliwal, V. K., Rathore, R. K., ... Pandey, C. M. (2014). Diffusion tensor tractography and neuropsychological assessment in patients with vitamin B_{12} deficiency. *Neuroradiology, 56*(2), 97–106.
Gupta, P. K., Gupta, R. K., Garg, R. K., Rai, Y., Roy, B., Pandey, C. M., ... Narayana, P. A. (2014). DTI correlates of cognition in conventional MRI of normal-appearing brain in patients with clinical features of subacute combined degeneration and biochemically proven vitamin B(12) deficiency. *AJNR. American Journal of Neuroradiology, 35*(5), 872–877.
Gupta, L., Gupta, R. K., Gupta, P. K., Malhotra, H. S., Saha, I., & Garg, R. K. (2016). Assessment of brain cognitive functions in patients with vitamin B_{12} deficiency using resting state functional MRI: A longitudinal study. *Magnetic Resonance Imaging, 34*(2), 191–196.
Gupta, S., Kim, S. Y., Artis, S., Molfese, D. L., Schumacher, A., Sweatt, J. D., ... Lubin, F. D. (2010). Histone methylation regulates memory formation. *Journal of Neuroscience, 30*(10), 3589–3599.
Hainsworth, A. H., Yeo, N. E., Weekman, E. M., & Wilcock, D. M. (2016). Homocysteine, hyperhomocysteinemia and vascular contributions to cognitive impairment and dementia (VCID). *Biochimica et Biophysica Acta, 1862*(5), 1008–1017.
Hannibal, L., Kim, J., Brasch, N. E., Wang, S., Rosenblatt, D. S., Banerjee, R., & Jacobsen, D. W. (2009). Processing of alkylcobalamins in mammalian cells: A role for the MMACHC (cblC) gene product. *Molecular Genetics and Metabolism, 97*(4), 260–266.
Hao, L., Ma, J., Zhu, J., Stampfer, M. J., Tian, Y., Willett, W. C., & Li, Z. (2007). Vitamin B-12 deficiency is prevalent in 35- to 64-year-old Chinese adults. *Journal of Nutrition, 137*(5), 1278–1285.
Hardlei, T. F., & Nexo, E. (2009). A new principle for measurement of cobalamin and corrinoids, used for studies of cobalamin analogs on serum haptocorrin. *Clinical Chemistry, 55*(5), 1002–1010.
Hathout, L., & El-Saden, S. (2011). Nitrous oxide-induced B(1)(2) deficiency myelopathy: Perspectives on the clinical biochemistry of vitamin B_{12}. *Journal of Neurological Sciences, 301*(1–2), 1–8. https://doi.org/10.1016/j.jns.2010.10.033.

Hay, G., Clausen, T., Whitelaw, A., Trygg, K., Johnston, C., Henriksen, T., & Refsum, H. (2010). Maternal folate and cobalamin status predicts vitamin status in newborns and 6-month-old infants. *Journal of Nutrition, 140*(3), 557–564.

Hay, G., Johnston, C., Whitelaw, A., Trygg, K., & Refsum, H. (2008). Folate and cobalamin status in relation to breastfeeding and weaning in healthy infants. *American Journal of Clinical Nutrition, 88*(1), 105–114.

Hazra, A. B., Tran, J. L., Crofts, T. S., & Taga, M. E. (2013). Analysis of substrate specificity in CobT homologs reveals widespread preference for DMB, the lower axial ligand of vitamin B(12). *Chemistry & Biology, 20*(10), 1275–1285.

Health Quality Ontario. (2013). Vitamin B_{12} and cognitive function: An evidence-based analysis. *Ontario Health Technology Assessment Series, 13*(23), 1–45.

Hill, M. H., Flatley, J. E., Barker, M. E., Garner, C. M., Manning, N. J., Olpin, S. E., ... Powers, H. J. (2013). A vitamin B-12 supplement of 500 μg/d for eight weeks does not normalize urinary methylmalonic acid or other biomarkers of vitamin B-12 status in elderly people with moderately poor vitamin B-12 status. *Journal of Nutrition, 143*(2), 142–147. https://doi.org/10.3945/jn.112.169193.

Hin, H., Clarke, R., Sherliker, P., Atoyebi, W., Emmens, K., Birks, J., ... Evans, J. G. (2006). Clinical relevance of low serum vitamin B_{12} concentrations in older people: The Banbury B_{12} study. *Age and Ageing, 35*, 412–422.

Hintikka, J., Tolmunen, T., Tanskanen, A., & Viinamaki, H. (2003). High vitamin B_{12} level and good treatment outcome may be associated in major depressive disorder. *BMC Psychiatry, 3*(1), 17.

Hodgkin, D. C., Kamper, J., Mackay, M., Pickworth, J., Trueblood, K. N., & White, J. G. (1956). Structure of vitamin B_{12}. *Nature, 178*(4524), 64–66.

Hoey, L., Strain, J. J., & McNulty, H. (2009). Studies of biomarker responses to intervention with vitamin B-12: A systematic review of randomized controlled trials. *The American Journal of Clinical Nutrition, 89*(6), 1981S–1996S.

Hooper, M., Hudson, P., Porter, F., & McCaddon, A. (2014). Patient journeys: Diagnosis and treatment of pernicious anaemia. *British Journal of Nursing, 23*(7), 376–381.

Hooshmand, B., Mangialasche, F., Kalpouzos, G., Solomon, A., Kareholt, I., Smith, A. D., ... Kivipelto, M. (2016). Association of vitamin B_{12}, folate, and sulfur amino acids with brain magnetic resonance imaging measures in older adults: A longitudinal population-based study. *JAMA Psychiatry, 73*(6), 606–613.

Hooshmand, B., Polvikoski, T., Kivipelto, M., Tanskanen, M., Myllykangas, L., Erkinjuntti, T., ... Solomon, A. (2013). Plasma homocysteine, Alzheimer and cerebrovascular pathology: A population-based autopsy study. *Brain, 136*(Pt 9), 2707–2716. https://doi.org/10.1093/brain/awt206.

Hooshmand, B., Solomon, A., Kareholt, I., Rusanen, M., Hanninen, T., Leiviska, J., ... Kivipelto, M. (2012). Associations between serum homocysteine, holotrans-cobalamin, folate and cognition in the elderly: A longitudinal study. *Journal of Internal Medicine, 271*(2), 204–212.

Huemer, M., Diodato, D., Schwahn, B., Schiff, M., Bandeira, A., Benoist, J. F., ... Dionisi-Vici, C. (2017). Guidelines for diagnosis and management of the cobalamin-related remethylation disorders cblC, cblD, cblE, cblF, cblG, cblJ and MTHFR deficiency. *Journal of Inherited Metabolic Disease, 40*(1), 21–48.

Hunt, A., Harrington, D., & Robinson, S. (2014). Vitamin B_{12} deficiency. *BMJ, 349*, g5226.

Hvas, A. M., Ellegaard, J., & Nexo, E. (2001). Vitamin B_{12} treatment normalizes metabolic markers but has limited clinical effect: A randomized placebo-controlled study. *Clinical Chemistry, 47*(8), 1396–1404.

Hvas, A. M., Juul, S., Lauritzen, L., Nexo, E., & Ellegaard, J. (2004). No effect of vitamin B-12 treatment on cognitive function and depression: A randomized placebo controlled study. *Journal of Affective Disorders, 81*(3), 269–273.

Jernerén, F., Elshorbagy, A. K., Oulhaj, A., Smith, S. M., Refsum, H., & Smith, A. D. (2015). Brain atrophy in cognitively impaired elderly: The importance of long-chain omega-3 fatty acids and B vitamin status in a randomized controlled trial. *American Journal of Clinical Nutrition, 102*(7), 215–221.

Jost, M., Fernandez-Zapata, J., Polanco, M. C., Ortiz-Guerrero, J. M., Chen, P. Y., Kang, G., … Drennan, C. L. (2015). Structural basis for gene regulation by a B_{12}-dependent photoreceptor. *Nature, 526*(7574), 536–541.

Kalita, J., Agarwal, R., Chandra, S., & Misra, U. K. (2013). A study of neurobehavioral, clinical psychometric, and P3 changes in vitamin B_{12} deficiency neurological syndrome. *Nutritional Neuroscience, 16*(1), 39–46.

Kalita, J., & Misra, U. K. (2015). Benefit of vitamin B-12 supplementation in asymptomatic elderly: A matter of endpoints. *American Journal of Clinical Nutrition, 102*(3), 529–530.

Kapiszewska, M., Kalemba, M., Wojciech, U., & Milewicz, T. (2005). Uracil misincorporation into DNA of leukocytes of young women with positive folate balance depends on plasma vitamin B_{12} concentrations and methylenetetrahydro-folate reductase polymorphisms. A pilot study. *Journal of Nutritional Biochemistry, 16*(8), 467–478.

Kim, J., Gherasim, C., & Banerjee, R. (2008). Decyanation of vitamin B_{12} by a trafficking chaperone. *Proceedings of the National Academy of Sciences of the United States of America, 105*(38), 14551–14554.

Kim, J., Hannibal, L., Gherasim, C., Jacobsen, D. W., & Banerjee, R. (2009). A human vitamin B_{12} trafficking protein uses glutathione transferase activity for processing alkylcobalamins. *Journal of Biological Chemistry, 284*(48), 33418–33424.

Kim, S., Lim, I. K., Park, G. H., & Paik, W. K. (1997). Biological methylation of myelin basic protein: Enzymology and biological significance. *International Journal of Biochemistry and Cell Biology, 29*(5), 743–751.

Kim, J. M., Stewart, R., Kim, S. W., Yang, S. J., Shin, I. S., & Yoon, J. S. (2008). Predictive value of folate, vitamin B_{12} and homocysteine levels in late-life depression. *British Journal of Psychiatry, 192*, 268–274.

Kirke, P. N., Molloy, A. M., Daly, L. E., Burke, H., Weir, D. G., & Scott, J. M. (1993). Maternal plasma folate and vitamin B_{12} are independent risk factors for neural tube defects. *Quarterly Journal of Medicine, 86*(11), 703–708.

Knight, B. A., Shields, B. M., Brook, A., Hill, A., Bhat, D. S., Hattersley, A. T., & Yajnik, C. S. (2015). Lower circulating B_{12} is associated with higher obesity and insulin resistance during pregnancy in a non-diabetic white British population. *PLoS One, 10*(8), e0135268.

Kobe, T., Witte, A. V., Schnelle, A., Grittner, U., Tesky, V. A., Pantel, J., … Floel, A. (2016). Vitamin B-12 concentration, memory performance, and hippocampal structure in patients with mild cognitive impairment. *American Journal of Clinical Nutrition, 103*, 1045–1054.

Koenig, V., Stanga, Z., Zerlauth, M., Bernasconi, L., Risch, M., Huber, A., & Risch, L. (2014). Prevalence of vitamin B(12) depletion and deficiency in Liechtenstein. *Public Health Nutrition, 17*(2), 241–247.

Koutmos, M., Gherasim, C., Smith, J. L., & Banerjee, R. (2011). Structural basis of multifunctionality in a vitamin B_{12}-processing enzyme. *Journal of Biological Chemistry, 286*(34), 29780–29787.

Krautler, B. (2005). Vitamin B_{12}: Chemistry and biochemistry. *Biochemical Society Transactions, 33*(Pt. 4), 806–810.

Krishnaveni, G. V., Hill, J. C., Veena, S. R., Bhat, D. S., Wills, A. K., Karat, C. L., … Fall, C. H. (2009). Low plasma vitamin B_{12} in pregnancy is associated with gestational 'diabesity' and later diabetes. *Diabetologia, 52*(11), 2350–2358.

Krishnaveni, G. V., & Yajnik, C. S. (2017). Developmental origins of diabetes—An Indian perspective. *European Journal of Clinical Nutrition, 71*, 865–869. https://doi.org/10.1038/ejcn.2017.87.

Kumar, T., Taneja, S., Sachdev, H. P. S., Refsum, H., Yajnik, C. S., Bhandari, N., & Strand, T. A. (2017). Supplementation of vitamin B_{12} or folic acid on hemoglobin concentration in children 6-36 months of age: A randomized placebo controlled trial. *Clinical Nutrition, 36*(4), 986–991.

Kuzminski, A. M., DelGiacco, E. J., Allen, R. H., Stabler, S. P., & Lindenbaum, J. (1998). Effective treatment of cobalamin deficiency with oral cobalamin. *Blood, 92*(4), 1191–1198.

Kvestad, I., Hysing, M., Shrestha, M., Ulak, M., Thorne-Lyman, A. L., Henjum, S., … Strand, T. A. (2017). Vitamin B-12 status in infancy is positively associated with development and cognitive functioning 5 y later in Nepalese children. *American Journal of Clinical Nutrition, 105*(5), 1122–1131.

Kvestad, I., Taneja, S., Kumar, T., Hysing, M., Refsum, H., Yajnik, C. S., … Vitamin, B. S. G. (2015). Vitamin B_{12} and folic acid improve gross motor and problem-solving skills in young North Indian children: A randomized placebo-controlled trial. *PLoS One, 10*(6), e0129915.

Kwok, T., Lee, J., Lam, L., & Woo, J. (2008). Vitamin B(12) supplementation did not improve cognition but reduced delirium in demented patients with vitamin B(12) deficiency. *Archives of Gerontology and Geriatrics, 46*(3), 273–282.

Kwok, T., Lee, J., Ma, R. C., Wong, S. Y., Kung, K., Lam, A., … Lam, L. (2016). A randomized placebo controlled trial of vitamin B_{12} supplementation to prevent cognitive decline in older diabetic people with borderline low serum vitamin B_{12}. *Clinical Nutrition (Edinburgh, Scotland), 36*, 1509–1515. https://doi.org/10.1016/j.clnu.2016.1010.1018.

Lai, J. S., Pang, W. W., Cai, S., Lee, Y. S., Chan, J. K., Shek, L. P., … Chong, M. F. (2017). High folate and low vitamin B_{12} status during pregnancy is associated with gestational diabetes mellitus. *Clinical Nutrition*. https://doi.org/10.1016/j.clnu.2017.03.022. [Epub ahead of print].

Leclerc, D., Wilson, A., Dumas, R., Gafuik, C., Song, D., Watkins, D., … Gravel, R. A. (1998). Cloning and mapping of a cDNA for methionine synthase reductase, a flavoprotein defective in patients with homocystinuria. *Proceedings of the National Academy of Sciences of the United States of America, 95*(6), 3059–3064.

Lerner-Ellis, J. P., Tirone, J. C., Pawelek, P. D., Dore, C., Atkinson, J. L., Watkins, D., … Rosenblatt, D. S. (2006). Identification of the gene responsible for methylmalonic aciduria and homocystinuria, cblC type. *Nature Genetics, 38*(1), 93–100.

Li, D., Mann, N. J., & Sinclair, A. J. (2006). A significant inverse relationship between concentrations of plasma homocysteine and phospholipid docosahexaenoic acid in healthy male subjects. *Lipids, 41*(1), 85–89.

Li, M. M., Yu, J. T., Wang, H. F., Jiang, T., Wang, J., Meng, X. F., … Tan, L. (2014). Efficacy of vitamins B supplementation on mild cognitive impairment and Alzheimer's disease: A systematic review and meta-analysis. *Current Alzheimer Research, 11*(9), 844–852.

Lindenbaum, J., Healton, E. B., Savage, D. G., Brust, J. C., Garrett, T. J., Podell, E. R., … Allen, R. H. (1988). Neuropsychiatric disorders caused by cobalamin deficiency in the absence of anemia or macrocytosis. *New England Journal of Medicine, 318*, 1720–1728.

Lövblad, K., Ramelli, G., Remonda, L., Nirkko, A. C., Ozdoba, C., & Schroth, G. (1997). Retardation of myelination due to dietary vitamin B_{12} deficiency: Cranial MRI findings. *Pediatric Radiology, 27*(2), 155–158.

Mandal, M., & Breaker, R. R. (2004). Gene regulation by riboswitches. *Nature Reviews Molecular Cell Biology, 5*(6), 451–463.

Martens, J. H., Barg, H., Warren, M. J., & Jahn, D. (2002). Microbial production of vitamin B_{12}. *Applied Microbiology and Biotechnology, 58*(3), 275–285.
Mathews, F. S., Gordon, M. M., Chen, Z., Rajashankar, K. R., Ealick, S. E., Alpers, D. H., & Sukumar, N. (2007). Crystal structure of human intrinsic factor: Cobalamin complex at 2.6-A resolution. *Proceedings of the National Academy of Sciences of the United States of America, 104*(44), 17311–17316.
McCaddon, A. (2006). Homocysteine and cognition—A historical perspective. *Journal of Alzheimer's Disease: JAD, 9*(4), 361–380.
McCaddon, A. (2013). Vitamin B_{12} in neurology and ageing; clinical and genetic aspects. *Biochimie, 95*(5), 1066–1076. https://doi.org/10.1016/j.biochi.2012.11.017.
McCaddon, A., & Miller, J. W. (2015). Assessing the association between homocysteine and cognition: Reflections on Bradford Hill, meta-analyses and causality. *Nutrition Reviews, 73*(10), 723–735.
McLean, E., de Benoist, B., & Allen, L. H. (2008). Review of the magnitude of folate and vitamin B_{12} deficiencies worldwide. *Food and Nutrition Bulletin, 29*(2 Suppl), S38–51.
Mera, P. E., St Maurice, M., Rayment, I., & Escalante-Semerena, J. C. (2007). Structural and functional analyses of the human-type corrinoid adenosyltransferase (PduO) from Lactobacillus reuteri. *Biochemistry, 46*(48), 13829–13836.
Metz, J. (1992). Cobalamin deficiency and the pathogenesis of nervous system disease. *Annual Review of Nutrition, 12*, 59–79.
Miles, L. M., Allen, E., Clarke, R., Mills, K., Uauy, R., & Dangour, A. D. (2017). Impact of baseline vitamin B_{12} status on the effect of vitamin B_{12} supplementation on neurologic function in older people: Secondary analysis of data from the OPEN randomised controlled trial. *European Journal of Clinical Nutrition, 71*, 1166–1172. https://doi.org/10.1038/ejcn.2017.7.
Miller, R. R., Leanza, C. M., Phillips, E. E., & Blacquire, K. D. (2003). Homocysteine-induced changes in brain membrane composition correlate with increased brain caspase-3 activities and reduced chick embryo viability. *Comparative Biochemistry and Physiology Part B, Biochemistry & Molecular Biology, 136*(3), 521–532.
Minot, G. R., & Murphy, L. P. (1926). Treatment of pernicious anaemia by special diet. *Journal of the American Medical Association, 87*, 470–476.
Mizrahi, E. H., Lubart, E., & Leibovitz, A. (2017). Low borderline levels of serum vitamin B_{12} may predict cognitive decline in elderly hip fracture patients. *The Israel Medical Association Journal, 19*(5), 305–308.
Molloy, A. M., Kirke, P. N., Troendle, J. F., Burke, H., Sutton, M., Brody, L. C., … Mills, J. L. (2009). Maternal vitamin B_{12} status and risk of neural tube defects in a population with high neural tube defect prevalence and no folic acid fortification. *Pediatrics, 123*(3), 917–923.
Moore, S. J., Lawrence, A. D., Biedendieck, R., Deery, E., Frank, S., Howard, M. J., … Warren, M. J. (2013). Elucidation of the anaerobic pathway for the corrin component of cobalamin (vitamin B_{12}). *Proceedings of the National Academy of Sciences of the United States of America, 110*(37), 14906–14911.
Moore, E., Mander, A., Ames, D., Carne, R., Sanders, K., & Watters, D. (2012). Cognitive impairment and vitamin B_{12}: A review. *International Psychogeriatrics*, 1–16. https://doi.org/10.1017/s1041610211002511.
Moore, A., Ryan, J., Watts, M., Pillay, I., Clinch, D., & Lyons, D. (2004). Orthostatic tolerance in older patients with vitamin B_{12} deficiency before and after vitamin B_{12} replacement. *Clinical Autonomic Research, 14*(2), 67–71.
Morkbak, A. L., Poulsen, S. S., & Nexo, E. (2007). Haptocorrin in humans. *Clinical Chemistry and Laboratory Medicine, 45*(12), 1751–1759.

Morris, M. S., Jacques, P. F., Rosenberg, I. H., & Selhub, J. (2007). Folate and vitamin B-12-status in relation to anemia, macrocytosis, and cognitive impairment in older Americans in the age of folic acid fortification. *American Journal of Clinical Nutrition*, *85*(1), 193–200.

Morris, M. S., Jacques, P. F., Rosenberg, I. H., & Selhub, J. (2010). Circulating unmetabolized folic acid and 5-methyltetrahydrofolate in relation to anemia, macrocytosis, and cognitive test performance in American seniors. *American Journal of Clinical Nutrition*, *91*, 1733–1744.

Mosconi, L., Murray, J., Davies, M., Williams, S., Pirraglia, E., Spector, N., … de Leon, M. J. (2014). Nutrient intake and brain biomarkers of Alzheimer's disease in at-risk cognitively normal individuals: A cross-sectional neuroimaging pilot study. *BMJ Open*, *4*(6), e004850.

Nielsen, M. J., Rasmussen, M. R., Andersen, C. B., Nexo, E., & Moestrup, S. K. (2012). Vitamin B_{12} transport from food to the body's cells—A sophisticated, multistep pathway. *Nature Reviews Gastroenterology & Hepatology*, *9*(6), 345–354.

Nilsson, M., Norberg, B., Hultdin, J., Sandstrom, H., Westman, G., & Lökk, J. (2005). Medical intelligence in Sweden. Vitamin B_{12}: Oral compared with parenteral? *Postgraduate Medical Journal*, *81*(953), 191–193.

Nizon, M., Ottolenghi, C., Valayannopoulos, V., Arnoux, J. B., Barbier, V., Habarou, F., … de Lonlay, P. (2013). Long-term neurological outcome of a cohort of 80 patients with classical organic acidurias. *Orphanet Journal of Rare Diseases*, *8*, 148.

Nomenclature of Corrinoids. (1976). Commission on Biochemical Nomenclature. *Pure and Applied Chemistry*, *48*, 495–502.

Nurk, E., Refsum, H., Bjelland, I., Drevon, C. A., Tell, G. S., Ueland, P. M., … Smith, A. D. (2012). Plasma free choline, betaine and cognitive performance: The Hordaland Health Study. *British Journal of Nutrition*, 1–9.

Obeid, R., Kasoha, M., Knapp, J. P., Kostopoulos, P., Becker, G., Fassbender, K., & Herrmann, W. (2007). Folate and methylation status in relation to phosphorylated tau protein(181P) and {beta}-amyloid(1-42) in cerebrospinal fluid. *Clinical Chemistry*, *53*(6), 1129–1136.

Oikonomidi, A., Lewczuk, P., Kornhuber, J., Smulders, Y., Linnebank, M., Semmler, A., & Popp, J. (2016). Homocysteine metabolism is associated with cerebrospinal fluid levels of soluble amyloid precursor protein and amyloid beta. *Journal of Neurochemistry*, *139*(2), 324–332.

O'Leary, F., Allman-Farinelli, M., & Samman, S. (2012). Vitamin B_{12} status, cognitive decline and dementia: A systematic review of prospective cohort studies. *British Journal of Nutrition*, *108*(11), 1948–1961.

Oner, T., Guven, B., Tavli, V., Mese, T., Yilmazer, M. M., & Demirpence, S. (2014). Postural orthostatic tachycardia syndrome (POTS) and vitamin B_{12} deficiency in adolescents. *Pediatrics*, *133*(1), e138–142.

Oulhaj, A., Jernerén, F., Refsum, H., Smith, A. D., & de Jager, C. A. (2016). Omega-3 fatty acid status enhances the prevention of cognitive decline by B vitamins in Mild Cognitive Impairment. *Journal of Alzheimer's Disease*, *50*(2), 547–557.

Padmanabhan, S., Jost, M., Drennan, C. L., & Elias-Arnanz, M. (2017). A new facet of vitamin B_{12}: Gene regulation by cobalamin-based photoreceptors. *Annual Review of Biochemistry*, *86*, 485–514.

Padovani, D., & Banerjee, R. (2009). A G-protein editor gates coenzyme B_{12} loading and is corrupted in methylmalonic aciduria. *Proceedings of the National Academy of Sciences of the United States of America*, *106*(51), 21567–21572.

Padovani, D., Labunska, T., Palfey, B. A., Ballou, D. P., & Banerjee, R. (2008). Adenosyltransferase tailors and delivers coenzyme B_{12}. *Nature Chemical Biology*, *4*(3), 194–196.

Parisi, F., Rousian, M., Koning, A. H., Willemsen, S. P., Cetin, I., Steegers, E. A., & Steegers-Theunissen, R. P. (2017). Periconceptional maternal biomarkers of one-carbon metabolism and embryonic growth trajectories: The Rotterdam Periconceptional Cohort (Predict Study). *Fertility and Sterility*, *107*, 691–698. https://doi.org/10.1016/j.fertnstert.2016.11.030.

Pawlak, R., Lester, S. E., & Babatunde, T. (2014). The prevalence of cobalamin deficiency among vegetarians assessed by serum vitamin B_{12}: A review of literature. *European Journal of Clinical Nutrition*, *68*(5), 541–548.

Pawlak, R., Parrott, S. J., Raj, S., Cullum-Dugan, D., & Lucus, D. (2013). How prevalent is vitamin B(12) deficiency among vegetarians? *Nutrition Reviews*, *71*(2), 110–117.

Payne, K. A., Quezada, C. P., Fisher, K., Dunstan, M. S., Collins, F. A., Sjuts, H., … Leys, D. (2015). Reductive dehalogenase structure suggests a mechanism for B_{12}-dependent dehalogenation. *Nature*, *517*(7535), 513–516.

Popp, J., Lewczuk, P., Linnebank, M., Cvetanovska, G., Smulders, Y., Kolsch, H., … Jessen, F. (2009). Homocysteine metabolism and cerebrospinal fluid markers for Alzheimer's disease. *Journal of Alzheimer's Disease: JAD*, *18*, 819–828.

Quadros, E. V. (2010). Advances in the understanding of cobalamin assimilation and metabolism. *British Journal of Haematology*, *148*(2), 195–204.

Ragsdale, S. W. (2008). Catalysis of methyl group transfers involving tetrahydrofolate and B (12). *Vitamins and Hormones*, *79*, 293–324.

Rajan, S., Wallace, J. I., Brodkin, K. I., Beresford, S. A., Allen, R. H., & Stabler, S. P. (2002). Response of elevated methylmalonic acid to three dose levels of oral cobalamin in older adults. *Journal of the American Geriatrics Society*, *50*(11), 1789–1795.

Refsum, H., & Smith, A. D. (2003). Low vitamin B-12 status in confirmed Alzheimer's disease as revealed by serum holotranscobalamin. *Journal of Neurology, Neurosurgery, and Psychiatry*, *74*(7), 959–961.

Reynolds, E. (2006). Vitamin B_{12}, folic acid, and the nervous system. *Lancet Neurology*, *5*(11), 949–960.

Rickes, E. L., Brink, N. G., Koniuszy, F. R., Wood, T. R., & Folkers, K. (1948). Crystalline Vitamin B_{12}. *Science*, *107*(2781), 396–397.

Rizzo, G., Lagana, A. S., Rapisarda, A. M., La Ferrera, G. M., Buscema, M., Rossetti, P., … Vitale, S. G. (2016). Vitamin B_{12} among vegetarians: Status, assessment and supplementation. *Nutrients*, *8*(12), 767.

Rogne, T., Tielemans, M. J., Chong, M. F., Yajnik, C. S., Krishnaveni, G. V., Poston, L., … Risnes, K. R. (2017). Associations of maternal vitamin B_{12} concentration in pregnancy with the risks of preterm birth and low birth weight: A systematic review and meta-analysis of individual participant data. *American Journal of Epidemiology*, *185*, 212–223. https://doi.org/10.1093/aje/kww212.

Rosenberg, I. H., & Miller, J. W. (1992). Nutritional factors in physical and cognitive functions of elderly people. *American Journal of Clinical Nutrition*, *55*, 1237S–1243S.

Roth, J. R., Lawrence, J. G., & Bobik, T. A. (1996). Cobalamin (coenzyme B_{12}): Synthesis and biological significance. *Annual Review of Microbiology*, *50*, 137–181.

Rutsch, F., Gailus, S., Miousse, I. R., Suormala, T., Sagne, C., Toliat, M. R., … Nurnberg, P. (2009). Identification of a putative lysosomal cobalamin exporter altered in the cblF defect of vitamin B_{12} metabolism. *Nature Genetics*, *41*(2), 234–239.

Scalabrino, G. (2009). The multi-faceted basis of vitamin B_{12} (cobalamin) neurotrophism in adult central nervous system: Lessons learned from its deficiency. *Progress in Neurobiology*, *88*(3), 203–220.

Scott, J. M., Dinn, J. J., Wilson, P., & Weir, D. G. (1981). Pathogenesis of subacute combined degeneration: A result of methyl group deficiency. *Lancet*, *318*(8242), 334–337.

Selhub, J., & Rosenberg, I. H. (2016). Excessive folic acid intake and relation to adverse health outcome. *Biochimie*, *126*, 71–78.

Selley, M. L. (2007). A metabolic link between S-adenosylhomocysteine and polyunsaturated fatty acid metabolism in Alzheimer's disease. *Neurobiology of Aging*, *28*(12), 1834–1839.

Shaikh, S. R., & Teague, H. (2012). N-3 fatty acids and membrane microdomains: From model membranes to lymphocyte function. *Prostaglandins, Leukotrienes and Essential Fatty Acids*, *87*(6), 205–208.

Sidhu, V. K., Huang, B. X., Desai, A., Kevala, K., & Kim, H. Y. (2016). Role of DHA in aging-related changes in mouse brain synaptic plasma membrane proteome. *Neurobiology of Aging*, *41*, 73–85.

Smith, E. L. (1948). Purification of anti-pernicious anaemia factors from liver. *Nature*, *161*(4095), 638.

Smith, A. D. (2008). The worldwide challenge of the dementias: A role for B vitamins and homocysteine? *Food and Nutrition Bulletin*, *29*, S143–172.

Smith, A. D. (2016). Hippocampus as a mediator of the role of vitamin B-12 in memory. *American Journal of Clinical Nutrition*, *103*, 959–960.

Smith, A. D., & Refsum, H. (2009). Vitamin B-12 and cognition in the elderly. *The American Journal of Clinical Nutrition*, *89*(Suppl), 707S–711S.

Smith, A. D., & Refsum, H. (2012). Do we need to reconsider the desirable blood level of vitamin B_{12}? *Journal of Internal Medicine*, *271*(2), 179–182.

Smith, A. D., & Refsum, H. (2016). Homocysteine, B vitamins, and cognitive impairment. *Annual Review of Nutrition*, *6*, 211–239.

Smith, A. D., & Refsum, H. (2017). Dementia prevention by disease-modification through nutrition. *Journal of the Prevention of Alzheimer's Disease*, *4*(3), 138–139.

Smith, A. D., Smith, S. M., de Jager, C. A., Whitbread, P., Johnston, C., Agacinski, G., … Refsum, H. (2010). Homocysteine-lowering by B vitamins slows the rate of accelerated brain atrophy in mild cognitive impairment. A randomized controlled trial. *PLoS One*, *5*(9), e12244.

Sofia, H. J., Chen, G., Hetzler, B. G., Reyes-Spindola, J. F., & Miller, N. E. (2001). Radical SAM, a novel protein superfamily linking unresolved steps in familiar biosynthetic pathways with radical mechanisms: Functional characterization using new analysis and information visualization methods. *Nucleic Acids Research*, *29*(5), 1097–1106.

Song, S. K., Sun, S. W., Ramsbottom, M. J., Chang, C., Russell, J., & Cross, A. H. (2002). Dysmyelination revealed through MRI as increased radial (but unchanged axial) diffusion of water. *NeuroImage*, *17*(3), 1429–1436.

Sontag, J. M., & Sontag, E. (2014). Protein phosphatase 2A dysfunction in Alzheimer's disease. *Frontiers in Molecular Neuroscience*, *7*, 16.

Spence, J. D. (2016). Metabolic B_{12} deficiency: A missed opportunity to prevent dementia and stroke. *Nutrition Research*, *36*(2), 109–116.

Stabler, S. P. (2013). Clinical practice. Vitamin B_{12} deficiency. *New England Journal of Medicine*, *368*(2), 149–160.

Stabler, S. P., & Allen, R. H. (2004). Vitamin B_{12} deficiency as a worldwide problem. *Annual Review of Nutrition*, *24*, 299–326.

Stabler, S. P., Allen, R. H., Savage, D. G., & Lindenbaum, J. (1990). Clinical spectrum and diagnosis of cobalamin deficiency. *Blood*, *76*, 871–881.

Stich, T. A., Yamanishi, M., Banerjee, R., & Brunold, T. C. (2005). Spectroscopic evidence for the formation of a four-coordinate Co2+ cobalamin species upon binding to the human ATP:cobalamin adenosyltransferase. *Journal of the American Chemical Society*, *127*(21), 7660–7661.

Strand, T. A., Taneja, S., Kumar, T., Manger, M. S., Refsum, H., Yajnik, C. S., & Bhandari, N. (2015). Vitamin B-12, folic acid, and growth in 6- to 30-month-old children: A randomized controlled trial. *Pediatrics*, *135*(4), e918–926. https://doi.org/10.1542/peds.2014-1848.

Strand, T. A., Taneja, S., Ueland, P. M., Refsum, H., Bahl, R., Schneede, J., ... Bhandari, N. (2013). Cobalamin and folate status predicts mental development scores in North Indian children 12-18 mo of age. *American Journal of Clinical Nutrition, 97*(2), 310–317.

Stubbe, J. (1994). Binding site revealed of nature's most beautiful cofactor. *Science, 266*(5191), 1663–1664.

Sukumar, N., Rafnsson, S. B., Kandala, N. B., Bhopal, R., Yajnik, C. S., & Saravanan, P. (2016). Prevalence of vitamin B-12 insufficiency during pregnancy and its effect on offspring birth weight: A systematic review and meta-analysis. *American Journal of Clinical Nutrition, 103*(5), 1232–1251.

Sukumar, N., Venkataraman, H., Wilson, S., Goljan, I., Selvamoni, S., Patel, V., & Saravanan, P. (2016). Vitamin B_{12} status among pregnant women in the UK and its association with obesity and gestational diabetes. *Nutrients, 8*(12), 768.

Szalkai, B., Grolmusz, V. K., & Grolmusz, V. I. (2017). Identifying combinatorial biomarkers by association rule mining in the CAMD Alzheimer's database. *Archives of Gerontology and Geriatrics, 73*, 300–307. https://doi.org/10.1016/j.archger.2017.08.006.

Taga, M. E., Larsen, N. A., Howard-Jones, A. R., Walsh, C. T., & Walker, G. C. (2007). BluB cannibalizes flavin to form the lower ligand of vitamin B_{12}. *Nature, 446*(7134), 449–453.

Taneja, S., Bhandari, N., Strand, T. A., Sommerfelt, H., Refsum, H., Ueland, P. M., ... Bhan, M. K. (2007). Cobalamin and folate status in infants and young children in a low-to-middle income community in India. *American Journal of Clinical Nutrition, 86*(5), 1302–1309.

Taneja, S., Strand, T. A., Kumar, T., Mahesh, M., Mohan, S., Manger, M. S., ... Bhandari, N. (2013). Folic acid and vitamin B-12 supplementation and common infections in 6-30-mo-old children in India: A randomized placebo-controlled trial. *American Journal of Clinical Nutrition, 98*(3), 731–737.

Tang, K. F., Li, Y. L., & Wang, H. Y. (2015). Quantitative assessment of maternal biomarkers related to one-carbon metabolism and neural tube defects. *Scientific Reports, 5*, 8510.

Tangney, C. C., Aggarwal, N. T., Li, H., Wilson, R. S., Decarli, C., Evans, D. A., & Morris, M. C. (2011). Vitamin B_{12}, cognition, and brain MRI measures: A cross-sectional examination. *Neurology*, 77(13), 1276–1282. https://doi.org/10.1212/WNL.0b013e3182315a33.

Tangney, C. C., Tang, Y., Evans, D. A., & Morris, M. C. (2009). Biochemical indicators of vitamin B_{12} and folate insufficiency and cognitive decline. *Neurology, 72*(4), 361–367.

Taskesen, M., Yaramis, A., Pirinccioglu, A. G., & Ekici, F. (2012). Cranial magnetic resonance imaging findings of nutritional vitamin B(12) deficiency in 15 hypotonic infants. *European Journal of Paediatric Neurology, 16*(3), 266–270.

Thomazeau, A., Bosch-Bouju, C., Manzoni, O., & Laye, S. (2017). Nutritional n-3 PUFA deficiency abolishes endocannabinoid gating of hippocampal long-term potentiation. *Cerebral Cortex, 27*(4), 2571–2579.

Toh, B. H., van Driel, I. R., & Gleeson, P. A. (1997). Pernicious anemia. *New England Journal of Medicine, 337*(20), 1441–1448.

Torsvik, I., Ueland, P. M., Markestad, T., & Bjorke-Monsen, A. L. (2013). Cobalamin supplementation improves motor development and regurgitations in infants: Results from a randomized intervention study. *American Journal of Clinical Nutrition, 98*(5), 1233–1240.

Torsvik, I. K., Ueland, P. M., Markestad, T., Midttun, O., & Bjorke Monsen, A. L. (2015). Motor development related to duration of exclusive breastfeeding, B vitamin status and B_{12} supplementation in infants with a birth weight between 2000-3000 g, results from a randomized intervention trial. *BMC Pediatrics, 15*, 218.

Troen, A. M., Shea-Budgell, M., Shukitt-Hale, B., Smith, D. E., Selhub, J., & Rosenberg, I. H. (2008). B-vitamin deficiency causes hyperhomocysteinemia and vascular cognitive impairment in mice. *Proceedings of the National Academy of Sciences of the United States of America, 105*, 12474–12479.

Tu, M. C., Lo, C. P., Chen, C. Y., & Huang, C. F. (2015). Correlation of Tc-99 m ethyl cysteinate dimer single-photon emission computed tomography and clinical presentations in patients with low cobalamin status. *BMC Neurology*, *15*, 251.

Tucker, K. L., Rich, S., Rosenberg, I., Jacques, P., Dallal, G. E., Wilson, P. W. F., & Selhub, J. (2000). Plasma vitamin B-12 concentrations relate to intake source in the Framingham Offspring Study. *American Journal of Clinical Nutrition*, *71*(2), 514–522.

Ueland, P. M., & Monsen, A. L. (2003). Hyperhomocysteinemia and B-vitamin deficiencies in infants and children. *Clinical Chemistry and Laboratory Medicine*, *41*(11), 1418–1426.

van Dyck, C. H., Lyness, J. M., Rohrbaugh, R. M., & Siegal, A. P. (2009). Cognitive and psychiatric effects of vitamin B_{12} replacement in dementia with low serum B_{12} levels: A nursing home study. *International Psychogeriatrics*, *21*(1), 138–147.

van Overbeek, E. C., Staals, J., & van Oostenbrugge, R. J. (2013). Vitamin B_{12} and progression of white matter lesions. A 2-year follow-up study in first-ever lacunar stroke patients. *PLoS One*, *8*(10), e78100.

Veber, D., & Scalabrino, G. (2015). Are PrP(C)s involved in some human myelin diseases? Relating experimental studies to human pathology. *Journal of Neurological Sciences*, *359*(1–2), 396–403.

Venkatramanan, S., Armata, I. E., Strupp, B. J., & Finkelstein, J. L. (2016). Vitamin B-12 and cognition in children. *Advances in Nutrition*, 7(5), 879–888.

Vey, J. L., & Drennan, C. L. (2011). Structural insights into radical generation by the radical SAM superfamily. *Chemical Reviews*, *111*(4), 2487–2506.

Victora, C. G., Bahl, R., Barros, A. J., Franca, G. V., Horton, S., Krasevec, J., … Rollins, N. C. (2016). Breastfeeding in the 21st century: Epidemiology, mechanisms, and lifelong effect. *Lancet*, *387*(10017), 475–490.

Vidal-Alaball, J., Butler, C. C., Cannings-John, R., Goringe, A., Hood, K., McCaddon, A., … Papaioannou, A. (2005). Oral vitamin B_{12} versus intramuscular vitamin B_{12} for vitamin B_{12} deficiency. *Cochrane Database of Systematic Reviews*(3), CD004655.

Vogiatzoglou, A., Oulhaj, A., Smith, A. D., Nurk, E., Drevon, C. A., Ueland, P. M., … Refsum, H. (2009). Determinants of plasma methylmalonic acid in a large population: Implications for assessment of vitamin B_{12} status. *Clinical Chemistry*, *55*, 2198–2206.

Vogiatzoglou, A., Refsum, H., Johnston, C., Smith, S. M., Bradley, K. M., de Jager, C., … Smith, A. D. (2008). Vitamin B_{12} status and rate of brain volume loss in community-dwelling elderly. *Neurology*, *71*, 826–832.

Vogiatzoglou, A., Smith, A. D., Nurk, E., Berstad, P., Drevon, C. A., Ueland, P. M., … Refsum, H. (2009). Dietary sources of vitamin B-12 and their association with plasma vitamin B-12 concentrations in the general population: The Hordaland Homocysteine Study. *American Journal of Clinical Nutrition*, *89*(4), 1078–1087.

Vogiatzoglou, A., Smith, A. D., Nurk, E., Drevon, C. A., Ueland, P. M., Vollset, S. E., … Refsum, H. (2013). Cognitive function in an elderly population: Interaction between vitamin B_{12} status, depression, and apolipoprotein E ε4: The Hordaland Homocysteine Study. *Psychosomatic Medicine*, *75*(1), 20–29.

Vry, M. S., Haerter, K., Kastrup, O., Gizewski, E., Frings, M., & Maschke, M. (2005). Vitamine-B_{12}-deficiency causing isolated and partially reversible leukoencephalopathy. *Journal of Neurology*, *252*(8), 980–982.

Warren, M. J., Raux, E., Schubert, H. L., & Escalante-Semerena, J. C. (2002). The biosynthesis of adenosylcobalamin (vitamin B_{12}). *Natural Product Reports*, *19*(4), 390–412.

Watkins, D., & Rosenblatt, D. S. (2013). Lessons in biology from patients with inborn errors of vitamin B_{12} metabolism. *Biochimie*, *95*(5), 1019–1022.

Wei, W., Liu, Y. H., Zhang, C. E., Wang, Q., Wei, Z., Mousseau, D. D., … Liu, G. P. (2011). Folate/vitamin-B_{12} prevents chronic hyperhomocysteinemia-induced tau hyperphosphorylation and memory deficits in aged rats. *Journal of Alzheimer's Disease*, *27*(3), 639–650.

Weikert, C., Dierkes, J., Hoffmann, K., Berger, K., Drogan, D., Klipstein-Grobusch, K., … Boeing, H. (2007). B vitamin plasma levels and the risk of ischemic stroke and transient ischemic attack in a German cohort. *Stroke, 38*(11), 2912–2918.

Werder, S. F. (2010). Cobalamin deficiency, hyperhomocysteinemia, and dementia. *Neuropsychiatric Disease and Treatment, 6*, 159–195.

Whipple, G. H., & Robscheit-Robbins, F. S. (1925). Favourable influence of liver, heart and skeletal muscle in diet on blood regeneration in anemia. *American Journal of Physiology, 72*, 408–418.

Widner, F. J., Lawrence, A. D., Deery, E., Heldt, D., Frank, S., Gruber, K., … Krautler, B. (2016). Total synthesis, structure, and biological activity of adenosylrhodibalamin, the non-natural rhodium homologue of coenzyme B_{12}. *Angewandte Chemie (International ed in English), 55*(37), 11281–11286.

Winkels, R. M., Brouwer, I. A., Clarke, R., Katan, M. B., & Verhoef, P. (2008). Bread cofortified with folic acid and vitamin B-12 improves the folate and vitamin B-12 status of healthy older people: A randomized controlled trial. *American Journal of Clinical Nutrition, 88*(2), 348–355.

Wong, C. W., Ip, C. Y., Leung, C. P., Leung, C. S., Cheng, J. N., & Siu, C. Y. (2015). Vitamin B_{12} deficiency in the institutionalized elderly: A regional study. *Experimental Gerontology, 69*, 221–225.

Wuerges, J., Garau, G., Geremia, S., Fedosov, S. N., Petersen, T. E., & Randaccio, L. (2006). Structural basis for mammalian vitamin B_{12} transport by transcobalamin. *Proceedings of the National Academy of Sciences of the United States of America, 103*(12), 4386–4391.

Xia, C., Makaretz, S. J., Caso, C., McGinnis, S., Gomperts, S. N., Sepulcre, J., … Dickerson, B. C. (2017). Association of in vivo [18F]AV-1451 tau PET imaging results with cortical atrophy and symptoms in typical and atypical Alzheimer disease. *JAMA Neurology, 74*(4), 427–436.

Yajnik, C. S., Deshpande, S. S., Jackson, A. A., Refsum, H., Rao, S., Fisher, D. J., … Fall, C. H. D. (2008). Vitamin B_{12} and folate concentrations during pregnancy and insulin resistance in the offspring: The Pune Maternal Nutrition Study. *Diabetologia, 51*, 29–38.

Yajnik, C. S., Deshpande, S. S., Lubree, H. G., Naik, S. S., Bhat, D. S., Uradey, B. S., … Yudkin, J. S. (2006). Vitamin B_{12} deficiency and hyperhomocysteinemia in rural and urban Indians. *The Journal of the Association of Physicians of India, 54*, 775–782.

Yamada, K., Gherasim, C., Banerjee, R., & Koutmos, M. (2015). Structure of human B_{12} trafficking protein CblD reveals molecular mimicry and identifies a new subfamily of nitro-FMN reductases. *Journal of Biological Chemistry, 290*(49), 29155–29166.

Zengin, E., Sarper, N., & Caki Kilic, S. (2009). Clinical manifestations of infants with nutritional vitamin B deficiency due to maternal dietary deficiency. *Acta Paediatrica, 98*(1), 98–102.

Zhang, Y., Hodgson, N. W., Trivedi, M. S., Abdolmaleky, H. M., Fournier, M., Cuenod, M., … Deth, R. C. (2016). Decreased brain levels of vitamin B_{12} in aging, autism and schizophrenia. *PLoS One, 11*(1).

Zhang, W., Li, Y., Wang, T. D., Meng, H. X., Min, G. W., Fang, Y. L., … Li, C. X. (2014). Nutritional status of the elderly in rural north China: A cross-sectional study. *The Journal of Nutrition, Health & Aging, 18*(8), 730–736.

Zhang, J., Liu, T. T., Zhang, W., Li, Y., Niu, X. Y., Fang, Y. L., … Li, C. X. (2016). Hyperhomocysteinemia is associated with vitamin B-12 deficiency: A cross-sectional study in a rural, elderly population of Shanxi China. *The Journal of Nutrition, Health & Aging, 20*(6), 594–601.

Zhang, D. M., Ye, J. X., Mu, J. S., & Cui, X. P. (2017). Efficacy of vitamin B supplementation on cognition in elderly patients with cognitive-related diseases. *Journal of Geriatric Psychiatry and Neurology, 30*(1), 50–59.

Zhuo, J. M., Wang, H., & Pratico, D. (2011). Is hyperhomocysteinemia an Alzheimer's disease (AD) risk factor, an AD marker, or neither? *Trends in Pharmacological Sciences, 32*(3), 562–571.

CHAPTER SEVEN

Dietary Vitamin C in Human Health

Matthew Granger, Peter Eck[1]
Department of Food and Human Nutritional Sciences, University of Manitoba, Winnipeg, MB, Canada
[1]Corresponding author: e-mail address: Peter.Eck@umanitoba.ca

Contents

Advances in Food and Nutrition Research, Volume 83
ISSN 1043-4526
https://doi.org/10.1016/bs.afnr.2017.11.006

Abstract

Vitamin C is essential to prevent scurvy in humans and is implicated in the primary prevention of common and complex diseases such as coronary heart disease, stroke, and cancer. This chapter reviews the latest knowledge about dietary vitamin C in human health with an emphasis on studies of the molecular mechanisms of vitamin C maintenance as well as gene–nutrient interactions modifying these relationships.

Epidemiological evidence indicates 5% prevalence for vitamin C deficiency and 13% prevalence for suboptimal status even in industrialized countries. The daily intake (dose) and the corresponding systemic concentrations (response) are related in a saturable relationship, and low systemic vitamin C concentrations in observational studies are associated with negative health outcomes.

However, there is no evidence that vitamin C supplementation impacts the risks for all-cause mortality, impaired cognitive performance, reduced quality of life, the development of eye diseases, infections, cardiovascular disease, and cancers. This might be related to the fact that prevention would not be realized by supplementation in populations already adequately supplied through dietary sources.

Recent genetic association studies indicate that the dietary intake might not be the sole determinant of systemic concentrations, since variations in genes participating in redox homeostasis and vitamin C transport had been associated with lowered plasma concentrations. However, impact sizes are generally low and these phenomena might only affect individual of suboptimal dietary supply.

1. INTRODUCTION

Vitamin C, existing in the two main forms of ascorbic and dehydroascorbic acid, is a ubiquitous metabolite in plants and animal. Eukaryotes, plants, fungi, and most animals can synthesize L-ascorbic acid (Drouin, Godin, & Page, 2011). Anthropoid primates, teleost fish, bats, passeriforme birds, and guinea pigs have lost this ability, and for these species it is an essential dietary component (Menniti, Knoth, & Diliberto, 1986; Ohta & Nishikimi, 1999). In humans, the ability to synthesize vitamin C was lost due to mutations in the L-gulonolactone oxidase (*GLO*) gene, that is responsible for catalyzing the synthesis of L-ascorbic acid from L-gulono-1,4-lactone, the last step in the ascorbic acid synthesis pathway in mammals (Nishikimi & Yagi, 1991). A profound lack of dietary supply will result in the deficiency disease scurvy, which can be fatal (Padayatty & Levine, 2016).

Due to the presence of scurvy, vitamin C's existence was known before its molecular discovery. In his 1753 work, *Treatise of the Scurvy*, James Lind noted that consumption of citrus fruit prevents scurvy, also known in Latin as scorbutus (Bartholomew, 2002). Hence, ascorbutus was the lack of scorbutus (scurvy), and the molecular name of ascorbic acid originates there (Grzybowski & Pietrzak, 2013). The water-soluble L-ascorbic acid was discovered by Albert Szent-Györgyi in 1928 and characterized as the anti-scorbutic factor by Szent-Györgyi and King in 1932 (King & Waugh, 1932; Svirbely & Szent-Györgyi, 1932). The chemical structure of ascorbic acid was deduced by Walter Norman Haworth in 1933 (Carpenter, 2012). Since then vitamin C has been extensively researched with speculations regarding its many beneficial functions in the maintenance of health and the curing of disease (Padayatty et al., 2003; Padayatty & Levine, 2016).

This chapter aims to update on the latest knowledge about dietary vitamin C in human health. The many epidemiological or dietary intervention studies scrutinizing the effects of vitamin C consumption and/or supplementation on physiological parameters, biomarkers, and clinical end points are not being reviewed in great detail, since they have been reviewed elsewhere. For reviews on technical issues of research on vitamin C, such as adequate sampling and sensitivity of assays refer to Levine, Wang, and Rumsey (1999), on the role as a physiologic antioxidant refer to Padayatty et al. (2003). The main emphasis is placed on studies of the molecular mechanisms of vitamin C maintenance as well as gene–nutrient interactions modifying these relationships.

2. VITAMIN C: BASIC PHYSIOLOGY

The biological functions of vitamin C revolve around its ability to alter its redox state, which enables it to function as a cofactor for eight known human enzymes and as a water-soluble antioxidant (Padayatty & Levine, 2016). At the physiological pH of 7.4, vitamin C exists as the ascorbic acid anion (Fig. 1; Padayatty et al., 2003; Padayatty & Levine, 2016).

The reduced ascorbic acid loses electrons sequentially, with the loss of one electron forming the ascorbic acid radical. Compared to other radical species it has a long half-life of many seconds to minutes (Buettner, 1993) and has been measured in blood and extracellular fluid samples (Chen et al., 2007). The oxidized form of vitamin C, dehydroascorbic acid, results from the loss of a second electron and can be recycled into the reduced form through enzyme mediated or reductive metabolic pathways (Dhariwal, Shirvan, & Levine, 1991; Dhariwal, Washko, & Levine, 1990; Wang et al., 1997). Dehydroascorbic acid has a half-life of only minutes, after which it undergoes hydrolytic ring rupture, the resulting 2,3-diketogulonic acid cannot reform its precursor and is unable to continue its role in

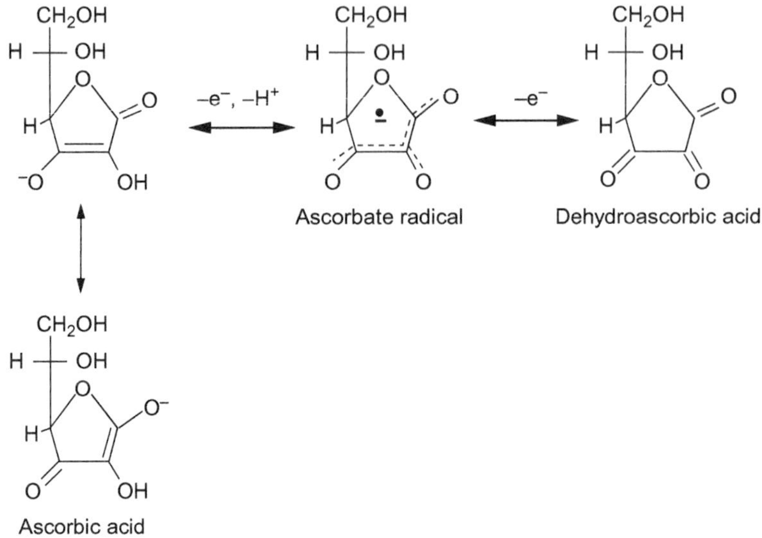

Fig. 1 The three states of vitamin C. The reduced ascorbic acid exists at the physiological pH of 7.4 as the ascorbic acid anion. The ascorbic acid radical results from the loss of one electron and is a stable radical species, indicated by the *dot*. Dehydroascorbic acid results from the loss of another electron and is the oxidized form of vitamin C, which can be recycled into ascorbic acid through reduction by glutathione or glutathione utilizing enzymes.

vitamin C metabolism. Ascorbic acid and dehydroascorbic acid utilize distinct pathways for cellular entry, while ascorbic acid utilizes sodium-dependent membrane transporters, dehydroascorbic acid utilizes facilitative glucose transporters.

Eight human enzymes are known to utilize ascorbic acid as a cofactor, three participate in collagen hydroxylation (Kivirikko & MyllylÄ, 1985; Peltonen, Halila, & Ryhänen, 1985; Peterkofsky, 1991; Prockop & Kivirikko, 1995) and two in carnitine biosynthesis (Dunn, Rettura, Seifter, & Englard, 1984). Of the three enzymes which participate in collagen hydroxylation, one is necessary for biosynthesis of the catecholamine norepinephrine (noradrenaline) (Kaufman, 1974; Levine et al., 1992), one is necessary for amidation of peptide hormones (Eipper, Milgram, Husten, Yun, & Mains, 1993; Eipper, Stoffers, & Mains, 1992), and one is involved in tyrosine metabolism (Englard & Seifter, 1986; Lindblad, Lindstedt, & Lindstedt, 1970). Details about these enzymes and their functional role are described by Padayatty and Levine (2016).

As a major water-soluble antioxidant in mammalian physiology, ascorbic acid scavenges potentially harmful oxidizing free radicals and can be irreversibly oxidized in this process, unless it is recycled (Frei, England, &

Ames, 1989; Frei, Stocker, England, & Ames, 1990), which leads to increased dietary requirements in situations of oxidative stress. Through redox sensing it also contributes to differential gene expression and mRNA translation, which could also contribute to the prevention of oxidative damage of intracellular proteins and DNA (Hitomi & Tsukagoshi, 1996; Padayatty et al., 2003; Qiao & May, 2011; Sram, Binkova, & Rossner, 2012). Plasma ascorbic acid contributes to the reduction of extracellular oxidants, increased endothelium-dependent vasodilatation, and reduced low-density lipoprotein oxidation (Ceriello et al., 2013; Polidori, Mecocci, Levine, & Frei, 2004; Richards, Crecelius, Larson, & Dinenno, 2015; Traber & Stevens, 2011).

In the intestinal tract ascorbic acid increases nontransferrin-mediated absorption of iron by reducing ferric iron (Fe^{3+}) to ferrous iron (Fe^{2+}) and also enhances transferrin-mediated uptake of iron via intracellular reduction of iron (Hallberg, Brune, & Rossander, 1989; Lane, Chikhani, Richardson, & Richardson, 2013a, 2013b; Lane & Richardson, 2014).

Related to the above described biological functions, vitamin C has a role in energy-yielding metabolism, collagen synthesis, nonheme iron absorption, and normal functioning of the nervous system. In spite of our knowledge on vitamin C physiology, significant uncertainties remain in the quest to link individual variability in vitamin C metabolism to improved and individualized recommendations. This chapter will update the latest knowledge on the genetic variability influencing vitamin C utilization and therefore recommendations.

It is well established that a severe dietary undersupply of vitamin C will result in scurvy, and many deficiency symptoms are reflected in the functions of ascorbic acid as a cofactor of known enzymes, such as defects in collagen leading to structural weakening in connective tissue (Padayatty & Levine, 2016). However, ascorbic acids role in the prevention or treatment of common and complex diseases is still uncertain. Even the widely held assumption that ascorbic acid is one of the major biological antioxidants and therefore has a prominent role in disease prevention has not been definitively validated (Padayatty et al., 2003; Padayatty & Levine, 2016).

3. CURRENT BENCHMARKS OF VITAMIN C STATUS AND DIETARY RECOMMENDATIONS

Plasma concentrations serve as the most readily available biomarker for vitamin C status. Values below 11 μmol/L specify deficiency coincide with the clinical symptoms of scurvy (Lykkesfeldt & Poulsen, 2010; Robitaille &

Hoffer, 2016). The highest concentrations observed in pharmacokinetic studies are between 70 and 80 μmol/L (Levine et al., 1996; Levine, Wang, Padayatty, & Morrow, 2001), seldom more than 100 μmol/L has been reported (Padayatty & Levine, 2016), and concentrations plateau in that range even during very high dietary supplementation. However, concentrations as low as 28 μmol/L are considered adequate (Hoffer, 2010), and consequently values between 11 and 28 μmol/L indicate marginal deficiency (often referred to as hypovitaminosis C), where scurvy is absent but the risk for chronic diseases is elevated.

The recommended dietary allowance (RDA) for vitamin C has been developed over the years, and the current recommendations in the United States and Canada are set to 90 mg/day for adult men and 75 mg/day for adult women. They are based on maintaining near-maximal neutrophil concentrations and to elicit antioxidant protection (Monsen, 2000). Recommendations stratified by age and metabolic status are summarized in Table 1.

RDAs differ widely across the world, however, which reflects the lack of underlying scientific evidence. For example, for adults in the United Kingdom 40 mg/day are recommended, while in France and Belgium the benchmark lies at 110 mg/day (Birlouez-Aragon, Fieux, Potier De Courcy, & Hercberg, 2001), and in Germany at 110 mg/day for males and at 95 mg/day for females (Bechthold, Leschik-Bonnet, Strohm, & Heseker, 2015).

Table 1 Recommended Dietary Allowances (RDAs) for Vitamin C in the United States and Canada

Age	Male (mg)	Female (mg)	Pregnancy (mg)	Lactation (mg)
0–6 Months	40[a]	40[a]		
7–12 Months	50[a]	50[a]		
1–3 Years	15	15		
4–8 Years	25	25		
9–13 Years	45	45		
14–18 Years	75	65	80	115
19+ Years	90	75	85	120
Smokers	Individuals who smoke require 35 mg/day more vitamin C than nonsmokers			

[a]Adequate intake (AI) is issued on a less validated scientific basis.
Recommended dietary allowances (RDAs) are issued if a strong scientific basis exists (Monsen, 2000).

Due to the large variability in RDAs worldwide, it has been suggested that vitamin C intakes above current RDAs could contribute to the prevention of chronic diseases, in particular cardiovascular diseases (CVDs)—principally coronary heart disease (CHD) and stroke—and certain cancers (Frei, Birlouez-Aragon, & Lykkesfeldt, 2012).

4. VITAMIN C STATUS IN THE GENERAL POPULATION

Studies describing the relationship between dietary intake and the plasma concentrations of vitamin C generally show that both low intake and plasma concentrations are common. Available data from the United States show mean plasma ascorbic acid concentrations of 48 μmol/L in males and 54.8 μmol/L in females (Schleicher, Carroll, Ford, & Lacher, 2009). However, 8.2% of males and 6% of females had plasma vitamin C concentrations below the deficiency threshold of 11 μmol/L. Among men, 18% of smokers had values below this threshold, while only 5.3% of nonsmokers had such low values. Among women, 15.3% of smokers and 4.2% of nonsmokers had similarly low values (Schleicher et al., 2009).

Around 5.5% of the Canadian general population who did not use supplements had deficient plasma vitamin C concentrations of less than 11 μmol/L (Fig. 2; Langlois, Cooper, & Colapinto, 2016), while approximately 11.6% of smokers, 5.8% overweight individuals had deficient vitamin C concentrations. About 13.5% of nonsupplement users had marginal deficient plasma vitamin C concentrations between 11 and 28 μmol/L, while 17.6% of smokers, 9% of nonsmokers, 17% of obese individuals, 8.2% of overweight individuals had marginal deficient plasma vitamin C concentrations (Langlois et al., 2016).

Even in industrialized countries, marginal deficiency or hypovitaminosis C can have a prevalence of about 15% of the general population (Lindblad, Tveden-Nyborg, & Lykkesfeldt, 2013), 30% of cigarette smokers (Pfeiffer, Sternberg, Schleicher, & Rybak, 2013; Schectman, Byrd, & Hoffmann, 1991), and 60% of hospitalized individuals (Evans-Olders, Eintracht, & Hoffer, 2009; Fain et al., 2003; Gan, Eintracht, & Hoffer, 2008; Gariballa & Forster, 2006; Hunt, Chakravorty, Annan, Habibzadeh, & Schorah, 1994; Teixeira, Carrie, Genereau, Herson, & Cherin, 2001; Wang et al., 2013; Zhang, Robitaille, Eintracht, & Hoffer, 2011). The exact health implications of marginal deficient vitamin C status remain unknown, but clinical symptoms may include fatigue or mood disruption (Crandon, Lund, & Dill, 1940; Wang et al., 2013; Zhang et al., 2011),

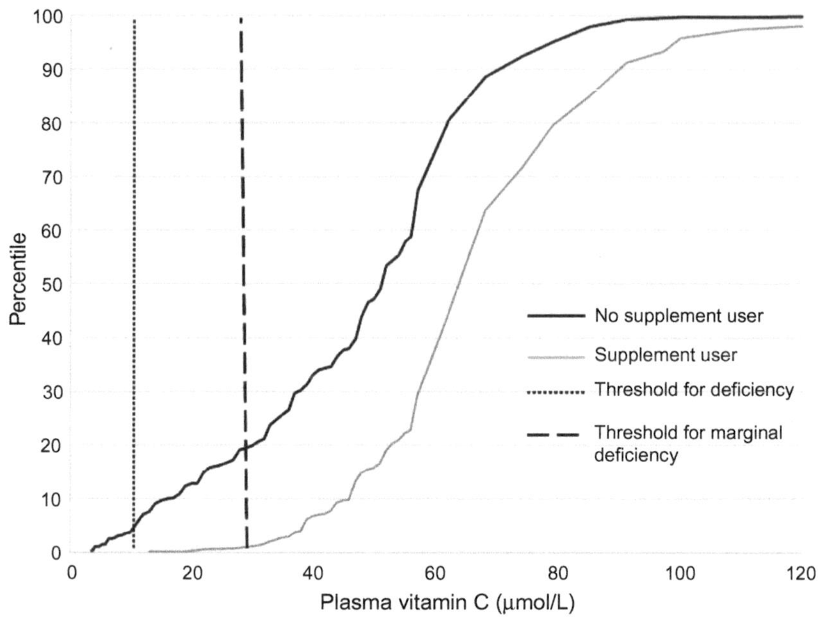

Fig. 2 Plasma vitamin C concentrations of Canadians. *Data sourced from the 2012/2013 Canadian Health Measures Survey (Langlois, K., Cooper, M., & Colapinto, C. K. (2016). Vitamin C status of Canadian adults: Findings from the 2012/2013 Canadian Health Measures Survey.* Health Reports, *27(5), 3–10).*

decreased immunity (Anthony & Schorah, 1982; Hemila & Louhiala, 2007; Hunt et al., 1994), impaired wound healing (Blass et al., 2013; Lund & Crandon, 1941; Sorensen et al., 2010), localized pain (Shibuya, Humphers, Agarwal, & Jupiter, 2013), and CVD (Frei et al., 2012; Juraschek, Guallar, Appel, & Miller, 2012; Padayatty & Levine, 2000; Rodrigo et al., 2013; Vita et al., 1998).

4.1 Associations of Vitamin C and Health Outcomes in Observational Studies

Since outcomes in major common and complex diseases are potentially modified by the vitamin C status, large-scale observational studies have examined these relationships. In the following sections the major observations for the most detrimental common and complex diseases, namely, CVD and cancers, are summarized. It should be noted that men with marginal deficient serum vitamin C concentrations below 28 μmol/L had a 57%

higher risk of all-cause mortality after 12–16 years of follow-up than men with the highest vitamin C concentrations above 74 μmol/L, creating the rationale to try to relate this to today's major health problems.

4.2 The Relation of Vitamin C to CVDs in Human Observational Studies

Many epidemiological studies assessed the association between dietary vitamin C intake and the risks of common and complex diseases. However, some of these observational studies did not consider biomarkers of vitamin C status such as plasma concentrations. The following section summarizes studies reporting on the biomarkers of vitamin C status, thereby eliminating the uncertainty of dietary assessments.

Serum vitamin C concentrations above 45 μmol/L (45.4 and 79.5 μmol/L) lowered the risk of CVD by about 25% compared to individuals with concentrations under 23 μmol/L (Simon, Hudes, & Tice, 2001). Similarly, a 33% lowered risk for CHDs was associated in subjects with mean plasma vitamin C concentrations of 77 μmol/L compared to those with 27 μmol/L (Boekholdt et al., 2006). Moreover, plasma vitamin C concentrations were inversely related to mortality from all causes and CVD (Khaw et al., 2001), where each 20 μmol/L increase in plasma vitamin C was associated with a 20% and 30% reduction in all-cause and CVD mortality, respectively. Concurring evidence was reported for associations of increased vitamin C plasma or serum concentrations (means ranging from 49.5 to 85.2 μmol/L) and decreased CVD risks (Langlois, Duprez, Delanghe, De Buyzere, & Clement, 2001; Nyyssonen, Parviainen, Salonen, Tuomilehto, & Salonen, 1997; Simon, Hudes, & Browner, 1998). Consistent with these findings, in young type 1 diabetic patients, poor vitamin C status was found to be associated with an increase in several early cardiovascular risk markers (Odermarsky, Lykkesfeldt, & Liuba, 2009).

A recent systematic review on the relationship between vitamin C and heart health concluded that in populations with already adequate vitamin C intake, further supplementation with vitamin C did not correlate with the risk of CVD (Moser & Chun, 2016). However, if dietary supply was suboptimal, risks for CVD were elevated (Moser & Chun, 2016).

Older adults with plasma vitamin C concentrations above 28 μmol/L had 30% less deaths from strokes (Gale, Martyn, Winter, & Cooper, 1995), and a 42% lower incidence of stroke occurred when plasma concentrations were above 66 μmol/L compared to individuals with vitamin C concentrations below 41 μmol/L (Myint et al., 2008). Similarly, subjects with vitamin C concentrations above 64 μmol/L had a 41% lower risk of stroke than those with vitamin C below 40 μmol/L (Yokoyama et al., 2000). Echoing these data, individuals with plasma concentrations below 28 μmol/L had a twofold increased risk of stroke compared to individuals with concentrations above 65.0 μmol/L.

Contrary to the positive effects reported, CVD mortality increased upon vitamin C supplementation in postmenopausal women with diabetes (Lee, Folsom, Harnack, Halliwell, & Jacobs, 2004), indicating that cases of specific pathologies should be viewed separate from the general population.

In conclusion, observational studies have produced mixed results regarding the relationship of vitamin C and CVDs. Generally, no relationship between vitamin C and CVD risk was observed at optimal plasma vitamin C concentrations, but suboptimal concentrations seem to be associated with elevated CVD risks.

4.3 The Relation of Vitamin C to Cancers in Human Observational Studies

More limited evidence exists for a relationship between vitamin C status and the prevention of certain cancers. In a case–control study, individuals with plasma concentrations above 51 μmol/L had a 45% lower risk of gastric cancer compared to individuals with concentrations below 29 μmol/L (Jenab et al., 2006). Men with serum vitamin C concentrations below 28 μmol/L had a 62% higher risk of cancer-related deaths after 12–16 years of follow-up than men above 73.8 μmol/L (Loria, Klag, Caulfield, & Whelton, 2000). Men with mean plasma vitamin C concentrations of 73 μmol/L had a 53% reduced cancer mortality compared to those of 21 μmol/L (Khaw et al., 2001).

Very low vitamin C status was also reported for cases of multiple myeloma (Sharma, Tripathi, Satyam, & Kumar, 2009) and unspecified advanced cancers, where it was associated with increased inflammation and lower survival (Mayland, Bennett, & Allan, 2005). However, in those case–control studies it was undetermined if reduced concentrations are the consequence of rather than the cause of the reported associations.

In conclusion, observational studies indicate that vitamin C plays a role in the prevention of gastric cancer, while the role in the prevention of other cancers is indicated, but requires further validation.

4.4 Human Intervention Studies Supplementing Vitamin C

The epidemiologic associations between adequate vitamin C status and decreased risk of CVD and cancers inspired controlled intervention studies to determine if a causal relationship exists. Those intervention studies regularly used vitamin C supplementation in combination with other "antioxidant vitamins" (most often vitamin E and β-carotene), or as part of a multivitamin–mineral mix, which would complicate the interpretation of positive findings. However, if intervention studies report no effect of vitamin C supplementations then they would be considered ineffective.

Many clinical studies have used interventions with vitamin C in relation to specific outcomes in patients populations, such as HIV (Guwatudde et al., 2015), type 2 diabetes (Shateri et al., 2016), depressive disorders (Sahraian, Ghanizadeh, & Kazemeini, 2015), and end-stage renal diseases (Biniaz, Sadeghi Shermeh, Ebadi, Tayebi, & Einollahi, 2014; Shateri et al., 2016). Those small and targeted studies will not be discussed in this chapter, which rather focuses on large-scale long-term studies on health outcomes in the general population. Within these intervention studies, cardiovascular outcomes and cancers had been a targeted most often.

4.5 Human Intervention Studies and Health Outcomes in Common and Complex Diseases

Major human intervention trials with vitamin C did not show improvements in all-cause mortality (Hercberg, Galan, Preziosi, et al., 2004), eye health (age-related macular degeneration, age-related cataract, lens opacity, or vision loss) (Age-Related Eye Disease Study Research Group, 2001; Christen et al., 2012), cognitive performance (Grodstein et al., 2013), infections (Avenell et al., 2005; Girodon et al., 1999; Graat, Schouten, & Kok, 2002), and overall quality of life (Avenell et al., 2005).

One clinical trial of moderate size reported the improvement of serum pepsinogen concentrations, which serve as a biomarker for the progression of gastric mucosal atrophy during *Helicobacter pylori* infection in a group highly supplemented with vitamin C (500 mg/day) vs the low-supplemented group (50 mg/day) (Sasazuki et al., 2003).

In conclusion, vitamin C supplementation did not modify all-cause mortality, cognitive performance, overall quality of life, the development of eye diseases, and infections.

4.6 Human Intervention Studies and Cardiovascular Outcomes

Major coronary events and fatal or nonfatal vascular events (2002) as well as incidences of myocardial infarction, stroke, coronary revascularization, or CVD death (Cook, Albert, Gaziano, et al., 2007) were not affected by supplementation with vitamins C + E + β-carotene. Similar supplementation with vitamins C + E + β-carotene + Zn + Se did not protect from ischemic CVD and all-cause mortality (Hercberg et al., 2004). Vitamin C combined with vitamin E did not reduce any cardiovascular events, myocardial infarction, stroke, or death from CVDs (Sesso, Buring, Christen, et al., 2008) and did not affect atherosclerotic lesions and carotid artery intima-media thickness (Salonen et al., 2003) or changes in minimum lumen diameter of coronary arteries (Waters, Alderman, Hsia, et al., 2002). Moreover, cerebrovascular diseases were not reduced upon daily supplementation with 14 vitamins and 12 minerals (Li et al., 1993).

Overall, the attempt to prove causal relationships in controlled clinical feeding trials has negative results for CVDs (Moser & Chun, 2016).

4.7 Human Intervention Studies and Cancer Outcomes

Vitamin C supplementation did not affect incidences or outcomes for major gastrointestinal cancers or overall cancer mortalities in the majority of studies. Specifically, supplementation with vitamin C and molybdenum (Blot et al., 1993), or vitamins C + E + β-carotene + zinc + selenium (Hercberg et al., 2004) did not reduce overall cancer incidence and disease-specific mortality. Similarly, interventions with vitamin C, vitamin E, β-carotene, and selenium did not reduce gastric cancer and overall cancer mortality (You et al., 2001). Daily supplementation with 14 vitamins and 12 minerals did reduce esophageal and gastric cancer deaths or total cancer incidences (Li et al., 1993). Occurrences of colorectal adenomas (Gaziano, Sesso, Christen, et al., 2012; Greenberg et al., 1994) or the progression of multifocal nonmetaplastic atrophy or intestinal metaplasia (Correa et al., 2000) were not correlated with multivitamin–mineral and β-carotene + vitamins C + E supplementation, respectively.

Vitamin C combined with vitamin E did not reduce the risks for prostate and total cancer (Gaziano, Glynn, Christen, et al., 2009); similarly, multivitamin supplementation did not reduce the risk for prostate, colorectal, and other site-specific cancers, however, reduced the risk of total cancer (Gaziano et al., 2012).

One solitary study reported that the positive association between red and processed meat intakes and breast cancer risk was eliminated upon supplementation with vitamins C + E + β-carotene + zinc + selenium (Pouchieu et al., 2014). Otherwise, all major intervention studies failed to prove a positive modification of cancer incidences and outcomes.

4.8 Human Intervention Studies and the Importance to Consider the Dose–Concentration Relationship for Vitamin C

Most evidence reported from large epidemiological cohorts, dietary intervention studies, as well as meta-analyses remain ambiguous, with many reporting that vitamin C supplementation had little or no effect on outcome measures (Moser & Chun, 2016; Padayatty & Levine, 2016). Specifically, most intervention studies do not show positive effects, while low systemic vitamin C concentrations in observational studies are associated with negative health outcomes. The explanation for this phenomenon might lay in the saturable relationship between daily intake (dose) and the corresponding systemic concentrations (response). In individuals with depleted plasma vitamin C, concentrations increase only modestly after small doses of up to 30 mg/day but increase more rapidly upon larger doses up to 100 mg/day reaching a plateau at about 200 mg/day (Fig. 3). Consequently, in studies where the control and treatment groups already have an adequate supply at baseline, plasma and cellular concentrations might not differ even upon interventions with high doses of vitamin C, eliminating the chance of

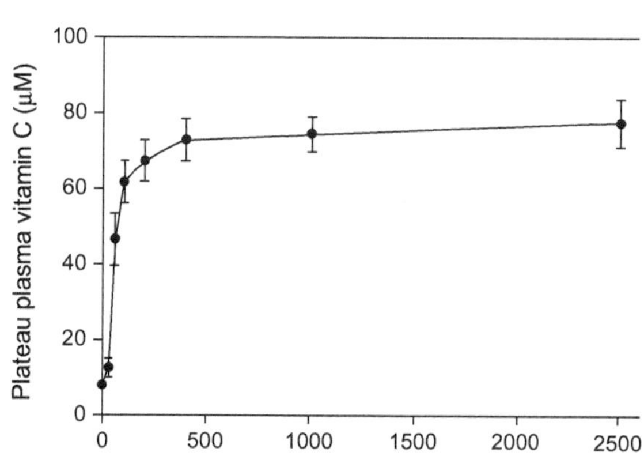

Fig. 3 Vitamin C concentrations as a function of the dose (Levine et al., 2001).

any demonstrable effects. Therefore, the effect of any supplementation would only be observed in groups undersupplied through the basal diet.

This phenomenon is exemplified by a recent meta-analysis on the effects of vitamin C supplementation on glycemic control, where overall no effects were observed on biomarkers of dysglycemia, as measured by glucose, HbA1c, and insulin concentrations (Ashor et al., 2017). However, in patients with type 2 diabetes, the high blood glucose concentrations were reduced upon vitamin C supplementation (Ashor et al., 2017), which could be attributed to improvements in the low vitamin C status of diabetic individuals (Chen et al., 2006; Tu et al., 2015).

5. GENETIC INFLUENCES ON VITAMIN C METABOLISM AND DISEASE PATHOLOGY

An individual's vitamin C status depends on the absorption, distribution, and metabolism of ascorbic acid where all of these processes may vary depending on genetic variations. Two genes encoding sodium-dependent ascorbic acid transporters have been identified, and polymorphisms in these have been associated with reduced systemic vitamin C concentrations and several diseases. The associations of polymorphisms in the ascorbic acid transporter genes to biomarkers of the vitamin C status and the disease associations will be discussed in separate sections, since it is currently unclear how they correlate. Polymorphisms in two glutathione *S*-transferases (GSTs) and haptoglobin had been associated with reduced vitamin C status and those will be summarized in separate sections; however, the disease associations for these genes will not be discussed, since their direct involvement in the ascorbic acid metabolism is unclear and correlations to the vitamin C status speculative at best.

5.1 Genetic Variations in the SLC23A1 Gene Associated to Altered Vitamin C Status

The *SLC23A1* gene encodes the sodium-dependent ascorbic acid transporter SVCT1 (SLC23A1) which plays a key role in the renal reabsorption of vitamin C (Daruwala, Song, Koh, Rumsey, & Levine, 1999; Tsukaguchi et al., 1999; Wang et al., 2000). Genetic variations in *SLC23A1* could lower the renal threshold for ascorbic acid reabsorption and therefore increase urinary losses and decrease steady-state plasma and body vitamin C concentrations (Corpe et al., 2010).

Five alleles of single nucleotide polymorphisms (SNPs, variations with frequencies over 1%) in the *SLC23A1* gene have been associated with lowered serum ascorbic acid concentrations; however, some genotypes showed mixed results with elevated concentrations in one study and decreased concentrations in others.

The average serum ascorbic acid concentrations were 5.3 μmol/L lower for SNP rs4257763-GG homozygotes in a setting indicating suboptimal values across this cohort (Cahill & El-Sohemy, 2009).

The SNP rs33972313-G, rs10063949-A, and rs6596473-C alleles were associated with 4.2, 2.9, and 2 μmol/L reductions in circulating concentrations of L-ascorbic acid, respectively (Timpson et al., 2010). These findings could not be replicated for the SNP rs10063949-A allele, but were replicated for the rs33972313-G and rs6596473-C alleles with associated reductions of 8.3 and 1 μmol/L, respectively (Timpson et al., 2010). A subsequent pooled analysis across five studies associated each additional rs33972313-G allele with a 6 μmol/L reduction in circulating vitamin C. In stark contrast to these results, the rs33972313-G allele was associated with a 3.1-μmol/L elevation in plasma vitamin C concentrations (Kobylecki, Afzal, Smith, & Nordestgaard, 2015), where GG homozygotes showed increases of 7.3 μmol/L and AG heterozygotes 4.1 μmol/L plasma vitamin C concentrations, respectively. Similarly, rs33972313-GG homozygotes had plasma ascorbic acid concentrations 6 μmol/L higher than those of GA heterozygotes, while AA homozygotes were found in this study (Duell et al., 2013).

SNP rs11950646-GG homozygotes had 8 μmol/L lower plasma vitamin C concentrations compared to AA heterozygotes, while a 3-μmol/L decrease in plasma vitamin C concentrations was observed in GA heterozygotes (Duell et al., 2013), indicating an allele dosage effect.

To conclude, emerging evidence indicates a role of genetic polymorphisms in the modulation of biomarkers of vitamin C status. However, the inconsistencies in the results need to be addressed in future controlled trials.

5.2 Genetic Variations in the SLC23A1 Gene Associated With Common and Complex Diseases

Several variations in the *SLC23A1* gene have been associated with increased risks of certain types of cancers, Crohn's disease, spontaneous preterm delivery, and aggressive periodontitis.

An 80% elevated risk of non-Hodgkin lymphoma was demonstrated for homozygotes for rs6596473-C/C and rs11950646-G/G (Skibola et al., 2008). A 150% increase in the risk for Crohn's disease was associated with the rs10063949-A/G heterozygotes and a 307% increase with G/G homozygotes, clearly showing an allele dosage effect (Amir Shaghaghi, Bernstein, Serrano Le, El-Gabalawy, & Eck, 2014). Certain haplotypes in *SLC23A1* gene were associated with increased risk of spontaneous preterm delivery (Erichsen et al., 2006). The rare allele of rs6596473 was associated with aggressive periodontitis (De Jong et al., 2014).

Overall, evidence of associations in the SLC23A1 gene with common and complex diseases is currently emerging but lacks enough depth to conclude on the validity.

5.3 Genetic Variations in the SLC23A2 Gene Associated to Altered Vitamin C Status

Three SNPs in the *SLC23A2* gene have been associated with lowered serum ascorbic acid concentrations. Plasma vitamin C concentrations were decreased by 5 μmol/L in SNP rs6053005-CC homozygotes and 4.3 μmol/L in CT heterozygotes compared to TT homozygotes (Duell et al., 2013). They were also decreased by 6 μmol/L in carrier of the SNP rs6133175-A allele (Duell et al., 2013). Moreover, rs1279386-GG homozygotes had approximately 1.1 μg/mL lower plasma vitamin C concentrations than the other genotypes (Zanon-Moreno et al., 2011).

Current evidence might indicate a role of *SLC23A2* polymorphism in the regulation of steady-state plasma vitamin C concentrations, but at present, research is only emerging thus it is premature to make any concrete conclusions.

5.4 Genetic Variations in the SLC23A2 Gene Associated With Common and Complex Diseases

Several variations in the SLC23A2 gene had been associated with the risk of six types of cancer, birth complications, CHD, and glaucoma.

Reduced risks for gastric cancer were reported for SNP rs12479919-AA homozygotes (Wright et al., 2009), and the haplotype for the common alleles of SNPs rs6139591 + rs2681116 + rs14147458 was inversely associated with gastric cancer (Wright et al., 2009). Similarly, the genotype rs6116569-C/T and the two haplotypes, CGTC (rs6052937, rs3787456, rs6116569, rs17339746) and ATC (rs6139587, rs6053005, rs2326576), in the *SLC23A2* gene were associated with the risk for gastric cancer (Duell et al., 2013).

The haplotype G-C for SNPs rs4987219 + rs1110277 was associated with a reduction in the risk of colorectal adenoma (Erichsen et al., 2008).

SNP rs12479919-CT heterozygosity interacted with SCARB1 (scavenger receptor class B 1) rs4765621 genotypes to elevate the risk for bladder cancer (Andrew et al., 2009).

The risks for non-Hodgkin lymphomas were elevated by 80% for genotypes rs6133175-GG, rs1715364-CC, rs1715385-AA, rs1776948-AA, and rs6139587-AA, as well as the two haplotypes rs1776948 + rs6139587-AA and rs1715385 + rs6133175 + rs1715364-AAC (Skibola et al., 2008). SNPs rs6133175-G and rs1776948-A elevated the risk for chronic lymphocytic leukemia by about 20% (Casabonne et al., 2017). The haplotype analysis of rs1715364 + rs6133175 supported the genotype results.

The risk of head and neck cancer following HPV16 infections was decreased for rs4987219-C/C homozygotes (Chen et al., 2009).

Elevated risks for spontaneous preterm birth were observed in carrier of the minor T allele of SNP rs6139591, showing 1.7- and 2.7-fold elevations for hetero- and homozygotes, respectively (Erichsen et al., 2006).

A 5.4-fold elevated risk of acute coronary syndrome was reported in women with the rs6139591-T/T genotype with low-dietary vitamin C (Dalgard et al., 2013). Moreover, women with the rs1776964-T/T genotype and high intake of vitamin C had a 3.4-fold increased risk of acute coronary syndrome, compared with C/C homozygotes with low intake. This might indicate that the genotype effects may not be completely compensated by a high dietary intake of vitamin C (Dalgard et al., 2013).

The risk for open-angle glaucoma was elevated by 1.7-fold in rs1279386-G/G homozygotes, potentially echoing the lower plasma vitamin C concentration observed for this genotype (Zanon-Moreno et al., 2011).

To conclude, reports on genetic associations of *SLC23A2* variations mostly consist of observation from single studies, which calls for replications in similar or larger studies in order to validate those results.

5.5 Genetic Variations in the GSTM1 and GSTT1 Genes Associated to Altered Vitamin C Status

GSTs are phase 2 enzymes which conjugate glutathione to xenobiotics for the purpose of detoxification. The two major isoforms, GSTT1 and GSTM1, are involved in oxidative stress pathways through the utilization and conjugation of glutathione (Block, Shaikh, Jensen, Volberg, & Holland, 2011), and their genes are deleted in ~20% and ~50% of the human population. The capacity of GSTO to recycle dehydroascorbic acid has been demonstrated (Linster & Van Schaftingen, 2007), but the participation of the GSTT1 and GSTM1 enzymes in the recycling of dehydroascorbic acid has not been proven. However, if they do not participate directly, they could spare the biochemical ascorbic acid consumption indirectly through recycling of other antioxidants.

For individuals carrying GSTT1 and GSTM1 null genotypes (**0/*0* gene is nonfunctional) a rare diet–gene interaction was reported: individuals had an increased risk of vitamin C deficiency when their dietary supply did not meet current recommendations (Cahill, Fontaine-Bisson, & El-Sohemy, 2009). Specifically, the risk of serum ascorbic acid deficiency in marginal vitamin C supply was elevated approximately 12-fold for the *GSTT1*0/*0* genotype but was only increased approximately twofold for carriers of the *GSTT1*1* allele. Moreover, the risk of serum ascorbic acid deficiency with marginal vitamin C intake was approximately fourfold elevated for the *GSTM1*0/*0* genotype, while it was approximately twofold for carriers of the *GSTM1*1* allele. Individuals with both nonfunctional genotypes had an approximately 15-fold elevated risk of serum ascorbic acid deficiency in marginal vitamin C supply compared to an approximately twofold risk for subjects with both functional genotypes. Serum ascorbic acid concentrations were decreased by approximately 10 μmol/L for individuals with both nonfunctional genotypes in marginal supply (Cahill et al., 2009).

Plasma vitamin C concentrations in *GSTM1*0/*0* homozygotes were decreased by 2.1 μmol/L in the general population and by 5.3 μmol/L in asbestos factory workers (Horska et al., 2011). In workers of a rock wool plant GSTT1**0/*0* homozygotes showed 9.5 μmol/L lower concentrations. Homozygotes for GSTM1/GSTT1 deletions had a vitamin C level that was 6.42 μmol/L lower.

In direct contrast to the above reported results, serum vitamin C concentrations were 5 μmol/L higher in GSTM**0/*0* homozygotes (Block et al., 2011). However, the dual deletions of GSTM1**0/*0* and GSTT**0/*0* were not associated with serum vitamin C concentrations. The association of vitamin C concentrations with the probability of being GSTM1**0/*0* was allele dosage dependent (Fig. 4), where subjects in the highest quartile of vitamin C concentrations were 2.6 times as likely to be GSTM1**0/*0*. SNPs did not modify any associations between dietary vitamin C intake and serum ascorbic acid, as reported in the previous paragraph (Block et al., 2011;

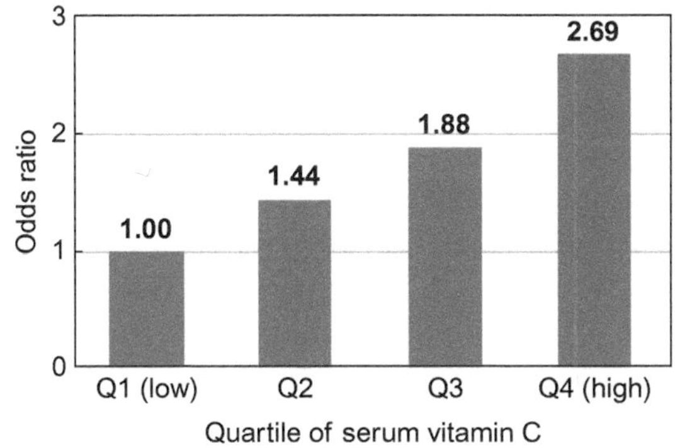

Fig. 4 Odds ratios for being GSTM1 null, by quartile (Q) of serum vitamin C, adjusted for age, sex, and BMI. 95% CIs: Q2 (0.80, 2.58), Q3 (1.04, 3.40), Q4 (1.46, 4.93). *P*-trend = 0.001. Median serum vitamin C: Q1 (41.46 μmol/L), Q2 (59.07 μmol/L), Q3 (67.59 μmol/L), Q4 (82.93 μmol/L) (Block et al., 2011).

Cahill et al., 2009). Furthermore, no associations with plasma vitamin C concentrations have been reported for GSTM1/T1 polymorphisms (Yuan et al., 2012).

Conflicting results of the associations of GSTT1 and GSTM1 null genotypes with circulating ascorbic acid concentrations do not allow deducing any causal relationships. In contrast to GSTO and glutaredoxin, the participation of the GSTT1 and GSTM1 enzymes in the recycling of dehydroascorbic acid has not been proven (Linster & Van Schaftingen, 2007); however, it is necessary to confirm the mechanistic relevance of any of the reported genetic variations.

5.6 Genetic Variations in the Haptoglobin Gene Associated to Altered Vitamin C Status

The haptoglobin protein binds free hemoglobin in the plasma, inhibits its oxidative activity, and therefore acts as an antioxidant. The haptoglobin gene allele Hp2 comprises a 1.7-kb partial duplication which does not have the same potency as the Hp1 allele, and its presence could therefore cause excessive metabolic consumption of ascorbic acid (Guthrie et al., 2014).

A 12-μmol/L decrease in serum ascorbate concentrations in the general population has been reported for Hp2-2 homozygotes (Delanghe et al., 1998), and a reduction of approximately 40 μmol/L was reported for male but not for female Hp2-2 homozygotes (Na et al., 2006). Similarly, serum ascorbic acid concentrations in HIV patients were approximately 11 μmol/L lower for Hp2-2 homozygotes (Delanghe et al., 1998). No differences in plasma ascorbic acid were observed between subjects with the Hp2-1 and Hp1-1 alleles (Delanghe et al., 1998; Na et al., 2006). In contrast, haptoglobin genotype was not associated with vitamin C concentrations in another cohort (Guthrie et al., 2014).

For individuals carrying the Hp2-2 genotype, a rare diet–gene interaction was reported: they had an increased risk of vitamin C deficiency when their dietary intake did not meet current recommendations (Cahill & El-Sohemy, 2010). Of the subjects who reported not meeting the RDA for vitamin C, those with the Hp2-2 genotype had 5.7 μmol/L lower average serum ascorbic acid concentrations than individuals with the Hp1 allele. Hp2-2 homozygotes had a 4.7-fold elevated risk for deficiency when they did not meet their RDA, while this was 1.7-fold for carriers of the Hp1 allele.

In conclusion, ascorbic acid oxidation might be increased in Hp2-2 homozygotes. However, the correlations with impaired iron status and lowered vitamin concentrations are unclear, and diet–gene interactions of haptoglobin alleles may extend beyond vitamin C regulation and should be examined with respect to the role of haptoglobin in iron status as well (Michels, Hagen, & Frei, 2013).

5.7 Conclusions From Large-Scale Observational/Intervention and Genetic Association Studies: Implications for Future Research

Around 5% of the general population in industrialized countries has deficient plasma vitamin C concentrations and about 13% have marginal deficient concentrations. Severe deficiency will lead to scurvy, while marginal deficiency is associated to elevated risks of all-cause mortality, CVD, and gastric cancer in observational studies.

Supplementation of Vitamin C did not reduce the risks for all-cause mortality, impaired cognitive performance, reduced quality of life, the development of eye diseases, infections, CVDs, and cancers. It is suggested that these results related to the fact that additional supplementation did not elevate systemic vitamin C level in individuals already in optimal supply.

Recent genetic association studies indicate that dietary intake is not the sole determinant of systemic vitamin C levels. Polymorphisms in two ascorbic acid transporter genes (*SLC23A1* and *SLC23A2*) are associated with lower system ascorbic acid concentrations. Common variations in three genes participating in the redox (*GSTM1* and *GSTT1*) and antioxidant metabolism (*haptoglobin*) are also associated with lower system ascorbic acid concentrations. The impact sizes are moderate, with reductions ranging between 5 and 10 μmol/L, which leads to the speculation that these variations would only be relevant in situations of suboptimal dietary intake, as defined by the current RDAs. However, although impact sizes of common polymorphism are expected to be low, epistatic interactions of multiple common variants can dramatically increase effect sizes (Abdullah, Jones, & Eck, 2015).

It is anticipated that future research will address these relations, specifically if epistatic effects of common polymorphism will predispose individuals to marginal systemic vitamin C concentrations even if they achieve current intake recommendations.

REFERENCES

Abdullah, M. M. H., Jones, P. J. H., & Eck, P. K. (2015). Nutrigenetics of cholesterol metabolism: Observational and dietary intervention studies in the postgenomic era. *Nutrition Reviews*, *73*(8), 523–543.

Age-Related Eye Disease Study Research Group. (2001). A randomized, placebo-controlled, clinical trial of high-dose supplementation with vitamins C and E and beta carotene for age-related cataract and vision loss: AREDS report no. 9. *Archives of Ophthalmology*, *119*(10), 1439–1452.

Amir Shaghaghi, M., Bernstein, C. N., Serrano León, A., El-Gabalawy, H., & Eck, P. (2014). Polymorphisms in the sodium-dependent ascorbate transporter gene SLC23A1 are associated with susceptibility to Crohn disease. *American Journal of Clinical Nutrition*, *99*, 378–383.

Andrew, A. S., Gui, J., Sanderson, A. C., Mason, R. A., Morlock, E. V., Schned, A. R., et al. (2009). Bladder cancer SNP panel predicts susceptibility and survival. *Human Genetics*, *125*(5–6), 527–539.

Anthony, H. M., & Schorah, C. J. (1982). Severe hypovitaminosis C in lung-cancer patients: The utilization of vitamin C in surgical repair and lymphocyte-related host resistance. *British Journal of Cancer*, *46*(3), 354–367.

Ashor, A. W., Werner, A. D., Lara, J., Willis, N. D., Mathers, J. C., & Siervo, M. (2017). Effects of vitamin C supplementation on glycaemic control: A systematic review and meta-analysis of randomised controlled trials. *European Journal of Clinical Nutrition*, *71*, 1371–1380. https://doi.org/10.1038/ejcn.2017.24.

Avenell, A., Campbell, M. K., Cook, J. A., Hannaford, P. C., Kilonzo, M. M., McNeill, G., et al. (2005). Effect of multivitamin and multimineral supplements on morbidity from infections in older people (MAVIS trial): Pragmatic, randomised, double blind, placebo controlled trial. *BMJ*, *331*(7512), 324–329.

Bartholomew, M. (2002). James Lind's treatise of the scurvy (1753). *Postgraduate Medicine Journal*, *78*(925), 695–696.

Bechthold, A., Leschik-Bonnet, E., Strohm, D., & Heseker, H. (2015). Updated 'reference values for the nutrient supply'. *Ernahrungs Umschau*, *62*(2).

Biniaz, V., Sadeghi Shermeh, M., Ebadi, A., Tayebi, A., & Einollahi, B. (2014). Effect of vitamin C supplementation on C-reactive protein levels in patients undergoing hemodialysis: A randomized, double blind, placebo-controlled study. *Nephro-Urology Monthly*, *6*(1), e13351.

Birlouez-Aragon, I., Fieux, B., Potier De Courcy, G., & Hercberg, S. (2001). *Vitamine C. Apports nutritionnels conseillés* (pp. 215–220). Paris: Tec et Doc. Lavoisier.
Blass, S. C., Goost, H., Burger, C., Tolba, R. H., Stoffel-Wagner, B., & Stehle, P. (2013). Extracellular micronutrient levels and pro-/antioxidant status in trauma patients with wound healing disorders: Results of a cross-sectional study. *Nutrition Journal, 12*, 157.
Block, G., Shaikh, N., Jensen, C. D., Volberg, V., & Holland, N. (2011). Serum vitamin C and other biomarkers differ by genotype of phase 2 enzyme genes GSTM1 and GSTT1. *The American Journal of Clinical Nutrition, 94*(3), 929–937.
Blot, W. J., Li, J. Y., Taylor, P. R., Guo, W., Dawsey, S., Wang, G. Q., et al. (1993). Nutrition intervention trials in Linxian, China: Supplementation with specific vitamin/mineral combinations, cancer incidence, and disease-specific mortality in the general population. *Journal of the National Cancer Institute, 85*(18), 1483–1492.
Boekholdt, S. M., Meuwese, M. C., Day, N. E., Luben, R., Welch, A., Wareham, N. J., et al. (2006). Plasma concentrations of ascorbic acid and C-reactive protein, and risk of future coronary artery disease, in apparently healthy men and women: The EPIC-Norfolk prospective population study. *British Journal of Nutrition, 96*(3), 516–522.
Buettner, G. R. (1993). The pecking order of free radicals and antioxidants: Lipid peroxidation, α-tocopherol, and ascorbate. *Archives of Biochemistry and Biophysics, 300*(2), 535–543.
Cahill, L. E., & El-Sohemy, A. (2009). Vitamin C transporter gene polymorphisms, dietary vitamin C and serum ascorbic acid. *Journal of Nutrigenetics and Nutrigenomics, 2*(6), 292–301.
Cahill, L. E., & El-Sohemy, A. (2010). Haptoglobin genotype modifies the association between dietary vitamin C and serum ascorbic acid deficiency. *The American Journal of Clinical Nutrition, 92*(6), 1494–1500.
Cahill, L. E., Fontaine-Bisson, B., & El-Sohemy, A. (2009). Functional genetic variants of glutathione S-transferase protect against serum ascorbic acid deficiency. *The American Journal of Clinical Nutrition, 90*(5), 1411–1417.
Carpenter, K. J. (2012). The discovery of vitamin C. *Annals of Nutrition and Metabolism, 61*(3), 259–264.
Casabonne, D., Gracia, E., Espinosa, A., Bustamante, M., Benavente, Y., Robles, C., et al. (2017). Fruit and vegetable intake and vitamin C transporter gene (SLC23A2) polymorphisms in chronic lymphocytic leukaemia. *European Journal of Nutrition, 56*(3), 1123–1133.
Ceriello, A., Novials, A., Ortega, E., Canivell, S., La Sala, L., Pujadas, G., et al. (2013). Vitamin C further improves the protective effect of glucagon-like peptide-1 on acute hypoglycemia-induced oxidative stress, inflammation, and endothelial dysfunction in type 1 diabetes. *Diabetes Care, 36*(12), 4104–4108.
Chen, Q., Espey, M. G., Sun, A. Y., Lee, J.-H., Krishna, M. C., Shacter, E., et al. (2007). Ascorbate in pharmacologic concentrations selectively generates ascorbate radical and hydrogen peroxide in extracellular fluid in vivo. *Proceedings of the National Academy of Sciences of the United States of America, 104*(21), 8749–8754.
Chen, H., Karne, R. J., Hall, G., Campia, U., Panza, J. A., Cannon, R. O., 3rd, et al. (2006). High-dose oral vitamin C partially replenishes vitamin C levels in patients with type 2 diabetes and low vitamin C levels but does not improve endothelial dysfunction or insulin resistance. *The American Journal of Physiology. Heart and Circulatory Physiology, 290*(1), H137–145.
Chen, A. A., Marsit, C. J., Christensen, B. C., Houseman, E. A., McClean, M. D., Smith, J. F., et al. (2009). Genetic variation in the vitamin C transporter, SLC23A2, modifies the risk of HPV16-associated head and neck cancer. *Carcinogenesis, 30*(6), 977–981.

Christen, W. G., Glynn, R. J., Sesso, H. D., Kurth, T., MacFadyen, J., Bubes, V., et al. (2012). Vitamins E and C and medical record-confirmed age-related macular degeneration in a randomized trial of male physicians. *Ophthalmology, 119*(8), 1642–1649.
Cook, N. R., Albert, C., Gaziano, M. J., et al. (2007). A randomized factorial trial of vitamins C and E and beta carotene in the secondary prevention of cardiovascular events in women: Results from the women's antioxidant cardiovascular study. *Archives of Internal Medicine, 167*(15), 1610–1618.
Corpe, C. P., Tu, H., Eck, P., Wang, J., Faulhaber-Walter, R., Schnermann, J., et al. (2010). Vitamin C transporter Slc23a1 links renal reabsorption, vitamin C tissue accumulation, and perinatal survival in mice. *Journal of Clinical Investigation, 120*(4), 1069–1083.
Correa, P., Fontham, E. T., Bravo, J. C., Bravo, L. E., Ruiz, B., Zarama, G., et al. (2000). Chemoprevention of gastric dysplasia: Randomized trial of antioxidant supplements and anti-helicobacter pylori therapy. *Journal of the National Cancer Institute, 92*(23), 1881–1888.
Crandon, J. H., Lund, C. C., & Dill, D. B. (1940). Experimental human scurvy. *New England Journal of Medicine, 223*, 353–369.
Dalgard, C., Christiansen, L., Vogel, U., Dethlefsen, C., Tjonneland, A., & Overvad, K. (2013). Variation in the sodium-dependent vitamin C transporter 2 gene is associated with risk of acute coronary syndrome among women. *PLoS One, 8*(8), e70421.
Daruwala, R., Song, J., Koh, W. S., Rumsey, S. C., & Levine, M. (1999). Cloning and functional characterization of the human sodium-dependent vitamin C transporters hSVCT1 and hSVCT2. *FEBS Letters, 460*(3), 480–484.
De Jong, T. M. H., Jochens, A., Jockel-Schneider, Y., Harks, I., Dommisch, H., Graetz, C., et al. (2014). SLC23A1 polymorphism rs6596473 in the vitamin C transporter SVCT1 is associated with aggressive periodontitis. *Journal of Clinical Periodontology, 41*(6), 531–540.
Delanghe, J. R., Langlois, M. R., Boelaert, J. R., Van Acker, J., Van Wanzeele, F., Van Der Groen, G., et al. (1998). Haptoglobin polymorphism, iron metabolism and mortality in HIV infection. *AIDS, 12*(9), 1027–1032.
Dhariwal, K. R., Shirvan, M., & Levine, M. (1991). Ascorbic acid regeneration in chromaffin granules: In situ kinetics. *Journal of Biological Chemistry, 266*(9), 5384–5387.
Dhariwal, K. R., Washko, P. W., & Levine, M. (1990). Determination of dehydroascorbic acid using high-performance liquid chromatography with coulometric electrochemical detection. *Analytical Biochemistry, 189*(1), 18–23.
Drouin, G., Godin, J. R., & Page, B. (2011). The genetics of vitamin C loss in vertebrates. *Current Genomics, 12*(5), 371–378.
Duell, E. J., Lujan-Barroso, L., Llivina, C., Muñoz, X., Jenab, M., Boutron-Ruault, M.-C., et al. (2013). Vitamin C transporter gene (SLC23A1 and SLC23A2) polymorphisms, plasma vitamin C levels, and gastric cancer risk in the EPIC cohort. *Genes & Nutrition, 8*(6), 549–560.
Dunn, W. A., Rettura, E., Seifter, E., & Englard, S. (1984). Carnitine biosynthesis from γ-butyrobetaine and from exogenous protein-bound 6-N-trimethyl-L-lysine by the perfused guinea pig liver. Effect of ascorbate deficiency on the in situ activity of γ-butyrobetaine hydroxylase. *Journal of Biological Chemistry, 259*(17), 10764–10770.
Eipper, B. A., Milgram, S. L., Husten, E. J., Yun, H. Y., & Mains, R. E. (1993). Peptidylglycine alpha-amidating monooxygenase: A multifunctional protein with catalytic, processing, and routing domains. *Protein Science, 2*(4), 489–497.
Eipper, B. A., Stoffers, D. A., & Mains, R. E. (1992). The biosynthesis of neuropeptides: Peptide alpha-amidation. *Annual Reviews in Neuroscience, 15*, 57–85.
Englard, S., & Seifter, S. (1986). The biochemical functions of ascorbic-acid. *Annual Review of Nutrition, 6*, 365–406.
Erichsen, H. C., Engel, S. A., Eck, P. K., Welch, R., Yeager, M., Levine, M., et al. (2006). Genetic variation in the sodium-dependent vitamin C transporters, SLC23A1, and

SLC23A2 and risk for preterm delivery. *American Journal of Epidemiology*, *163*(3), 245–254.

Erichsen, H. C., Peters, U., Eck, P., Welch, R., Schoen, R. E., Yeager, M., et al. (2008). Genetic variation in sodium-dependent vitamin C transporters SLC23A1 and SLC23A2 and risk of advanced colorectal adenoma. *Nutrition and Cancer*, *60*(5), 652–659.

Evans-Olders, R., Eintracht, S., & Hoffer, L. J. (2009). Metabolic origin of hypovitaminosis C in acutely hospitalized patients. *Nutrition*, *26*(11 – 12), 1070–1074.

Fain, O., Paries, J., Jacquart, B., Moel, G., Kettaneh, A., & Stirnemann, J. (2003). Hypovitaminosis C in hospitalized patients. *European Journal of Internal Medicine*, *14*(7), 419–425.

Frei, B., Birlouez-Aragon, I., & Lykkesfeldt, J. (2012). Authors' perspective: What is the optimum intake of vitamin C in humans? *Critical Reviews in Food Science & Nutrition*, *52*(9), 815–829.

Frei, B., England, L., & Ames, B. N. (1989). Ascorbate is an outstanding antioxidant in human blood plasma. *Proceedings of the National Academy of Sciences of the United States of America*, *86*(16), 6377–6381.

Frei, B., Stocker, R., England, L., & Ames, B. N. (1990). Ascorbate: The most effective antioxidant in human blood plasma. In I. Emerit, L. Packer, & C. Auclair (Eds.), *Antioxidants in therapy and preventive medicine* (pp. 155–163). Boston, MA: Springer.

Gale, C. R., Martyn, C. N., Winter, P. D., & Cooper, C. (1995). Vitamin-C and risk of death from stroke and coronary heart-disease in cohort of elderly people. *British Medical Journal*, *310*(6994), 1563–1566.

Gan, R., Eintracht, S., & Hoffer, L. J. (2008). Vitamin C deficiency in a university teaching hospital. *Journal of the American College of Nutrition*, *27*(3), 428–433.

Gariballa, S., & Forster, S. (2006). Effects of acute-phase response on nutritional status and clinical outcome of hospitalized patients. *Nutrition*, *22*(7–8), 750–757.

Gaziano, J., Glynn, R. J., Christen, W. G., et al. (2009). Vitamins E and C in the prevention of prostate and total cancer in men: The physicians' health study II randomized controlled trial. *JAMA*, *301*(1), 52–62.

Gaziano, J., Sesso, H. D., Christen, W. G., et al. (2012). Multivitamins in the prevention of cancer in men: The physicians' health study II randomized controlled trial. *JAMA*, *308*(18), 1871–1880.

Girodon, F., Galan, P., Monget, A. L., Boutron-Ruault, M. C., Brunet-Lecomte, P., Preziosi, P., et al. (1999). Impact of trace elements and vitamin supplementation on immunity and infections in institutionalized elderly patients—A randomized controlled trial MIN. VIT. AOX. geriatric network. *Archives of Internal Medicine*, *159*(7), 748–754.

Graat, J. M., Schouten, E. G., & Kok, F. J. (2002). Effect of daily vitamin E and multivitamin-mineral supplementation on acute respiratory tract infections in elderly persons: A randomized controlled trial. *JAMA*, *288*(6), 715–721.

Greenberg, E. R., Baron, J. A., Tosteson, T. D., Freeman, D. H., Jr., Beck, G. J., Bond, J. H., et al. (1994). A clinical trial of antioxidant vitamins to prevent colorectal adenoma. Polyp prevention study group. *New England Journal of Medicine*, *331*(3), 141–147.

Grodstein, F., O'Brien, J., Kang, J. H., Dushkes, R., Cook, N. R., Okereke, O., et al. (2013). Long-term multivitamin supplementation and cognitive function in men: A randomized trial. *Annals of Internal Medicine*, *159*(12), 806–814.

Grzybowski, A., & Pietrzak, K. (2013). Albert Szent-Gyorgyi (1893–1986): The scientist who discovered vitamin C. *Clinical Dermatology*, *31*(3), 327–331.

Guthrie, P. A. I., Abdollahi, M. R., Gaunt, T., Lawlor, D. A., Ben-Shlomo, Y., Gallacher, J., Smith, G.D., Day, I.N.M. and Rodriguez, S. (2014). Haptoglobin duplicon, hemoglobin, and vitamin C: Analyses in the British women's heart and health study and caerphilly prospective study. Disease Markers. 2014. ID529456.

Guwatudde, D., Wang, M., Ezeamama, A. E., Bagenda, D., Kyeyune, R., Wamani, H., et al. (2015). The effect of standard dose multivitamin supplementation on disease progression in HIV-infected adults initiating HAART: A randomized double blind placebo-controlled trial in Uganda. *BMC Infectious Diseases*, *15*, 348.

Hallberg, L., Brune, M., & Rossander, L. (1989). The role of vitamin C in iron absorption. *International Journal for Vitamin & Nutrition Research Supplement*, *30*, 103–108.

Hemila, H., & Louhiala, P. (2007). Vitamin C may affect lung infections. *Journal of the Royal Society of Medicine*, *100*(11), 495–498.

Hercberg, S., Galan, P., Preziosi, P., et al. (2004). The SU.VI.MAX study: A randomized, placebo-controlled trial of the health effects of antioxidant vitamins and minerals. *Archives of Internal Medicine*, *164*(21), 2335–2342.

Hitomi, K., & Tsukagoshi, N. (1996). Role of ascorbic acid in modulation of gene expression. In J. R. Harris (Ed.), *Subcellular biochemistry: Ascorbic acid: Biochemistry and biomedical cell biology* (pp. 41–56). Boston, MA, US: Springer.

Hoffer, L. J. (2010). Re: Vitamin C deficiency in a population of young Canadian adults. *American Journal of Epidemiology*, *171*(3), 387.

Horska, A., Mislanova, C., Bonassi, S., Ceppi, M., Volkovova, K., & Dusinska, M. (2011). Vitamin C levels in blood are influenced by polymorphisms in glutathione S-transferases. *European Journal of Nutrition*, *50*(6), 437–446.

Hunt, C., Chakravorty, N. K., Annan, G., Habibzadeh, N., & Schorah, C. J. (1994). The clinical effects of vitamin C supplementation in elderly hospitalised patients with acute respiratory infections. *International Journal for Vitamin & Nutrition Research*, *64*, 212–219.

Jenab, M., Riboli, E., Ferrari, P., Sabate, J., Slimani, N., Norat, T., et al. (2006). Plasma and dietary vitamin C levels and risk of gastric cancer in the European Prospective Investigation into Cancer and Nutrition (EPIC-EURGAST). *Carcinogenesis*, *27*(11), 2250–2257.

Juraschek, S. P., Guallar, E. L., Appel, L. J., & Miller, E. R. (2012). Effects of vitamin C supplementation on blood pressure: A meta-analysis of randomized controlled trials. *The American Journal of Clinical Nutrition*, *95*(5), 1079–1088.

Kaufman, S. (1974). Dopamine-beta-hydroxylase. *Journal of Psychiatric Research*, *11*, 303–316.

Khaw, K. T., Bingham, S., Welch, A., Luben, R., Wareham, N., Oakes, S., et al. (2001). Relation between plasma ascorbic acid and mortality in men and women in EPIC-Norfolk prospective study: A prospective population study. *Lancet*, *357*(9257), 657–663.

King, C. G., & Waugh, W. A. (1932). The chemical nature of vitamin C. *Science*, *75*(1944), 357–358.

Kivirikko, K. I., & MyllylÄ, R. (1985). Post-translational processing of procollagens. *Annals of the New York Academy of Sciences*, *460*, 187–201.

Kobylecki, C. J., Afzal, S., Smith, G. D., & Nordestgaard, B. G. (2015). Genetically high plasma vitamin C, intake of fruit and vegetables, and risk of ischemic heart disease and all-cause mortality: A Mendelian randomization study. *American Journal of Clinical Nutrition*, *101*(6), 1135–1143.

Lane, D. J. R., Chikhani, S., Richardson, V., & Richardson, D. (2013a). Transferrin iron uptake is stimulated by ascorbate *via* an intracellular reductive mechanism. *American Journal of Hematology*, *88*(5), E189–E190.

Lane, D. J. R., Chikhani, S., Richardson, V., & Richardson, D. R. (2013b). Transferrin iron uptake is stimulated by ascorbate via an intracellular reductive mechanism. *Biochimica et Biophysica Acta-Molecular Cell Research*, *1833*(6), 1527–1541.

Lane, D. J., & Richardson, D. R. (2014). The active role of vitamin C in mammalian iron metabolism: Much more than just enhanced iron absorption!. *Free Radical Biology and Medicine*, *75*, 69–83.

Langlois, K., Cooper, M., & Colapinto, C. K. (2016). Vitamin C status of Canadian adults: Findings from the 2012/2013 Canadian health measures survey. *Health Reports*, *27*(5), 3–10.

Langlois, M., Duprez, D., Delanghe, J., De Buyzere, M., & Clement, D. L. (2001). Serum vitamin C concentration is low in peripheral arterial disease and is associated with inflammation and severity of atherosclerosis. *Circulation*, *103*(14), 1863–1868.

Lee, D.-H., Folsom, A. R., Harnack, L., Halliwell, B., & Jacobs, D. R. (2004). Does supplemental vitamin C increase cardiovascular disease risk in women with diabetes? *The American Journal of Clinical Nutrition*, *80*(5), 1194–1200.

Levine, M., Conry-Cantilena, C., Wang, Y., Welch, R. W., Washko, P. W., Dhariwal, K. R., et al. (1996). Vitamin C pharmacokinetics in healthy volunteers: Evidence for a recommended dietary allowance. *Proceedings of the National Academy of Sciences of the United States of America*, *93*(8), 3704–3709.

Levine, M., Dhariwal, K. R., Washko, P., Welch, R., Wang, Y. H., Cantilena, C. C., et al. (1992). Ascorbic acid and reaction kinetics in situ: A new approach to vitamin requirements. *Journal of Nutritional Science and Vitaminology*, *38*, 169–172.

Levine, M., Wang, Y., Padayatty, S. J., & Morrow, J. (2001). A new recommended dietary allowance of vitamin C for healthy young women. *Proceedings of the National Academy of Sciences of the United States of America*, *98*(17), 9842–9846.

Levine, M., Wang, Y., & Rumsey, S. C. (1999). Analysis of ascorbic acid and dehydroascorbic acid in biological samples. *Methods in Enzymology*, *299*, 65–76.

Li, J. Y., Taylor, P. R., Li, B., Dawsey, S., Wang, G. Q., Ershow, A. G., et al. (1993). Nutrition intervention trials in linxian, China: Multiple vitamin/mineral supplementation, cancer incidence, and disease-specific mortality among adults with esophageal dysplasia. *Journal of the National Cancer Institute*, *85*(18), 1492–1498.

Lindblad, B., Lindstedt, G., & Lindstedt, S. (1970). Mechanism of enzymic formation of homogentisate from *p*-hydroxyphenylpyruvate. *Journal of the American Chemical Society*, *92*(25), 7446–7449.

Lindblad, M., Tveden-Nyborg, P., & Lykkesfeldt, L. (2013). Regulation of vitamin C homeostasis during deficiency. *Nutrients*, *5*(8), 2860–2879.

Linster, C. L., & Van Schaftingen, E. (2007). Vitamin C: Biosynthesis, recycling and degradation in mammals. *FEBS Journal*, *274*(1), 1–22.

Loria, C. M., Klag, M. J., Caulfield, L. E., & Whelton, P. K. (2000). Vitamin C status and mortality in US adults. *The American Journal of Clinical Nutrition*, *72*(1), 139–145.

Lund, C. C., & Crandon, J. H. (1941). Ascorbic acid and human wound healing. *Annals of Surgery*, *114*(4), 776–790.

Lykkesfeldt, J., & Poulsen, H. E. (2010). Is vitamin C supplementation beneficial? Lessons learned from randomised controlled trials. *British Journal of Nutrition*, *103*(9), 1251–1259.

Mayland, C. R., Bennett, M. I., & Allan, K. (2005). Vitamin C deficiency in cancer patients. *Palliative Medicine*, *19*(1), 17–20.

Menniti, F. S., Knoth, J., & Diliberto, E. J., Jr. (1986). Role of ascorbic acid in dopamine beta-hydroxylation. The endogenous enzyme cofactor and putative electron donor for cofactor regeneration. *Journal of Biological Chemistry*, *261*(36), 16901–16908.

Michels, A. J., Hagen, T. M., & Frei, B. (2013). Human genetic variation influences vitamin C homeostasis by altering vitamin C transport and antioxidant enzyme function. *Annual Review of Nutrition*, *33*(1), 45–70.

Monsen, E. R. (2000). Dietary reference intakes for the antioxidant nutrients: Vitamin C, vitamin E, selenium, and carotenoids. *Journal of the American Dietetic Association*, *100*(6), 637–640.

Moser, M. A., & Chun, O. K. (2016). Vitamin C and heart health: A review based on findings from epidemiologic studies. *International Journal of Molecular Sciences*, *17*(8), 1328.

Myint, P. K., Luben, R. N., Welch, A. A., Bingham, S. A., Wareham, N. J., & Khaw, K. T. (2008). Plasma vitamin C concentrations predict risk of incident stroke over 10 y in 20 649 participants of the European prospective investigation into cancer-Norfolk prospective population study. *The American Journal of Clinical Nutrition, 87*(1), 64–69.

Na, N., Delanghe, J. R., Taes, Y. E. C., Torck, M., Baeyens, W. R. G., & Ouyang, J. (2006). Serum vitamin C concentration is influenced by haptoglobin polymorphism and iron status in Chinese. *Clinica Chimica Acta, 365*(1–2), 319–324.

Nishikimi, M., & Yagi, K. (1991). Molecular basis for the deficiency in humans of gulonolactone oxidase, a key enzyme for ascorbic acid biosynthesis. *The American Journal of Clinical Nutrition, 54*(6 Suppl), 1203S–1208S.

Nyyssonen, K., Parviainen, M. T., Salonen, R., Tuomilehto, J., & Salonen, J. T. (1997). Vitamin C deficiency and risk of myocardial infarction: Prospective population study of men from eastern Finland. *BMJ, 314*(7081), 634–638.

Odermarsky, M., Lykkesfeldt, J., & Liuba, P. (2009). Poor vitamin C status is associated with increased carotid intima-media thickness, decreased microvascular function, and delayed myocardial repolarization in young patients with type 1 diabetes. *American Journal of Clinical Nutrition, 90*(2), 447–452.

Ohta, Y., & Nishikimi, M. (1999). Random nucleotide substitutions in primate nonfunctional gene for L-gulono-gamma-lactone oxidase, the missing enzyme in L-ascorbic acid biosynthesis. *Biochimica et Biophysica Acta, 1472*(1–2), 408–411.

Padayatty, S. J., Katz, A., Wang, Y. H., Eck, P., Kwon, O., Lee, J. H., et al. (2003). Vitamin C as an antioxidant: Evaluation of its role in disease prevention. *Journal of the American College of Nutrition, 22*(1), 18–35.

Padayatty, S. J., & Levine, M. (2000). Vitamin C and myocardial infarction: The heart of the matter. *The American Journal of Clinical Nutrition, 71*(5), 1027–1028.

Padayatty, S. J., & Levine, M. (2016). Vitamin C: The known and the unknown and Goldilocks. *Oral Diseases, 22*(6), 463–493.

Peltonen, L., Halila, R., & Ryhänen, L. (1985). Enzymes converting procollagens to collagens. *Journal of Cellular Biochemistry, 28*(1), 15–21.

Peterkofsky, B. (1991). Ascorbate requirement for hydroxylation and secretion of procollagen: Relationship to inhibition of collagen synthesis in scurvy. *The American Journal of Clinical Nutrition, 54*(6 Suppl), 1135S–1140S.

Pfeiffer, C. M., Sternberg, M. R., Schleicher, R. L., & Rybak, M. E. (2013). Dietary supplement use and smoking are important correlates of biomarkers of water-soluble vitamin status after adjusting for sociodemographic and lifestyle variables in a representative sample of U.S. adults. *Journal of Nutrition, 143*, 957S–965S.

Polidori, M. C., Mecocci, P., Levine, M., & Frei, B. (2004). Short-term and long-term vitamin C supplementation in humans dose-dependently increases the resistance of plasma to ex vivo lipid peroxidation. *Archives of Biochemistry and Biophysics, 423*(1), 109–115.

Pouchieu, C., Deschasaux, M., Hercberg, S., Druesne-Pecollo, N., Latino-Martel, P., & Touvier, M. (2014). Prospective association between red and processed meat intakes and breast cancer risk: Modulation by an antioxidant supplementation in the SU.VI. MAX randomized controlled trial. *International Journal of Epidemiology, 43*(5), 1583–1592.

Prockop, D. J., & Kivirikko, K. I. (1995). Collagens: Molecular biology, diseases, and potentials for therapy. *Annual Review of Biochemistry, 64*, 403–434.

Qiao, H., & May, J. M. (2011). Regulation of the human ascorbate transporter SVCT2 exon 1b gene by zinc-finger transcription factors. *Free Radical Biology and Medicine, 50*(9), 1196–1209.

Richards, J. C., Crecelius, A. R., Larson, D. G., & Dinenno, F. A. (2015). Acute ascorbic acid ingestion increases skeletal muscle blood flow and oxygen consumption via local vasodilation during graded handgrip exercise in older adults. *American Journal of Physiology. Heart and Circulatory Physiology, 309*(2), H360–H368.

Robitaille, L., & Hoffer, L. J. (2016). A simple method for plasma total vitamin C analysis suitable for routine clinical laboratory use. *Nutrition Journal*, *15*(1), 40.
Rodrigo, R., Korantzopoulos, P., Cereceda, M., Asenjo, J., Zamorano, J., Villalabeitia, E., et al. (2013). A randomized controlled trial to prevent post-operative atrial fibrillation by antioxidant reinforcement. *Journal of the American College of Cardiology*, *62*(16), 1457–1465.
Sahraian, A., Ghanizadeh, A., & Kazemeini, F. (2015). Vitamin C as an adjuvant for treating major depressive disorder and suicidal behavior, a randomized placebo-controlled clinical trial. *Trials*, *16*, 94.
Salonen, R. M., Nyyssönen, K., Kaikkonen, J., Porkkala-Sarataho, E., Voutilainen, S., Rissanen, T. H., et al. (2003). Six-year effect of combined vitamin C and E supplementation on atherosclerotic progression. *The Antioxidant Supplementation in Atherosclerosis Prevention (ASAP) Study*, *107*(7), 947–953.
Sasazuki, S., Sasaki, S., Tsubono, Y., Okubo, S., Hayashi, M., Kakizoe, T., et al. (2003). The effect of 5-year vitamin C supplementation on serum pepsinogen level and *Helicobacter pylori* infection. *Cancer Science*, *94*(4), 378–382.
Schectman, G., Byrd, J. C., & Hoffmann, R. (1991). Ascorbic acid requirements for smokers: Analysis of a population survey. *The American Journal of Clinical Nutrition*, *53*(6), 1466–1470.
Schleicher, R. L., Carroll, M. D., Ford, E. S., & Lacher, D. A. (2009). Serum vitamin C and the prevalence of vitamin C deficiency in the United States: 2003–2004 National health and nutrition examination survey (NHANES). *The American Journal of Clinical Nutrition*, *90*(5), 1252–1263.
Sesso, H. D., Buring, J. E., Christen, W. G., et al. (2008). Vitamins E and C in the prevention of cardiovascular disease in men: The physicians' health study II randomized controlled trial. *JAMA*, *300*(18), 2123–2133.
Sharma, A., Tripathi, M., Satyam, A., & Kumar, L. (2009). Study of antioxidant levels in patients with multiple myeloma. *Leukemia & Lymphoma*, *50*(5), 809–815.
Shateri, Z., Ali Keshavarz, S., Hosseini, S., Chamari, M., Hosseini, M., & Nasli, E. (2016). Effect of Vitamin C supplementation on blood pressure level in type 2 diabetes mellitus: A randomized, double-blind, placebo-controlled trial. *Biosciences Biotechnology Research Asia*, *13*(1), 279–286.
Shibuya, N., Humphers, J. M., Agarwal, M. R., & Jupiter, D. C. (2013). Efficacy and safety of high-dose vitamin C on complex regional pain syndrome in extremity trauma and surgery—Systematic review and meta-analysis. *Journal of Foot and Ankle Surgery*, *52*(1), A7–A8.
Simon, J. A., Hudes, E. S., & Browner, W. S. (1998). Serum ascorbic acid and cardiovascular disease prevalence in U.S. adults. *Epidemiology*, *9*(3), 316–321.
Simon, J. A., Hudes, E. S., & Tice, J. A. (2001). Relation of serum ascorbic acid to mortality among US adults. *Journal of the American College of Nutrition*, *20*(3), 255–263.
Skibola, C. F., Bracci, P. M., Halperin, E., Nieters, A., Hubbard, A., Paynter, R. A., et al. (2008). Polymorphisms in the estrogen receptor 1 and vitamin C and matrix metalloproteinase gene families are associated with susceptibility to lymphoma. *PLoS One*, *3*(7), e2816.
Sorensen, L. T., Toft, B. G., Rygaard, J., Ladelund, S., Paddon, M., & James, T. (2010). Effect of smoking, smoking cessation, and nicotine patch on wound dimension, vitamin C, and systemic markers of collagen metabolism. *Surgery*, *148*(5), 982–990.
Sram, R. J., Binkova, B., & Rossner, P., Jr. (2012). Vitamin C for DNA damage prevention. *Mutation Research/Fundamental and Molecular Mechanisms of Mutagenesis*, *733*(1–2), 39–49.
Svirbely, J. L., & Szent-Györgyi, A. (1932). The chemical nature of vitamin C. *Biochemical Journal*, *26*(3), 865–870.

Teixeira, A., Carrie, A. S., Genereau, T., Herson, S., & Cherin, P. (2001). Vitamin C deficiency in elderly hospitalized patients. *American Journal of Medicine, 111*(6), 502.

Timpson, N. J., Forouhi, N. G., Brion, M. J., Harbord, R. M., Cook, D. G., Johnson, P., et al. (2010). Genetic variation at the SLC23A1 locus is associated with circulating concentrations of L-ascorbic acid (vitamin C): Evidence from 5 independent studies with >15,000 participants. *The American Journal of Clinical Nutrition, 92*(2), 375–382.

Traber, M. G., & Stevens, J. F. (2011). Vitamins C and E: Beneficial effects from a mechanistic perspective. *Free Radical Biology & Medicine, 51*(5), 1000–1013.

Tsukaguchi, H., Tokui, T., Mackenzle, B., Berger, U. V., Chen, X. Z., Wang, Y., et al. (1999). A family of mammalian Na+−dependent L-ascorbic acid transporters. *Nature, 399*(6731), 70–75.

Tu, H., Li, H., Wang, Y., Niyyati, M., Wang, Y., Leshin, J., et al. (2015). Low red blood cell vitamin C concentrations induce red blood cell fragility: A link to diabetes via glucose, glucose transporters, and dehydroascorbic acid. *eBioMedicine, 2*(11), 1735–1750.

Vita, J. A., Keaney, J. F., Raby, K. E., Morrow, J. D., Freedman, J. E., & Lynch, S. (1998). Low plasma ascorbic acid independently predicts the presence of an unstable coronary syndrome. *Journal of the American College of Cardiology, 31*(5), 980–986.

Wang, Y., Liu, X. J., Robitaille, L., Eintracht, S., Macnamara, E., & Hoffer, L. J. (2013). Effects of vitamin C and vitamin D administration on mood and distress in acutely hospitalized patients. *The American Journal of Clinical Nutrition, 98*(3), 705–711.

Wang, Y. X., Mackenzie, B., Tsukaguchi, H., Weremowicz, S., Morton, C. C., & Hediger, M. A. (2000). Human vitamin C (L-ascorbic acid) transporter SVCT1. *Biochemical and Biophysical Research Communications, 267*(2), 488–494.

Wang, Y., Russo, T. A., Kwon, O., Chanock, S., Rumsey, S. C., & Levine, M. (1997). Ascorbate recycling in human neutrophils: Induction by bacteria. *Proceedings of the National Academy of Sciences of the United States of America, 94*(25), 13816–13819.

Waters, D. D., Alderman, E. L., Hsia, J., et al. (2002). Effects of hormone replacement therapy and antioxidant vitamin supplements on coronary atherosclerosis in postmenopausal women: A randomized controlled trial. *JAMA, 288*(19), 2432–2440.

Wright, M. E., Andreotti, G., Lissowska, J., Yeager, M., Zatonski, W., Chanock, S. J., et al. (2009). Genetic variation in sodium-dependent ascorbic acid transporters and risk of gastric cancer in Poland. *European Journal of Cancer, 45*(10), 1824–1830.

Yokoyama, T., Date, C., Kokubo, Y., Yoshiike, N., Matsumura, Y., & Tanaka, H. (2000). Serum vitamin C concentration was inversely associated with subsequent 20-year incidence of stroke in a Japanese rural community—The Shibata study. *Stroke, 31*(10), 2287–2294.

You, W. C., Chang, Y. S., Heinrich, J., Ma, J. L., Liu, W. D., Zhang, L., et al. (2001). An intervention trial to inhibit the progression of precancerous gastric lesions: Compliance, serum micronutrients and S-allyl cysteine levels, and toxicity. *European Journal of Cancer Prevention, 10*(3), 257–263.

Yuan, L. H., Meng, L. P., Ma, W. W., Li, S., Feng, J. F., Yu, H. L., et al. (2012). The role of glutathione S-transferase M1 and T1 gene polymorphisms and fruit and vegetable consumption in antioxidant parameters in healthy subjects. *British Journal of Nutrition, 107*(6), 928–933.

Zanon-Moreno, V., Ciancotti-Olivares, L., Asencio, J., Sanz, P., Ortega-Azorin, C., Pinazo-Duran, M. D., et al. (2011). Association between a SLC23A2 gene variation, plasma vitamin C levels, and risk of glaucoma in a Mediterranean population. *Molecular Vision, 17*(322–24), 2997–3004.

Zhang, M., Robitaille, L., Eintracht, S., & Hoffer, L. J. (2011). Vitamin C provision improves mood in acutely hospitalized patients. *Nutrition, 27*(5), 530–533.

FURTHER READING

Food Standard Agency. (2007). FSA nutrient and food based guidelines for UK institutions. Retrieved October 09, 2017, from https://www.food.gov.uk/sites/default/files/multimedia/pdfs/nutrientinstitution.pdf.

Heart Protection Study Collaborative Group. (2002). MRC/BHF Heart Protection Study of antioxidant vitamin supplementation in 20,536 high-risk individuals: A randomised placebo-controlled trial. *The Lancet, 360*(9326), 23–33.

Levine, M., Dhariwal, K. R., Washko, P. W., Butler, J. D., Welch, R. W., Wang, Y., et al. (1991). Ascorbic acid and in situ kinetics: A new approach to vitamin requirements. *American Journal of Clinical Nutrition, 54*(Suppl. 6), 1157S–1162S.

Smirnoff, N., Conklin, P. L., & Loewus, F. A. (2001). Biosynthesis of ascorbic acid in plants: A renaissance. *Annual Review of Plant Physiology and Plant Molecular Biology, 52*(1), 437–467.

CHAPTER EIGHT

Water-Soluble Vitamin E—Tocopheryl Phosphate

Jean-Marc Zingg[1]
Miller School of Medicine, University of Miami, Miami, FL, United States
[1]Corresponding author: e-mail address: jaz42@miami.edu

Contents

Advances in Food and Nutrition Research, Volume 83
ISSN 1043-4526
https://doi.org/10.1016/bs.afnr.2017.12.007

Abstract

The hydrophobicity of vitamin E poses transport and metabolic challenges to regulate its bioavailability and to prevent its accumulation in lipid-rich tissues such as adipose tissue, brain, and liver. Water-soluble precursors of vitamin E (α-tocopherol, αT), such as its esters with acetate (αTA), succinate (αTS), or phosphate (αTP), have increased solubility in water and stability against reaction with free radicals, but they are rapidly converted during their uptake into the lipid-soluble vitamin E. Therefore, the bioavailability of these precursors as intact molecules is low; nevertheless, at least for αTS and αTP, the recent research has revealed unique regulatory effects on signal transduction and gene expression and the modulation of cellular events ranging from proliferation, survival/apoptosis, lipid uptake and metabolism, phagocytosis, long term potentiation, cell migration, telomere maintenance, and angiogenesis. Moreover, water-soluble derivatives of vitamin E including some based on αTP are increasingly used as components of nanocarriers for enhanced and targeted delivery of drugs and other molecules (vitamins, including αT and αTP itself, vitamin D3, carnosine, caffeine, docosahexaenoic acid (DHA), insulin) and cofactors such as coenzyme Q10. In this review, the chemical characteristics, transport, metabolic pathways, and molecular mechanisms of action of αTP in cells and tissues are summarized and put into perspective with its possible role in the prevention of a number of diseases.

1. INTRODUCTION

There are four fat-soluble vitamins, vitamin A (retinol), vitamin D (cholecalciferol), vitamin K (phylloquinone and menaquinones), and vitamin E (tocopherols and tocotrienols), and their hydrophobicity poses transport and metabolic challenges to regulate their bioavailability and to prevent their accumulation in cells of lipid-rich tissues such as adipose tissue, brain, and liver. At high concentrations, these vitamins can be absorbed by passive diffusion mechanisms and bulk endocytosis, and at least in intestinal epithelia cells a certain degree of competition between these vitamins suggests common mechanisms of uptake (Goncalves, Margier, et al., 2014; Goncalves et al., 2015; Traber, 2008). However, more recent research suggests that at the lower doses found in the diet and the body, their transport and distribution are selective and regulated involving specific recognition by carrier proteins, transporters, and receptors, and possibly facilitated by intracellular metabolic trapping by enzymatic modification as it occurs with fatty acids (Pownall & Moore, 2014). Part of this regulation occurs by selective and regulated uptake and transport in cells, plasma and tissue, specific binding to and activation by enzymes, structural proteins and transcription factors, selective metabolism, and selective secretion/excretion. In particular, vitamin A but also vitamin D levels in the body are tightly regulated not only by their transport but also by their conversion to their most active forms to prevent negative and potential toxic effects of their accumulation (Borel, Caillaud, & Cano, 2015; Kono & Arai, 2015; Reboul & Borel, 2011). In contrast, for vitamin E no serious toxic biological effects due to accumulation in tissues have been described. Although some symptoms such as increased anticoagulant activity due to hypervitaminosis E have been reported in animal models (Abdo, Rao, Montgomery, Dinowitz, & Kanagalingam, 1986; Jack Yang & Desai, 1977), in humans, diseases are mostly due to deficiencies of vitamin E (Traber, 2014; Ulatowski & Manor, 2013). Nevertheless, the uptake and transport of vitamin E are highly sophisticated and selective, in part to prevent accumulation but mostly because it is geared toward retaining mainly one of the eight natural vitamin E analogues, α-tocopherol (αT), by means of selective enrichment and by preventing its metabolism occurring with the other seven analogues of vitamin E (Traber, 2013).

αT is one member of a group of eight major natural analogues of vitamin E (α-, β-, γ-, δ-tocopherol and tocotrienols), which are all hydrophobic and occur as *RRR*-side-chain stereoisomers (Fig. 1) (Traber & Arai, 1999; Ulatowski & Manor, 2013). These eight vitamin E analogues share more or less similar chemical antioxidant activity, but nevertheless distinguish themselves by clear individual physicochemical and biological effects at a molecular level in humans and higher animals. Most importantly, only αT is enriched in plasma 10- to 100-fold, from about 0.3–2.5 μM (as measured for the non-αT analogues) to an average of 23 μM (Meydani, Koga, & Ali, 2001; Zingg, 2007b). This enrichment is mostly the consequence of selective retention, protection, and incorporation into very low density lipoproteins (VLDL) of *RRR*-α-tocopherol by the liver α-tocopherol transfer protein (α-TTP), and of the enhanced metabolic degradation of the other tocopherols (β-, γ-, and δ-), of excess αT, and of all tocotrienols by cytochrome P450 enzymes and their subsequent elimination (reviewed in Traber, 2013; Wu & Croft, 2007). As a consequence of enrichment, the αT analogue is mostly considered as the vitamin E analogue with the essential function (Azzi, 2017), regardless of the eight analogues having essentially equal chemical antioxidant activities. Accordingly, the biological

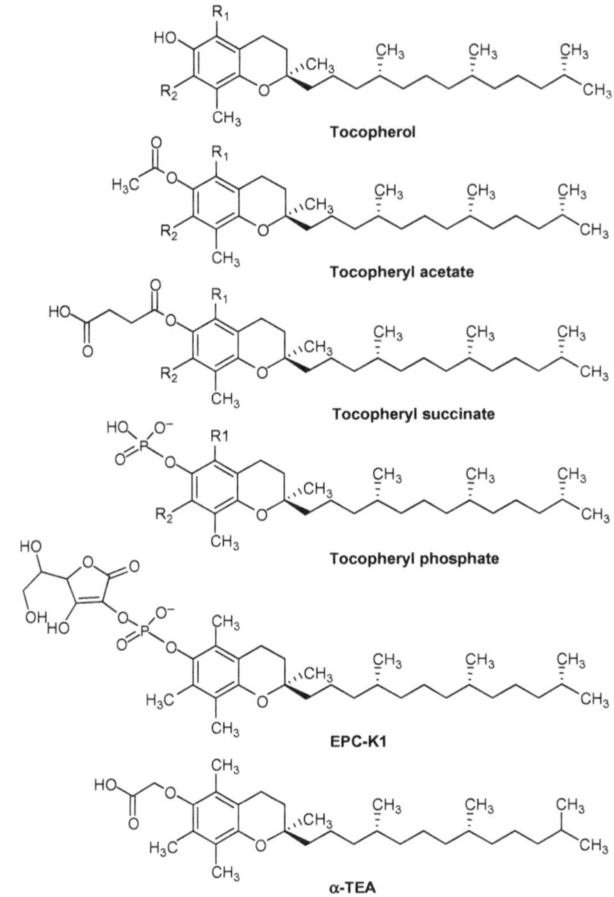

R_1	R_2	
CH_3	CH_3	α
CH_3	H	β
H	CH_3	γ
H	H	δ

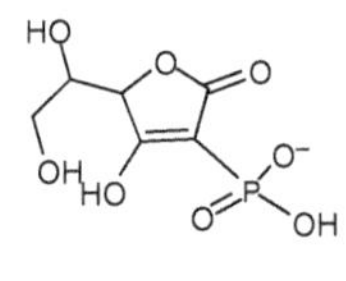

Fig. 1 Structure of vitamin E and water-soluble vitamin E analogues. Shown are tocopherol, tocopheryl acetate, tocopheryl succinate, and tocopheryl phosphate, having four differently methylated residues R1 and R2, giving their alpha-, beta-, gamma-, or delta-analogues. EPC-K1 is a composite molecule between α-tocopherol and L-ascorbate, linked by a phosphodiester bond; shown is also L-ascorbyl phosphate that can be generated by hydrolysis of EPC-K1. α-TEA is an ether-linked acetic acid analogue of α-tocopherol.

effectiveness in vivo of the four forms of vitamin E (tocopherols) has been established as $\alpha \gg \gamma > \delta > \beta$ (Burton & Traber, 1990), with αT as the only form being recognized by the Food and Nutrition Board of the Institute of Medicine (IOM) to meet human requirements (Galli et al., 2016; IOS, 2000).

The interaction with the food matrix, the formation of micelles with bile/lipids in the intestinal tract, and the specific interaction with a number of proteins such as membrane transporters, plasma transporters, lipoproteins, and lipid transport proteins (LTP) that act as biological extra- and intracellular carriers can be seen to be designed to overcome the inherent water insolubility of vitamin E and to possibly also protect it from oxidative and metabolic destruction. These proteins regulate uptake and transport, distribution to tissues, cells and subcellular sites in the body, and the metabolism to a number of water-soluble urinary vitamin E metabolites such as the carboxyethylhydroxychromanes (CEHC) that have the specific role in removing too much vitamin E from the body (Birringer, 2010; Zhao et al., 2010). In contrast, synthetic water-soluble vitamin E analogues (mainly α-tocopheryl acetate (αTA), α-tocopheryl succinate (αTS), and α-tocopheryl phosphate (αTP)) (Fig. 1) are rapidly converted to the non-modified hydrophobic αT during uptake and therefore show low bioavailability as intact molecules but are valuable precursors of the natural αT. These water-soluble derivatives of vitamin E have been mainly synthesized for commercial purpose as supplements since they have increased shelf-life due to their resistance against oxidation. As supplement, vitamin E is mostly available in different forms: a mixture of natural tocopherols and tocotrienols (extracted from natural sources), natural RRR-αT, synthetic αT, which is an equimolar racemic mixture of eight stereoisomers (*all rac*-αT), or a mixture of the synthetic water-soluble tocopheryl esters (mainly αTA or αTS). In this chapter, the chemical and biological characteristics of these water-soluble analogues of vitamin E and in particular of water-soluble αTP that was only recently recognized to occur naturally are reviewed.

2. WATER-SOLUBLE VITAMIN ANALOGUES

The ability to act as antioxidants renders the natural vitamin E analogues inherently unstable due to reaction with reactive oxygen species (ROS); thus, several stabilized vitamin E derivatives (e.g., αTA, αTS, αTP, α-TEA (α-tocopheryloxyacetic acid, or 2,5,7,8-tetramethyl-2*R*-(4*R*,8*R*,12-trimethyltridecyl)chroman-6-yloxy acetic acid), EPC-K1 (the phosphodiester of vitamins E and C (L-ascorbic acid), L-ascorbic acid 2-[3,4-dihydro-2,5,7,8-tetramethyl-2-(4,8,12-trimethytridecyl)-2H-1-benzopyran-6-yl-hydrogen-phosphate] potassium salt), and others) have been synthesized for usage in dietary supplements, cosmetics, antiaging, and anticancer agents (Fig. 1) (reviewed in Birringer, EyTina, Salvatore, & Neuzil, 2003; Cerecetto & Lopez, 2007; Neuzil, Dong, et al., 2007; Neuzil, Tomasetti, et al., 2007; Zingg, 2007a, 2007b, 2007c). In addition to being chemically more stable, the water solubility, transport, metabolism, and cellular activities of these derivatives are also different. Most of these tocopherol derivatives are modified at the 6-hydroxyl group of the chromanol ring and therefore cannot act chemically as antioxidants. These derivatives are usually prepared using αT, since this is the analogue that is selectively enriched by the liver α-TTP and therefore has the highest biological vitamin E activity (Burton & Traber, 1990), and αTA is the most often used analogue in food supplements and cosmetic products. Whereas these derivatives may have advantages for processing, storage, and absorption, they can be considered to be provitamins since they are readily converted to the natural parent form by intestinal or less efficiently by epidermal esterases (with the exception of the ether analogue α-TEA which is more stable), and only the nonesterified tocopherols are efficiently taken up and appear in plasma within hours and thus ultimately perform the same function in the body as the natural αT (Birringer, 2010; Cheeseman et al., 1995; Horwitt, Elliott, Kanjananggulpan, & Fitch, 1984; Lauridsen, Hedemann, & Jensen, 2001; Mathias, Harries, Peters, & Muller, 1981; Muller, Manning, Mathias, & Harries, 1976; Nakamura, Aoyama, Fujita, & Katsui, 1975; Nierenberg, Lester, & Colacchio, 1987; Zhao et al., 2010). αTP appears to be an exception, being more resistant to hydrolysis and being able to be absorbed as such to some extent (Gianello, Hall, Kennepohl, Libinaki, & Ogru, 2007; Gianello et al., 2005; Zingg, Libinaki, et al., 2010).

As discussed in the following, some of these vitamin E derivatives in their unconverted form, such as αTS, αTP, EPC-K1, or α-TEA, can act as completely novel compounds, are differently transported, have their own effects on cellular signaling and gene expression, reduce inflammation and atherosclerosis, and at higher concentrations exert potent growth-inhibitory and apoptotic activities in a wide spectrum of in vitro and in vivo cancer models (including prostate, breast, neuroblastoma, mesothelioma, etc.) (Birringer et al., 2003; Cerecetto & Lopez, 2007; Neuzil, Dong, et al., 2007; Neuzil, Tomasetti, et al., 2007; Zingg, 2007b). Whereas these molecular and cellular effects are well studied for αTS and α-TEA, less is known for αTP, despite the fact that it has been first synthesized and evaluated almost 80 years ago (Karrer & Bussmann, 1940).

3. EARLY STUDIES WITH α-TOCOPHERYL PHOSPHATE

The phosphorylated form of αT, αTP (Fig. 1), was first synthesized in the early 1940s and later evaluated in several experimental systems (Karrer & Bussmann, 1940; Zhukova, Zakharova, Sokolova, Sarycheva, & Evstigneeva, 1983; Zingg, Meydani, & Azzi, 2010). These early studies mostly focused on the cellular effects in vitro and in animal models and were often based on the usage of crude homogenates and high concentrations of αTP that may exert nonspecific activities. Since the biochemical knowledge and methodology were less advanced at that time, the specificity and relevance of these effects are difficult to evaluate and may need to be reassessed using today's techniques. These early biochemical in vitro studies revealed that αTP modulated many enzymes by various mechanisms as already reviewed (Zingg, Meydani, et al., 2010). Therefore, in the present review, only the more recently described regulatory effects are addressed, and when possible put into context with the observations gained in these early studies.

Some of the early effects were suggested to be the results of Ca^{2+} removal (Ames, 1947; Govier & Jetter, 1948; Nason & Lehman, 1955) since αTP forms an insoluble salt with Ca^{2+} (Ames, 1947), as it does with a number of other cations such as sodium, potassium, lithium, ammonium, diethylamine, and diethanolamine (Solmssen & Lee, 1949). Interestingly, the ability of αTP to form a Ca^{2+} salt and to increase intracellular Ca^{2+} was recently shown to contribute to increase the moisture-retaining ability in stratum corneum after topical application to the skin of hairless mice (Kato & Takahashi, 2012). In these experiments, αTP increased ceramide levels by increasing the expression of the ceramide synthetases and serine palmitoyltransferases (SPTLC1 and SPTLC2), without affecting ceramidases (ASAH1 and ASAH2), leading to induction of keratinocyte differentiation as monitored by increased expression of the differentiation markers transglutaminase 1, cytokeratin 10, involucrin, and loricrin (Kato & Takahashi, 2012). Only about 3.1%–4.8% of αTP was converted to αT within 24 h, suggesting that these effects are due to the intact αTP (Kato & Takahashi, 2012). Interestingly, in the human keratinocyte cell line (NCTC 2544), αT and γT were also able to increase transglutaminase 1 expression involved in terminal keratinocyte differentiation via stimulating peroxisome proliferator receptor gamma (PPARγ) (De Pascale et al., 2006), and in view of the results with mouse skin, it appears possible that this could involve de novo synthesis of αTP or γTP. Nevertheless, since other studies suggested that αTP is rapidly dephosphorylated in the skin (Nakayama, Katoh, Tsuzuki, & Kobayashi, 2003; Nishio et al., 2011), it remains to be proven whether these effects are mostly due to the intact αTP or due to αT.

In addition to the enzymes described early on, more recent research has shown that αTP inhibits glutathione S-transferase omega (Sampayo-Reyes & Zakharyan, 2006), and cAMP and cGMP phosphodiesterases

(Sakai, Okano, Makino, & Tsudzuki, 1976) possibly leading to increased levels of cellular cAMP (Zingg, Azzi, & Meydani, 2017). Further, it stimulates rat liver phenylalanine hydroxylase (PAH) which converts L-phenylalanine to L-tyrosine (Abita, Parniak, & Kaufman, 1984; Brase & Westfall, 1972), although another study reported also inhibition (Zhukova et al., 1983). Moreover, the treatment in human brain microvascular endotheliocytes (HBMEs) with 150 μM αTP reduced their cell size, decreased cellular levels of free radicals, and decreased telomere loss by retaining telomerase activity leading to an increased cellular life span (Tanaka, Moritoh, & Miwa, 2007). Interestingly, in these studies, conversion of αTP to αT was strongly reduced in old cells, suggesting age-dependent regulatory effects of αTP. In contrast to that, telomerase activity is inhibited by αTA in ovarian cancer cells (Bermudez, Ahmadi, Lowell, & Kruk, 2007).

Early animal studies showed that parenteral administration of αTP to rabbits had a higher preventive activity against muscular dystrophy induced by vitamin E deficiency than αT (Eppstein & Morgulis, 1941; Hove & Harris, 1947). Also in rabbits, αTP inhibited the higher rate of oxygen consumption as well as the augmented activity of succinoxidase observed in vitamin E-deficient muscles, whereas αT had no effect (Houchin, 1942; Houchin & Mattill, 1942). However, some of these experiments could not be repeated (Basinski & Hummel, 1947; Hummel & Basinski, 1948) and a direct role of αTP as intermediate in oxidative phosphorylation was not supported experimentally (Scott, 1965). Interestingly, when compared to αTA, αTP was more effective when administered parenterally, but was equally effective when administered orally (Eppstein & Morgulis, 1941), what is possibly due to different rates of hydrolysis of parentally administrated αTA and αTP and specific regulatory effects of intact αTP. At that time, it was proposed that αTP inhibits the enzyme diphosphopyridine nucleotidase (DPN), leading to inhibition of mitochondrial succinoxidase by the promotion of oxaloacetate formation by malic dehydrogenase (Govier & Jetter, 1948), what was in subsequent studies explained by its action as a surface-active anion similar to sodium dodecylsulfate (SDS), preventing the interaction of succinic dehydrogenase with cytochrome *c* (Rabinovitz & Boyer, 1950). In contrast to that, in later studies succinate–cytochrome *c* reductase was found to be activated by αTP and αTS after inhibition by digitonin, whereas the diphosphopyridine nucleotide (DPNH) reductase and cytochrome *c* oxidase were inhibited by αTP and αTS with and without digitonin (Detwiler, Garrett, & Nason, 1966).

Interestingly, recent research into the apoptotic mechanisms of αTP and αTS may offer additional explanations to these early findings with mitochondrial enzymes (Section 4).

After these initial early studies most work on the function and health benefits of vitamin E either focused on the natural tocopherols/tocotrienols or the synthetic racemic mixtures, and animal studies were mostly done with αTA, since this is the analogue that is most often used as a supplement. More recently, research on αTP regained interest after it was shown to occur at low levels naturally in foodstuff such as chocolate and cheese and in animals as well as in human tissues (liver, fat) after developing a novel isolation and detection method (Section 5) (Gianello et al., 2005; Ogru et al., 2003). The natural presence of αTP in the human body prompts a number of questions regarding its formation, transport, metabolism, biological function, and relevance that are currently beginning to be addressed (Negis et al., 2005). These questions and the more recent molecular and cellular activities of αTP as a natural compound and active lipid mediator involved in signal transduction and gene expression capable of exerting regulatory effects at a cellular level have been reviewed (Eskin, 2013; Ghayour-Mobarhan, Saghiri, Ferns, & Sahebkar, 2014; Zingg, Meydani, et al., 2010; Zingg, Meydani, & Azzi, 2012) and are discussed in detail in the following sections.

4. INDUCTION OF APOPTOSIS BY α-TOCOPHERYL PHOSPHATE

Recent research into the apoptotic molecular mechanisms of αTP and αTS may offer additional explanations to the above described early findings on regulatory effects on mitochondrial enzymes. Whereas at low sub-apoptotic concentrations these analogues can influence signal transduction and gene expression and show antiinflammatory and antiatherosclerotic effects, at higher concentrations, these analogues can induce apoptosis by acting as "mitocans" by influencing selectively the mitochondria of cancer cells, and normal cells are less affected (Fig. 2) (reviewed in Kline, Yu, & Sanders, 2004; Kluckova, Bezawork-Geleta, Rohlena, Dong, & Neuzil, 2013; Neuzil, 2002; Neuzil, Dong, et al., 2007; Neuzil, Tomasetti, et al., 2007; Prasad, Kumar, Yan, Hanson, & Cole, 2003). At least for αTS, the apoptotic effects in cancer cells are well studied and involve increased transport of αTS to mitochondria, the inhibition of complex II (succinate dehydrogenase (SDH) or succinate:ubiquinone oxidoreductase (SQR)) in the mitochondrial electron transport chain leading to increased levels of

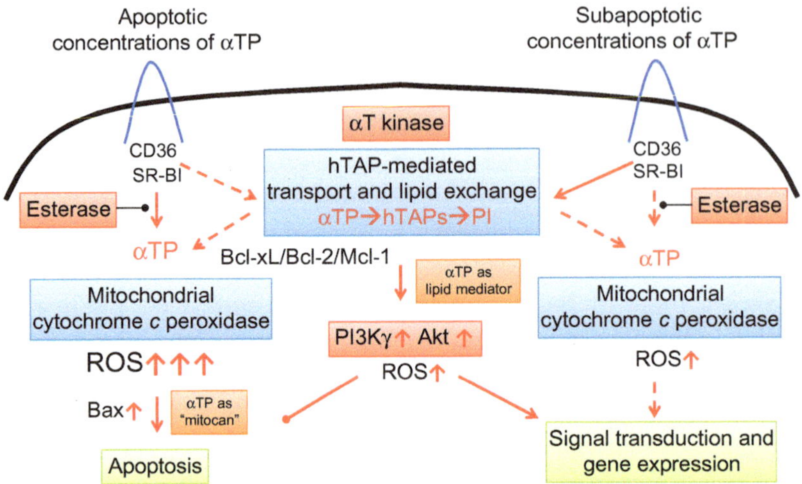

Fig. 2 Model for the induction of apoptosis by α-tocopheryl phosphate. High concentrations of αTP can act as a "mitocan" by converting the mitochondrial cytochrome *c*-mediated electron transport activity into a peroxidase activity leading to the production apoptotic concentrations of reactive oxygen species (ROS). The level of proteins involved in αTP uptake (CD36, SR-BI), distribution (hTAPs), synthesis (αT kinase), hydrolysis (phosphatase/esterase), and response (Bcl-xL/Bcl-2/Mcl-1 and Bax) may determine the cell type-specific sensitivity to αTP. At lower concentrations of αTP, these proteins and the lower levels of ROS may instead of inducing apoptosis modulate signal transduction (e.g., PI3K, Akt) and gene expression (e.g., VEGF).

ROS, and the inhibition of antiapoptotic Bcl-xL/Bcl-2/Mcl-1 proteins by interfering with their interaction with proapoptotic Bax (Neuzil, Dong, Rohlena, Truksa, & Ralph, 2013; Swettenham, Witting, Salvatore, & Neuzil, 2005). Moreover, similar to cardiolipin, apoptotic concentrations of αTP and αTS induce partial unfolding of cytochrome *c* and convert it from an electron-transferring protein into a peroxidase resulting in selective cardiolipin peroxidation and the induction of apoptosis (Fig. 2) (Samhan-Arias, Tyurina, & Kagan, 2011; Yanamala et al., 2014). Interestingly, in early studies and at concentrations that exceed succinic dehydrogenase, αTP has been proposed to act on an associated system such as the cytochrome *c* reductase (Basinski & Hummel, 1947). Cytochrome *c* acts in the mitochondrial electron transfer chain between complexes III and IV, and it appears possible that at subapoptotic concentrations of αTP and αTS (and possibly also at cell physiological concentrations of αTP), its partial unfolding may shift mitochondrial electron transfer into the production of low levels of free radicals relevant for signal transduction and gene expression as observed in cultured cells with αTP (Fig. 2) (Zingg, Libinaki, et al., 2010).

Many synthetic vitamin E derivatives have been synthesized and most of them induce apoptosis at higher concentrations (reviewed in Birringer et al., 2003; Cerecetto & Lopez, 2007; Neuzil, Dong, et al., 2007; Neuzil, Tomasetti, et al., 2007; Zingg, 2007b). For αTP and αTS, their apoptotic activity may also be influenced by modulating their intracellular and mitochondrial concentrations by regulated cellular import, uptake into mitochondria, metabolism, and export. Whereas αT in cell culture is not toxic or apoptotic until high concentrations, αTP is cytotoxic to THP-1 monocytes, murine MG-63 melanoma cells, and HT116 colorectal cancer cells, but in these cells, cytotoxicity and apoptosis are only observed at high concentrations (>50 μM) (Dolfi et al., 2013; Munteanu et al., 2004; Negis, Meydani, Zingg, & Azzi, 2007; Rezk, van der Vijgh, Bast, & Haenen, 2007). No apoptosis but protective effects were reported in HBMEs at a concentration of αTP of up to 150 μM (Tanaka et al., 2007), suggesting that the apoptotic potency of αTP may be influenced by cell-type-specific differences in uptake, metabolism, and cellular retention of the intact molecule by transport proteins and the level and presence of specific target proteins such as Bcl-xL/Bcl-2/Mcl-1/Bax and cytochrome *c* (Neuzil et al., 2013; Swettenham et al., 2005).

Several proteins have been identified capable of influencing the cellular levels and activity of αTP, including the organic anion transporters (OATs) (Negis et al., 2007), the scavenger receptors (CD36, SR-BI) (Zingg et al., 2017), intracellular phosphatases/esterases (Gianello et al., 2005; Kagan, Bakalova, Serbinova, & Stoytchev, 1990; Nakayama et al., 2003; Negis et al., 2005; Nishio et al., 2011), transport proteins such as α-TTP, and the three human tocopherol-associated proteins (hTAP1, hTAP2, hTAP3, or SEC14L2, SEC14L3, SEC14L4, respectively) (Zingg, Azzi, & Meydani, 2015; Zingg, Libinaki, Meydani, & Azzi, 2014). The hTAP proteins bind and increase the cellular uptake of αT, αTP, and αTS (Neuzil, Dong, Wang, & Zingg, 2006; Ni et al., 2005; Panagabko et al., 2003; Zingg et al., 2015, 2014), stimulate the in vitro transport of αT to mitochondria (Zingg, Kempna, et al., 2008), and induce mitochondria-mediated apoptosis by αTS in hTAP1-overexpressing mesothelioma cells, most likely resulting from increased transport to mitochondrial proteins such as Bcl-xL/Bcl-2/Mcl-1, cytochrome *c*, mitochondrial complex II, or possibly ATP synthase (Dong et al., 2008; Kluckova et al., 2013; Neuzil et al., 2006; Neuzil, Dyason, et al., 2007; Ni et al., 2005; Shiau et al., 2006; Swettenham et al., 2005; Wen & Dhar, 2016; Yanamala et al., 2014). The activity may also be influenced by the levels and presence of specific enzymes and cellular structures that mediate the signaling of αTP, as it has been demonstrated for αT and γT with protein kinase C alpha (PKCα) or the phosphatase PHLPP2 (Huang et al., 2013; McCary et al., 2012).

In contrast, a decrease in apoptosis and increased survival by αTP treatment was observed in HBMEs (Tanaka et al., 2007) and in a rat ischemia/reperfusion model, in which αTP normalized reduced phosphorylation of Akt, p44/42 mitogen-activated kinase β (MAPK), p38 MAPKβ, and NF-κB binding, but decreased phosphorylation of Src and MAPKα (Mukherjee, Lekli, Das, Azzi, & Das, 2008; Wu et al., 2017). These events are most likely due to increased resistance of normal cells to αTP when compared to cancer cells as it was observed with αTS (Swettenham et al., 2005), and the efficient conversion to αT and the consequent generally lower subapoptotic concentrations of αTP in tissues in vivo able to modulate protective pathways of signal transduction and gene expression such as PI3K/Akt (Fig. 2). In fact, at subapoptotic concentrations αTP stimulated PI3K/Akt activity and ROS production that in THP-1 monocytes is mostly the result of activation of PI3Kγ, and αT could antagonize αTP in these events (Zingg, Libinaki, et al., 2010). At supapoptotic concentrations of αTP, ROS production was inhibited by the PI3K-inhibitor wortmannin and by mitochondrial inhibitors (rotenone, antimycin, TTFA), but only wortmannin inhibited also Akt activation (Zingg, Libinaki, et al., 2010), suggesting that Akt was not directly activated by ROS produced from the inhibition of mitochondrial electron transfer. Nevertheless, some generation of regulatory ROS from αTP-mediated conversion of cytochrome *c* into a peroxidating enzyme producing H_2O_2 cannot be completely excluded at present time (Fig. 2) (Samhan-Arias et al., 2011; Yanamala et al., 2014). In response to αT, the protective PI3K/Akt pathway has been observed to be both activated (Numakawa et al., 2006; Royer et al., 2009; Vejux et al., 2009) or inhibited after stimulation with oxLDL (Kempna, Reiter, Arock, Azzi, & Zingg, 2004; Lee et al., 2009; Munteanu et al., 2006), indicating that the cellular response may depend on the applied concentration of αTP and degree of interconversion of αT and αTP in a given tissue and cell type.

5. METHODS OF ISOLATION AND DETECTION OF α-TOCOPHERYL PHOSPHATE

The relatively late identification of αTP as a natural form of vitamin E is most likely due to three reasons. First, αTP occurs at much lower concentration than αT: due to the low amounts (nmol/g) present in plants and animal tissues, αTP may have escaped most early less sensitive detection methods such as thin-layer chromatography (TLC) or UV absorbance. Second, αTP escapes classical sample workups commonly used for αT: whereas αT as a hydrophobic molecule segregates efficiently into organic phases with solvents such as hexane, αTP does not. With calculated pK_a values of 6.07 and 1.64, αTP is amphiphilic and is expected to carry two negative charges with physiological conditions and mostly occurs in solution as disodium salt (Birringer, 2010). Third, the increased chemical stability of αTP to hydrolysis when compared to other often extracted analogues of vitamin E such as αTA and αTS that made it necessary to develop new methods of extraction and detection of αTP as intact form.

5.1 Extraction of αTP

In most cases, standard analytical methods for the extraction of vitamin E from complex matrices such as plants or animal tissues involve a hydrolysis step of triglycerides employing sodium or potassium hydroxide (2M NaOH or KOH, respectively) to liberate tocopherols and to ensure that α-tocopheryl esters, such as added αTA or αTS, are converted to free αT and made soluble in organic solvents. However, αTP is not hydrolyzed in this step but simply converts the molecule into the disodium or dipotassium salts. These sodium and potassium salts of αTP are highly water soluble and relatively insoluble in organic solvents, which results in αTP remaining in the aqueous layer after extraction of the free αT with organic solvents. As a solution to this, effective extraction of αTP into the organic phase occurs when the alkaline aqueous solution containing the water-soluble disodium or dipotassium salt is acidified by addition of concentrated HCl, the αTP salt is converted back to the free acid, and αTP is readily extracted into the hexane phase (Gianello et al., 2005). Moreover, when αTP is prepared from tissues, the method uses a preparation of lipids initially extracted with dichloromethane (Gianello et al., 2005; Ogru et al., 2003).

5.2 Detection of αTP

Detection of extracted αTP has been achieved by using TLC, high-performance liquid chromatography (HPLC), electrospray mass spectrometry (ESMS), liquid chromatography MS (LCMS), and LCMS/MS (Gianello et al., 2005; Negis et al., 2007; Ogru et al., 2003; Zingg, Libinaki, et al., 2010). Electrochemical detection or gas chromatography

MS (GCMS) is not applicable, since αTP lacks redox activity and is inert to chemical derivatization (Birringer, 2010). However, fluorescence detection of αTP was used in several studies with a detection limit of 10 ng/mL (Gianello et al., 2005; Negis et al., 2006, 2007; Ogru et al., 2003). For measuring αTP in plasma samples, a rapid protocol involves extraction of αTP by mixing 200 μL of plasma with 20 μL of glacial acetic acid and subsequently with 180 μL of isopropanol. Samples are sonicated for 5 min and then centrifuged and supernatants are analyzed by HPLC and fluorescence detection with peaks at an excitation of 297 nm and emission of 319 nm (Zingg, Libinaki, et al., 2010).

5.3 Detection of Naturally Formed αTP

Whereas these extraction and detection methods gave initial evidence for the natural presence of low amounts of αTP in plants, foodstuff, and tissues, further evidence for the synthesis of αTP from αT and of αT from αTP was required to demonstrate a possibly functional and physiological importance. In cultured cells, the formation of αTP from αT was demonstrated in situ in digitonin-permeabilized human primary cardiac smooth muscle cells (HCA-SMC) and human mast cells (HMC-1) using the synthesis from αT plus radioactive ^{32}P-ATP, a reaction that is stimulated by the tocopherol-associated proteins (hTAP) (Negis et al., 2005; Zingg et al., 2014). In animals and similar also in cultured NIH-3T3-L1 preadipocytes, de novo synthesis of αTP was demonstrated using young, male rats dosed orally for 24 h with ^{14}C-αT, and labeled ^{14}C-αTP was extracted from the liver and analyzed by HPLC, with labeled αTP representing about 1% of the total αTP pool (Gianello et al., 2005). For the dephosphorylation reaction an αTP phosphatase or esterase activity has been detected in several experimental systems (Gianello et al., 2005; Kagan et al., 1990; Nakayama et al., 2003; Negis et al., 2005; Nishio et al., 2011). Although these experiments give evidence for the presence of enzymatic activities able to synthesize αTP from αT and ATP, and to convert it back, to date, no enzymes have been identified or cloned specifically catalyzing these phosphorylation or dephosphorylation reactions.

Based on these experiments, only small amounts of αTP are formed from αT by a putative αT kinase in cultured cells, plasma, and animal tissues, indicating that αTP is not a storage form of vitamin E, but more likely may act as a cofactor and/or an active lipid mediator able to regulate cellular events by modulating signal transduction and gene expression. Similar to other phosphorylated lipids such as phosphatidylinositol phosphate (PIP) and others, the local synthesis of αTP at subcellular sites may generate sufficient amounts to exert its specific cellular action, and the addition of αTP (or PIPs) to the medium of cultured cells or as supplement in the diet may not necessarily reflect exactly how it acts in the cells of the body (reviewed in Zingg, Meydani, et al., 2010). Nevertheless, it appears possible that the cellular levels of αTP are locally increased under certain conditions, e.g., by stimulating import via increased CD36/SR-BI expression during inflammation and atherosclerosis (Zingg et al., 2017), by reducing the activity of the αTP phosphatase during aging (Tanaka et al., 2007), or by stimulating the phosphorylation of αT via hTAP1-mediated lipid exchange as a result of increased hTAP1 expression and transport to the mitochondria in cancer cells (Neuzil et al., 2006; Ni et al., 2005; Zingg, Kempna, et al., 2008), situations in which αTP may act as a natural endogenous lipid mediator (Sections 7 and 8) and at higher concentrations induce apoptosis by acting as an endogenous "mitocan" (Section 4) (reviewed in (Kluckova et al., 2013; Zingg, Meydani, et al., 2010).

6. BIOAVAILABILITY OF α-TOCOPHERYL PHOSPHATE

αTP occurs in low amounts in foods (nmol/g of extracted material), but the phosphorylated forms of β-, γ-, and δ-tocopherols/tocotrienols have so far not yet been detected, because they are either not formed or they are rapidly hydrolyzed and their endogenous level is too low to detect (Gianello et al., 2005; Nakayama et al., 2003; Negis et al., 2005; Ogru et al., 2003; Zingg, Libinaki, et al., 2010). The amounts of αTP in foods are variable and generally much lower than the free αT (Ogru et al., 2003), but food-derived αTP may contribute to the amounts of αTP (~0.1 μg/g) detected in animal tissues (including humans) that has been measured to be about 100 times lower than that of αT (~10 μg/g) (Gianello et al., 2005). Accordingly, supplementation of the diet of rats with αTP results in increased appearance of αTP and αT in liver and adipose tissue (Ogru et al., 2003), and no significant toxicity is observed in several animal models, indicating that the amounts taken up from the diet remain low (Gianello et al., 2007; Libinaki, Ogru, Gianello, Bolton, & Geytenbeek, 2005). In plasma of unsupplemented rats, minipigs, and humans (Zingg, Libinaki, et al., 2010), αTP levels are similar as measured in unsupplemented rabbits, having a basal plasma αTP level of 0.19 μM that increased to 29.4 μM after supplementation with 1.33 g/kg of body weight of αTP for 4 weeks suggesting mechanisms of transport and retention of the intact form (Negis et al., 2006). Supplementation of rats, minipigs, and humans for 12 weeks with PhosphaE (TPM™), a mixture of pure αTP, di-αTP, and αT, increased the plasma αTP level by about 5- to 10-fold (Zingg, Libinaki, et al., 2010). To date it is unknown whether the same route of uptake is taken by αTP as demonstrated for αT or αTA and where exactly and to what degree αTP is hydrolyzed (Brisson et al., 2008; Rigotti, 2007). At least in THP-1 human monocytes/macrophages, cellular uptake of αTP involves OATs (Negis et al., 2007), and/or binding and selective transport by the CD36/FAT scavenger receptor/fatty acid transporter as well as the SR-BI scavenger receptor, whereas uptake of αT occurs more by clathrin-mediated endocytosis, possibly also indicating that the different route of uptake of αTP and αT can influence their regulatory effects on signal transduction and gene expression (Zingg et al., 2017).

Several gene polymorphisms and alternative splicing have been detected in the CD36 and SR-BI genes, with some of them relevant for the plasma and tissue levels of vitamin E, but it remains to be investigated whether they can affect the ability of these receptors to modulate signal transduction and gene expression in response to αT and αTP (Andersen, Lenhard, Whatling, Eriksson, & Odeberg, 2006; Borel & Desmarchelier, 2016; Borel et al., 2007; Cheung et al., 2007; Lecompte et al., 2011; Ma et al., 2004; Zingg, Ricciarelli, Andorno, & Azzi, 2002). A similar regulatory influence of polymorphisms on signaling and gene expression may also occur with other transporters and metabolic enzymes relevant for the vitamin E levels in plasma and tissues such as ATP binding cassette transporter A1 and G1 (ABCA1, ABCG1) (Frikke-Schmidt, Nordestgaard, Jensen, Steffensen, & Tybjaerg-Hansen, 2008; Goncalves, Roi, Nowicki, Niot, & Reboul, 2014; Mustacich, Leonard, Devereaux, Sokol, & Traber, 2006; Oram, Vaughan, & Stocker, 2001), Niemann-Pick-C1-like 1 (NPC1L1) (Narushima, Takada, Yamanashi, & Suzuki, 2008), afamin (Voegele et al., 2002), albumin (Fanali, Fasano, Ascenzi, Zingg, & Azzi, 2013; Oram et al., 2001), cytochrome P450 4F2 (Bardowell, Stec, & Parker, 2010), or α-TTP and hTAP1/2/3 (Borel, Desmarchelier, Nowicki, Bott, & Tourniaire, 2015; Brigelius-Flohe et al., 2002; Doring, Rimbach, & Lodge, 2004; Jialal & Devaraj, 2003; Kono & Arai, 2015; Mocchegiani et al., 2014; Zingg, Azzi, & Meydani, 2008). These gene polymorphisms in vitamin E transport and regulatory genes may ultimately translate into altered risks for diseases in which vitamin E is known to have a preventive role, such as atherosclerosis (Love-Gregory et al., 2014), prostate cancer (Major et al., 2014; Major et al., 2011), nonalcoholic hepatic steatosis (NASH) (Athinarayanan et al., 2014), and metabolic syndrome (Noel et al., 2010), but also explain the difficulty to reach clear evidence for beneficial effects in vitamin E supplementation studies (Borel, Desmarchelier, et al., 2015; Mocchegiani et al., 2014).

7. MODULATION OF SIGNAL TRANSDUCTION AND GENE EXPRESSION BY α-TOCOPHERYL PHOSPHATE

The natural occurrence of low levels of αTP suggests that it is not a storage form but rather may have some role as a transport form, a structural membrane component, a cofactor, or an active lipid mediator or "second messenger" capable of exerting regulatory effects at a cellular level (Zingg, Meydani, et al., 2010). As an intact molecule, αTP has per se no chemical antioxidant activity since it is phosphorylated at the chromanol –OH group, but it may do so after dephosphorylation to αT (Tanaka et al., 2007). Nevertheless, αTP was suggested to reduce oxidative stress by preventing the propagation of free radicals in membranes from one polyunsaturated fatty acid to another by changing the membrane properties, or possibly by interference with their enzymatic generation (Rezk, Haenen, Van Der Vijgh, & Bast, 2004; Tanaka et al., 2007). In contrast to that, a modest increase of ROS production has been detected at subapoptotic concentrations of αTP with a possible role in modulating signal transduction (Zingg, Libinaki, et al., 2010), whereas higher levels of ROS are produced with apoptotic concentrations of αTP (Fig. 2; Section 4; Yanamala et al., 2014). As reviewed in the following, many of the cellular effects observed are unique to αTP and not seen by αT (Table 1), suggesting that they are not the result of formation of αT, but rather the action as an intact molecule that modulates signal transduction and gene expression by nonantioxidant molecular mechanisms. Nevertheless, it cannot be completely excluded at this time that to a certain degree the higher potency of αTP may be due to a better uptake of the molecule and its subsequent intracellular hydrolysis, providing more αT to sensitive regulatory sites. The regulatory effects of αT on signal transduction and gene expression have been recently reviewed (Mocchegiani et al., 2014; Rimbach, Moehring, Huebbe, & Lodge, 2010; Zingg, 2007a, 2015) and are therefore not specifically addressed here.

Table 1 A Selection of Recently Described Cellular Regulatory Effects of αTP

Activity and Measurements	Proposed Mechanisms	Cells	References
Inhibition of cell proliferation and normalization of oxLDL-stimulated CD36 scavenger receptor expression by αTPM™ (a mixture αTP, di-αTP, and αT), and also by pure αTP	Reduction of CD36 mRNA and protein expression, of cellular CD36 surface exposition, increase of CD36 internalization and signaling	THP-1 monocytes and rat aortic smooth muscle cells (RASMCs)	Munteanu et al. (2004); Negis et al. (2007); Ricciarelli, Zingg, and Azzi (2000); Zingg et al. (2017)
Inhibition of CD36 and SR-BI scavenger receptor surface exposition, oxLDL uptake, and bacterial phagocytosis	Binding and transport by CD36 and SR-BI and induction of their internalization	THP-1 monocytes/macrophages	Munteanu et al. (2004); Ricciarelli et al. (2000); Zingg et al. (2015, 2017)
Stimulation of the PI3K and/or PKB/Akt signaling pathway, a pathway that is downregulated by αT	Action as an active "lipid mediator" by hTAP-mediated lipid exchange	THP-1 monocytes and cultured cortical neurons	Kempna et al. (2004); Munteanu et al. (2006); Numakawa et al. (2006)
Stimulation of angiogenesis	Induction of vascular endothelial growth factor (VEGF) expression by stimulation of CD36/PI3Kγ/Akt and low levels of ROS production, CREB, and Hif1α activation	THP-1 monocytes and HEK293 cells	Zingg et al. (2017); Zingg et al. (2014)
Inhibition of cell proliferation and lipid accumulation in preadipocytes, whereas in differentiated adipocytes, lipid uptake is stimulated. These processes may be relevant in the regulation of excess fat accumulation and in prevention of lipotoxicity	Activation of a transcriptional gene set (insulin-induced gene 1 (INSIG1), sestrin 2 (SESN2), and tribbles homolog 3 (TRB3)) in preadipocytes, but inhibition of the same gene set in differentiated adipocytes	THP-1 monocytes, mouse NIH-3T3-L1 preadipocytes and differentiated adipocytes	Lirangi, Meydani, Zingg, and Azzi (2012); Zingg et al. (2017); Zingg et al. (2014)

Inhibition of gap junctional intercellular communication and toxicity	No effect on connexin 43 phosphorylation and cell proliferation	Rat liver epithelial cells (IAR203)	Chaumontet, Bex, Veran, and Martel (2008)
Induction of apoptosis at high concentrations	Possible action as an endogenous "mitocan" (Section 4)	THP-1 monocytes and MG-63 melanoma cells	Munteanu et al. (2004); Negis et al. (2007); Rezk et al. (2007); Yanamala et al. (2014)
Induction of apoptosis, inhibition of cell proliferation, and reduction of colony formation	γTP and δTP had markedly higher effects that αTP	HCT116 colorectal cancer cells	Dolfi et al. (2013)
Protection against oxidative damage	Attenuation of Ca^{2+} influx and via a genomic action	Neurons	Crouzin et al. (2007); de Jesus Ferreira et al. (2005)
Protection against ethyl methanesulfonate (EMS)-induced cell death and lipid peroxidation, protection against lethal doses of gamma irradiation	Possibly by increasing the tissue αT levels	Cultured rat hepatocytes	Fariss (1997)
Protection against ultraviolet-induced damage	Prevention of photodamage, DNA degradation and lipid peroxidation	Cultured mouse skin	Nakayama et al. (2003)
Protection against methamphetamine- and morphine-induced toxicity	Reduction of free radical production	Mouse brain and lung	Ito, Mori, Kanazawa, and Sawaguchi (2007)
Activation of phenylalanine hydroxylase (PAH)	Binding/conformational change	Rat liver	Abita et al. (1984)
Inhibition of glutathione S-transferase omega	Noncompetitive inhibition	In vitro	Sampayo-Reyes and Zakharyan (2006)

8. MOLECULAR MECHANISMS OF α-TOCOPHERYL PHOSPHATE ACTION

The molecular mechanisms by which αT and αTP act as signaling molecules have recently been reviewed and may involve several pathways (Azzi, Meydani, Meydani, & Zingg, 2016; Zingg, 2015; Zingg, Meydani, et al., 2010). For being a signaling molecule, αTP is expected to act differently or with higher potency than αT, and recent research has indeed detected regulatory effects that are unique or more pronounced with αTP (Table 1). Although regulatory effects of αTP on a number of enzymes have been reported early on (Ames & Risley, 1949), they may be nonspecific and occur only in vitro and not explain the increased activity of αTP when compared to αT or αTA observed in some animal studies (reviewed in Zingg, Meydani, et al., 2010). The levels of αTP used in most in vitro experiments are often higher than the levels found in plasma and tissues (Negis et al., 2006); hence, the in vitro cellular response may not necessarily reflect physiological effects. However, the cellular concentrations may substantially differ from the added one, for example, as a result of precipitation of αTP by $CaCl_2$ and $MgSO_4$ in the cell culture media, suggesting that the amount of αTP transferred inside the cells is more relevant and should be measured. Moreover, a higher local concentration of αTP with cellular importance may be achieved in the in vivo situation (as it also occurs with PIPs), for example, by local production of αTP inside the cells by an αT kinase induced only in specific cells or circumstances, by inhibition of an αTP phosphatase/esterase, or by cellular uptake, enrichment, and compartmentalization of αTP at specific subcellular sites. Thus, the biological function and molecular action of αTP in vivo are not completely clear to date, but as discussed in the following recent research suggests that at physiological concentrations αTP can affect enzymes and membrane properties in a specific manner and modulate signal transduction and gene expression by a number of mechanisms.

8.1 Alteration of Membrane Properties by αTP and of Protein–Membrane Interactions

The negative charge of the phosphate group of αTP makes it more similar to phosphorylated messenger lipids and thus possibly better able to modulate specific and nonspecific protein–membrane interactions (reviewed in Zingg, 2006, 2015; Zingg & Azzi, 2006; Zingg, Meydani, et al., 2010).

αTP may play a similar "second messenger" role as known for the phosphorylated forms of PIPs that influence membrane curvature and serve as subcellular address by attracting enzymes such as kinases, phosphatases, and NADPH oxidases to the plasma membrane, leading to their activation/inactivation. In fact, αT and αTP affect signaling and gene expression by influencing the interaction of specific enzymes with the membrane, such as PI3K, Akt, PHLPP1, PP2A, and PKC (Fig. 3) (Huang et al., 2013; Kempna et al., 2004; Munteanu et al., 2006; Zingg, 2015; Zingg, Libinaki, et al., 2010). Moreover, protein domains may exist able to bind specifically to αTP in the plasma membrane, and binding may either activate their own, yet to be identified, signaling cascade or interfere with the signaling of other enzymes.

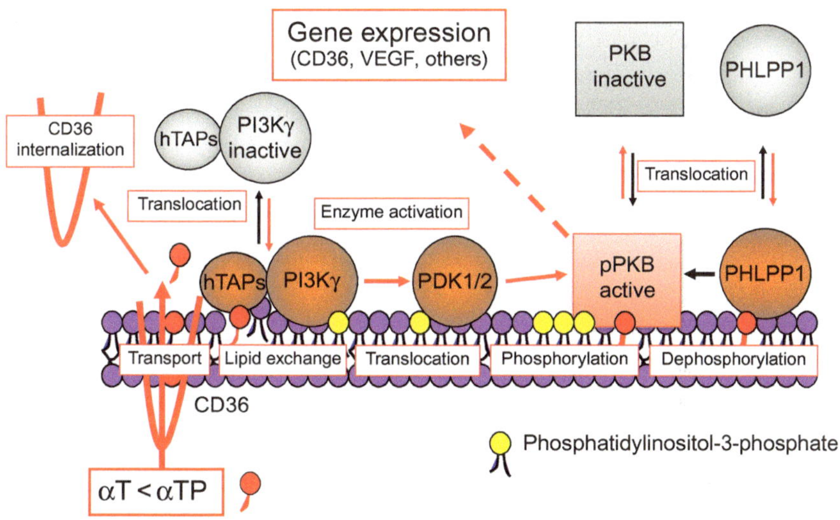

Fig. 3 Model for signal transduction by α-tocopheryl phosphate. Binding of αTP to CD36 leads to αTP uptake and CD36 internalization, whereas αT is less recognized. Within the cells, binding of αTP to hTAPs leads to dissociation of an inactive hTAPs/PI3Kγ heterodimer and/or αTP-mediated lipid exchange enabling phosphorylation of phosphatidylinositol (PI) by PI3Kγ. The higher level of phosphorylated PI (PIP), αT, and αTP leads to translocation and activation of PI3Kγ and PKB/Akt. Activated PKB/Akt becomes dephosphorylated and inactivated by the phosphatase PHLPP1, which translocates to the plasma membrane in response to αT. Note that as a consequence of CD36 internalization, the binding, transport, or signaling of other ligands such as oxLDL, bacteria, virus, parasites, cholesterol crystals, and fatty acids is reduced (Fig. 4). *Adapted from Zingg, J. M. (2015). Vitamin E: A role in signal transduction.* Annual Review of Nutrition, 35, *135–173.*

8.2 Formation of Salts With αTP but Not αT

For long it has been recognized that in contrast to αT, αTP can form a number of different salts (Solmssen & Lee, 1949). In particular, the formation of an insoluble Ca^{2+} salt has been linked to regulatory effects of αTP, either by changing the extracellular concentration of Ca^{2+} or by transporting Ca^{2+} into the cells (Ames, 1947; Govier & Jetter, 1948; Kato & Takahashi, 2012; Nason & Lehman, 1955). Likewise, the formation of an organic salt of αTP with serotonin has been postulated to facilitate transfer into the hydrophobic phase and therefore may readily insert into the plasma membrane (Dombro, Bradham, Campbell, & Woolley, 1961).

8.3 Differential Binding of αTP and αT to Proteins and Promotion of Lipid Exchange

Several proteins such as the three hTAPs can bind both αT and αTP, and binding of αT or αTP differently influences their activity. The in vitro activity of PI3Kγ and PI3Kα is inhibited by hTAP1 indicating the formation of an inactive hTAP1/PI3Kγ heterodimer (Kempna et al., 2003; Ni et al., 2005; Zingg et al., 2015, 2014). The binding of PI to hTAP1 is reversed by αT and more so by αTP leading to stimulation of PI3Kγ activity, suggesting that αT and αTP promote the dissociation of the inactive complex and/or the release of sequestered PI from hTAP1 for subsequent presentation to the kinase by means of a heterotypic lipid exchange mechanism (Fig. 3) (Zingg et al., 2015, 2014). Moreover, in a homotypic lipid exchange reaction, the αT phosphorylation activity is stimulated by recombinant hTAP1 which binds αT and αTP, thus facilitating the transport and presentation of the substrate (αT) to the putative αT kinase and/or the removal of the product (αTP) away from it, thus increasing the enzyme's catalytic turnover (Zingg et al., 2014). In these experiments the αT phosphorylation activity was assayed in the presence of permeabilized cells in situ possibly allowing also heterotypic ligand exchange when encountering the cellular plasma membranes in a similar manner as observed with other SEC14-like proteins (Ghosh & Bankaitis, 2011). The differential modulation of the PI3K/Akt signal transduction pathway via hTAP-mediated lipid exchange may be at the basis for the observed differential regulatory effects of αT and αTP on gene expression, e.g., by transporting these ligands to specific enzymes such as cytosolic PI3Kγ, or to membrane sites accessible for regulating PI3K/PKB/PHLPP1 (Huang et al., 2013, 2009; Munteanu et al., 2006; Zingg, Libinaki, et al., 2010).

Analogous lipid-exchange mechanisms with PI/PC and PI/αT were recently visualized with the crystal structure of the closest SEC14p homolog—the *Saccharomyces cerevisiae* Sfh1a (Schaaf et al., 2008), as well as proposed for human α-TTP, respectively (Bankaitis, Mousley, & Schaaf, 2010; Ghosh & Bankaitis, 2011; Ile, Schaaf, & Bankaitis, 2006; Kono & Arai, 2015; Kono et al., 2013). The cellular response to αT and αTP may depend on whether they enable catalysis of PI by PI3K or PI4K, the conversion of αT to αTP by a putative αT kinase, or vice versa by an αTP phosphatase, or influence the membrane composition in a given tissue and cell type (Zingg, Libinaki, et al., 2010). It remains to be investigated whether αT/αTP can influence other enzymes as well, such as PI4K, phospholipases, squalene epoxidase, fatty acid synthase, and choline-phosphate cytidylyltransferases via the modulation of hTAPs-mediated lipid exchange (Kempna et al., 2003; Mokashi, Singh, & Porter, 2005; Ni et al., 2005; Panagabko et al., 2003; Shibata et al., 2001).

8.4 Differential Uptake, Transport, and Metabolism of αTP and αT

As a hydrophobic molecule αT occurs mainly in hydrophobic membranes, bound to transport proteins, lipoproteins (chylomicrons, VLDL, LDL, HDL), afamin, albumin or in the gastrointestinal tract emulsified in lipid droplets, and it is therefore taken up via lipoprotein receptors or by endocytosis. In contrast, αTP is water soluble and therefore may follow a different route of uptake and end up at different subcellular sites. Accordingly, in cultured monocytes, αTP appears to be taken up similar to other fatty acids by binding to CD36 and SR-BI that possibly channel it via two tunnels to the cell membrane and/or by OATs (Fig. 3), whereas αT is more taken up by clathrin-dependent endocytosis (Negis et al., 2007; Zingg et al., 2017). Thus, the different regulatory effects of αTP and αT on signal transduction and gene expression may be the consequence of their different cellular uptake, their different molecular action on different subcellular targets, and their different chemical ability to scavenge free radicals, and possibly to the intracellular hydrolysis of αTP by phosphatases/esterases that may lead to higher levels of intracellular αT at specific sites when compared to treatment directly with αT (Gianello et al., 2005; Kagan et al., 1990; Nakayama et al., 2003; Negis et al., 2005). However, at least in THP-1 cells, only 10% of the added αTP was hydrolyzed and even an αT concentration 10-fold higher than that produced by αTP hydrolysis was not effective in inhibiting cell proliferation (Negis et al., 2007). Therefore, a preferential direct interaction of αTP with specific proteins and cellular structures may occur, as described with αTS binding to mitochondrial proteins such as Bcl-xL/Bcl-2/Mcl-1, cytochrome *c*, mitochondrial complex II (succinate dehydrogenase (SDH) or succinate:ubiquinone oxidoreductase (SQR)), or ATP synthase (Dong et al., 2008; Kluckova et al., 2013; Neuzil et al., 2006; Neuzil, Dyason, et al., 2007; Ni et al., 2005; Shiau et al., 2006; Swettenham et al., 2005; Wen & Dhar, 2016; Yanamala et al., 2014).

8.5 Differential Reduction of CD36 and SR-BI Surface Exposition by αTP and αT

The CD36 scavenger receptor binds and imports a variety of ligands at the cell surface and mediates their regulatory effects on signal transduction and gene expression (Fig. 3). Therefore, compounds that can influence CD36 exposition at the cellular surface may modulate a number of subsequent cellular events in response to these ligands. So far, the reduction of CD36 by αT and more so αTP has been linked to a number of cellular events, including the inhibition of oxLDL uptake and of oxLDL-induced PKB/Akt activation, the accumulation of atherogenic lipids and foam cell formation, reduction of cell proliferation of human THP-1 monocytes and rat aortic smooth muscle cells (RASMCs), and inhibition of bacterial phagocytosis (Devaraj, Hugou, & Jialal, 2001; Munteanu et al., 2006, 2004; Ricciarelli et al., 2000; Zingg et al., 2017; Zingg, Libinaki, et al., 2010). Mechanistically, the binding of αT and αTP to CD36 may trigger conformational changes and the internalization and disappearance of CD36 from the cell surface, thus lowering the cellular response to other CD36 ligands (Jay, Chen, Paz, Hung, & Hamilton, 2015; Zingg, 2015; Zingg et al., 2017). The recent crystal structure of the CD36 ectodomain identified two fatty acids (palmitic and stearic) within two tunnels in the protein (Hsieh et al., 2016; Neculai et al., 2013; Pepino, Kuda, Samovski, & Abumrad, 2014), and αTP may be channeled via these tunnels or possibly binds and blocks them. As a consequence of αTP binding the response to a number of other ligands of CD36 may be affected, including ligands such as fatty acids (Pohl, Ring, Korkmaz, Ehehalt, & Stremmel, 2005), ceramide (Luan & Griffiths, 2006), myristic acid (Isenberg et al., 2007), and anionic phospholipids such as phosphatidylserine, phosphatidylinositol, sphingomyelin, phosphatidylcholine, and lysophosphatidic acid (Engelmann & Wiedmann, 2010; Rigotti, Acton, & Krieger, 1995).

At the molecular level, the mechanisms regulating internalization and reduction of cell surface exposition of CD36 after binding of αTP may involve ligand-induced phosphorylation and conformational changes (Jay et al., 2015), CD36 ubiquitination (Smith, Su, El-Maghrabi, Stahl, & Abumrad, 2008), additional proteins (such as dynamin (Sun et al., 2007), tropomyosin (Bondeva et al., 1998; Naga Prasad, Jayatilleke, Madamanchi, & Rockman, 2005), or actin (Collins et al., 2009)), and/or signal transduction after binding to CD36 (reviewed in Febbraio & Silverstein, 2007). The degree of exposition of CD36 at the cellular surface may also be influenced by the rate of transfer of CD36 to the plasma membrane and of CD36 maturation. In skeletal muscle, upregulation of FOXO1 plays a role in increasing expression and membrane exposition of CD36 (Bastie et al., 2005; Nahle et al., 2008), whereas preliminary results suggest that overexpression of FOXO3a also increases CD36 exposition at the membrane.

9. ANIMAL AND HUMAN STUDIES WITH α-TOCOPHERYL PHOSPHATE

Several animal studies with αTP were performed early on and some of them showed increased activity of αTP when compared to αTA (Table 2) (reviewed in Zingg, Meydani, et al., 2010). In rabbits, parenteral administration of αTP had a higher preventive activity against muscular dystrophy induced by vitamin E deficiency than αT (Eppstein & Morgulis, 1941; Hove & Harris, 1947; Karrer & Bussmann, 1940). Possible reasons for the higher activity of αTP were investigated, and it was found that αTP inhibited the augmented activity of succinoxidase as well as the higher rate of oxygen consumption observed in vitamin E-deficient muscles, whereas αT had no effect (Houchin, 1942; Houchin & Mattill, 1942). Interestingly, in contrast to αTP, αTA was not effective when administered parenterally, but was equally effective when administered orally (Eppstein & Morgulis, 1941), suggesting differences in uptake and hydrolysis of αTP and αTA as well. In more recent animal studies, supplementation of the diet of rats with αTP increased the deposition of αTP and αT in liver and adipose tissue, without exerting any significant toxicity in several animal models (Libinaki et al., 2005; Ogru et al., 2003). In a model of hypercholesterolemic rabbits and in apoE knockout mice, atherosclerosis progression and inflammation is more efficiently prevented by αTP when compared to αT, most likely resulting from reduction of CD36 scavenger receptor overexpression and of inhibition of several proinflammatory cytokines (IL1β, IL6, IL8, PAI1, TNFα, and CRP) (Libinaki et al., 2010, 2017; Negis et al., 2006).

Table 2 A Selection of Regulatory Effects of αTP Observed in Animal Studies

Animal Studies and Measurements	Proposed Mechanisms	Animal Treatment	References
Increase of prothrombin clotting time	Antiproteolytic activity of αTP	Intraperitoneal injection of αTP to rats; in vitro plasma coagulation time	Zierler, Grob, and Lilienthal (1948)
Metabolism of carbohydrates and suppression of glycogenolysis in skeletal muscle	Modulation of phosphoglucomutase system	Injection into rat skeletal muscle	Quaife (1954); Zierler, Andres, Levy, Anderson, and Lilienthal (1953); Zierler, Folk, and Lilienthal (1953); Zierler, Levy, and Lilienthal (1953)
Prevention of muscular dystrophy	Not analyzed	Vitamin E-deficient rabbits	Eppstein and Morgulis (1941); Hove and Harris (1947)
Induction of hippocampal long-term potentiation	Antioxidant	Tissue slices isolated from guinea pig hippocampal CA1 pyramidal neurons	Xie and Sastry (1993)
Regulatory effects on synaptic glutaminergic and GABAergic transmission	Differential interaction with the cannabinoid receptor (CB1R) system by αT and αTP in a nonantioxidant manner, inhibition of Forskolin-induced Erk1/2 phosphorylation, reversion of hypothermia induced by CB1R agonist	Primary neuronal cultures from rat hippocampus	Crouzin et al. (2011)
Protection against ultraviolet-induced damage	Conversion to αT and antioxidant	Mouse skin	Nakayama et al. (2003)
Protection against methamphetamine- and morphine-induced toxicity	Antioxidant	Mice	Ito et al. (2007)

Prolongation of the therapeutic action of barbiturates, prevention of the agitation phase, and prolongation of thiopental-induced depression	Not analyzed	Mice and rats	Giarman, Bowers, Quie, and Hampton (1954); Mirsky and Giarman (1955)
Prevention of atherosclerosis	Reduction of CD36 overexpression, foam cell formation, and inhibition of inflammatory cytokine production	Hypercholesterolemic rabbits, apoE knockout mice	Libinaki, Vinh, Tesanovic-Klajic, Widdop, and Gaspari (2017); Negis et al. (2006)
Reduction of inflammatory cytokines	Reduction CD36 at the cellular surface and of lipid uptake, possibly reduced NLRP3 inflammasome activation, and of IL1β expression in response to CD36-facilitated cholesterol crystal formation	Hypercholesterolemic rabbits	Kagan and Horng (2013); Libinaki et al. (2010); Negis et al. (2006); Sheedy et al. (2013)
Protection against myocardial and brain ischemia/reperfusion injury	Modulation of signal transduction and gene expression	Heart and brain of rats	Mukherjee et al. (2008); Wu et al. (2017)
Potentiation of cell death and lipid peroxidation after ischemia/reperfusion injury	Increased intracellular free $[Ca^{2+}]$ in an age-dependent manner, increased production of ROS, increased intracellular $[Ca^{2+}]$, lipid peroxidation, and cell death	Immature cerebellar granule cells subjected to ischemia/reoxygenation	Dyatlov, Makovetskaia, Leonhardt, Lawrence, and Carpenter (1998)

10. POSSIBLE MECHANISMS OF DISEASE PREVENTION BY α-TOCOPHERYL PHOSPHATE

As outlined in the following, the regulatory effects on signal transduction and gene expression in vitro and in animal studies and in particular the inhibition of CD36 cell surface exposition by αT, and more so by αTP and EPC-K1 may affect a number of cellular events of ligands that require CD36 for their action and therefore influence a number of cellular processes and diseases including inflammation, atherosclerosis, insulin resistance, diabetes, angiogenesis/vasculogenesis, infections with virus, bacteria, and parasites, cancer, and metastasis (Fig. 4) (Azzi et al., 2016; Mocchegiani et al., 2014).

10.1 Atherosclerosis

As an antioxidant in the lipid phase, vitamin E can prevent the formation of proatherogenic oxidized low-density lipoproteins (oxLDL) and thus prevent atherosclerosis. In addition, vitamin E (αT) was one of the first compounds reported to affect the uptake and signaling of the CD36 ligand oxLDL by reducing CD36 at the cellular surface and by inhibiting oxLDL-induced PKB/Akt activation, thus preventing the accumulation of atherogenic lipids and foam cell formation in monocytes/macrophages and vascular smooth muscle cells (Devaraj et al., 2001; Munteanu et al., 2006; Ricciarelli et al., 2000). Interestingly, the phosphorylated form of αT, αTP, was more potent than αT in reducing oxLDL uptake, as well as in reducing the proliferation of human THP-1 monocytes and RASMCs, events that most likely involve the increased ability of αTP to reduce CD36 surface exposition and/or of CD36-mediated signaling (Munteanu et al., 2004; Zingg et al., 2017; Zingg, Libinaki, et al., 2010). Moreover, since common bacteria, such as the periodontal pathogen *Porphyromonas gingivalis*, or *Chlamydia pneumoniae*, and viruses such as the cytomegalovirus have been suggested to accelerate the development of atherosclerosis (Gibson et al., 2004; Gibson, Yumoto, Takahashi, Chou, & Genco, 2006; Grayston, Kuo, Campbell, Wang, & Jackson, 1997; Gura, 1998; Lalla et al., 2003), it appears possible that downregulation of CD36 or other scavenger receptors by αTP may prevent their atherogenic action in the vascular wall in a similar manner as it was observed in hypercholesterolemic rabbits and apoE knockout mice (Libinaki et al., 2010, 2017; Negis et al., 2006).

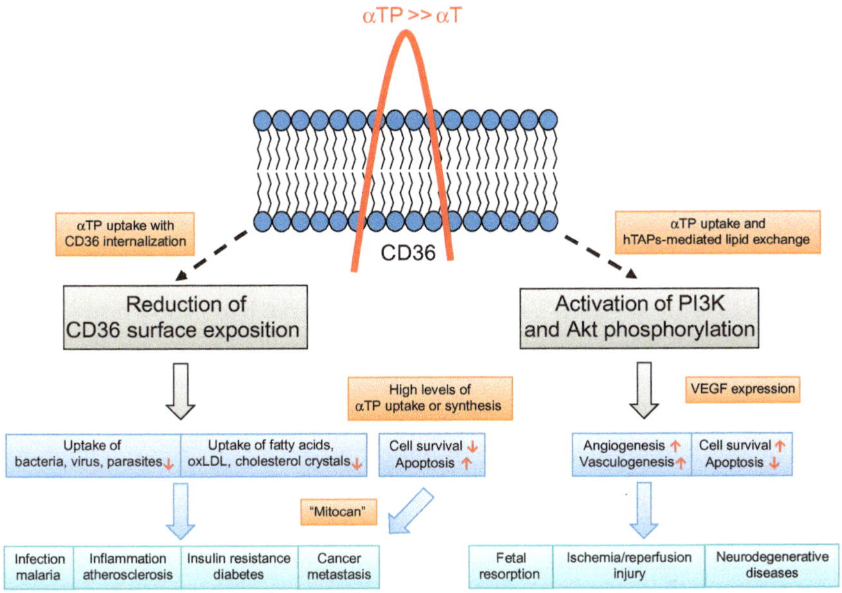

Fig. 4 Model for the prevention of diseases by modulation of CD36 activity by sub-apoptotic concentrations of α-tocopheryl phosphate. αT and more so αTP lead to reduction of CD36 cell surface exposition, by triggering a conformational change leading to induction of CD36 internalization. The consequent reduced response to other CD36 ligands such as bacteria, virus, parasites, fatty acids, oxLDL, and cholesterol crystals may lead to a reduction of infection, inflammation, atherosclerosis, insulin resistance, lipotoxicity, cancer, and metastasis. Uptake of αTP by CD36 leads also to activation of PI3K/Akt signal transduction via hTAP-mediated lipid exchange and increased expression of the vascular endothelial growth factor (VEGF). The consequent increased angiogenesis, vasculogenesis, and cell survival may lead to prevention of fetal resorption, ischemia/reperfusion injury, and neurodegenerative diseases. At high concentrations, αTP may prevent cancer and metastasis by acting as "mitocan" or by inducing lipotoxicity by reducing CD36-mediated fatty acids transport into the mitochondria and their subsequent metabolism (Fig. 2).

10.2 Inflammation

In hypercholesterolemic rabbits, αTP had antiinflammatory effects as evidenced by a reduction of several inflammatory cytokines (IL1β, IL6, IL8, PAI1, TNFα, and CRP), and the prevention of inflammation may contribute to the antiatherosclerotic effects (Libinaki et al., 2010; Negis et al., 2006). Since CD36 has been implicated in CD36-facilitated cholesterol crystal formation and subsequent NLRP3 inflammasome activation, the reduction of CD36 at the cellular surface and of lipid uptake by αTP may contribute to the antiinflammatory effects of αTP at the initiating early stage (Libinaki et al., 2010; Negis et al., 2006; Sheedy et al., 2013).

10.3 Cancer and Metastasis

A possible role of vitamin E in cancer prevention has been investigated in a number of studies, but so far no clear consensus has been reached (Negis, Zingg, Libinaki, Meydani, & Azzi, 2009). At the molecular and cellular level, vitamin E may inhibit tumor formation by influencing signal transduction and gene expression, inhibiting cell proliferation, preventing lipid peroxidation and of DNA damage, inducing apoptosis, and some of the natural and synthetic tocopherol and tocotrienol analogues and derivatives including γT, δT, αTP, αTS, α-TEA, and EPC-K1 have been shown to be more active in modulating some of these events (reviewed in Birringer et al., 2003; Cerecetto & Lopez, 2007; Neuzil, Dong, et al., 2007; Neuzil, Tomasetti, et al., 2007; Zingg, 2007b). Interestingly, it was recently shown that metastatic cells upregulate CD36 to import more fatty acids as energy source and that inhibition of CD36 induces lipotoxicity by accumulating unmetabolized lipids since they are not taken up into mitochondria and metabolized (Pascual et al., 2017). Based on these results the reduction of CD36 exposition at the cellular/mitochondrial surface by αTP may have anticancer and antimetastatic activities. In fact, at least in THP-1 monocytic leukemia cells, αTP at nontoxic doses lowers the intracellular lipid levels (Zingg et al., 2017), and it remains to be investigated whether higher toxic αTP doses involve lipotoxic mechanisms as well as the apoptotic action as a "mitocan" (Section 4).

In HT-1080 human fibrosarcoma and in DUMS-16 human dermal fibroblasts, αTP exerts antiinvasive activities through a decrease in cell motility, repression of adhesion to the extracellular matrix, reduction of free radicals production, cell morphological changes, and especially alteration of cortactin distribution in invadopodia that dispatch signal transduction for tumor invasion and the formation of vesicles (Saitoh, Yumoto, & Miwa, 2009). Interestingly, also EPC-K1 was able to inhibit metastasis of rat hepatocellular carcinoma cells (AH-60C) to the liver and reduced the production of lipid peroxides (Hirai et al., 1997), what in view of the regulatory effects of EPC-K1 on CD36 surface exposition and lipid uptake, may also involve inhibition of CD36 (Zingg et al., 2017). Increased uptake of free fatty acids via CD36 also promotes epithelial–mesenchymal transition relevant for the progression of a number of tumor cells such as hepatocellular carcinoma (Nath, Li, Roberts, & Chan, 2015) and high glucose-induced renal tubular epithelial cells (HK-1) (Hou et al., 2015), events that possibly can be inhibited by αTP via downregulation of CD36.

10.4 Infection

CD36 recognizes a variety of bacteria and interacts with bacterial lipids together with Toll-like receptors 2 and 6, leading to the modulation of signal transduction important for native immunity and for inflammatory processes in response to bacterial pathogens (Areschoug & Gordon, 2009; Baranova et al., 2008; Canton, Neculai, & Grinstein, 2013). In cultured cells, CD36 mediates uptake and phagocytosis of *Escherichia coli* and *Staphylococcus aureus* (Philips, Rubin, & Perrimon, 2005), and the activation of the PI3K/Akt signaling pathway is important for the internalization of these bacteria by endothelial (Oviedo-Boyso et al., 2011) and epithelial cells (Wang et al., 2013). In THP-1 monocytes/macrophages, αTP triggered a decrease of surface CD36 and lowered the uptake of fluorescent *S. aureus* bioparticles (Munteanu et al., 2004; Zingg et al., 2017; Zingg, Libinaki, et al., 2010), which may contribute to decreased phagocytosis previously observed in guinea pigs with high concentrations of vitamin E (De la Fuente, Carazo, Correa, & Del Rio, 2000). The inhibitory effect of αTP on phagocytosis may have some benefits against infection with facultative intracellular pathogens such as *S. aureus*, by facilitating their killing by antibiotics by keeping them extracellular before they are able to evade the immune system (Fraunholz & Sinha, 2012), as demonstrated in a mouse sepsis model (Leelahavanichkul et al., 2012), and in wounds infected with methicillin-infected *S. aureus* (Pierpaoli et al., 2011; Provinciali et al., 2011). However, a certain level of CD36 expression may be important for a correct response to bacteria, since mice with a homozygous CD36 mutation render them hypersusceptible to *S. aureus* infection (Stuart et al., 2005), albeit this response to infection also depends on the site of infection (Blanchet, Jouvion, Fitting, Cavaillon, & Adib-Conquy, 2014) and timing (Sharif et al., 2013). Since CD36 is also a major receptor for the malaria parasite *Plasmodium falciparum* by recognizing erythrocytes infected with *P. falciparum*, the reduction of CD36 by αTP may influence the susceptibility to severe and fatal malaria infection (Franke-Fayard et al., 2005; Patel et al., 2007). Moreover, since the uptake and replication of hepatitis C virus (HCV) are modulated by CD36 (Cheng et al., 2016), SR-BI (Cocquerel, Voisset, & Dubuisson, 2006), PI3K/PI4K (Bishe, Syed, & Siddiqui, 2012), and SEC14L2/TAP1 (Saeed et al., 2015), αT and αTP may modulate the HCV infectious cycle by influencing these proteins.

10.5 Prevention of Lipotoxicity and NASH

Internalization of CD36 by vitamin E and reduction of cellular lipid accumulation may contribute to the protective effects of αT and αTP against lipotoxicity occurring during high-fat-induced development of atherosclerosis and NASH (Munteanu & Zingg, 2007; Negis et al., 2006; Ozer et al., 2006; Ricciarelli et al., 2000; Sanyal et al., 2010). Regulatory effects of αTP on CD36 and plasma and tissue lipid levels may be cell type and organ specific as demonstrated in studies in which the level of expression of CD36 has been modified, e.g., by overexpressing CD36 or CD36 knockout mice (Bonen et al., 2007; Koonen et al., 2007). In hepatocytes, CD36 signaling resulted in inhibition of cholesterol synthesis by HMG-CoA reductase via activation of the PPARγ/PGC-1α pathway and INSIG1/2 expression with consequent reduction of cellular neutral lipid and cholesterol levels (Rodrigue-Way et al., 2013), events that may contribute to the preventive effects of vitamin E against NASH (Sanyal et al., 2010). Likewise, in THP-1 monocytes and in preadipocytes, αTP induced INSIG1 leading to decreased lipid accumulation, whereas in differentiated adipocytes it was inhibited, and it remains to be investigated whether this occurs via engaging CD36 receptor activity (Lirangi et al., 2012; Zingg et al., 2017; Zingg, Libinaki, et al., 2010). Only a slight reduction of plasma lipids was observed in mice supplemented with αTP (Libinaki et al., 2010), possibly since most of αTP is dephosphorylated during uptake (Nishio et al., 2011), and it remains to be resolved whether stronger effects can be achieved if more intact αTP could be made bioavailable.

10.6 Insulin Resistance, Metabolic Syndrome, and Diabetes

A high-fat diet has been causally linked with insulin resistance as a result of accumulation of intracellular free fatty acids which impair the signal transduction from the insulin receptor to Akt mainly as a consequence of activation of PTEN, p38, JNK, PKC, NF-κB, and ATF-2 (Boucher, Kleinridders, & Kahn, 2014; Kennedy & Kashyap, 2011; Liang, Han, Senokuchi, & Tall, 2007; Wang et al., 2006). CD36 deficiency increases insulin sensitivity in muscle but induces insulin resistance in the liver in mice, and insulin resistance was the result of the uptake of fatty acids into cells and tissues rather than the presence of fatty acids in plasma (Goudriaan et al., 2003). Thus, since αT and more so αTP reduce CD36 surface exposition and intracellular lipid accumulation, it may prevent the development of insulin resistance. In addition to that, αTP may alleviate insulin resistance by activating the CD36/PI3K/Akt-signaling pathway via hTAP/SEC14L2-mediated lipid exchange in an insulin/insulin receptor-independent manner (Zingg et al., 2014; Zingg et al., 2015, 2017). In fact, a high dose of vitamin E transiently reduced insulin resistance and associated parameters in overweight subjects (Manning et al., 2004).

10.7 Modulation of Angiogenesis and Vasculogenesis

Binding and internalization of αTP by CD36 increase PI3Kγ/Akt activity via tocopherol-associated protein 1 (hTAP1/SEC14L2)-mediated lipid exchange and lead to increased expression of the vascular endothelial growth factor (VEGF) (Zingg et al., 2017, 2014; Wu et al., 2017). The induction of VEGF expression by αT and more so by αTP suggests that it may be involved in angiogenesis and vasculogenesis, important for a number of diseases in which VEGF is reduced including atherosclerosis, diabetes, neurodegenerative disease, wound healing, and ischemia/reperfusion injury, and it may prevent fetal resorption that occurs during vitamin E deficiency and defines the essential activity of vitamin E (reviewed in Zingg, Libinaki, et al., 2010; Zingg, Meydani, et al., 2010; Zingg et al., 2012).

10.8 Depression and Neurodegenerative Disease

In tissue slices isolated from guinea pig hippocampal CA1 pyramidal neurons, an induction of hippocampal long-term potentiation by αTP was seen, possibly indicating some effects on memory and learning (Xie & Sastry, 1993). αTP also prolonged the therapeutic action of barbiturates, prevented the agitation phase (Giarman et al., 1954), and exerted a pronounced prolonging effect on thiopental-induced depression in mice and rats, but not in rabbits and dogs (Mirsky & Giarman, 1955). Moreover, some of the neurological symptoms of vitamin E deficiency, like the increased permeability of capillaries during encephalomalacia in the chick, could be the result of impaired formation of a vascular network due to low VEGF levels in the brain that possibly could be prevented by αTP (Dam & Glavind, 1942). The antidepressant-like effects observed with αT or αTP treatment (Lobato et al., 2010; Manosso et al., 2013) could also be explained by induction of VEGF (Nowacka & Obuchowicz, 2011), or by increasing serotonin production by stimulating tryptophan hydroxylase 1 (Tph-1) expression (Yano et al., 2015). By similar mechanisms, the modulation of VEGF expression and angiogenesis by αTP (Zingg et al., 2017, 2014;

Wu et al., 2017) could play a role in other diseases, such as age-dependent macular degeneration or Alzheimer's disease (Ohno-Matsui, 2011). A reduction of CD36 by αTP may reduce astrocyte activation and astroglial scar formation as it occurs during a stroke (Bao et al., 2012).

11. POTENTIAL USES OF α-TOCOPHERYL PHOSPHATE AND OF ITS ANALOGS AND DERIVATIVES

As an active lipid mediator affecting signaling and gene expression with cellular effects on inflammation, lipid metabolism, phagocytosis, and apoptosis, increased delivery of αTP and αT to tissues may have applications for prevention and treatment of diseases ranging from atherosclerosis, dyslipidemia, diabetes, inflammation, infection, neurodegenerative diseases, skin diseases, and various cancers. However, since similar to αTA and other αT esters, αTP is to a large degree converted to αT during uptake via the gastrointestinal tract and less in topical applications to skin, αTP may act mostly as a precursor of αT (Section 2) (Ben-Shabat, Kazdan, Beit-Yannai, & Sintov, 2013; Nakayama et al., 2003; Negis et al., 2007; Nishio et al., 2011). When high concentrations of intact αTP can be achieved, it may be more toxic than intact αTA due to its more-pronounced properties as a surface-active anion similar to SDS (Ben-Shabat et al., 2013; Rabinovitz & Boyer, 1950). Nevertheless, the amphiphilic water-soluble characteristics may also be of advantage in certain situations, e.g., in diseases with mutated carrier proteins or chaperones such as α-TTP, hTAP1/2/3, NPC1L1, or lipoproteins that are required as transporter for the lipophilic αT (Kono & Arai, 2015; Ulatowski & Manor, 2013). A certain increase of delivery of intact αTP to cells in tissues may therefore be useful and may be accomplished by formation of nanocarrier complexes of αTP able to enhance the delivery of αTP to and across cell membranes. In this approach, nanocarriers may act as chemical vehicles for molecules such as αT and αTP and act by protecting them and enhancing their transport, exchange, and metabolic conversion, and by stimulating enzymes (e.g., PI3Kγ)-dependent lipid reactions, functions that are otherwise performed by lipoproteins and LTP (e.g., α-TTP, NPC1L1, or hTAP1/2/3) (Jay & Hamilton, 2016; Kono & Arai, 2015; Kono et al., 2013; Zingg et al., 2015, 2014). Alternatively or in combination with supplementation, higher tissue levels and activity of αTP may be achieved by direct injection as demonstrated in early studies (Section 3), or possibly by stimulating the αT kinase or inhibiting the αTP phosphatase/esterase, e.g., via stimulation of hTAP1/2/3-mediated lipid exchange (Fig. 5) (Zingg et al., 2015, 2014). For applications in biotechnology, it can also be envisioned that the facilitation of lipid exchange by hTAP1/2/3 and/or nanocarriers can serve to catalyze lipid reactions not only at nanoscale for lipid transfer and signaling in cells but also on a larger scale.

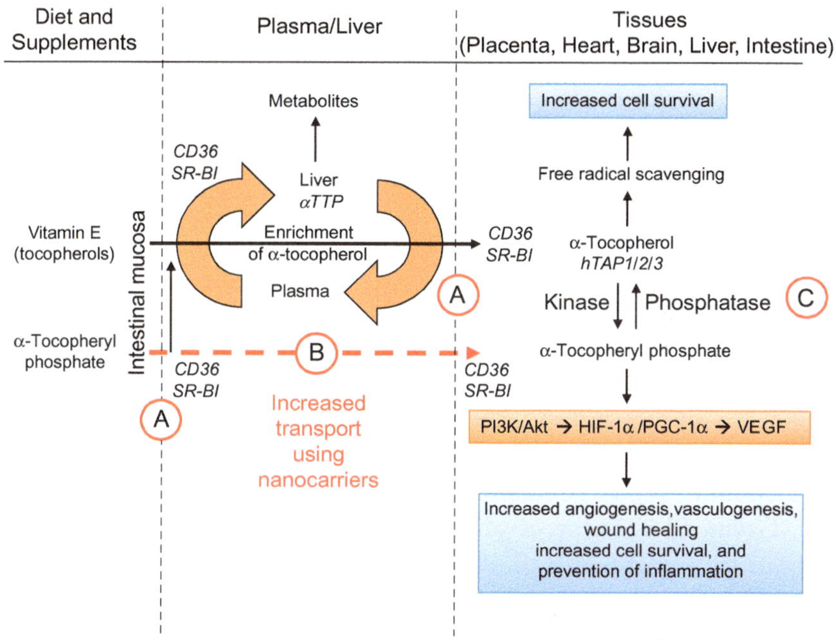

Fig. 5 Transport and distribution of vitamin E and α-tocopheryl phosphate. The four tocopherols are taken up in the intestine and only the α-tocopherol analogue is selectively enriched in the plasma by the liver α-TTP, whereas the other analogues are metabolized and secreted. In tissues, αT either acts as a scavenger of reactive oxygen species (ROS) and increases cell survival, or it may become phosphorylated locally to αTP which acts as an active lipid mediator influencing signal transduction and gene expression. αTP can stimulate VEGF expression, possibly leading to increased cell survival, angiogenesis, vasculogenesis, and wound healing, and to a reduction of inflammation. In addition to direct injection/infusion, the concentration of αTP in tissues can hypothetically be increased by stimulating expression of transport proteins such as CD36 and/or SR-BI (A), by means of nanocarriers (B), or by stimulating the αT kinase, e.g., via hTAP1/2/3-mediated lipid exchange or by inhibiting the αTP phosphatase/esterase (C). At high concentrations, αTP may act similar to αTS as endogenous "mitocan" able to induce apoptosis in cancer cells (Fig. 2). *Adapted from Zingg, J. M., Meydani, M., & Azzi, A. (2012). Alpha-tocopheryl phosphate—An activated form of vitamin E important for angiogenesis and vasculogenesis?* Biofactors, 38*(1), 24–33.*

Vitamin E derivatives-based nanomedicines may have high potential for drug delivery, in particular to improve the water solubility of hydrophobic drugs and to enhance their therapeutic efficiency (reviewed in Duhem, Danhier, & Preat, 2014). So far, strategies to use vitamin E derivatives were mostly focused on αTS or the water-soluble vitamin E polyethylene glycol-succinate that can act as prodrugs, micelles, liposomes, and nanoparticles; on ESeroS-GS containing αT and glutathione with succinic acid as the core; or on sugar-based amphiphilic macromolecules complexes with αT that acted as antiatheroinflammatory nanotherapeutics (Duan et al., 2011; Lewis et al., 2016; Wei et al., 2003; Zhang, Tan, & Feng, 2012). In an alternative approach with mitochondrial-targeted αTS nanoparticles (based on poly(lactic-co-glycolic acid)-polyethylene glycol (PLGA-b-PEG-OH)), increased delivery and induction of apoptosis by inhibiting antiapoptotic Bcl-xL/Bcl-2/Mcl-1 and activation of complex V (ATP synthase) were observed (Wen & Dhar, 2016).

Another strategy is based on the mixture of αT, αTP, and di-tocopheryl phosphate ester (known as TPM™) that can form self-assembling liposomes able to act as efficient nanocarriers for delivery of molecules across the skin, such as drugs (e.g., oxycodone, caffeine) and other molecules (vitamins, including αT and αTP itself, vitamin D3, the omega-3 fatty acid docosahexaenoic acid (DHA), carnosine, insulin) and cofactors such as coenzyme Q10 (Gavin, El-Tamimy, Keah, & Boyd, 2017; Libinaki & Gavin, 2017; Mozafari, Javanmard, & Raji, 2017; Smith et al., 2015). Increased delivery to mitochondria has been achieved using triphenylphosphonium-modified vitamin E (αT and αTS) or coenzyme Q10 (Smith, Porteous, Gane, & Murphy, 2003). For αTP, the levels in mitochondria are not known, but similar to αT and αTS, it binds to and therefore may be transported to the mitochondria by hTAP1/2/3 (Neuzil et al., 2006; Ni et al., 2005; Panagabko et al., 2003; Zingg, Kempna, et al., 2008) after which it is possibly transferred into mitochondria via the CD36 scavenger receptor (Holloway, Luiken, Glatz, Spriet, & Bonen, 2008; Zingg et al., 2015, 2017, 2014).

Related phosphorylated vitamin E analogues include phosphoglitazone (troglitazone-phosphate) (Little et al., 2008) and EPC-K1, a composite molecule between vitamin E (α-tocopherol) and vitamin C (L-ascorbate), linked by a phosphodiester bond (Fig. 1). EPC-K1 is most likely partially converted to αTP or L-ascorbyl phosphate spontaneously or by enzymes in vivo, as suggested by in vitro stability studies showing αTP as a hydrolytic decomposition product (Ohba, Hori, Kadowaki, Takamatsu, & Matsuoka, 1994).

In contrast to αTP, EPC-K1 can act via its enolic hydroxyl group as a potent scavenger for both hydrophilic and hydrophobic radicals, including hydroxyl radicals, superoxide, peroxynitrites, as well as alkyl and lipid radicals (Wei et al., 1999). Moreover, EPC-K1 can chelate Cu^{2+} and Fe^{2+}, thus reducing free radical generation via the Fenton reaction (Tomita, Kashima, & Tsujimoto, 2000). Similar to αTP (Mukherjee et al., 2008; Wu et al., 2017), EPC-K1 was able to protect against ischemia–reperfusion injury during transplantation or vessel occlusion in several studies, e.g., in muscle (Hirose, Yamaga, Ide, Tanoue, & Takagi, 1997), lung (Nagahiro et al., 1997), liver (Yagi et al., 1997; Yagi, Sakagami, Nakagawa, Takaishi, & Orita, 1992), and brain (Kato, Yanaka, Nagase, Hirayama, & Nose, 2003), and during myocardial infarction (Kuribayashi, Katori, Majima, & Yoshida, 1996; Yamada et al., 1998) possibly also as a result of stimulating VEGF expression (Zingg et al., 2017, 2014; Wu et al., 2017). Since the combinatorial treatment with vitamin E and C was more potent against atherosclerosis in several studies, EPC-K1 may also show beneficial effects after hydrolysis into αTP and L-ascorbic acid, although this has not yet been tested (Abudu, Miller, Attaelmannan, & Levinson, 2004; Antoniades, Tousoulis, Tentolouris, Toutouzas, & Stefanadis, 2003; Chan, 1993; May, 1999).

12. CONCLUSIONS

In recent years, research on the water-soluble phosphorylated form of αT, αTP, has revealed unique molecular and cellular activities that are different from that of αT, αTA, αTS, α-TEA, and EPC-K1. Moreover, the natural occurrence of low levels of αTP in plants, foodstuff, and tissues prompts a number of questions on the possible cellular and physiological functions of αTP. At least in cultured cells, αTP has unique transport pathways, affects signal transduction and gene expression, and modulates apoptosis/survival, inflammation, lipid uptake and metabolism, bacterial phagocytosis, angiogenesis, cell migration, long-term potentiation, and telomere length, and the specific molecular mechanisms of its action are beginning to be discovered. These molecular and cellular activities of αTP may help to explain some of the more pronounced regulatory effects of αTP when compared to αT or αTA observed in animal studies over the almost 80 years since its first synthesis, including the prevention of muscle dystrophy, atherosclerosis, inflammation, and its antidepressant effects.

REFERENCES

Abdo, K. M., Rao, G., Montgomery, C. A., Dinowitz, M., & Kanagalingam, K. (1986). Thirteen-week toxicity study of D-alpha-tocopheryl acetate (vitamin E) in Fischer 344 rats. *Food and Chemical Toxicology*, *24*(10–11), 1043–1050.

Abita, J. P., Parniak, M., & Kaufman, S. (1984). The activation of rat liver phenylalanine hydroxylase by limited proteolysis, lysolecithin, and tocopherol phosphate. Changes in conformation and catalytic properties. *Journal of Biological Chemistry*, *259*, 14560–14566.

Abudu, N., Miller, J. J., Attaelmannan, M., & Levinson, S. S. (2004). Vitamins in human arteriosclerosis with emphasis on vitamin C and vitamin E. *Clinica Chimica Acta*, *339*, 11–25.

Ames, S. R. (1947). Effect of calcium on the inhibition of the succinic oxidase system by D-alpha-tocopheryl phosphate. *Journal of Biological Chemistry*, *503*, 503–512.

Ames, S. R., & Risley, H. A. (1949). Effects of tocopherols and their phosphates on enzyme systems. *Annals of the New York Academy of Sciences*, *52*, 149155.

Andersen, M., Lenhard, B., Whatling, C., Eriksson, P., & Odeberg, J. (2006). Alternative promoter usage of the membrane glycoprotein CD36. *BMC Molecular Biology*, 7, 8.

Antoniades, C., Tousoulis, D., Tentolouris, C., Toutouzas, P., & Stefanadis, C. (2003). Oxidative stress, antioxidant vitamins, and atherosclerosis. From basic research to clinical practice. *Herz*, *28*, 628–638.

Areschoug, T., & Gordon, S. (2009). Scavenger receptors: Role in innate immunity and microbial pathogenesis. *Cellular Microbiology*, *11*(8), 1160–1169.

Athinarayanan, S., Wei, R., Zhang, M., Bai, S., Traber, M. G., Yates, K., et al. (2014). Genetic polymorphism of cytochrome P450 4F2, vitamin E level and histological response in adults and children with nonalcoholic fatty liver disease who participated in PIVENS and TONIC clinical trials. *PLoS One*, *9*(4), e95366.

Azzi, A. (2017). Many tocopherols, one vitamin E. *Molecular Aspects of Medicine*. https://doi.org/10.1016/j,mam.2017.06.004.

Azzi, A., Meydani, S. N., Meydani, M., & Zingg, J. M. (2016). The rise, the fall and the renaissance of vitamin E. *Archives of Biochemistry and Biophysics*, *595*, 100–108. https://doi.org/10.1016/j.abb.2015.11.010.

Bankaitis, V. A., Mousley, C. J., & Schaaf, G. (2010). The Sec14 superfamily and mechanisms for crosstalk between lipid metabolism and lipid signaling. *Trends in Biochemical Sciences*, *35*(3), 150–160.

Bao, Y., Qin, L., Kim, E., Bhosle, S., Guo, H., Febbraio, M., et al. (2012). CD36 is involved in astrocyte activation and astroglial scar formation. *Journal of Cerebral Blood Flow & Metabolism*, *32*(8), 1567–1577.

Baranova, I. N., Kurlander, R., Bocharov, A. V., Vishnyakova, T. G., Chen, Z., Remaley, A. T., et al. (2008). Role of human CD36 in bacterial recognition, phagocytosis, and pathogen-induced JNK-mediated signaling. *Journal of Immunology*, *181*(10), 7147–7156.

Bardowell, S. A., Stec, D. E., & Parker, R. S. (2010). Common variants of cytochrome P450 4F2 exhibit altered vitamin E-{omega}-hydroxylase specific activity. *Journal of Nutrition*, *140*(11), 1901–1906.

Basinski, D. H., & Hummel, J. P. (1947). Further observations on the succinic dehydrogenase system and the effects of tocopherol esters. *Journal of Biological Chemistry*, *167*(2), 339–343.

Bastie, C. C., Nahle, Z., McLoughlin, T., Esser, K., Zhang, W., Unterman, T., et al. (2005). FoxO1 stimulates fatty acid uptake and oxidation in muscle cells through CD36-dependent and -independent mechanisms. *Journal of Biological Chemistry*, *280*(14), 14222–14229.

Ben-Shabat, S., Kazdan, Y., Beit-Yannai, E., & Sintov, A. C. (2013). Use of alpha-tocopherol esters for topical vitamin E treatment: Evaluation of their skin permeation and metabolism. *The Journal of Pharmacy and Pharmacology, 65*(5), 652–658.

Bermudez, Y., Ahmadi, S., Lowell, N. E., & Kruk, P. A. (2007). Vitamin E suppresses telomerase activity in ovarian cancer cells. *Cancer Detection and Prevention, 31*(2), 119–128.

Birringer, M. (2010). Analysis of vitamin E metabolites in biological specimen. *Molecular Nutrition & Food Research, 54*(5), 588–598.

Birringer, M., EyTina, J. H., Salvatore, B. A., & Neuzil, J. (2003). Vitamin E analogues as inducers of apoptosis: Structure-function relation. *British Journal of Cancer, 88*, 1948–1955.

Bishe, B., Syed, G., & Siddiqui, A. (2012). Phosphoinositides in the hepatitis C virus life cycle. *Viruses, 4*(10), 2340–2358.

Blanchet, C., Jouvion, G., Fitting, C., Cavaillon, J. M., & Adib-Conquy, M. (2014). Protective or deleterious role of scavenger receptors SR-A and CD36 on host resistance to Staphylococcus aureus depends on the site of infection. *PLoS One, 9*(1), e87927.

Bondeva, T., Pirola, L., Bulgarelli-Leva, G., Rubio, I., Wetzker, R., & Wymann, M. P. (1998). Bifurcation of lipid and protein kinase signals of PI3Kgamma to the protein kinases PKB and MAPK. *Science, 282*, 293–296.

Bonen, A., Han, X. X., Habets, D. D., Febbraio, M., Glatz, J. F., & Luiken, J. J. (2007). A null mutation in skeletal muscle FAT/CD36 reveals its essential role in insulin- and AICAR-stimulated fatty acid metabolism. *American Journal of Physiology. Endocrinology and Metabolism, 292*(6), E1740–1749.

Borel, P., Caillaud, D., & Cano, N. J. (2015). Vitamin D bioavailability: State of the art. *Critical Reviews in Food Science and Nutrition, 55*(9), 1193–1205.

Borel, P., & Desmarchelier, C. (2016). Genetic variations involved in Vitamin E status. *International Journal of Molecular Sciences, 17*(12), E2094. https://doi.org/10.3390/ijms17122094.

Borel, P., Desmarchelier, C., Nowicki, M., Bott, R., & Tourniaire, F. (2015). Can genetic variability in alpha-tocopherol bioavailability explain the heterogeneous response to alpha-tocopherol supplements? *Antioxidants & Redox Signaling, 22*, 669–678. https://doi.org/10.1089/ars.2014.6144.

Borel, P., Moussa, M., Reboul, E., Lyan, B., Defoort, C., Vincent-Baudry, S., et al. (2007). Human plasma levels of vitamin E and carotenoids are associated with genetic polymorphisms in genes involved in lipid metabolism. *Journal of Nutrition, 137*(12), 2653–2659.

Boucher, J., Kleinridders, A., & Kahn, C. R. (2014). Insulin receptor signaling in normal and insulin-resistant states. *Cold Spring Harbor Perspectives in Biology, 6*(1), a009191. https://doi.org/10.1101/cshperspect.a009191.

Brase, D. A., & Westfall, T. C. (1972). Stimulation of rat liver phenylalanine hydroxylase activity by derivatives of vitamin E. *Biochemical and Biophysical Research Communications, 48*(5), 1185–1191.

Brigelius-Flohe, R., Kelly, F. J., Salonen, J. T., Neuzil, J., Zingg, J. M., & Azzi, A. (2002). The European perspective on vitamin E: Current knowledge and future research. *American Journal of Clinical Nutrition, 76*, 703–716.

Brisson, L., Castan, S., Fontbonne, H., Nicoletti, C., Puigserver, A., & Ajandouz el, H. (2008). Alpha-tocopheryl acetate is absorbed and hydrolyzed by Caco-2 cells comparative studies with alpha-tocopherol. *Chemistry and Physics of Lipids, 154*(1), 33–37.

Burton, G. W., & Traber, M. G. (1990). Vitamin E: Antioxidant activity, biokinetics, and bioavailability. *Annual Review of Nutrition, 10*, 357–382.

Canton, J., Neculai, D., & Grinstein, S. (2013). Scavenger receptors in homeostasis and immunity. *Nature Reviews Immunology, 13*(9), 621–634.

Cerecetto, H., & Lopez, G. V. (2007). Antioxidants derived from vitamin E: An overview. *Mini-Reviews in Medicinal Chemistry, 7*, 315–338.

Chan, A. C. (1993). Partners in defense, vitamin E and vitamin C. *Canadian Journal of Physiology and Pharmacology*, *71*, 725–731.
Chaumontet, C., Bex, V., Veran, F., & Martel, P. (2008). The vitamin E analog tocopherol succinate strongly inhibits gap junctional intercellular communication in rat liver epithelial cells (IAR203). *The Journal of Nutritional Biochemistry*, *19*(4), 263–268.
Cheeseman, K. H., Holley, A. E., Kelly, F. J., Wasil, M., Hughes, L., & Burton, G. (1995). Biokinetics in humans of RRR-alpha-tocopherol: The free phenol, acetate ester, and succinate ester forms of vitamin E. *Free Radical Biology & Medicine*, *19*, 591–598.
Cheng, J. J., Li, J. R., Huang, M. H., Ma, L. L., Wu, Z. Y., Jiang, C. C., et al. (2016). CD36 is a co-receptor for hepatitis C virus E1 protein attachment. *Scientific Reports*, *6*, 21808. https://doi.org/10.1038/srep21808.
Cheung, L., Andersen, M., Gustavsson, C., Odeberg, J., Fernandez-Perez, L., Norstedt, G., et al. (2007). Hormonal and nutritional regulation of alternative CD36 transcripts in rat liver—A role for growth hormone in alternative exon usage. *BMC Molecular Biology*, *8*, 60.
Cocquerel, L., Voisset, C., & Dubuisson, J. (2006). Hepatitis C virus entry: Potential receptors and their biological functions. *Journal of General Virology*, *87*(Pt. 5), 1075–1084.
Collins, R. F., Touret, N., Kuwata, H., Tandon, N. N., Grinstein, S., & Trimble, W. S. (2009). Uptake of oxidized low density lipoprotein by CD36 occurs by an actin-dependent pathway distinct from macropinocytosis. *Journal of Biological Chemistry*, *284*(44), 30288–30297.
Crouzin, N., de Jesus Ferreira, M. C., Cohen-Solal, C., Aimar, R. F., Vignes, M., & Guiramand, J. (2007). Alpha-tocopherol-mediated long-lasting protection against oxidative damage involves an attenuation of calcium entry through TRP-like channels in cultured hippocampal neurons. *Free Radical Biology & Medicine*, *42*(9), 1326–1337.
Crouzin, N., de Jesus Ferreira, M. C., Cohen-Solal, C., M'Kadmi, C., Bernad, N., Martinez, J., et al. (2011). Alpha-tocopherol and alpha-tocopheryl phosphate interact with the cannabinoid system in the rodent hippocampus. *Free Radical Biology & Medicine*, *51*(9), 1643–1655.
Dam, H., & Glavind, J. (1942). Factors influencing capillary permeability in the vitamin E deficient chick. *Science*, *96*(2488), 235–236.
de Jesus Ferreira, M. C., Crouzin, N., Barbanel, G., Cohen-Solal, C., Recasens, M., Vignes, M., et al. (2005). A transient treatment of hippocampal neurons with alpha-tocopherol induces a long-lasting protection against oxidative damage via a genomic action. *Free Radical Biology & Medicine*, *39*(8), 1009–1020.
De la Fuente, M., Carazo, M., Correa, R., & Del Rio, M. (2000). Changes in macrophage and lymphocyte functions in guinea-pigs after different amounts of vitamin E ingestion. *British Journal of Nutrition*, *84*(1), 25–29.
De Pascale, M. C., Bassi, A. M., Patrone, V., Villacorta, L., Azzi, A., & Zingg, J. M. (2006). Increased expression of transglutaminase-1 and PPARgamma after vitamin E treatment in human keratinocytes. *Archives of Biochemistry and Biophysics*, *447*, 97–106.
Detwiler, T. C., Garrett, R. H., & Nason, A. (1966). Effects of digitonin and tocopherol on bovine heart muscle reduced diphosphopyridine nucleotide- and succinate-cytochrome c reductase and cytochrome c oxidase. *Journal of Biological Chemistry*, *241*(7), 1621–1631.
Devaraj, S., Hugou, I., & Jialal, I. (2001). Alpha-tocopherol decreases CD36 expression in human monocyte-derived macrophages. *Journal of Lipid Research*, *42*, 521–527.
Dolfi, S. C., Yang, Z., Lee, M. J., Guan, F., Hong, J., & Yang, C. S. (2013). Inhibitory effects of different forms of tocopherols, tocopherol phosphates, and tocopherol quinones on growth of colon cancer cells. *Journal of Agricultural and Food Chemistry*, *61*(36), 8533–8540.
Dombro, R. S., Bradham, L. S., Campbell, N. K., & Woolley, D. W. (1961). Serotonin receptors. III. The role of salt formation. *Biochimica et Biophysica Acta*, *54*, 516–524.

Dong, L. F., Low, P., Dyason, J. C., Wang, X. F., Prochazka, L., Witting, P. K., et al. (2008). Alpha-tocopheryl succinate induces apoptosis by targeting ubiquinone-binding sites in mitochondrial respiratory complex II. *Oncogene, 27*(31), 4324–4335.

Doring, F., Rimbach, G., & Lodge, J. K. (2004). In silico search for single nucleotide polymorphisms in genes important in vitamin E homeostasis. *IUBMB Life, 56*, 615–620.

Duan, W., Zhou, J., Zhang, S., Zhao, K., Zhao, L., Ogata, K., et al. (2011). ESeroS-GS modulates lipopolysaccharide-induced macrophage activation by impairing the assembly of TLR-4 complexes in lipid rafts. *Biochimica et Biophysica Acta, 1813*(5), 772–783.

Duhem, N., Danhier, F., & Preat, V. (2014). Vitamin E-based nanomedicines for anti-cancer drug delivery. *Journal of Controlled Release, 182*, 33–44.

Dyatlov, V. A., Makovetskaia, V. V., Leonhardt, R., Lawrence, D. A., & Carpenter, D. O. (1998). Vitamin E enhances Ca(2+)-mediated vulnerability of immature cerebellar granule cells to ischemia. *Free Radical Biology & Medicine, 25*(7), 793–802.

Engelmann, B., & Wiedmann, M. K. (2010). Cellular phospholipid uptake: Flexible paths to coregulate the functions of intracellular lipids. *Biochimica et Biophysica Acta, 1801*(6), 609–616.

Eppstein, S. H., & Morgulis, S. (1941). Factors influencing the onset and cure of nutritional muscular dystrophy. *Journal of Nutrition, 23*, 473–482.

Eskin, M. (2013). Vitamin E and the role of water-soluble alpha-tocopheryl phosphate. *Vitamins & Minerals, 3*(1), 1000e123.

Fanali, G., Fasano, M., Ascenzi, P., Zingg, J.-M., & Azzi, A. (2013). α-Tocopherol binding to human serum albumin. *Biofactors, 39*(3), 294–303.

Fariss, M. W. (1997). *Anionic tocopherol esters as antioxidants and cytoprotectants* (5th ed.). New York: Marcel Dekker.

Febbraio, M., & Silverstein, R. L. (2007). CD36: Implications in cardiovascular disease. *International Journal of Biochemistry & Cell Biology, 39*(11), 2012–2030.

Franke-Fayard, B., Janse, C. J., Cunha-Rodrigues, M., Ramesar, J., Buscher, P., Que, I., et al. (2005). Murine malaria parasite sequestration: CD36 is the major receptor, but cerebral pathology is unlinked to sequestration. *Proceedings of the National Academy of Sciences of the United States of America, 102*, 11468–11473.

Fraunholz, M., & Sinha, B. (2012). Intracellular Staphylococcus aureus: Live-in and let die. *Frontiers in Cellular Infection and Microbiology, 2*, 43. https://doi.org/10.3389/fcimb.2012.00043.

Frikke-Schmidt, R., Nordestgaard, B. G., Jensen, G. B., Steffensen, R., & Tybjaerg-Hansen, A. (2008). Genetic variation in ABCA1 predicts ischemic heart disease in the general population. *Arteriosclerosis, Thrombosis, and Vascular Biology, 28*(1), 180–186.

Galli, F., Azzi, A., Birringer, M., Cook-Mills, J. M., Eggersdorfer, M., Frank, J., et al. (2016). Vitamin E: Emerging aspects and new directions. *Free Radical Biology & Medicine, 102*, 16–36.

Gavin, P. D., El-Tamimy, M., Keah, H. H., & Boyd, B. J. (2017). Tocopheryl phosphate mixture (TPM) as a novel lipid-based transdermal drug delivery carrier: Formulation and evaluation. *Drug Delivery and Translation Research, 7*, 53–65. https://doi.org/10.1007/s13346-016-0331-x.

Ghayour-Mobarhan, M., Saghiri, Z., Ferns, G., & Sahebkar, A. (2014). Alpha-tocopheryl phosphate as a bioactive derivative of vitamin E: A review of the literature. *Journal of Dietary Supplements, 12*(4), 359–372.

Ghosh, R., & Bankaitis, V. A. (2011). Phosphatidylinositol transfer proteins: Negotiating the regulatory interface between lipid metabolism and lipid signaling in diverse cellular processes. *Biofactors, 37*(4), 290–308.

Gianello, R., Hall, W. C., Kennepohl, E., Libinaki, R., & Ogru, E. (2007). Subchronic oral toxicity study of mixed tocopheryl phosphates in rats. *International Journal of Toxicology, 26*(5), 475–490.

Gianello, R., Libinaki, R., Azzi, A., Gavin, P. D., Negis, Y., Zingg, J. M., et al. (2005). Alpha-tocopheryl phosphate: A novel, natural form of vitamin E. *Free Radical Biology & Medicine, 39*, 970–976.

Giarman, N. J., Bowers, G. N., Jr., Quie, P. G., & Hampton, L. J. (1954). Potentiation of certain barbiturates by alpha-tocopherol phosphate. *Archives Internationales de Pharmacodynamie et de Therapie, 97*(3–4), 473–482.

Gibson, F. C., 3rd, Yumoto, H., Takahashi, Y., Chou, H. H., & Genco, C. A. (2006). Innate immune signaling and Porphyromonas gingivalis-accelerated atherosclerosis. *Journal of Dental Research, 85*, 106–121.

Gibson, F. C., 3rd, Hong, C., Chou, H. H., Yumoto, H., Chen, J., Lien, E., et al. (2004). Innate immune recognition of invasive bacteria accelerates atherosclerosis in apolipoprotein E-deficient mice. *Circulation, 109*, 2801–2806.

Goncalves, A., Margier, M., Roi, S., Collet, X., Niot, I., Goupy, P., et al. (2014). Intestinal scavenger receptors are involved in vitamin K1 absorption. *Journal of Biological Chemistry, 289*(44), 30743–30752.

Goncalves, A., Roi, S., Nowicki, M., Dhaussy, A., Huertas, A., Amiot, M. J., et al. (2015). Fat-soluble vitamin intestinal absorption: Absorption sites in the intestine and interactions for absorption. *Food Chemistry, 172*, 155–160.

Goncalves, A., Roi, S., Nowicki, M., Niot, I., & Reboul, E. (2014). Cluster-determinant 36 impacts on vitamin E postprandial response. *Molecular Nutrition & Food Research, 58*(12), 2297–2300.

Goudriaan, J. R., Dahlmans, V. E., Teusink, B., Ouwens, D. M., Febbraio, M., Maassen, J. A., et al. (2003). CD36 deficiency increases insulin sensitivity in muscle, but induces insulin resistance in the liver in mice. *Journal of Lipid Research, 44*(12), 2270–2277.

Govier, W. M., & Jetter, N. S. (1948). The effect of α-tocopheryl phosphate on diphosphopyridine nucleotidase. *Science, 107*(2771), 146–147.

Grayston, J. T., Kuo, C. C., Campbell, L. A., Wang, S. P., & Jackson, L. A. (1997). *Chlamydia pneumoniae* and cardiovascular disease. *Cardiologia, 42*, 1145–1151.

Gura, T. (1998). Infections: A cause of artery-clogging plaques? *Science, 281*, 35, 37.

Hirai, T., Yoshimoto, A., Iwata, T., Yamashita, Y., Kuwahara, M., & Toge, T. (1997). Enhancing effect of thoraco-laparotomy on liver metastasis and the role played by active oxygens in its mechanism. *Surgery Today, 27*(11), 1040–1045.

Hirose, J., Yamaga, M., Ide, J., Tanoue, M., & Takagi, K. (1997). Reduced ischemia-reperfusion injury in muscle. Experiments in rats with EPC-K1, a new radical scavenger. *Acta Orthopaedica Scandinavica, 68*, 369–373.

Holloway, G. P., Luiken, J. J., Glatz, J. F., Spriet, L. L., & Bonen, A. (2008). Contribution of FAT/CD36 to the regulation of skeletal muscle fatty acid oxidation: An overview. *Acta Physiologica (Oxford), 194*(4), 293–309.

Horwitt, M. K., Elliott, W. H., Kanjananggulpan, P., & Fitch, C. D. (1984). Serum concentrations of alpha-tocopherol after ingestion of various vitamin E preparations. *American Journal of Clinical Nutrition, 40*(2), 240–245.

Hou, Y., Wu, M., Wei, J., Ren, Y., Du, C., Wu, H., et al. (2015). CD36 is involved in high glucose-induced epithelial to mesenchymal transition in renal tubular epithelial cells. *Biochemical and Biophysical Research Communications, 468*(1–2), 281–286.

Houchin, O. B. (1942). The in vitro effect of alpha-tocopherol and its phosphate derivative on oxidation in muscle tissue. *Journal of Biological Chemistry, 146*, 313–321.

Houchin, O. B., & Mattill, H. A. (1942). The influence of parenteral administration of alpha-tocopherol phosphate on the metabolic process in dystrophic muscle. *Journal of Biological Chemistry, 146*, 309–312.

Hove, E. L., & Harris, P. L. (1947). Relative activity of the tocopherols in curing muscular dystrophy in rabbits. *Journal of Nutrition, 33*, 95–106.

Hsieh, F. L., Turner, L., Bolla, J. R., Robinson, C. V., Lavstsen, T., & Higgins, M. K. (2016). The structural basis for CD36 binding by the malaria parasite. *Nature Communications*, *7*, 12837.

Huang, P. H., Chuang, H. C., Chou, C. C., Wang, H., Lee, S. L., Yang, H. C., et al. (2013). Vitamin E facilitates the inactivation of the kinase Akt by the phosphatase PHLPP1. *Science Signaling*, *6*(267), ra19.

Huang, P. H., Wang, D., Chuang, H. C., Wei, S., Kulp, S. K., & Chen, C. S. (2009). Alpha-tocopheryl succinate and derivatives mediate the transcriptional repression of androgen receptor in prostate cancer cells by targeting the PP2A-JNK-Sp1-signaling axis. *Carcinogenesis*, *30*(7), 1125–1131.

Hummel, J. P., & Basinski, D. H. (1948). The in vitro effect of tocopherol phosphate on the respiration of muscle from normal and dystrophic rabbits. *Journal of Biological Chemistry*, *172*(2), 417–420.

Ile, K. E., Schaaf, G., & Bankaitis, V. A. (2006). Phosphatidylinositol transfer proteins and cellular nanoreactors for lipid signaling. *Nature Chemical Biology*, *2*, 576–583.

IOS. (2000). *Dietary reference intakes for vitamin C, vitamin E, selenium, and carotenoids*. (pp. 186–283). Washington, DC: National Academy Press.

Isenberg, J. S., Jia, Y., Fukuyama, J., Switzer, C. H., Wink, D. A., & Roberts, D. D. (2007). Thrombospondin-1 inhibits nitric oxide signaling via CD36 by inhibiting myristic acid uptake. *Journal of Biological Chemistry*, *282*(21), 15404–15415.

Ito, S., Mori, T., Kanazawa, H., & Sawaguchi, T. (2007). Differential effects of the ascorbyl and tocopheryl derivative on the methamphetamine-induced toxic behavior and toxicity. *Toxicology*, *240*(1–2), 96–110.

Jack Yang, N. Y., & Desai, I. D. (1977). Effect of high levels of dietary vitamin E on hematological indices and biochemical parameters in rats. *Journal of Nutrition*, *107*(8), 1410–1417.

Jay, A. G., Chen, A. N., Paz, M. A., Hung, J. P., & Hamilton, J. A. (2015). CD36 binds oxidized low density lipoprotein (LDL) in a mechanism dependent upon fatty acid binding. *Journal of Biological Chemistry*, *290*(8), 4590–4603.

Jay, A. G., & Hamilton, J. A. (2016). The enigmatic membrane fatty acid transporter CD36: New insights into fatty acid binding and their effects on uptake of oxidized LDL. *Prostaglandins, Leukotrienes and Essential Fatty Acids*. https://doi.org/10.1016/j.plefa.2016.05.005.

Jialal, I., & Devaraj, S. (2003). Antioxidants and atherosclerosis: Don't throw out the baby with the bath water. *Circulation*, *107*, 926–928.

Kagan, V. E., Bakalova, R. A., Serbinova, E. E., & Stoytchev, T. S. (1990). Fluorescence measurements of incorporation and hydrolysis of tocopherol and tocopheryl esters in biomembranes. *Methods in Enzymology*, *186*, 355–367.

Kagan, J. C., & Horng, T. (2013). NLRP3 inflammasome activation: CD36 serves double duty. *Nature Immunology*, *14*(8), 772–774.

Karrer, P., & Bussmann, G. (1940). D,L-alpha-Tocopherol-phosphorsäure-ester. *Helvetica Chimica Acta*, *23*, 1137–2238.

Kato, E., & Takahashi, N. (2012). Improvement by sodium DL-alpha-tocopheryl-6-O-phosphate treatment of moisture-retaining ability in stratum corneum through increased ceramide levels. *Bioorganic & Medicinal Chemistry*, *20*(12), 3837–3842.

Kato, N., Yanaka, K., Nagase, S., Hirayama, A., & Nose, T. (2003). The antioxidant EPC-K1 ameliorates brain injury by inhibiting lipid peroxidation in a rat model of transient focal cerebral ischaemia. *Acta Neurochirurgica (Wien)*, *145*, 489–493. discussion 493.

Kempna, P., Reiter, E., Arock, M., Azzi, A., & Zingg, J. M. (2004). Inhibition of HMC-1 mast cell proliferation by vitamin E: Involvement of the protein kinase B pathway. *Journal of Biological Chemistry*, *279*, 50700–50709.

Kempna, P., Zingg, J. M., Ricciarelli, R., Hierl, M., Saxena, S., & Azzi, A. (2003). Cloning of novel human SEC14p-like proteins: Cellular localization, ligand binding and functional properties. *Free Radical Biology & Medicine*, *34*, 1458–1472.

Kennedy, D. J., & Kashyap, S. R. (2011). Pathogenic role of scavenger receptor CD36 in the metabolic syndrome and diabetes. *Metabolic Syndrome and Related Disorders*, *9*(4), 239–245.

Kline, K., Yu, W., & Sanders, B. G. (2004). Vitamin E and breast cancer. *Journal of Nutrition*, *134*, 3458S–3462S.

Kluckova, K., Bezawork-Geleta, A., Rohlena, J., Dong, L., & Neuzil, J. (2013). Mitochondrial complex II, a novel target for anti-cancer agents. *Biochimica et Biophysica Acta*, *1827*(5), 552–564.

Kono, N., & Arai, H. (2015). Intracellular transport of fat-soluble vitamins A and E. *Traffic*, *16*(1), 19–34.

Kono, N., Ohto, U., Hiramatsu, T., Urabe, M., Uchida, Y., Satow, Y., et al. (2013). Impaired alpha-TTP-PIPs interaction underlies familial vitamin E deficiency. *Science*, *340*, 1106–1110. https://doi.org/10.1126/science.1233508.

Koonen, D. P., Jacobs, R. L., Febbraio, M., Young, M. E., Soltys, C. L., Ong, H., et al. (2007). Increased hepatic CD36 expression contributes to dyslipidemia associated with diet-induced obesity. *Diabetes*, *56*(12), 2863–2871.

Kuribayashi, Y., Katori, M., Majima, M., & Yoshida, K. (1996). Inhibitory effects of a phosphate diester of alpha-tocopherol and ascorbic acid (EPC-K1) on myocardial infarction in rats. *International Journal of Tissue Reactions*, *18*, 73–79.

Lalla, E., Lamster, I. B., Hofmann, M. A., Bucciarelli, L., Jerud, A. P., Tucker, S., et al. (2003). Oral infection with a periodontal pathogen accelerates early atherosclerosis in apolipoprotein E-null mice. *Arteriosclerosis, Thrombosis, and Vascular Biology*, *23*, 1405–1411.

Lauridsen, C., Hedemann, M. S., & Jensen, S. K. (2001). Hydrolysis of tocopheryl and retinyl esters by porcine carboxyl ester hydrolase is affected by their carboxylate moiety and bile acids. *Journal of Nutritional Biochemistry*, *12*, 219–224.

Lecompte, S., Szabo de Edelenyi, F., Goumidi, L., Maiani, G., Moschonis, G., Widhalm, K., et al. (2011). Polymorphisms in the CD36/FAT gene are associated with plasma vitamin E concentrations in humans. *American Journal of Clinical Nutrition*, *93*(3), 644–651.

Lee, H. J., Ju, J., Paul, S., So, J. Y., DeCastro, A., Smolarek, A., et al. (2009). Mixed tocopherols prevent mammary tumorigenesis by inhibiting estrogen action and activating PPAR-gamma. *Clinical Cancer Research*, *15*(12), 4242–4249.

Leelahavanichkul, A., Bocharov, A. V., Kurlander, R., Baranova, I. N., Vishnyakova, T. G., Souza, A. C., et al. (2012). Class B scavenger receptor types I and II and CD36 targeting improves sepsis survival and acute outcomes in mice. *Journal of Immunology*, *188*(6), 2749–2758.

Lewis, D. R., Petersen, L. K., York, A. W., Ahuja, S., Chae, H., Joseph, L. B., et al. (2016). Nanotherapeutics for inhibition of atherogenesis and modulation of inflammation in atherosclerotic plaques. *Cardiovascular Research*, *109*(2), 283–293.

Liang, C. P., Han, S., Senokuchi, T., & Tall, A. R. (2007). The macrophage at the crossroads of insulin resistance and atherosclerosis. *Circulation Research*, *100*(11), 1546–1555.

Libinaki, C. P., & Gavin, P. D. (2017). Changes in the bioavailability of omega-3 (DHA) through alpha-tocopheryl phosphate mixture (TPM) after oral administration in rats. *Nutrients*, *8*(9), 1042. https://doi.org/10.3390/nu9091042.

Libinaki, R., Ogru, E., Gianello, R., Bolton, L., & Geytenbeek, S. (2005). Evaluation of the safety of mixed tocopheryl phosphates (MTP)—A formulation of alpha-tocopheryl phosphate plus alpha-di-tocopheryl phosphate. *Food and Chemical Toxicology*, *44*, 916–932.

Libinaki, R., Tesanovic, S., Heal, A., Nikolovski, B., Vinh, A., Widdop, R. E., et al. (2010). The effect of tocopheryl phosphate on key biomarkers of inflammation: Implication in the reduction of atherosclerosis progression in a hypercholesterolemic rabbit model. *Clinical and Experimental Pharmacology & Physiology*, *37*(5–6), 587–592.

Libinaki, R., Vinh, A., Tesanovic-Klajic, S., Widdop, R., & Gaspari, T. (2017). The effect of tocopheryl phosphates (TPM) on the development of atherosclerosis in apolipoprotein-E deficient mice. *Clinical and Experimental Pharmacology & Physiology*, *44*(Suppl. 1), 107–116. https://doi.org/10.1111/1440-1681.12821.

Lirangi, M., Meydani, M., Zingg, J. M., & Azzi, A. (2012). Alpha-tocopheryl-phosphate regulation of gene expression in preadipocytes and adipocytes. *Biofactors*, *38*(6), 450–457.

Little, P. J., Ballinger, M. L., Survase, S., Osman, N., Ogru, E., Geytenbeek, S., et al. (2008). Phosphorylated troglitazone activates PPARgamma and inhibits vascular smooth muscle cell proliferation and proteoglycan synthesis. *Journal of Cardiovascular Pharmacology*, *51*(3), 274–279.

Lobato, K. R., Cardoso, C. C., Binfare, R. W., Budni, J., Wagner, C. L., Brocardo, P. S., et al. (2010). Alpha-tocopherol administration produces an antidepressant-like effect in predictive animal models of depression. *Behavioural Brain Research*, *209*(2), 249–259.

Love-Gregory, L., Sherva, R., Schappe, T., Qi, J. S., McCrae, J., Klein, S., et al. (2014). Common CD36 SNPs reduce protein expression and may contribute to a protective atherogenic profile. *Human Molecular Genetics*, *20*(1), 193–201.

Luan, Y., & Griffiths, H. R. (2006). Ceramides reduce CD36 cell surface expression and oxidised LDL uptake by monocytes and macrophages. *Archives of Biochemistry and Biophysics*, *450*(1), 89–99.

Ma, X., Bacci, S., Mlynarski, W., Gottardo, L., Soccio, T., Menzaghi, C., et al. (2004). A common haplotype at the CD36 locus is associated with high free fatty acid levels and increased cardiovascular risk in Caucasians. *Human Molecular Genetics*, *13*(19), 2197–2205.

Major, J. M., Yu, K., Weinstein, S. J., Berndt, S. I., Hyland, P. L., Yeager, M., et al. (2014). Genetic variants reflecting higher vitamin e status in men are associated with reduced risk of prostate cancer. *Journal of Nutrition*, *144*(5), 729–733.

Major, J. M., Yu, K., Wheeler, W., Zhang, H., Cornelis, M. C., Wright, M. E., et al. (2011). Genome-wide association study identifies common variants associated with circulating vitamin E levels. *Human Molecular Genetics*, *20*(19), 3876–3883.

Manning, P. J., Sutherland, W. H., Walker, R. J., Williams, S. M., De Jong, S. A., Ryalls, A. R., et al. (2004). Effect of high-dose vitamin E on insulin resistance and associated parameters in overweight subjects. *Diabetes Care*, *27*(9), 2166–2171.

Manosso, L. M., Neis, V. B., Moretti, M., Daufenbach, J. F., Freitas, A. E., Colla, A. R., et al. (2013). Antidepressant-like effect of alpha-tocopherol in a mouse model of depressive-like behavior induced by TNF-alpha. *Progress in Neuro-Psychopharmacology & Biological Psychiatry*, *46*, 48–57.

Mathias, P. M., Harries, J. T., Peters, T. J., & Muller, D. P. (1981). Studies on the in vivo absorption of micellar solutions of tocopherol and tocopheryl acetate in the rat: Demonstration and partial characterization of a mucosal esterase localized to the endoplasmic reticulum of the enterocyte. *Journal of Lipid Research*, *22*, 829–837.

May, J. M. (1999). Is ascorbic acid an antioxidant for the plasma membrane? *The FASEB Journal*, *13*, 995–1006.

McCary, C. A., Yoon, Y., Panagabko, C., Cho, W., Atkinson, J., & Cook-Mills, J. M. (2012). Vitamin E isoforms directly bind PKCalpha and differentially regulate activation of PKCalpha. *Biochemical Journal*, *441*(1), 189–198.

Meydani, M., Koga, T., & Ali, S. (2001). Vitamin E deficiency. In *Encyclopedia of life sciences* (pp. 1–6). Nature Publishing Group.

Mirsky, J. H., & Giarman, N. J. (1955). Studies on the potentiation of thiopental. *The Journal of Pharmacology and Experimental Therapeutics*, *114*(2), 240–249.

Mocchegiani, E., Costarelli, L., Giacconi, R., Malavolta, M., Basso, A., Piacenza, F., et al. (2014). Vitamin E-gene interactions in aging and inflammatory age-related diseases: Implications for treatment. A systematic review. *Ageing Research Reviews*, *14*, 81–101.

Mokashi, V., Singh, D. K., & Porter, T. D. (2005). Supernatant protein factor stimulates HMG-CoA reductase in cell culture and in vitro. *Archives of Biochemistry and Biophysics*, *433*, 474–480.

Mozafari, M. R., Javanmard, R., & Raji, M. (2017). Tocosome: Novel drug deliver system containing phospholipids and tocopheryl phosphates. *International Journal of Pharmaceutics*, *528*(1–2), 381–382.

Mukherjee, S., Lekli, I., Das, M., Azzi, A., & Das, D. K. (2008). Cardioprotection with alpha-tocopheryl phosphate: Amelioration of myocardial ischemia reperfusion injury is linked with its ability to generate a survival signal through Akt activation. *Biochimica et Biophysica Acta*, *1782*(9), 498–503.

Muller, D. P., Manning, J. A., Mathias, P. M., & Harries, J. T. (1976). Studies on the intestinal hydrolysis of tocopheryl esters. *International Journal for Vitamin and Nutrition Research*, *46*, 207–210.

Munteanu, A., Taddei, M., Tamburini, I., Bergamini, E., Azzi, A., & Zingg, J. M. (2006). Antagonistic effects of oxidized low density lipoprotein and {alpha}-tocopherol on CD36 scavenger receptor expression in monocytes: Involvement of protein kinase B and peroxisome proliferator-activated receptor-{gamma}. *Journal of Biological Chemistry*, *281*, 6489–6497.

Munteanu, A., & Zingg, J. M. (2007). Cellular, molecular and clinical aspects of vitamin E on atherosclerosis prevention. *Molecular Aspects of Medicine*, *28*, 538–590.

Munteanu, A., Zingg, J. M., Ogru, E., Libinaki, R., Gianello, R., West, S., et al. (2004). Modulation of cell proliferation and gene expression by alpha-tocopheryl phosphates: Relevance to atherosclerosis and inflammation. *Biochemical and Biophysical Research Communications*, *318*, 311–316.

Mustacich, D. J., Leonard, S. W., Devereaux, M. W., Sokol, R. J., & Traber, M. G. (2006). Alpha-tocopherol regulation of hepatic cytochrome P450s and ABC transporters in rats. *Free Radical Biology & Medicine*, *41*, 1069–1078.

Naga Prasad, S. V., Jayatilleke, A., Madamanchi, A., & Rockman, H. A. (2005). Protein kinase activity of phosphoinositide 3-kinase regulates beta-adrenergic receptor endocytosis. *Nature Cell Biology*, 7(8), 785–796.

Nagahiro, I., Aoe, M., Yamashita, M., Date, H., Andou, A., & Shimizu, N. (1997). EPC-K1 is effective in lung preservation in an ex vivo rabbit lung perfusion model. *Annals of Thoracic Surgery*, *63*, 954–959.

Nahle, Z., Hsieh, M., Pietka, T., Coburn, C. T., Grimaldi, P. A., Zhang, M. Q., et al. (2008). CD36-dependent regulation of muscle FoxO1 and PDK4 in the PPAR delta/beta-mediated adaptation to metabolic stress. *Journal of Biological Chemistry*, *283*(21), 14317–14326.

Nakamura, T., Aoyama, Y., Fujita, T., & Katsui, G. (1975). Studies on tocopherol derivatives: V. Intestinal absorption of several D,1-3,4-3H2-alpha-tocopheryl esters in the rat. *Lipids*, *10*(10), 627–633.

Nakayama, S., Katoh, E. M., Tsuzuki, T., & Kobayashi, S. (2003). Protective effect of alpha-tocopherol-6-O-phosphate against ultraviolet B-induced damage in cultured mouse skin. *Journal of Investigative Dermatology*, *121*, 406–411.

Narushima, K., Takada, T., Yamanashi, Y., & Suzuki, H. (2008). Niemann-pick C1-like 1 mediates alpha-tocopherol transport. *Molecular Pharmacology*, *74*(1), 42–49.

Nason, A., & Lehman, I. R. (1955). Tocopherol as an activator of cytochrome C reductase. *Science*, *122*(3157), 19–22.

Nath, A., Li, I., Roberts, L. R., & Chan, C. (2015). Elevated free fatty acid uptake via CD36 promotes epithelial-mesenchymal transition in hepatocellular carcinoma. *Scientific Reports*, *5*, 14752.

Neculai, D., Schwake, M., Ravichandran, M., Zunke, F., Collins, R. F., Peters, J., et al. (2013). Structure of LIMP-2 provides functional insights with implications for SR-BI and CD36. *Nature*, *504*(7478), 172–176.

Negis, Y., Aytan, N., Ozer, N., Ogru, E., Libinaki, R., Gianello, R., et al. (2006). The effect of tocopheryl phosphates on atherosclerosis progression in rabbits fed with a high cholesterol diet. *Archives of Biochemistry and Biophysics*, *450*, 63–66.

Negis, Y., Meydani, M., Zingg, J. M., & Azzi, A. (2007). Molecular mechanism of alpha-tocopheryl-phosphate transport across the cell membrane. *Biochemical and Biophysical Research Communications*, *359*(2), 348–353.

Negis, Y., Zingg, J. M., Libinaki, R., Meydani, M., & Azzi, A. (2009). Vitamin E and cancer. *Nutrition and Cancer*, *61*(6), 875–878.

Negis, Y., Zingg, J. M., Ogru, E., Gianello, R., Libinaki, R., & Azzi, A. (2005). On the existence of cellular tocopheryl phosphate, its synthesis, degradation and cellular roles: A hypothesis. *IUBMB Life*, *57*, 23–25.

Neuzil, J. (2002). Alpha-tocopheryl succinate epitomizes a compound with a shift in biological activity due to pro-vitamin-to-vitamin conversion. *Biochemical and Biophysical Research Communications*, *293*, 1309–1313.

Neuzil, J., Dong, L. F., Ramanathapuram, L., Hahn, T., Chladova, M., Wang, X. F., et al. (2007). Vitamin E analogues as a novel group of mitocans: Anti-cancer agents that act by targeting mitochondria. *Molecular Aspects of Medicine*, *28*(5–6), 607–645.

Neuzil, J., Dong, L. F., Rohlena, J., Truksa, J., & Ralph, S. J. (2013). Classification of mitocans, anti-cancer drugs acting on mitochondria. *Mitochondrion*, *13*(3), 199–208.

Neuzil, J., Dong, L. F., Wang, X. F., & Zingg, J. M. (2006). Tocopherol-associated protein-1 accelerates apoptosis induced by alpha-tocopheryl succinate in mesothelioma cells. *Biochemical and Biophysical Research Communications*, *343*, 1113–1117.

Neuzil, J., Dyason, J. C., Freeman, R., Dong, L. F., Prochazka, L., Wang, X. F., et al. (2007). Mitocans as anti-cancer agents targeting mitochondria: Lessons from studies with vitamin E analogues, inhibitors of complex II. *Journal of Bioenergetics*, *39*(1), 65–72.

Neuzil, J., Tomasetti, M., Zhao, Y., Dong, L. F., Birringer, M., Wang, X. F., et al. (2007). Vitamin E analogs, a novel group of 'mitocans,' as anti-cancer agents: The importance of being redox-silent. *Molecular Pharmacology*, *71*, 1185–1199.

Ni, J., Wen, X., Yao, J., Chang, H. C., Yin, Y., Zhang, M., et al. (2005). Tocopherol-associated protein suppresses prostate cancer cell growth by inhibition of the phosphoinositide 3-kinase pathway. *Cancer Research*, *65*, 9807–9816.

Nierenberg, D. W., Lester, D. C., & Colacchio, T. A. (1987). Determination of tocopherol and tocopherol acetate concentrations in human feces using high-performance liquid chromatography. *Journal of Chromatography*, *413*, 79–89.

Nishio, K., Ishida, N., Saito, Y., Ogawa-Akazawa, Y., Shichiri, M., Yoshida, Y., et al. (2011). Alpha-tocopheryl phosphate: Uptake, hydrolysis, and antioxidant action in cultured cells and mouse. *Free Radical Biology & Medicine*, *50*(12), 1794–1800.

Noel, S. E., Lai, C. Q., Mattei, J., Parnell, L. D., Ordovas, J. M., & Tucker, K. L. (2010). Variants of the CD36 gene and metabolic syndrome in Boston Puerto Rican adults. *Atherosclerosis*, *211*(1), 210–215.

Nowacka, M. M., & Obuchowicz, E. (2011). Vascular endothelial growth factor (VEGF) and its role in the central nervous system: A new element in the neurotrophic hypothesis of antidepressant drug action. *Neuropeptides*, *46*(1), 1–10.

Numakawa, Y., Numakawa, T., Matsumoto, T., Yagasaki, Y., Kumamaru, E., Kunugi, H., et al. (2006). Vitamin E protected cultured cortical neurons from oxidative stress-induced cell death through the activation of mitogen-activated protein kinase and phosphatidylinositol 3-kinase. *Journal of Neurochemistry*, *97*, 1191–1202.

Ogru, E., Gianello, R., Libinaki, R., Smallridge, A., Bak, R., & Geytenbeck, S.et al.(2003). *Vitamin E phosphate: An endogenous form of vitamin E*. Bologna, Italy: Medimond S.R.L.

Ohba, M., Hori, Y., Kadowaki, E., Takamatsu, T., & Matsuoka, M. (1994). Stability studies on L-ascorbic acid DL-alpha-tocopherol phosphoric acid diester potassium salt (EPC-K). *Yakugaku Zasshi*, *114*, 514–522.

Ohno-Matsui, K. (2011). Parallel findings in age-related macular degeneration and Alzheimer's disease. *Progress in Retinal and Eye Research*, *30*(4), 217–238.

Oram, J. F., Vaughan, A. M., & Stocker, R. (2001). ATP-binding cassette transporter A1 mediates cellular secretion of alpha-tocopherol. *Journal of Biological Chemistry*, *276*(43), 39898–39902.

Oviedo-Boyso, J., Cortes-Vieyra, R., Huante-Mendoza, A., Yu, H. B., Valdez-Alarcon, J. J., Bravo-Patino, A., et al. (2011). The phosphoinositide-3-kinase-Akt signaling pathway is important for Staphylococcus aureus internalization by endothelial cells. *Infection and Immunity*, *79*(11), 4569–4577.

Ozer, N. K., Negis, Y., Aytan, N., Villacorta, L., Ricciarelli, R., Zingg, J. M., et al. (2006). Vitamin E inhibits CD36 scavenger receptor expression in hypercholesterolemic rabbits. *Atherosclerosis*, *184*(1), 15–20.

Panagabko, C., Morley, S., Hernandez, M., Cassolato, P., Gordon, H., Parsons, R., et al. (2003). Ligand specificity in the CRAL-TRIO protein family. *Biochemistry*, *42*, 6467–6474.

Pascual, G., Avgustinova, A., Mejetta, S., Martin, M., Castellanos, A., Attolini, C. S., et al. (2017). Targeting metastasis-initiating cells through the fatty acid receptor CD36. *Nature*, *541*(7635), 41–45.

Patel, S. N., Lu, Z., Ayi, K., Serghides, L., Gowda, D. C., & Kain, K. C. (2007). Disruption of CD36 impairs cytokine response to Plasmodium falciparum glycosylphosphatidylinositol and confers susceptibility to severe and fatal malaria in vivo. *Journal of Immunology*, *178*, 3954–3961.

Pepino, M. Y., Kuda, O., Samovski, D., & Abumrad, N. A. (2014). Structure-function of CD36 and importance of fatty acid signal transduction in fat metabolism. *Annual Review of Nutrition*, *34*, 281–303.

Philips, J. A., Rubin, E. J., & Perrimon, N. (2005). Drosophila RNAi screen reveals CD36 family member required for mycobacterial infection. *Science*, *309*, 1251–1253.

Pierpaoli, E., Cirioni, O., Barucca, A., Orlando, F., Silvestri, C., Giacometti, A., et al. (2011). Vitamin E supplementation in old mice induces antimicrobial activity and improves the efficacy of daptomycin in an animal model of wounds infected with methicillin-resistant *Staphylococcus aureus*. *Journal of Antimicrobial Chemotherapy*, *66*(9), 2184–2185.

Pohl, J., Ring, A., Korkmaz, U., Ehehalt, R., & Stremmel, W. (2005). FAT/CD36-mediated long-chain fatty acid uptake in adipocytes requires plasma membrane rafts. *Molecular Biology of the Cell*, *16*(1), 24–31.

Pownall, H., & Moore, K. (2014). Commentary on fatty acid wars: The diffusionists versus the translocatists. *Arteriosclerosis, Thrombosis, and Vascular Biology*, *34*(5), e8–9.

Prasad, K. N., Kumar, B., Yan, X. D., Hanson, A. J., & Cole, W. C. (2003). Alpha-tocopheryl succinate, the most effective form of vitamin E for adjuvant cancer treatment: A review. *Journal of the American College of Nutrition*, *22*, 108–117.

Provinciali, M., Cirioni, O., Orlando, F., Pierpaoli, E., Barucca, A., Silvestri, C., et al. (2011). Vitamin E improves the in vivo efficacy of tigecycline and daptomycin in an animal model of wounds infected with meticillin-resistant *Staphylococcus aureus*. *Journal of Medical Microbiology*, *60*(Pt 12), 1806–1812.

Quaife, M. L. (1954). Fat-soluble vitamins. *Annual Review of Biochemistry*, *23*, 215–244.

Rabinovitz, M., & Boyer, P. D. (1950). The inhibition of the succinoxidase system by alpha-tocopheryl phosphate and sodium dodecyl sulfate. *Journal of Biological Chemistry, 183*, 111–121.

Reboul, E., & Borel, P. (2011). Proteins involved in uptake, intracellular transport and basolateral secretion of fat-soluble vitamins and carotenoids by mammalian enterocytes. *Progress in Lipid Research, 50*(4), 388–402.

Rezk, B. M., Haenen, G. R., Van Der Vijgh, W. J., & Bast, A. (2004). The extraordinary antioxidant activity of vitamin E phosphate. *Biochimica et Biophysica Acta, 1683*, 16–21.

Rezk, B. M., van der Vijgh, W. J., Bast, A., & Haenen, G. R. (2007). Alpha-tocopheryl phosphate is a novel apoptotic agent. *Frontiers in Bioscience, 12*, 2013–2019.

Ricciarelli, R., Zingg, J. M., & Azzi, A. (2000). Vitamin E reduces the uptake of oxidized LDL by inhibiting CD36 scavenger receptor expression in cultured aortic smooth muscle cells. *Circulation, 102*, 82–87.

Rigotti, A. (2007). Absorption, transport, and tissue delivery of vitamin E. *Molecular Aspects of Medicine, 28*(5-6), 423–436.

Rigotti, A., Acton, S. L., & Krieger, M. (1995). The class B scavenger receptors SR-BI and CD36 are receptors for anionic phospholipids. *Journal of Biological Chemistry, 270*, 16221–16224.

Rimbach, G., Moehring, J., Huebbe, P., & Lodge, J. K. (2010). Gene-regulatory activity of alpha-tocopherol. *Molecules, 15*(3), 1746–1761.

Rodrigue-Way, A., Caron, V., Bilodeau, S., Keil, S., Hassan, M., Levy, E., et al. (2013). Scavenger receptor CD36 mediates inhibition of cholesterol synthesis via activation of the PPARgamma/PGC-1α pathway and Insig1/2 expression in hepatocytes. *FASEB Journal, 28*(4), 1910–1923.

Royer, M. C., Lemaire-Ewing, S., Desrumaux, C., Monier, S., Pais de Barros, J. P., Athias, A., et al. (2009). 7-Ketocholesterol incorporation into sphingolipid/cholesterol-enriched (lipid raft) domains is impaired by vitamin E: A specific role for alpha-tocopherol with consequences on cell death. *Journal of Biological Chemistry, 284*(23), 15826–15834.

Saeed, M., Andreo, U., Chung, H. Y., Espiritu, C., Branch, A. D., Silva, J. M., et al. (2015). SEC14L2 enables pan-genotype HCV replication in cell culture. *Nature, 524*, 471–475.

Saitoh, Y., Yumoto, A., & Miwa, N. (2009). Alpha-tocopheryl phosphate suppresses tumor invasion concurrently with dynamic morphological changes and delocalization of cortactin from invadopodia. *International Journal of Oncology, 35*(6), 1277–1288.

Sakai, T., Okano, T., Makino, H., & Tsudzuki, T. (1976). Activation of cyclic AMP phosphodiesterase by a new vitamin E derivative. *Journal of Cyclic Nucleotide Research, 2*, 163–170.

Samhan-Arias, A. K., Tyurina, Y. Y., & Kagan, V. E. (2011). Lipid antioxidants: Free radical scavenging versus regulation of enzymatic lipid peroxidation. *Journal of Clinical Biochemistry and Nutrition, 48*(1), 91–95.

Sampayo-Reyes, A., & Zakharyan, R. A. (2006). Tocopherol esters inhibit human glutathione S-transferase omega. *Acta Biochimica Polonica, 53*(3), 547–552.

Sanyal, A. J., Chalasani, N., Kowdley, K. V., McCullough, A., Diehl, A. M., Bass, N. M., et al. (2010). Pioglitazone, vitamin E, or placebo for nonalcoholic steatohepatitis. *New England Journal of Medicine, 362*(18), 1675–1685.

Schaaf, G., Ortlund, E. A., Tyeryar, K. R., Mousley, C. J., Ile, K. E., Garrett, T. A., et al. (2008). Functional anatomy of phospholipid binding and regulation of phosphoinositide homeostasis by proteins of the sec14 superfamily. *Molecular Cell, 29*(2), 191–206.

Scott, P. M. (1965). Chemical studies concerning the possible role of chromanyl phosphates and quinones in oxidative phosphorylation. *Journal of Biological Chemistry, 240*, 1374–1380.

Sharif, O., Matt, U., Saluzzo, S., Lakovits, K., Haslinger, I., Furtner, T., et al. (2013). The scavenger receptor CD36 downmodulates the early inflammatory response while enhancing bacterial phagocytosis during pneumococcal pneumonia. *Journal of Immunology, 190*(11), 5640–5648.

Sheedy, F. J., Grebe, A., Rayner, K. J., Kalantari, P., Ramkhelawon, B., Carpenter, S. B., et al. (2013). CD36 coordinates NLRP3 inflammasome activation by facilitating intracellular nucleation of soluble ligands into particulate ligands in sterile inflammation. *Nature Immunology, 14*(8), 812–820.

Shiau, C. W., Huang, J. W., Wang, D. S., Weng, J. R., Yang, C. C., Lin, C. H., et al. (2006). Alpha-tocopheryl succinate induces apoptosis in prostate cancer cells in part through inhibition of BCL-XL/BCL-2 function. *Journal of Biological Chemistry, 281*, 11819–11825.

Shibata, N., Arita, M., Misaki, Y., Dohmae, N., Takio, K., Ono, T., et al. (2001). Supernatant protein factor, which stimulates the conversion of squalene to lanosterol, is a cytosolic squalene transfer protein and enhances cholesterol biosynthesis. *Proceedings of the National Academy of Sciences of the United States of America, 98*, 2244–2249.

Smith, R. A. J., Porteous, C. M., Gane, A. M., & Murphy, M. P. (2003). Delivery of bioactive molecules to mitochondria in vivo. *Proceedings of the National Academy of Sciences of the United States of America, 100*, 5407–5412.

Smith, J., Su, X., El-Maghrabi, R., Stahl, P. D., & Abumrad, N. A. (2008). Opposite regulation of CD36 ubiquitination by fatty acids and insulin: Effects on fatty acid uptake. *Journal of Biological Chemistry, 283*(20), 13578–13585.

Smith, M. T., Wyse, B. D., Edwards, S. R., El-Tamimy, M., Gaetano, G., & Gavin, P. (2015). Topical application of a novel oxycodone gel formulation (tocopheryl phosphate mixture) in a rat model of peripheral inflammatory pain produces localized pain relief without significant systemic exposure. *Journal of Pharmaceutical Sciences, 104*(7), 2388–2396.

Solmssen, U. V., & Lee, J. (1949). US Patent No. 2,457,932. United States Patent Office.

Stuart, L. M., Deng, J., Silver, J. M., Takahashi, K., Tseng, A. A., Hennessy, E. J., et al. (2005). Response to *Staphylococcus aureus* requires CD36-mediated phagocytosis triggered by the COOH-terminal cytoplasmic domain. *Journal of Cell Biology, 170*, 477–485.

Sun, B., Boyanovsky, B. B., Connelly, M. A., Shridas, P., van der Westhuyzen, D. R., & Webb, N. R. (2007). Distinct mechanisms for OxLDL uptake and cellular trafficking by class B scavenger receptors CD36 and SR-BI. *Journal of Lipid Research, 48*(12), 2560–2570.

Swettenham, E., Witting, P. K., Salvatore, B. A., & Neuzil, J. (2005). Alpha-tocopheryl succinate selectively induces apoptosis in neuroblastoma cells: Potential therapy of malignancies of the nervous system? *Journal of Neurochemistry, 94*, 1448–1456.

Tanaka, Y., Moritoh, Y., & Miwa, N. (2007). Age-dependent telomere-shortening is repressed by phosphorylated alpha-tocopherol together with cellular longevity and intracellular oxidative-stress reduction in human brain microvascular endotheliocytes. *Journal of Cellular Biochemistry, 102*(3), 689–703.

Tomita, T., Kashima, M., & Tsujimoto, Y. (2000). Characterization of the activity of L-ascorbic acid 2-[3,4-dihydro-2,5,7,8-tetramethyl-2-(4,8,12-trimethyltridecyl)-2H-1-benzopyran-6-yl-hydrogen phosphate] potassium salt in hydroxyl radical elimination. *Chemical & Pharmaceutical Bulletin (Tokyo), 48*, 330–333.

Traber, M. G. (2008). Vitamin E and K interactions—A 50-year-old problem. *Nutrition Reviews, 66*(11), 624–629.

Traber, M. G. (2013). Mechanisms for the prevention of vitamin E excess. *Journal of Lipid Research, 54*(9), 2295–2306. https://doi.org/10.1194/jlr.R032946.

Traber, M. G. (2014). Vitamin E inadequacy in humans: Causes and consequences. *Advances in Nutrition, 5*(5), 503–514.

Traber, M. G., & Arai, H. (1999). Molecular mechanisms of vitamin E transport. *Annual Review of Nutrition, 19*, 343–355.
Ulatowski, L., & Manor, D. (2013). Vitamin E trafficking in neurologic health and disease. *Annual Review of Nutrition, 33*, 87–103.
Vejux, A., Guyot, S., Montange, T., Riedinger, J. M., Kahn, E., & Lizard, G. (2009). Phospholipidosis and down-regulation of the PI3-K/PDK-1/Akt signalling pathway are vitamin E inhibitable events associated with 7-ketocholesterol-induced apoptosis. *Journal of Nutritional Biochemistry, 20*(1), 45–61.
Voegele, A. F., Jerkovic, L., Wellenzohn, B., Eller, P., Kronenberg, F., Liedl, K. R., et al. (2002). Characterization of the vitamin E-binding properties of human plasma afamin. *Biochemistry, 41*, 14532–14538.
Wang, J. H., Zhang, K., Wang, N., Qiu, X. M., Wang, Y. B., & He, P. (2013). Involvement of phosphatidylinositol 3-kinase/Akt signaling pathway in beta1 integrin-mediated internalization of *Staphylococcus aureus* by alveolar epithelial cells. *Journal of Microbiology, 51*(5), 644–650.
Wang, X. L., Zhang, L., Youker, K., Zhang, M. X., Wang, J., LeMaire, S. A., et al. (2006). Free fatty acids inhibit insulin signaling-stimulated endothelial nitric oxide synthase activation through upregulating PTEN or inhibiting Akt kinase. *Diabetes, 55*(8), 2301–2310.
Wei, T., Chen, C., Li, F., Zhao, B., Hou, J., Xin, W., et al. (1999). Antioxidant properties of EPC-K1: A study on mechanisms. *Biophysical Chemistry, 77*, 153–160.
Wei, T., Zhao, X., Hou, J., Ogata, K., Sakaue, T., Mori, A., et al. (2003). The antioxidant ESeroS-GS inhibits NO production and prevents oxidative stress in astrocytes. *Biochemical Pharmacology, 66*(1), 83–91.
Wen, R., & Dhar, S. (2016). *Turn up the cellular power generator with vitamin E analogue formulation*. Royal Society of Chemistry.
Wu, J. H., & Croft, K. D. (2007). Vitamin E metabolism. *Molecular Aspects of Medicine, 28*(5–6), 437–452.
Wu, Z., Zhong, X., Meng, L., Fang, X., He, Y., Li, D., et al. (2017). Alpha-tocopherol, especially alpha-tocopherol phosphate, exerts antiapoptotic and angiogenic effects on rat bone marrow-derived endothelial progenitor cells under high-glucose and hypoxia conditions. *Journal of Vascular Surgery*. https://doi.org/10.1016/j.jvs.2017.02.051.
Xie, Z., & Sastry, B. R. (1993). Induction of hippocampal long-term potentiation by alpha-tocopherol. *Brain Research, 604*, 173–179.
Yagi, T., Sadamori, H., Nakagawa, H., Ishikawa, T., Sasaki, H., Ohta, K., et al. (1997). Hepatropic soluble vitamin E derivative EPC-K1 prevents warm ischemia/reperfusion injury of non-heart-beating donor liver transplantation in pigs. *Transplantation Proceedings, 29*, 1390–1393.
Yagi, T., Sakagami, K., Nakagawa, H., Takaishi, Y., & Orita, K. (1992). A newly developed hydroxyl radical scavenger, EPC-K1 can improve the survival of swine warm ischemia-damaged transplanted liver grafts. *Transplant International, 5*(Suppl. 1), S420–423.
Yamada, S., Yashita, T., Yamaguchi, K., Kyuki, K., Sakaue, T., & Ogata, K. (1998). Effect of EPC-K1 on myocardial infarct size in a rat model with coronary artery occlusion and reperfusion. *General Pharmacology, 31*, 165–167.
Yanamala, N., Kapralov, A. A., Djukic, M., Peterson, J., Mao, G., Klein-Seetharaman, J., et al. (2014). Structural re-arrangement and peroxidase activation of cytochrome c by anionic analogues of vitamin E: Tocopherol-succinate and tocopherol-phosphate. *Journal of Biological Chemistry, 289*(47), 32488–32498.
Yano, J. M., Yu, K., Donaldson, G. P., Shastri, G. G., Ann, P., Ma, L., et al. (2015). Indigenous bacteria from the gut microbiota regulate host serotonin biosynthesis. *Cell, 161*(2), 264–276.
Zhang, Z., Tan, S., & Feng, S. S. (2012). Vitamin E TPGS as a molecular biomaterial for drug delivery. *Biomaterials, 33*(19), 4889–4906.

Zhao, Y., Lee, M. J., Cheung, C., Ju, J. H., Chen, Y. K., Liu, B., et al. (2010). Analysis of multiple metabolites of tocopherols and tocotrienols in mice and humans. *Journal of Agricultural and Food Chemistry, 58*(8), 4844–4852.

Zhukova, E. E., Zakharova, E. I., Sokolova, N. A., Sarycheva, I. K., & Evstigneeva, R. P. (1983). *Esters of DL-alpha-tocopherol and phosphoric acid.* Plenum Publishing Corporation.

Zierler, K. L., Andres, R., Levy, R. I., Anderson, H. M., & Lilienthal, J. L., Jr. (1953). On the mechanism of action of alpha-tocopheryl phosphate, with special reference to carbohydrate metabolism of striated muscle. II. Effect on the capacity of rat diaphragm to dissimilate hexose phosphates. *Bulletin of the Johns Hopkins Hospital, 92*(1), 32–39.

Zierler, K. L., Folk, B. P., & Lilienthal, J. L., Jr. (1953). On the mechanism of action of alpha-tocopheryl phosphate, with special reference to carbohydrate metabolism of striated muscle. I. Modification of epinephrine effect, hyperlactacidemia, by alpha-tocopheryl phosphate in the rat. *Bulletin of the Johns Hopkins Hospital, 92*(1), 26–30.

Zierler, K. L., Grob, D., & Lilienthal, J. L., Jr. (1948). On the antithrombic and antiproteolytic activity of alpha tocopheryl phosphate. *The American Journal of Physiology, 153*(1), 127–132.

Zierler, K. L., Levy, R. I., & Lilienthal, J. L., Jr. (1953). On the mechanism of action of alpha-tocopheryl phosphate, with special reference to carbohydrate metabolism of striated muscle. *Bulletin of the Johns Hopkins Hospital, 92*(1), 41–46.

Zingg, J. M. (2006). In *Modulation of signal transduction by vitamin E. Paper presented at the XIII congress of the society for free radical research international, Davos.*

Zingg, J. M. (2007a). Modulation of signal transduction by vitamin E. *Molecular Aspects of Medicine, 28*, 481–506.

Zingg, J. M. (2007b). Molecular and cellular activities of vitamin E analogues. *Mini Reviews in Medicinal Chemistry, 7*, 543–558.

Zingg, J. M. (2007c). Vitamin E: An overview of major research directions. *Molecular Aspects of Medicine, 28*, 400–422.

Zingg, J. M. (2015). Vitamin E: A role in signal transduction. *Annual Review of Nutrition, 35*, 135–173.

Zingg, J. M., & Azzi, A. (2006). Molecular activities of vitamin E. In M. S. Meskin, W. R. Bidlack, & R. K. Randoph (Eds.), *Phytochemicals: Nutrient-gene interactions* (pp. 175–206). Taylor & Francis.

Zingg, J. M., Azzi, A., & Meydani, M. (2008). Genetic polymorphisms as determinants for disease-preventive effects of vitamin E. *Nutrition Reviews, 66*(7), 406–414.

Zingg, J. M., Azzi, A., & Meydani, M. (2015). Induction of VEGF expression by alpha-tocopherol and alpha-tocopheryl phosphate via PI3Kgamma/PKB and hTAP1/SEC14L2-mediated lipid exchange. *Journal of Cellular Biochemistry, 116*, 398–407.

Zingg, J. M., Azzi, A., & Meydani, M. (2017). Alpha-tocopheryl phosphate induces VEGF expression via CD36/PI3Kgamma in THP-1 monocytes. *Journal of Cellular Biochemistry, 118*, 1855–1867. https://doi.org/10.1002/jcb.25871.

Zingg, J. M., Kempna, P., Paris, M., Reiter, E., Villacorta, L., Cipollone, R., et al. (2008). Characterization of three human sec14p-like proteins: α-Tocopherol transport activity and expression pattern in tissues. *Biochimie, 90*(11–12), 1703–1715.

Zingg, J. M., Libinaki, R., Lai, C. Q., Meydani, M., Gianello, R., Ogru, E., et al. (2010). Modulation of gene expression by alpha-tocopherol and alpha-tocopheryl phosphate in THP-1 monocytes. *Free Radical Biology and Medicine, 49*(12), 1989–2000.

Zingg, J. M., Libinaki, R., Meydani, M., & Azzi, A. (2014). Modulation of phosphorylation of tocopherol and phosphatidylinositol by hTAP1/SEC14L2-mediated lipid exchange. *PLoS One, 9*(7), e101550, PONE-D-14-10912 [pii].

Zingg, J. M., Meydani, M., & Azzi, A. (2010). Alpha-tocopheryl phosphate—An active lipid mediator? *Molecular Nutrition & Food Research, 54*(5), 1–14.

Zingg, J. M., Meydani, M., & Azzi, A. (2012). Alpha-tocopheryl phosphate—An activated form of vitamin E important for angiogenesis and vasculogenesis? *Biofactors*, *38*(1), 24–33.
Zingg, J. M., Ricciarelli, R., Andorno, E., & Azzi, A. (2002). Novel 5' exon of scavenger receptor CD36 is expressed in cultured human vascular smooth muscle cells and atherosclerotic plaques. *Arteriosclerosis, Thrombosis, and Vascular Biology*, *22*, 412–417.

FURTHER READING

Zingg, J. M. (2010). Alpha-tocopheryl phosphate—An active lipid mediator? *Molecular Biology Reports*, *54*(5), 679–692.

Kolibri 360 GmbH
Ferdinand-Jühlke-Straße 7
99095 Erfurt, Deutschland
produktsicherheit@kolibri360.de